ZDENĚK KOPAL

*Department of Astronomy, University of Manchester*

# AN INTRODUCTION
# TO THE STUDY OF THE
# MOON

SPRINGER-SCIENCE+BUSINESS MEDIA, B.V.

ISBN 978-94-011-7547-0      ISBN 978-94-011-7545-6 (eBook)
DOI 10.1007/978-94-011-7545-6

1966

Additional material to this book can be downloaded from http://extras.springer.com.

Sunset over the lunar landscape in Oceanus Procellarum, in the neighbourhood of the crater Flamsteed, as televised to the Earth by the first American soft-lander Surveyor 1 on June 13, 1966 (reproduced by courtesy of the Jet Propulsion Laboratory, California Institute of Technology, and of the U.S. National Aeronautics and Space Administration).

# PREFACE

After several decades spent in astronomical semi-obscurity, the Moon has of late suddenly emerged as an object of considerable interest to students of astronomy as well as of other branches of natural science and technology; and the reasons for this are indeed of historical significance. For the Moon has now been destined to be the first celestial body outside the confines of our own planet to be reconnoitered at a close range by means of spacecraft built and sent out by human hand for this purpose. At the time of writing, not less than ten such spacecraft of American as well as Russian origin landed already on different parts of the lunar surface; and some of these provided remarkable records of its detail structure to a spatial resolution increased thousandfold over that attained so far from our ground-based facilities.

A renewed interest in our satellite, stemming from this source, on the part of the students of many branches of science and technology has also underlined the need for presenting the gist of our present knowledge in this field in the form that could serve as an introduction to the study of the Moon not only for astronomers, but also for serious students from other branches of science or technology. There is, to be sure, no lack of literature on the Moon today – perhaps the contrary is true – but very little of it has presented its subject as integral part of the underlying physical sciences, and as another example of the applicability of their general methods for its study. This represents a task, and a challenge, which we shall attempt to meet.

The aim of the present volume is thus very literally expressed by its title: namely, to provide an introduction to the study of the Moon, based on the broad background of astronomy, physics and chemistry with which the reader will be expected to have some familiarity. The emphasis throughout will be on the development of basic methods from (whenever possible) the first principles, rather than on a mere enumeration of facts; for a proper understanding of the methods of lunar research should, in the long run, prove more rewarding for the reader than a simple account of latest facts, the arsenal of which will change from year to year.

The four parts of which this book consists have been organized in accordance with this plan. The first contains a survey of our present knowledge of the motion of the Moon and dynamics of the Earth-Moon system, obtained mainly by astronomical methods. The second part will be concerned with the internal structure of our satellite; and the third, with the lunar surface features and their interpretation. In the latter, we shall firmly keep in mind that the surface of the Moon must be regarded as a "boundary condition" of all internal processes discussed in Part II, as well as an "impact counter" of all external collisions which disfigured the face of the Moon since the time of its origin; for in no other way can an interpretation of its prolific surface markings possess any meaning. The fourth and concluding part of the volume will deal with the radiation reaching us from the Moon in all domains of the spectrum; and its bearing on the microstructure of the lunar surface on a scale much smaller than that resolvable by telescopic means not only from the Earth, but also from orbiting spacecraft.

As the reader can easily verify by a glance at its table of contents, the present volume attempts to cover almost the entire field of lunar studies; but the depth to which this will be done will vary somewhat from chapter to chapter – depending not the least on the previous coverage of their subjects in existing literature in the English language. For instance, the motion of the Moon in space will be outlined in Chapter 2 only briefly for the sake of completeness, as the student of this subject is well served by the available literature in both its fundamental and expository aspects. The same is also the case with the photometry of lunar eclipses, which has in recent years been extensively dealt with, e.g., by Link. On the other hand, this is not true yet of the motion of the Moon around its centre of gravity (on which the only available monograph by Habibullin exists in the Russian language); and for this reason, more space has been given to it in the present volume.

The subject of Part II – internal structure of the Moon – has not so far been monographically treated in any language. This should explain the attention which we have paid to it in this book; and its author hopes that the reader may find our present introduction to it adequate. In Part III – concerned with the topography of our satellite and an interpretation of its surface features – the reader's attention will likewise be focussed on the methods of research rather than on any descriptive account of the lunar landscape – mainly because such accounts have already filled many a book of both old and recent date; and there is no need of repetition. On the other hand, the last part of our volume (concerned with lunar photometry and its interpretation) will bring astronomy and physics in common focus more closely than has been true of any other part of this book; and its concluding chapters – summarizing, in fact, the entire contents of this volume – will draw on the interpretation of observed facts by all methods known to human science.

Each chapter of this book is, moreover, accompanied with bibliographical notes containing additional references and other points which would have made the main text too discursive. Such bibliographies are by no means complete – for any attempt at completeness would have occupied with bibliographical material a volume exceeding considerably in size this whole book. Special and more complete bibliographies of lunar research have in recent years been compiled by several writers*; and their existence renders any repetition superfluous. Besides a large part of older literature, even though of historical significance, may be devoid of other interest for a modern critical reader of the subject. One of the aims of the present volume (and of the selection of references quoted herein) has been to assist the user in the choice of collateral reading by an attempt to separate the chaff from the wheat. However, the limitations of space alone collaborated also in keeping the list of references as given on pp.

440–457 to its present size. The latter can, therefore, claim neither completeness nor uniqueness in selection; but any significant omissions could easily be rectified by using the present bibliography as a guide.

The references are almost solely to investigations published in books or journals that should be freely available to any interested reader; references to mere abstracts or reports of limited circulation have been kept to a minimum. In order to introduce the subject to the reader as it stands today, we have selected the references mainly from the literature that has appeared in the past ten years. Exception to this has been made only in Chapter 15 ('Mapping of the Moon') where historical interests are in the forefront; but elsewhere these have been relegated to the 'Bibliographical Notes'.

In conclusion, it is scarcely necessary to emphasise that, in the years to come, our acquaintance with our only natural satellite is going to become very much more intimate than it has been as long as astronomical telescopes represented our principal link of communications; and many issues still in suspense should be resolved in the near future by direct exploration by means of spacecraft of increasing effectiveness. As a result of these changing means of research, the Moon may soon cease to be treated as an astronomical object, and be professionally annexed by other sciences more intimately connected with direct exploration. The present volume may, perhaps, be the last monograph devoted to our satellite to come out from the pen of an astronomer. Nevertheless, it is equally certain that the present unprecedented advances in space research of the solar system have been made possible only on the basis of all knowledge which astronomers gathered so far largely from our terrestrial vantage point – a knowledge which, by itself, represents a remarkable achievement of the human mind – and which we aim in the present volume to introduce to the technical student of the subject.

The text of this book has originated in the courses on different aspects of lunar studies which the present writer has given to graduate students at the University of Manchester (and other institutions) in the past several years. A sabbatical leave in 1964–65, which he spent as honorary research fellow at the Boeing Scientific Research Laboratories in Seattle, Washington, offered a welcome chance for committing their contents to a written form. May I take this opportunity to express my sincere appreciation to Mr. G. L. Hollingsworth, Director of the Boeing Scientific Research Laboratories, to Dr. Burton H. Colvin, Head of the Mathematics Research Laboratory, to

---

* Cf., e.g., J. W. SALISBURY, R. A. VAN TASSEL, J. E. M. ADLER, R. T. DODD, Jr., and V. G. SMALLEY, 'Bibliography of Lunar and Planetary Research, 1960–63' (AFCRL Notes No. 684, 62–676, 63–903, 64–885), Bedford, Mass., 1961–64; A. C. MASON, 'Bibliography of the Moon', Military Geol. Branch, U.S. Army, Washington, 1960; A. POGO, 'Annotated Bibliography of Physical Observations of the Moon 1910–1960', Cal. Inst. of Tech., Pasadena 1961; or L. R. MAGNOLIA and J. R. TREW, 'The Lunar Problem – Bibliography' (STL/AB 61–5110–40), Space Technology Laboratories, Los Angeles, 1961.

For earlier literature the reader can consult the relevant sections of the yearly volumes of *Astronomische Jahresberichte* which have appeared since 1900.

A good many references to earlier literature – but, unfortunately, mixing freely chaff with wheat – can be found also in FIELDER's *Structure of the Lunar Surface* (Pergamon Press, London, 1961) or BALDWIN's *Measure of the Moon* (Univ. of Chicago Press, 1963).

Dr. John C. Noyes, Head of the Astro-Geophysics Research, as well as to all other friends for placing all facilities of these laboratories so generously at my disposal at all stages of this work, and for their unfailing courtesy and cooperation. A large part of the difficult MS was expertly typed for the press by Miss Linda Baqui in Seattle; while, at Manchester, the unfailing assistance of Miss Ellen B. Finlay in all editorial matters has been – as always – invaluable.

Many illustrations accompanying Part III of this book have been secured in the course of an extensive Manchester Programme in lunar photography at the Observatoire du Pic-du-Midi, supported since 1959 by the Air Force Cambridge Research Laboratories and the Aeronautical Chart and Information Center of the U.S. Air Force. May I, at this opportunity, express once more to these institutions our sincere appreciation of the unfailing cooperation with which they have met all our needs.

Last, but not least, I wish to thank Professor C. de Jager, Editor of the *Astrophysics and Space Science Library*, for inviting me to publish this book in his series; and to the D. Reidel Publishing Company for the care which they have bestowed on the printing of the entire volume.

ZDENĚK KOPAL

*"Greenfield", Parkway,*
*Wilmslow, Cheshire,*
*December, 1965*

# TABLE OF CONTENTS

PART THREE

*TOPOGRAPHY OF THE MOON*

PART FOUR

*LUNAR RADIATION AND SURFACE STRUCTURE*

# PART ONE

# MOTION OF THE MOON AND DYNAMICS
# OF THE EARTH-MOON SYSTEM

## CHAPTER 1

# BASIC FACTS: DISTANCE, SIZE, AND MASS

The aim of the present introductory chapter will be to introduce to the reader the Moon as a celestial body, and to review its basic properties such as the distance, size, and mass, the knowledge of which will constitute a necessary prerequisite for proper understanding of the structure and other physical characteristics of the Moon to be discussed in the main part of this book. To the astronomer the Moon has indeed been a friend of old standing; and at least a rudimentary knowledge of its motion goes very far back in the history of mankind on this planet; for since prehistory times the waxing and waning of lunar phases and the light changes accompanying them provided the first astronomical basis for the reckoning of the time. Whenever we go sufficiently far back in the history of almost any primitive civilization, we find them invariably dependent on the lunar, rather than the solar, calendar: the month became a unit of time long before the concept of the year emerged from accumulating observations; and the Moon as the graceful carrier of this knowledge thus gained entrance, as a female deity, to the pantheons of most ancient nations.

The first accurate observers of the Moon's motion and of the length of the month were the Babylonians; but the scientific study of the Moon as a celestial body did not really commence before the advent of the Greeks. The Ionian philosopher THALES (632–546 B.C.) is generally credited to have been the first to recognize the lunar phases as being due to varying illumination by the Sun; and ANAXAGORAS (500–450 B.C.) recognized the true nature of lunar eclipses. Not long thereafter, HIPPARCHOS (between 150–130 B.C.) determined the *distance* separating us from the Moon, by an ingenious geometrical method based on the phenomena of lunar eclipses which we shall explain later (Chapter 5), and determined this distance from observations in terms of the dimensions of the terrestrial globe correctly within a few per cent!

In more recent times, astronomers have re-determined this distance very much more accurately by triangulation – the work of the Greenwich and Cape of Good Hope Observatories is classic in this connection – and found it to vary between 364 400 and 406 730 km each month (due to the eccentricity of lunar relative orbit). The mean distance of the Moon amounts very closely to 384 400 km, and is equal to 60 268 times the Earth's equatorial radius of 6378.17 km (which would, therefore, be seen from the center of the Moon at an angle of $\pi_{\mathbb{C}} = 57'2''.46$ or $3422''.46$, called the Moon's *mean horizontal parallax*), or 0.00257 times the mean distance separating us from the Sun (the latter being, on the average, 389 times as far away from us as the Moon). The distance to the Moon represents, therefore, less than one per cent of the distance separating us from our next two nearest celestial neighbours – the planets

Fig. 1-1.   Change in the apparent diameter of the Moon between perigee (left) and apogee (right), due to the eccentricity of the lunar orbit.

Venus and Mars – even at the time of their closest approach. Light traverses this distance in 1.28 seconds; and a spaceship which can disengage itself from the gravitational field of the Earth can reach the Moon after a flight of 65–70 hours.

Quite recently, the distance to the Moon has been re-measured by a laboratory (non-astronomical) method relying on a knowledge of the velocity of propagation of electromagnetic waves through empty space (nowadays known to better than one part in a million). This method consists of sending out short radar pulses in the direction of the Moon (cf. Chapter 21) and of the accurate timing of the return of their echoes. Strictly speaking, this method measures the instantaneous distance between the transmitting-receiving antenna on Earth and the reflecting portion of the lunar surface; but the time variation of this slant range permits to separate the size of the two bodies from the distance of their centers. In 1959, Yaplee and his collaborators (BURTON, CRAIG, and YAPLEE, 1959) re-determined in this way the Earth-Moon distance and found its mean value to be $384402 \pm 1$ km, in close accord with previous astronomical determinations. The distance separating us from the Moon at any time can, therefore, be nowadays regarded as known within approximately one kilometer; and further refinements of the radio-echo techniques (coupled with a better knowledge of the exact form of the lunar globe) are potentially capable of reducing the remaining uncertainty by a factor of 5–10.

The *apparent diameter* of the lunar disc in the sky has since ancient times been known to be close to half a degree (Archimedes and Hipparchos adopted for it the round value of "720th part of the zodiac"), and varies somewhat on account of the varying distance of the Moon from the observer on the surface of the Earth as well as on the position in its relative orbit. When the topocentric observations are freed from the parallactic effect and reduced to the center of the Earth (for the requisite geometry, cf. Chapter 3), the mean geocentric apparent diameter of the lunar disc is found to amount to $1865''.2$ – oscillating by $204''.8$ between perigee and apogee (see Figure 1-1) – which at the mean distance of 384402 km corresponds to a mean radius of the lunar globe of 1738 km. The Moon is, therefore, little more than one-quarter in size of the Earth. Its surface covers an area of 37.96 million km$^2$; and its volume amounts to 21.99 milliard cubic kilometers (i.e., approximately 2.03 per cent of that of the Earth).

The next quantity of basic importance for the understanding of the fundamental characteristics of our satellite is its *mass*. The mass of the Moon – like that of any other celestial body – can be determined only from the effects of its attraction on another body of known mass; and in the case of the Moon, this will be our Earth (or, quite recently, an impinging spaceship). There are, in principle, three ways by which this can be done: namely, by a study of the effects of the Moon on (a) the orbital motion of the Earth in space; (b) the axial rotation of the Earth; (c) the infall of a spacecraft.

To take up the case (a) first, the oft-repeated statement that the Earth revolves around the Sun in an ellipse is not sufficiently precise; for, in actual fact, it is the center of gravity of the Earth-Moon system which describes (approximately) this curve, while both the Earth and the Moon revolve around this common center of gravity in orbits which are exactly alike in form, but whose absolute dimensions are inversely proportional to their masses.

The Earth does not, therefore, roll around the Sun in an ellipse; it rather wobbles about it in the period of one (synodic) month; and so does, of course, the Moon – only more so. In point of fact, inasmuch as the rate of free fall towards the Sun at the mean distance of the Earth is twice as large as that of the fall of the Moon towards the Earth, the curvature of the absolute orbit of the center of gravity of the Earth-Moon system around the Sun is twice as large as that of the lunar orbit around the Earth. It follows, therefore, that the absolute orbit of the Moon in space is always concave relative to the Sun – at the time of the new Moon it is merely smaller than at full Moon – but its curvature never changes sign. This particular characteristic of its space motion renders our Moon unique among all satellites of the solar system, as it is also in several other dynamical respects which we shall discuss later on.

But let us return to the monthly wobbling of the Earth in space due to the presence of the Moon. The necessary result of such a wobbling must be a slight alternate eastward and westward apparent displacement, on the celestial sphere, of every astronomical object viewed from the Earth – as compared with the place where this object would be seen if the Earth had no satellite and moved around the Sun alone. In the case of the stars (or of more distant planets) this displacement is quite insensible; but it can be well identified in the apparent motion of the Sun – or, better still, of one of the asteroids which can approach the Earth at a close range.

In more precise terms, the geocentric amplitude $L$ of this "lunar inequality" is related with the ratio $\mu$ of the mass of the Moon to that of the Earth, the solar parallax $\pi_\odot$ and the lunar horizontal parallax $\pi_{\mathbb{C}}$ by means of the simple equation

$$L = \frac{\mu}{1+\mu} \frac{\sin \pi_\odot}{\sin \pi_{\mathbb{C}}} \tag{1-1}$$

NEWCOMB (1895) deduced from the observations of the Sun the value of $L$ to be equal to $6''.456 \pm 0''.012$; and subsequent work based on the measured positions of the asteroid Eros led to $L = 6''.4305 \pm 0''.0031$ (HINKS, 1909); $6''.4378 \pm 0''.0012$ (SPENCER JONES, 1941); and $6''.4429 \pm 0''.0015$ (DELANO, 1950). BROUWER and CLEMENCE (1961) adopted the weighted mean of $L = 6''.4378 \pm 0''.0023$ as the most probable value of the constant of lunar inequality, which combined with the lunar horizontal parallax $\pi_{\mathbb{C}} = 3422''.46 \pm 0''.02$ already quoted and $\pi_\odot = 8''.79415 \pm 0''.00005$ as recently determined from the radar rangings of Venus (MUHLEMAN, HOLDRIDGE, and BLOCK, 1962) leads to a value of $\mu^{-1} = 81.33 \pm 0.03$ for the mass-ratio Earth:Moon.*

The method (b) by which the mass of the Moon can be independently ascertained from the motion of the Earth is based on the interpretation of the observed rates at which the axis of rotation of our planet (inclined to its orbital plane by $66°.5$) precesses and nutates in space under the combined gravitational attraction of the Sun and the Moon. The method leads to a less accurate determination of $\mu$ because of the uncertainty inherent in our knowledge of the constant of nutation (influenced as it is by the internal structure of the Earth); but confirms that the Earth is approximately $81.3$ times as massive as the Moon.

Quite recently, these astronomical determinations of the Earth:Moon mass-ratios were largely superseded by more direct determinations of the Moon's mass from the perturbations exerted by it on the motion of man-made spacecraft. Thus from the perturbations suffered by the American Mariner 2 at the commencement of its journey to Venus in 1922 ANDERSON and NULL (1963) obtained the value of $\mu^{-1} = 81.3015 \pm$

---

* BROUWER and CLEMENCE (1961), using an older value of $\pi_\odot = 8''.7984 \pm 0''.0006$ for the solar parallax, obtained a slightly larger value of $81.37 \pm 0.03$ for $\mu^{-1}$.

TABLE 1-1

Values of $Gm_{\text{☾}}$

| Spacecraft | $Gm_{\text{☾}}$ in km³ sec⁻² | $\mu^{-1}$ |
|---|---|---|
| Ranger 6: | $4902.62 \pm 0.14$ | $81.3037 \pm 0.0023$ |
| Ranger 7: | $4902.58 \pm 0.15$ | $81.3043 \pm 0.0026$ |
| Ranger 8: | $4902.64 \pm 0.12$ | $81.3033 \pm 0.0018$ |
| Ranger 9: | $4902.75 \pm 0.20$ | $81.3015 \pm 0.0030$ |
| Mariner 4: | $4902.72 \pm 0.20$ | $81.3020 \pm 0.0030$ |

0.0033; while similar perturbations exhibited since by Rangers 6–9 as well as by Mariner 4 led to the following values of the product $Gm_{\text{☾}}$ (where $G$ stands for the constant of gravitation), which were communicated to the author by W. L. SJOGREN of the Jet Propulsion Laboratory, California Institute of Technology, in advance of publication (See Table 1-1). As it is known that, for the Earth,

$$Gm_{\oplus} = 398601 \pm 1 \text{ km}^3 \text{ sec}^{-2}, \qquad (1\text{-}2)$$

the corresponding values of the Earth:Moon mass-ratios $\mu^{-1}$ are as given in the last column of the foregoing tabulation – and their mean value of 81.303 can be regarded as correct within half a unit of the last place.

This latter value of $\mu$ renders the mean distance of the Earth's center from the common center of gravity of the Earth-Moon system equal to $384402/(1+\mu^{-1})=4671$ km, and places it well inside the mantle of the terrestrial globe (though nearer to its surface than to the center). Moreover, since the value of the gravitation constant $G=(6.668\pm0.005)\times10^{-8}$ cm³/g sec², the absolute masses of the Earth and of the Moon result as $m_{\oplus}=5.978\times10^{27}$g and $m_{\text{☾}}=7.353\times10^{25}$g, respectively; their uncertainties of the order of one part in a thousand are almost solely due to the limits of our present knowledge of the gravitation constant.

The lunar *mass* of $7.35\times10^{25}$g or over 73 trillion tons may loom large in comparison with all terrestrial standards; but on cosmic scales it constitutes but a relatively tiny speck. And neither is the *mean density* of the lunar globe at all unusual; for dividing the mass just found by the lunar volume of $2.199\times10^{25}$ cm³, we find its mean density to be $\rho=3.34$ g/cm³ – i.e., only a little more than the density of common granite rocks of the Earth's crust (2.78 g/cm³), and considerably less than the mean density of the terrestrial globe (5.54 g/cm³). The *gravitational acceleration* $Gm_{\text{☾}}/r_{\text{☾}}^2$ on the lunar surface is, therefore, only some 162 cm/sec² (i.e., less than one-sixth of the terrestrial one); and the *velocity of escape* $(2Gm_{\text{☾}}/r_{\text{☾}})^{\frac{1}{2}}$ from the lunar gravitational field is close to 2.38 km/sec* – in comparison with its terrestrial value of 11.2 km/sec. Many

* For its more exact definition, cf. Chapter 6.

authors have dwelt on the advantages of this fact for economy of effort in lunar environment – that much less muscular work would be required there to lift weights or
throw stones at a distance. However, it should also be stressed that, by the same
reason, our own weight would work less effectively for us if we wished to use it to
compress anything, or to drive a shovel into the ground by stepping upon it.

## CHAPTER 2

# THE MOTION OF THE MOON IN SPACE

Since time immemorial, the observations of the motion of the Moon in the sky have constituted one of the principal preoccupations of astronomers for countless generations. The reasons, originally, have been utilitarian rather than scientific; as the apparent motion of the Moon was so intimately connected with the calendar and time keeping for agricultural and cultic purposes. Until the advent of modern astronomy in relatively recent times, the observers had no way of understanding why the Moon kept moving in the sky in so complicated a manner (of which they discovered many essential features); and it was not until Newton's discovery of the law of universal gravitation that the theory of the lunar orbit could gradually be developed on a logical basis.

In more precise terms, the principal objective of the lunar theory is to find the Moon's motion in space under the gravitational attraction of the Earth, the Sun, as well as of all other bodies (planets) which may act upon it. If planetary action were ignored, and the Moon's mass disregarded in comparison with those of the Sun and the Earth (the latter two being treated as mass-points), the differential equations governing the Moon's motion can be written down at once in the simple form

$$\frac{d^2x}{dt^2} + \frac{Gm_\oplus x}{r^3} = \frac{\partial R}{\partial x}, \tag{2-1}$$

$$\frac{d^2y}{dt^2} + \frac{Gm_\oplus y}{r^3} = \frac{\partial R}{\partial y}, \tag{2-2}$$

$$\frac{d^2z}{dt^2} + \frac{Gm_\oplus z}{r^3} = \frac{\partial R}{\partial z}, \tag{2-3}$$

where $x$, $y$, $z$ are the rectangular coordinates of the Moon in a system whose origin coincides with the Earth's center (such that $x^2 + y^2 + z^2 = r^2$); $m_\oplus$ is the Earth's mass; and the disturbing function $R$ arising from the attraction of the Sun on the Moon can be expressed as

$$R = Gm_\odot \left\{ \frac{1}{\Delta} - \frac{xx' + yy' + zz'}{r'^3} \right\}, \tag{2-4}$$

where $x'$, $y'$ $z'$ $(x'^2 + y'^2 + z'^2 = r'^2)$ are the coordinates of the Sun; $m_\odot$, the Sun's mass; and

$$\Delta^2 = (x - x')^2 + (y - y')^2 + z^2 \tag{2-5}$$

if the orbit of the Earth around the Sun has been identified with the $xy$-plane.

If, to a first approximation, the solar attraction too were ignored, the solution of the problem would be well known; and the relative orbit of the Moon around the Earth would be an *ellipse* of fixed position in space. In actual fact – as was found by Kepler seventy-seven years before the publication of Newton's *Principia*, the orbit of the Moon is only approximately an ellipse – of mean eccentricity $e = 0.05490$ – and its plane is inclined to the ecliptic (i.e., the $xy$-plane) by an angle $i$ which, at present (1964) is equal to $5°8'43''.4$. The actual lunar orbit departs, however, from a stationary ellipse, and its form is distorted not only by the attraction of the Sun (or, more precisely, the difference of the solar attraction on the Moon and the Earth), but also (though to a much smaller extent) by the oblateness of the Earth and the attraction of other planets.

The principal effect of solar perturbations is to cause the Keplerian ellipse of the lunar orbit to rotate slowly in space, in such a way that the *line of apsides* joining the points at which the Moon is nearest to the Earth (perigee) and farthest away from it (apogee) *advances* (i.e., moves toward the East) at a rate of $146427''.9$ per annum (all other non-solar perturbations acting upon the apsidal line will add only $8''.3$ to its secular motion), corresponding to a complete revolution of the apsidal line in 3232.57 days or 8.85 years. In the meantime, the *line of the nodes* (i.e., the intersection of the lunar orbit with the ecliptic) *recedes* (moving West) at an annular rate of $69679''.4$ – i.e., little less than half the rate at which the apsides advance – corresponding to a complete revolution in 6798.36 rays or approximately 18.61 years.

This motion of the plane of the lunar orbit in space has several important consequences; and those of most immediate interest concern the *orbital period* of our satellite. If we define this period as an interval of time after which the Moon will have made a complete revolution with respect to the celestial sphere – the so-called *sideric month* – this is known with great precision to be equal (in 1964) to 27.32166150 mean solar days or approximately 27 days, 7 hours, 43 minutes and 11.5 sec. During the time of one sideric month the Sun has, however, moved eastward by approximately one-twelfth of the entire circle; and, consequently, after the lapse of one sideric month the Moon would not have returned to the same phase. As a result, the time interval which elapses between two successive identical phases of the Moon – the so-called *synodic month* – is longer than the sideric one, and equal to 29.5305883 days or 29 days, 12 hours, 44 minutes and 2.8 seconds; this is the period after which the Moon returns to the same phase.* Should we, however, define a month as a time interval after which the Moon will return to the same place in its relative orbit around the Earth, this so-called *anomalistic month* will likewise be longer than the sideric one (though shorter than the synodic month) because of the secular advance of the apsides, and is equal to 27.5545503 days or 27 days, 13 hours, 18 minutes and 37.4 seconds; while a time interval between two successive nodal passages – the so-called *draconic month* – is shortest of them all (because the nodes recede) and equal to 27.212220 days or 27 days, 5 hours, 5 minutes and 35.8 seconds.

* It is of interest to note that this period was determined already by Hipparchos correctly to one second.

After having surveyed thus the effects of the principal *secular* perturbations of the lunar orbit – the advance of the apsides and regression of the nodes – caused by the solar attraction, let us turn to the *periodic* perturbations due to the same cause. The next important periodic perturbation in the Moon's motion is the *evection*, due to the effect of solar attraction on the eccentricity of the lunar orbit. If the apsidal line is parallel with the direction towards the Sun (i.e., when the Moon is "full" or "new" at perigee and apogee, or vice versa) the attraction of the Sun tends to elongate the ellipse of the relative lunar orbit and increase its eccentricity. If, on the other hand, the apsidal line is perpendicular to the line joining the positions of the "full" and "new" Moon – a line to which the astronomers refer as syzygies – solar attraction will tend to widen the ellipse and diminish its eccentricity. In all, this phenomenon recurring in the period of 31 807 days, can displace the position of the Moon in longitude by as much as $1°16'26''.4$, and cause the apparent orbital eccentricity (the mean value of which is equal to 0.0549) to fluctuate between 0.0432 and 0.0666. The displacement in longitude due to evection, which can amount to more than five apparent semi-diameters of the Moon, was discovered already in antiquity and known to Hipparchos.

The second most important periodic perturbation – the so-called *variation* – arises from the periodic interference of the attraction of the Sun and the Earth. Because of their relative positions, the gravitational attraction of these bodies acts in concert if the Moon is approaching the "full" or "new" phase, but against each other if the Moon is receding from these phases; the effect will vanish at both quarters when the two forces are perpendicular. The period of this variation will, therefore, be equal to one-half of the synodic month; and although its action can displace the position of the Moon in longitude by as much as $39'29''.9$ (i.e., by more than its apparent semi-diameter), it remained unnoticed by all Greek astronomers including Ptolemy – perhaps because it vanishes (when changing sign) at both full and new Moon and does not affect the times of solar or lunar eclipses (with the predictions of which the Greeks were so much concerned). It was apparently first noticed by the medieval Arab astronomer ABDUL-WEFA (939–998 A.D.) and established by Tycho DE BRAHE (1546–1601). Another periodic perturbation of the Moon's motion – third in order of importance by its magnitude – is the so-called *annual inequality*, due to the eccentricity of the terrestrial orbit around the Sun. Its amplitude attains $11'8''.9$ in longitude, and it fluctuates in the period of one anomalistic year. It was discovered by KEPLER (1571–1630).

All lunar perturbations described in previous paragraphs are sufficiently large to be noticed by the naked eye; and all of them were discovered in the days of pre-telescopic astronomy. A proper understanding of their causes was, of course, impossible while NEWTON laid down the foundations of the celestial mechanics in his *Principia* (1687). The first analytical theory of the Moons' motion was attempted by Newton himself, who succeeded in accounting for the existence of the principal secular as well as periodic perturbations of the Moon's motion enumerated earlier in this section.*

* In the 1687 edition of the *Principia,* Newton accounted for only about a half of the secular advance of lunar perigee produced by the solar action. However, an examination of Newton's unpublished

Even so great a man found, however, the problems posed by the lunar theory more than a match for his ingenuity, and is said to have told his friend Halley that it "made his head ache and kept him awake to often that would think of it no more".

Newton certainly was not the only one who must have experienced this feeling; and the great problem which he opened up continued to exercise the genius of the best minds of many generations up to our own time. In the 18th century, the development of the lunar theory was almost entirely the work of five men: L. EULER (1707–1783), A. C. CLAIRAUT (1713–1765), J. D'ALEMBERT (1717–1783), J. L. LAGRANGE (1736–1813) and P. S. LAPLACE (1749–1827). Their work was, to a large extent, stimulated by a general demand prevalent at that time for accurate lunar tables for the use of navigators in determining their position at sea; and this (together with the fact that the motion of the Moon presented the best test for Newton's theory of gravitation) induced the British Government, as well as several academic and scientific societies, to offer very substantial prizes for lunar tables which would agree satisfactorily with the observations.

In response to such needs, Clairaut and d'Alembert in 1747 presented (on the same day) to the Academy of Sciences in Paris their memoires on the motion of the Moon, which in 1749 Clairaut improved by explaining more completely the motion of its perigee (as was done also by Euler and d'Alembert a little later). Clairaut won the prize offered by the St. Petersburg Academy of Sciences for his *Théorie de la Lune* in 1752; while Euler published a year later a lunar theory which agreed with the observations so well that he was granted a reward of £ 3000 by the British Government. In 1772, Euler published his second lunar theory which represented a further improvement of his earlier work. In the meantime, Lagrange laid down the foundations on which much of the 19th century work on the Moon was based; while Laplace made one of his most important contributions to celestial mechanics in 1787, when he explained (in the third volume of his *Traité de mécanique céleste*) the cause of the secular acceleration of the Moon's mean motion.

All these investigators (with the exception of Euler) and, following them, DAMOISEAU (1827) and PLANA (1832) carried out their work in rectangular coordinates, in terms of the Moon's true longitude as the independent variable; while LUBBOCK (1830–1834) and DE PONTÉCOULANT (1846) developed a similar theory in polar coordinates. On the other hand, EULER (1753) and later POISSON (1835), HANSEN (1862–1864) and DELAUNAY (1860, 1867) employed for the lunar theory the method of the "variation of constants" of elliptic motion. Hansen's work was largely numerical; but by its completeness it became the foundation of lunar ephemerides in nautical almanacs for many years; while Delaunay developed a literal theory of the Moon's motion in canonical variables, which led to analytic expansions with many hundreds of periodic terms (each representing a separate perturbation).

In 1877, a notable advance in further improvement of the lunar theory was made

posthumous papers contained in the Portsmouth Collection revealed, in 1872, that later Newton accounted for almost the entire motion of the perigee (within 8% of it observed value) by solar perturbations of the second order.

by HILL, who employed a rotating system of coordinates. While the first attempt to use such coordinates was already made by Euler in his second theory (1772) just over a century before Hill, Euler let the z-axis of his rectangular coordinate system (perpendicular to the ecliptic) rotate with the Moon's mean angular velocity, while in Hill's work the coordinate system rotates with the radius-vector Sun-Earth. In more specific terms, while all investigators preceding Hill departed in one way or another from our equations (2-1 to 2-5) and thus used the ellipse as an approximate solution, Hill employed for the same purpose the *variational orbit* defined by the following equations of the restricted problem of three bodies,

$$\frac{d^2 X}{dt^2} - 2n\frac{dY}{dt} = \frac{\partial U}{\partial X}, \tag{2-6}$$

$$\frac{d^2 Y}{dt^2} + 2n\frac{dX}{dt} = \frac{\partial U}{\partial Y}, \tag{2-7}$$

$$\frac{d^2 Z}{dt^2} = \frac{\partial U}{\partial Z}, \tag{2-8}$$

in which the Moon is considered as a mass particle moving in gravitational field of the revolving dipole Sun-Earth.

In the foregoing equations, $X$, $Y$, $Z$ stand for the rectangular coordinates of the Moon (with origin at the center of the Earth, and the $X$-axis in the direction of the Sun, while the $XY$-plane coincides with the ecliptic), rotating in space with an angular velocity $n$; and the potential function

$$U = \tfrac{1}{2}n^2(X^2 + Y^2) + \frac{G(m_\oplus + m_\text{☾})}{r} + R \tag{2-9}$$

where $R$ denotes the solar disturbing function as defined by Equation (2-4); so that, more explicitly,

$$\begin{aligned} U = \tfrac{1}{2}n^2(X^2 + Y^2) &+ \frac{G(m_\oplus + m_\text{☾})}{\sqrt{X^2 + Y^2 + Z^2}} \\ &+ Gm_\odot\left\{\frac{1}{\sqrt{(X-X')^2 + Y^2 + Z^2}} - \frac{X}{X'^2}\right\} \end{aligned} \tag{2-10}$$

where $X'$ denotes the distance of the Sun along the $X$-axis.

The principal advantage of the foregoing system over that underlying Euler's work of 1772 is the fact that, unlike Euler's system, Equations (2-6)–(2-8) admit of a certain integral in a closed form; for if we multiply them successively by $2(dX/dt)$, $2(dY/dt)$, $2(dZ/dt)$ and add, the so resulting an equation can be readily integrated to yield the celebrated Jacobian *vis-viva integral* of the form

$$\left(\frac{dX}{dt}\right)^2 + \left(\frac{dY}{dt}\right)^2 + \left(\frac{dZ}{dt}\right)^2 = V^2 = 2U - C \tag{2-11}$$

where $V$ denotes the velocity of the particle, and $C$ is a constant.

In applying this system to the theory of the Moon's motion, Hill ignored the eccentricity $e=0.01675$ of the terrestial orbit around the Sun to render $n$ as well as $X'$ constant, neglected the inclination $i=5°8'43''$ of the lunar orbit to the ecliptic to set $Z=0$, and also the ratio $X/X'$ as small; departures from these conditions being treated subsequently as perturbations of his variational orbit. This method has become the basis of all subsequent work on the Moon's motion, at the hands of E. W. BROWN (between 1897–1908) and D. BROUWER, who in collaboration with G. M. CLEMENCE and W. J. ECKERT converted subsequently the underlying theory into most precise existing numerical tables of the Moon's motion valid for long intervals of time – so precise that this motion has now universally been adopted as the fundamental astronomical clock marking the passage of the time. It is true that crystal or molecular clocks in the laboratory can now measure short intervals of time with greater precision than that attainable by any kind of astronomical measurements; but an inherent lack of stability of such devices over long intervals renders our Moon still the most reliable guardian of the time. In this respect, the Moon has retained its traditional role from the dawn of civilization till the present, and is likely to retain it for a long time to come.

The importance of our only natural satellite in this connection can thus scarcely be overestimated; and it should be realized that the Moon has retained this importance only because its motion has been so thoroughly understood – thus making possible a sufficiently accurate conversion of the observed positions into time. In point of fact, our astronomical time is defined now as the independent variable in the lunar equations of motion; and the observed motion of the Moon thus serves as a handle of the universal clock.

If, in spite of this, the theory of the motion of the Moon in space has been dealt with in this section only in a summary fashion, the reason is the fact that the reader desirous of getting more fully acquainted with this subject is already well served by the existing literature. Brown's *Introductory Treatise on the Lunar Theory* (Cambridge, 1896) was recently reprinted in paperback form (Dover, New York, 1960); and Tisserand's *Lunar Theory* (being Volume III of his *Traité de mécanique céleste*, Paris, 1894) became likewise recently availbale in a modern reprint (Gauthier-Villars, Paris, 1960). For a shorter account of the lunar theory the reader can, moreover, consult with profit Chapters 12–14 of the *Methods of Celestial Mechanics* by Brouwer and Clemence (Academic Press, New York, 1961); while for a still more concise survey of the field cf. Brouwer and Hori's Chapter 1 in the *Physics and Astronomy of the Moon* (ed. by Z. Kopal, Acad. Press, New York and London, 1962).

There is, however, another reason which should make it legitimate to present only an outline of the theory of the Moon's motion in an introductory volume to the study of our satellite as an astronomical body: and that is the fact the motion of the Moon in space is virtually independent of any physical properties of the lunar globe as such – the coupling between the two being extremely weak. For instance, of the total annual recession of the Moon's node due to all causes – which amounts to $-69679''.36$ – only about $-0''.16$ arises from the distortion of the figure of the Moon; and of the total amount $146435''.21$ of the annual advance of the perigee, only $+0''.03$ goes back to the same cause; while the relativistic corrections to both amount to $0''.02$ per annum (cf. DE SITTER, 1917). On account of the smallness (and also very nearly spherical distribution) of its mass, the Moon has virtually no say about its motion through space, being completely at the mercy of outside forces which it cannot influence; and has to suffer the ignominy of being treated as a nondescript mass-particle without any individuality of its own. For this reason, the mathematically beautiful and powerful theories of the motion of the Moon in space are of only auxiliary interest for the actual studies of the Moon to which we wish to introduce the reader in this volume; all data that we may need in this connection are available in tabular form, to the requisite precision, in existing astronomical almanacs.

When, however, we turn to another kind of the Moon's motion: namely, one which the lunar globe performs about its own center of gravity, the situation becomes completely different; and in spite of the smallness of such motions they will prove a powerful source of information on the distribution of mass inside the lunar globe – in fact, the only such source we possess so far – so that its theory merits a closer analysis which will be given in the next chapters.

## CHAPTER 3

# ROTATION OF THE MOON; OPTICAL LIBRATIONS

The revolution of the Moon around the Earth, and with the Earth around the Sun, are not the only motions performed by our satellites. As has been realized in the earliest days of lunar observations from the fact that the Moon exhibit to us on Earth (almost) always the same face, it must also rotate with a uniform angular velocity about an axis fixed in space and inclined but little to its orbital plane.

Toward the end of the 17th century, J. D. CASSINI from a long series of observations deduced three empirical laws respecting lunar rotation, which bear his name and can be expressed as follows:

(a) The Moon rotates eastward, about a fixed axis, with constant angular velocity and in a period equal to one sideric month*

(b) The inclination $I$ of the Moon's axis of rotation to the ecliptic remains constant; and

(c) The poles of the Moon's axis of rotation, of the ecliptic, and of the lunar orbit lie in the same plane, and on one great circle, in that order (cf. Figure 3-1).

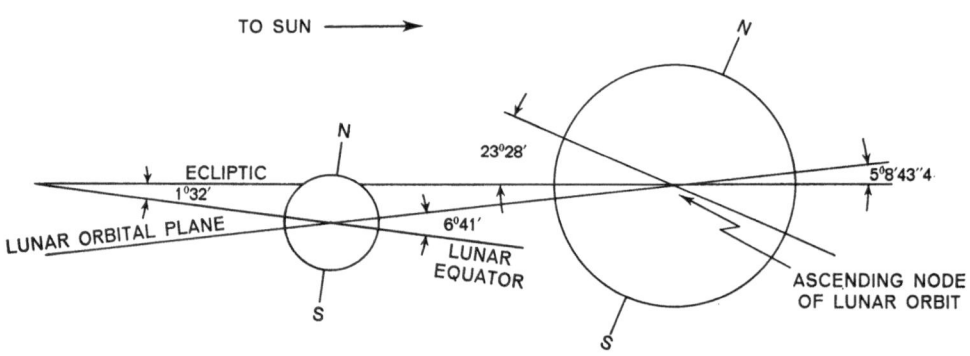

RELATIVE ORIENTATION-MOON, EARTH, AND ECLIPTIC.

Fig. 3-1.

---

* This synchronism is now known (cf. BANACHIEWICZ, 1950) to be accurate within at least 0.1 sec (e.g., approximately 1 part in 30 million); and may have been so for an astronomically long time.

The angle $i$ between the Moon's orbit and the ecliptic is known with great exactitude to be equal to $5°8'43''.4$ (Brown). The inclination $I$ of the lunar equator to the ecliptic is known with somewhat lesser precision to amount to $I = 1°32'1'' \pm 7''$ (KOZIEL, 1964). Hence, in accordance with Cassini's third law, the inclination of the Moon's equator to its orbital plane should be constant and equal to the sum of $I + i = 6°40'44''$. These facts entail, in turn, certain interesting consequence which manifest themselves to us as the *optical librations* of our satellite.

The oft-repeated statement that the Moon exhibits to the Earth always the same face on account of the synchronism prevailing between rotation and revolution is only approximately true, but cannot be exact for several reasons. The first is the fact that, if the axial rotation of the Moon is uniform (Cassini's first law), the angular velocity of its revolution in an elliptic orbit (varying as it does with inverse square of the radius vector, in accordance with Kepler's second law) is not; being sometimes ahead, and at other times behind, the orbital motion. The effect of this must be an alternating periodic angular displacement of lunar objects in longitude by as much as $7°57'$ about the center of the Moon – a phenomenon known as the *optical libration in longitude* (cf. Figure 3-2).

Secondly, as the lunar axis of rotation is not perpendicular to the orbital plane, but its inclination deviates from $90°$ by $I + i$ (Cassini's second and third law), we can see sometimes more of one polar region and, at other times, more of the other in the course of each month. This phenomenon give rise to an *optical libration in latitude* by $6°51'$ (Figure 3-2). Again, when the Moon is rising for the observer on the Earth we look over its upper edge, seeing a little more of that part of the Moon than if we were observing it from the Earth's center; and when the Moon is setting the converse is true. This *diurnal libration* (not of the Moon, to be sure, but of the observer) amounts to $\pi_{\mathbb{C}} = 57'2''.6$, and superposes upon all other librations to enable us to see considerably more than one-half of the lunar surface from the Earth. On the whole, not less than $59\%$ of the entire lunar globe can be seen at one time or another from the Earth; only $41\%$ being permanently invisible, and an equal amount never disappearing; the remainder of $18\%$ being alternately visible and invisible. We may add that, on October 1959, the cameras aboard the Russian third space station unveiled also the essential features of a major part ($28\%$) of the far side of the Moon (cf. Figures 16-2 and 16-3). The relative position of Lunik 3 at the time of the photography was such that not all of the lunar far side was exposed to the camera, and approximately $13\%$ of it remained uncharted till July 20, 1965, when another Russian space probe (Zond 6) succeeded (cf. Figure 16-4) in recording for us all but a small fraction of the remainder (LIPSKY, 1965).

The actual magnitude of lunar optical librations at anytime is easy to compute from known elements of the Moon's motion. Let, in what follows, $b$ and $l$ denote the selenographic latitude and longitude of the Earth's center (or, what is the same, the selenographic latitude and longitude of the apparent center of the lunar disc as would be seen from the center of the Earth). These optical (geocentric) libration angles can be evaluated from the equations

Fig. 3-2. A comparison of two aspects of the full-moon face, as influenced by the optical librations of our satellite, and recorded on photographs taken on 1961 December 21, 21$^h$ 18$^m$ 20$^s$ UT (libration constants: $l = +5°.7$, $b = +4°.5$) |–| left |–| and on 1962 April 20, 3$^h$ 45$^m$ 18$^s$ UT (libration constants: $l = -6°.6$, $b = -4°.1$) |–| right.

$$\left.\begin{array}{ll}
\cos b \cos (l + l_{\mathbb{C}} - \Omega + \varDelta) = & - \cos (\alpha_{\mathbb{C}} - \Omega') \cos \delta_{\mathbb{C}}, \\
\cos b \sin (l + l_{\mathbb{C}} - \Omega + \varDelta) = & - \sin (\alpha_{\mathbb{C}} - \Omega') \cos \delta_{\mathbb{C}} \cos i \\
& \quad - \sin \delta_{\mathbb{C}} \sin i, \\
\sin b \qquad\qquad\qquad = & + \sin (\alpha_{\mathbb{C}} - \Omega') \cos \delta_{\mathbb{C}} \sin i \\
& \quad - \sin \delta_{\mathbb{C}} \cos i,
\end{array}\right\} \qquad (3\text{-}1)$$

in terms of the instantaneous right-ascension ($\alpha_{\mathbb{C}}$) and declination ($\delta_{\mathbb{C}}$) of the Moon; or from an alternative set of equations

$$\begin{array}{ll}
\cos b \cos (l + l_{\mathbb{C}} - \Omega) = & \cos (\lambda_{\mathbb{C}} - \Omega) \cos \beta_{\mathbb{C}}, \\
\cos b \sin (l + l_{\mathbb{C}} - \Omega) = & \sin (\lambda_{\mathbb{C}} - \Omega) \cos \beta_{\mathbb{C}} \cos I \\
& \quad - \sin \beta_{\mathbb{C}} \sin I, \\
\sin b \qquad\qquad\qquad = & - \sin (\lambda_{\mathbb{C}} - \Omega) \cos \beta_{\mathbb{C}} \sin I \\
& \quad - \sin \beta_{\mathbb{C}} \cos I,
\end{array} \qquad (3\text{-}2)$$

in terms of the Moon's ecliptical latitude $\beta_{\mathbb{C}}$ and longitude $\lambda_{\mathbb{C}}$, which are related with the respective equatorial coordinates $\alpha_{\mathbb{C}}$ and $\beta_{\mathbb{C}}$ by means of the well-known equations

$$\left.\begin{array}{ll}
\cos \beta_{\mathbb{C}} \cos \lambda_{\mathbb{C}} = \cos \delta_{\mathbb{C}} \cos \alpha_{\mathbb{C}} \\
\cos \beta_{\mathbb{C}} \sin \lambda_{\mathbb{C}} = \sin \delta_{\mathbb{C}} \sin \varepsilon + \cos \delta_{\mathbb{C}} \sin \alpha_{\mathbb{C}} \cos \varepsilon \\
\sin \beta_{\mathbb{C}} \qquad = \sin \delta_{\mathbb{C}} \cos \varepsilon - \cos \delta_{\mathbb{C}} \sin \alpha_{\mathbb{C}} \sin \varepsilon
\end{array}\right\} \qquad (3\text{-}3)$$

yielding

$$\left.\begin{array}{l}
\tan \lambda_{\mathbb{C}} = \cos (N - \varepsilon) \sec N \tan \alpha_{\mathbb{C}}, \\
\tan \beta_{\mathbb{C}} = \tan (N - \varepsilon) \sin \lambda_{\mathbb{C}},
\end{array}\right\} \qquad (3\text{-}4)$$

with

$$\tan N = \tan \delta_{\mathbb{C}} \csc \alpha_{\mathbb{C}}, \qquad (3\text{-}5)$$

where $\Omega$ denotes the longitude of the ascending node of the lunar orbit on the ecliptic; $\Omega'$, the longitude of the ascending node of the lunar equator on the terrestrial equator (both $\Omega$ and $\Omega'$ being measured from the vernal equinox); $l$, the mean longitude of the Moon in its orbit; $\varepsilon$, the obliquity of the ecliptic; while $\varDelta$ is the angular distance between the terrestrial equator and the ecliptic measured along the lunar equator; the latter two angles being defined by the equations

$$\sin \Omega' = - \sin I \csc i \sin \Omega \qquad (3\text{-}6)$$

and

$$\sin \varDelta = - \sin \varepsilon \csc i \sin \Omega, \qquad (3\text{-}7)$$

respectively.

The values of $b$ and $l$ resulting from the solution of the systems (3-1) or (3-2) represent *geocentric* librations. Their *topocentric* values $b'$, $l'$ (taking account of the diurnal libration) can however be obtained from the same system of Equations (3-1) or (3-2) in which *the geocentric coordinates* $\alpha_{\mathbb{C}}$, $\delta_{\mathbb{C}}$ *or* $\lambda_{\mathbb{C}}$, $\beta_{\mathbb{C}}$ *of the Moon have been replaced by their respective topocentric values* (i.e., their apparent values at the actual place and time of observation). If primes denote such topocentric values of the respective coordinates at a distance $\rho$ from the center of the Earth and geocentric latitude $\varphi$, with

$\Theta$ denoting the sidereal time of observation, while $r$ and $r'$ stand for the geocentric ($r$) and topocentric ($r'$) distance of the Moon's center, the exact equations relating the (instantaneous) topocentric right ascension $\alpha'_{\mathbb{C}}$ and declination $\delta'_{\mathbb{C}}$ with their geocentric values (as tabulated in any standard ephemeris) are of the form

$$
\left.
\begin{aligned}
r' \cos \delta'_{\mathbb{C}} \cos \alpha'_{\mathbb{C}} &= r \cos \delta_{\mathbb{C}} \cos \alpha_{\mathbb{C}} - \rho \cos \varphi \cos \Theta \\
r' \cos \delta'_{\mathbb{C}} \sin \alpha'_{\mathbb{C}} &= r \cos \delta_{\mathbb{C}} \sin \alpha_{\mathbb{C}} - \rho \cos \varphi \sin \Theta \\
r' \sin \delta'_{\mathbb{C}} &= r \sin \delta_{\mathbb{C}} - \rho \sin \varphi,
\end{aligned}
\right\}
\tag{3-8}
$$

which can be solved to yield the differences

$$
\tan (\alpha'_{\mathbb{C}} - \alpha_{\mathbb{C}}) = \frac{\rho \cos \varphi \sin \pi \sec \delta_{\mathbb{C}} \sin (\alpha - \Theta)}{1 - \rho \cos \varphi \sin \pi \sec \delta \cos (\alpha - \Theta)}
\tag{3-9}
$$

and

$$
\tan (\delta'_{\mathbb{C}} - \delta_{\mathbb{C}}) = \frac{\rho \sin \varphi \sin \pi \csc \Gamma \sin (\Gamma - \delta_{\mathbb{C}})}{\rho \sin \varphi \sin \pi \csc \Gamma \cos (\Gamma - \delta_{\mathbb{C}}) - 1},
\tag{3-10}
$$

where the auxiliary angle $\Gamma$ is defined by the equation

$$
\tan \Gamma = \tan \varphi \cos \frac{\alpha'_{\mathbb{C}} - \alpha_{\mathbb{C}}}{2} \sec \left( \Theta - \frac{\alpha'_{\mathbb{C}} + \alpha_{\mathbb{C}}}{2} \right),
\tag{3-11}
$$

and where the geocentric radius-vector $r$ has been expressed in terms of the lunar equatorial horizontal parallax $\pi_{\mathbb{C}}$ by means of the relation

$$
\frac{\rho}{r} = \sin \pi_{\mathbb{C}}.
\tag{3-12}
$$

Moreover once the angles defined by the foregoing equations have been evaluated, the ratio of the topocentric and geocentric distance of the Moon results from

$$
\frac{r'}{r} = \frac{\sin (\delta_{\mathbb{C}} - \Gamma)}{\sin (\delta'_{\mathbb{C}} - \Gamma)},
\tag{3-13}
$$

so that the *augmentation* of the apparent semi-diameter $s'$ of the lunar disk, as seen by the observer on the surface of the Earth, over its geocentric semi-diameter $s$ is given by the equation

$$
\frac{s'}{s} = \frac{r}{r'}.
\tag{3-14}
$$

As the observer on Earth can never be at a greater distance from the Moon than $r$ if our satellite is to be visible above the horizon, it follows that $s' > s$.

All equations given so far in this section are exact, and should permit us to evaluate $b, l$ and $b', l'$ or the ratios $s'/s$ or $r'/r$ to the same precision to which the underlying data are known. If, however, highest accuracy is not required, the system (3-2) of equations can be solved approximately to yield

$$b = -\beta_{\mathbb{C}} - B + \tan^2 \tfrac{1}{2} I \sin 2 (B + \beta_{\mathbb{C}}) + \ldots, \tag{3-15}$$

where we have abbreviated

$$\tan B = \tan I \sin (\lambda_{\mathbb{C}} - \Omega); \tag{3-16}$$

or, correctly to the squares of the small angle $I = 1°32'$, the geocentric optical libration in latitude can be obtained from

$$b + \beta_{\mathbb{C}} = -I \sin (\lambda_{\mathbb{C}} - \Omega) + \tfrac{1}{4} I^2 \sin 2 \beta_{\mathbb{C}} + \ldots; \tag{3-17}$$

while, similarly, the optical libration in longitude

$$\begin{aligned} l &= \lambda_{\mathbb{C}} - l_{\mathbb{C}} + b \ I \cos (\lambda_{\mathbb{C}} - \Omega) + \tfrac{1}{4} I^2 \sin 2 (\lambda_{\mathbb{C}} - \Omega) + \ldots \\ &= \lambda_{\mathbb{C}} - l_{\mathbb{C}} - \beta_{\mathbb{C}} I \cos (\lambda_{\mathbb{C}} - \Omega) - \tfrac{1}{4} I^2 \sin 2 (\lambda_{\mathbb{C}} - \Omega) + \ldots \end{aligned} \right\} \tag{3-18}$$

The corrections necessary to reduce the geocentric values of $b$ and $l$ to the topocentric (primed) ones can be evaluated by taking advantage of the fact that, if highest accuracy is not required, the solution (3-9) and (3-10) of Equations (3-8) can be approximated by the expressions

$$\alpha_{\mathbb{C}}' - \alpha_{\mathbb{C}} = -\rho \cos \varphi \sin \pi_{\mathbb{C}} \sec \delta_{\mathbb{C}} \sin \Theta + \ldots, \tag{3-19}$$

$$\delta_{\mathbb{C}}' - \delta_{\mathbb{C}} = +\rho \sin \pi_{\mathbb{C}} (\sin \varphi \cos \delta_{\mathbb{C}} - \cos \varphi \sin \delta_{\mathbb{C}} \cos \Theta) + \ldots, \tag{3-20}$$

and, similarly, (3-14) can be approximated by

$$s' = s (1 + \rho \sin \pi \cos z) \propto s (1 + 0.0166 \cos z), \tag{3-21}$$

where $z$ denotes the zenith distance of the Moon at the time and place of observation; and, as such, can be evaluated from known data with the aid of the formula

$$\cos z = \sin \varphi \sin \delta_{\mathbb{C}}' + \cos \varphi \cos \delta_{\mathbb{C}}' \cos (\alpha_{\mathbb{C}}' - \Theta). \tag{3-22}$$

The topocentric values $b'$ and $l'$ then follow from the equations

$$\sin b' = \cos \sigma \sin b + \sin \sigma \cos b \cos (Q - C) \tag{3-23}$$

and

$$\sin (l' - l) = -\sin \sigma \sin (Q - C) \sec b', \tag{3-24}$$

where $\sigma$ denotes the selenocentric angle between the observer and the center of the Earth; and, as such, is given by the equation

$$\tan \sigma = \frac{\sin \gamma}{\csc \pi_{\mathbb{C}} - \cos \gamma}, \tag{3-25}$$

where $\gamma$ is the geocentric angle between the observer and the Moon (i.e., geocentric "zenith distance" of the Moon, obtainable from Equation (3-22) in which geocentric coordinates $\alpha_{\mathbb{C}}$, $\delta_{\mathbb{C}}$ are used in place of topocentric ones); $Q$, the azimuth of the observer measured at the sub-lunar point; and $C$ the position angle of the Moon's central meridian (measured positively eastward from the north). The latter two angles are, in turn, defined by the equations

$$\left.\begin{array}{l} \sin Q = \cos \varphi \csc \gamma \sin (\alpha_{\mathbb{C}} - \Theta), \\ \cos Q = (\sin \varphi - \cos \gamma \sin \delta_{\mathbb{C}}) \sec \delta_{\mathbb{C}} \csc \gamma, \end{array}\right\} \qquad (3\text{-}26)$$

and

$$\left.\begin{array}{l} \sin C = - \sin i \sec b \cos (\alpha_{\mathbb{C}} - \Omega'), \\ \quad = + \sin i \sec \delta_{\mathbb{C}} \cos (l + l' - \Omega + \varDelta). \end{array}\right\} \qquad (3\text{-}27)$$

Since, however, $\pi_{\mathbb{C}} < 1°$ and, therefore,

$$\sigma = \pi_{\mathbb{C}} \sin \gamma, \qquad (3\text{-}28)$$

Equations (23) and (24) can be readily approximated (cf., e.g., ATKINSON, 1951) by

$$b' - b = \pi_{\mathbb{C}} \sin \gamma \cos (Q - C) \qquad (3\text{-}29)$$

and

$$l' - l = \pi_{\mathbb{C}} \sin \gamma \sin (Q - C) \sec b'. \qquad (3\text{-}30)$$

Moreover, because (on account of the smallness of $i$), $C$ is likewise a small angle whose squares and higher powers can be ignored (so that $\cos C \sim 1$) an insertion from (3-26) and (3-27) in (3-29) and (3-30) reveals that, with an accuracy sufficient for many purposes,

$$b' - b = \pi_{\mathbb{C}} \{\sin \varphi - \sin \delta_{\mathbb{C}} \cos \gamma \qquad (3\text{-}31)$$
$$+ \cos \varphi \sin i \sin (\alpha_{\mathbb{C}} - \Theta) \cos (l + l_{\mathbb{C}} - \Omega + \varDelta)\} \sec \delta_{\mathbb{C}}$$

and

$$l' - l = \pi_{\mathbb{C}} \{\cos b \cos \varphi \sin (\alpha_{\mathbb{C}} - \Theta) \qquad (3\text{-}32)$$
$$+ (\sin \varphi - \sin \delta_{\mathbb{C}} \cos \gamma) \sin i \sec \delta_{\mathbb{C}} \cos (\alpha_{\mathbb{C}} - \Omega')\} \sec^2 b.$$

The reader will doubtless appreciate now the fact that – even in their simplest form as given above – the corrections necessary to convert the geocentric librations of the Moon into topocentric librations are not exactly simple. While the geocentric values $b$, $l$ of the Moon's optical libration, as defined by Equations (3-1) or (3-2), can usually be taken with sufficient precision from the existing ephemerides, their reductions to the topocentric values appropriate for a particular place and time of observation (or the converse operation) must be performed by the observer himself, with the aid of such formulae as given above. Fortunately, the absolute amounts of the differences $b' - b$ or $l' - l$ are small (always less than a degree) and need not, therefore, be usually computed to too many significant places.

The optical librations of the Moon represent phenomena of some magnitude, and were detected early in the history of lunar studies. The first one to have noticed the periodic displacement of lunar spots alternately toward the eastern and western limb appears to have been Galileo GALILEI, who in his *Dialogues on the Two Great World*

*Systems* had Salviati voice two possible causes: namely, the libration in latitude and diurnal libration – while the bulk of the phenomenon he observed was due to the libration in longitude. This libration was recognized as such by RICCIOLI and HEVELIUS between 1638–41. In 1654, in a letter to Riccioli, Hevelius expressed first the opinion that the true cause of the observed libration in longitude is the non-uniformity of the motion of the Moon in its orbit (implicit, in fact, in Kepler's second law known since 1609). The correct explanation of all three optical librations of the Moon in terms of the characteristics of its orbit was advanced in the third volume of his *Principia* by NEWTON (who, incidentally, was the first to introduce the term "optical libration" in this connection). Later in the same volume Newton, in discussing the problem of the figure of the Moon, mentioned also the possibility of its physical librations caused by the attraction of the Earth.

The announcement, in 1693, of Cassini's laws as stated at the commencement of this section followed the publication of the *Principia* by only six years; but as neither Jean Domenico Cassini himself, nor his son Jacques ever published the observations from which the three laws were derived, these did not exert their full impact on the contemporary scientific thought until they were later confirmed by MAYER (1748–49) and LALANDE (1763). For Newton, following Hevelius, the axis of rotation of the Moon was still perpendicular to its orbital plane (i.e., he took $I=0$). Cassini was the first to establish that this inclination was finite, and adopted for it a value of $I=2\frac{1}{2}°$, which Lalande reduced to $1°43'$.

By that time it has been realized that Cassini's laws, far from being accidental, must represent at least approximate integrals of the equations of the problem of the Moon's motion about its centre of mass, in much the same way as Kepler's laws anticipated empirically certain closed integrals of the problem of two bodies; and – in the contemporary words of Tobias Mayer – "anyone who could show the natural cause of this connection would be a happy and famous man, who would justly earn laurels for a new and great discovery". The dynamical problem of the Moon's rotation arising from the existence of Cassini's laws was felt to constitute such a challenge that the Paris Academy of Sciences offered a special prize for its solution.

Both Euler and d'Alembert responded (rather unsuccessfully) to this challenge; but the laurels of which Mayer spoke were really carried away by Lagrange who won the Academy's prize in 1764; and who in two classical papers (the first submitted to the Paris Academy in 1768; the second – much more important – to the Berlin Academy in 1780) presented so complete a solution to the problem that subsequent investigations of Laplace (1798) or Poisson only filled in the details.

## CHAPTER 4

# MOTION OF THE MOON ABOUT ITS CENTER OF GRAVITY; PHYSICAL LIBRATIONS

Let $x$, $y$, $z$ represent a set of selenocentric rectangular coordinates, the axes of which coincide with the principal axes of inertia of the Moon, and their origin with the center of gravity of its mass. Let, moreover, these axes share the mean rotation of our satellite (i.e., be fixed in its globe and share its motion – the body axes); while

$$A = \int (y^2 + z^2) \, dm, \tag{4-1}$$

$$B = \int (x^2 + z^2) \, dm, \tag{4-2}$$

$$C = \int (x^2 + y^2) \, dm, \tag{4-3}$$

represent the moments of inertia about the $x$, $y$, $z$-axes (the products of inertia being by definition, zero). If so, the well-known Eulerian equations of motion of a solid body about its center of gravity assume the forms

$$A\dot{\omega}_x - (B - C)\omega_y\omega_z = F_x, \tag{4-4}$$

$$B\dot{\omega}_y - (C - A)\omega_z\omega_x = F_y, \tag{4-5}$$

$$C\dot{\omega}_z - (A - B)\omega_x\omega_y = F_z, \tag{4-6}$$

where $\omega_{x,y,z}$ stand for the angular velocities of rotation about the respective axes; and $F_{x,y,z}$, for the components of the forces acting on the Moon from outside. The velocity components $\omega_{x,y,z}$ are, in turn, expressible in terms of the corresponding rotations of the Eulerian angles $\theta$, $\varphi$, $\psi$ (cf. Figure 4-1) as

$$\omega_x = -\dot{\psi} \sin\theta \sin\varphi - \dot{\theta} \cos\varphi, \tag{4-7}$$

$$\omega_y = -\dot{\psi} \sin\theta \cos\varphi + \dot{\theta} \sin\varphi, \tag{4-8}$$

$$\omega_z = \dot{\psi} \cos\theta \qquad\qquad + \dot{\varphi}, \tag{4-9}$$

where (as well as in (4-1) to (4-3)) the dots denote ordinary differentiation with respect to the time.

Let us consider first what would happen if the Moon were alone in space, so that the right-hand sides of the equations (4-1)–(4-3) would vanish. If so, and if, by way of abbreviation, we introduce the quantities

$$\frac{C - B}{A} = \alpha, \quad \frac{C - A}{B} = \beta, \quad \frac{B - A}{C} = \gamma, \tag{4-10}$$

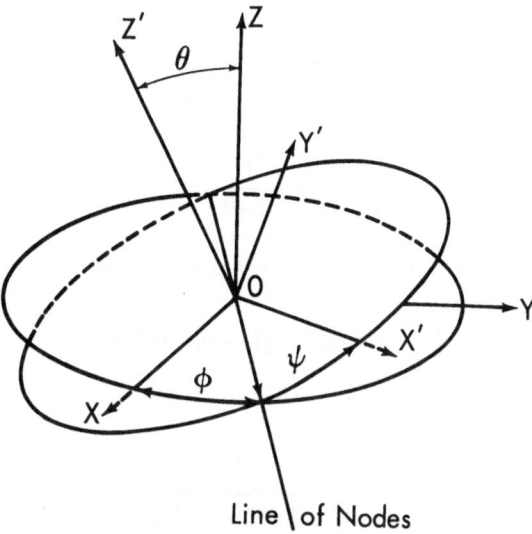

Fig. 4-1.   Definition of Eulerian angles.

constrained by the definitions (4-1)–(4-3) of $A$, $B$, and $C$ to satisfy the relation

$$\alpha - \beta + \gamma = \alpha\beta\gamma , \tag{4-11}$$

our system of equations becomes homogeneous and of the form

$$\left.\begin{array}{l}\dot\omega_x + \alpha\omega_y\omega_z = 0 , \\ \dot\omega_y - \beta\omega_x\omega_z = 0 , \\ \dot\omega_z + \gamma\omega_x\omega_y = 0 ;\end{array}\right\} \tag{4-12}$$

and these can be integrated as they stand. For observe that if we multiply these three equations successively by $\omega_x$, $\omega_y$, $\omega_z$ and add, the outcome can be readily integrated to yield

$$A\omega_x^2 + B\omega_y^2 + C\omega_z^2 = 2T , \tag{4-13}$$

where $T$ denotes the (constant) kinetic energy of motion. Moreover, if we multiply next the same equations by $A\omega_x$, $B\omega_y$ and $C\omega_z$ and add, we similarly find that

$$A^2\omega_x^2 + B^2\omega_y^2 + C^2\omega_z^2 = G^2 , \tag{4-14}$$

where $G$ denotes another constant of integration.

In order to complete the solution, let us set

$$\left.\begin{array}{l}\omega_x = \Omega\sqrt{\alpha\sigma}\,\cos\chi , \\ \omega_y = \Omega\sqrt{\beta\sigma}\,\sin\chi , \\ \omega_z = \Omega\sqrt{1 - \gamma\sigma\,\sin^2\chi} ,\end{array}\right\} \tag{4-15}$$

where we have abbreviated

$$\Omega^2 = \frac{G^2 - 2AT}{C(C - A)} \tag{4-16}$$

and

$$\sigma = \frac{C(2CT - G^2)}{(C - B)(G^2 - 2AT)} = \frac{2T - C\Omega^2}{C - B}. \tag{4-17}$$

If so, the differential equation defining $\chi$ will prove to be

$$\left(\frac{d\chi}{dt}\right)^2 = \Omega^2 \alpha \beta (1 - \gamma\sigma \sin^2 \chi), \tag{4-18}$$

and this readily integrates into

$$\sqrt{\alpha\beta}(t - t_0)\Omega = \int_0^{\chi} \frac{d\chi}{\sqrt{1 - \gamma\sigma \sin^2 \chi}} = F(\chi, \sqrt{\gamma\sigma}), \tag{4-19}$$

where $t_0$ is the third integration constant, and $F(\chi, k)$ denotes an incomplete elliptic integral of the first kind, an inversion of which renders $\chi$ a doubly-periodic function of the time.

If, however, the modulus $k = \sqrt{\gamma\sigma}$ of this integral is small, one can expand the integrand on the right-hand side of (4-19) in a binomial series and integrate term-by-term, finding that

$$\chi = u + \tfrac{1}{8}k^2 \sin 2u + \ldots, \tag{4-20}$$

where

$$u = \frac{\sqrt{\alpha\beta}(t - t_0)\Omega}{1 + \tfrac{1}{4}k^2}. \tag{4-21}$$

Hence,

$$\sin \chi = \sin u + \tfrac{1}{4}k^2 \sin 2u \cos u + \ldots, \tag{4-22}$$

$$\cos \chi = \cos u - \tfrac{1}{4}k^2 \sin 2u \sin u + \ldots, \tag{4-23}$$

and, therefore,

$$\omega_x = \Omega \sqrt{\alpha\sigma} \cos u (1 - k^2 \sin^2 u + \ldots), \tag{4-24}$$

$$\omega_y = \Omega \sqrt{\beta\sigma} \sin u (1 + k^2 \sin^2 u + \ldots), \tag{4-25}$$

and

$$\omega_z = \Omega (1 - \tfrac{1}{2}k^2 \sin^2 u + \ldots) \tag{4-26}$$

$$= \Omega (1 - \tfrac{1}{4}k^2 + \tfrac{1}{4}k^2 \cos 2u + \ldots).$$

Accordingly, to this order of approximation, the angular velocities $\omega_x$ and $\omega_y$ about the $x$ and $y$-axes should oscillate with an amplitude of $\Omega \sqrt{\alpha\sigma}$ and $\Omega \sqrt{\beta\sigma}$ and a frequency of

$$\frac{\Omega \sqrt{\alpha\beta}}{1 + \tfrac{1}{4}k^2};$$

while $\omega_z$ should oscillate around its mean value of $\Omega(1-\frac{1}{4}k^2)$ with an amplitude of $\frac{1}{4}k^2\Omega$ and twice the above frequency. As the constants $G$ and $T$ (and, therefore, $\sigma$) as well as $t_0$ are wholly arbitrary, our configuration can obviously oscillate about all three axes with an arbitrary amplitude and phase – hence the name of *free oscillations* – but the period of such oscillations should bear to the mean period of rotation about the $z$-axis the ratio of

$$\frac{1-(\frac{1}{4}k^2)^2}{\sqrt{\alpha\beta}},$$

or essentially $(\alpha\beta)^{-\frac{1}{2}}$, since $k^4$ becomes ignorable within the framework of our approximation.

After having established this preliminary result, let us return to the full-dress system of Equations (4-1)–(4-3) and allow the motion of the Moon about its own center of gravity be disturbed by the gravitational attraction of the Earth. If so, and if the action of the latter can be regarded as that of a mass-point*, the components of force on the right-hand sides of the Equations (4-1)–(4-3) assume the deceptively simple forms

$$F_x = \frac{3Gm_\oplus}{R^5}(C-B)\,y_E z_E, \tag{4-27}$$

$$F_y = \frac{3Gm_\oplus}{R^5}(A-C)\,z_E x_E, \tag{4-28}$$

$$F_z = \frac{3Gm_\oplus}{R^5}(B-A)\,x_E y_E, \tag{4-29}$$

where $G$ stands for the gravitational constant; $m_\oplus$, the mass of the Earth; $R$, the radius-vector of the Moon's relative orbit (i.e., the distance between the centers of the two bodies); and $x_E$, $y_E$, $z_E$ the selenographic coordinates of the Earth's center in the rotating system of the Moon's body axes.

We term the apparent simplicity of the Equations (4-27)–(4-29) deceptive; since in order to formulate them in the form suitable for eventual solution we must first express the coordinates $x_E$, $y_E$, $z_E$ of the Earth, referred to the Moon's principal axes of inertia, as explicit function of the time; and the whole complexity of the problem is, in fact, stored in this operation.

In order to develop the requisite geometry in steps which the reader could follow with ease, let us depart from the equations

$$\left.\begin{array}{rl}
x_E = & \xi\cos\varphi + \eta\cos\theta\sin\varphi - \zeta\sin\theta\sin\varphi, \\
y_E = & -\xi\sin\varphi + \eta\cos\theta\cos\varphi - \zeta\sin\theta\cos\varphi, \\
z_E = & +\eta\sin\theta + \zeta\cos\theta,
\end{array}\right\} \tag{4-30}$$

---

* Terms arising from the terrestrial oblateness on the right-hand sides of equations (4-4)–(4-6) would be of the order of $Gm_\oplus/R^8$ and, therefore, quite insensible. The same would be true of the terms representing the solar attraction on the Moon.

where $\xi$, $\eta$, $\zeta$ are selenocentric rectangular coordinates of the Earth at the time when $\theta = \varphi = 0$. Let, moreover, $L$, $B$ denote the true geocentric longitude and latitude of the Moon. As the selenocentric latitude of the Earth is then equal to $-B$, and seleno-centric longitude of the Earth becomes $L + 180°$, it follows that

$$\left.\begin{array}{l} \xi = -R \cos B \cos (L - \psi) = R \cos B \cos v, \\ \eta = -R \cos B \sin (L - \psi) = R \cos B \sin v, \\ \zeta = -R \sin B, \end{array}\right\} \quad (4\text{-}31)$$

where $R$ denotes the radius-vector in its relative orbit around the Earth, and

$$v = L + 180° - \psi \quad (4\text{-}32)$$

is the true selenocentric longitude of the Earth referred to the lunar equator.

In order to express $B$ in terms of $v$, let $A_\oplus$, $D_\oplus$ denote the selenocentric longitude and latitude of the Earth referred to the ecliptic ($A_\oplus$, like $L$, being measured from the vernal equinox). If so, the two sets of coordinates $(L, B)$ and $(A_\oplus, D_\oplus)$ are related by

$$\left.\begin{array}{l} \cos B \cos (L - \psi) = \cos D_\oplus \cos (A_\oplus - \Omega), \\ \cos B \sin (L - \psi) = \cos D_\oplus \sin (A_\oplus - \Omega) \cos i - \sin D_\oplus \sin i, \\ \sin B \qquad = \cos D_\oplus \sin (A_\oplus - \Omega) \sin i + \sin D_\oplus \cos i, \end{array}\right\} \quad (4\text{-}33)$$

where $i$ denotes, as before, the angle between the terrestrial and lunar equators, and $\Omega$ is the longitude of the ascending node of the Moon's orbit. Since, however, the Earth always remains on the ecliptic, by definition $D_\oplus = 0$; and if so, a division of the third equation of the foregoing set by the second leads to a relation

$$\tan B = \tan i \sin (L - \psi) = -\tan i \sin v, \quad (4\text{-}34)$$

which enables us to express $B$ in terms of $v$ and $i$.

But let us return now to equations (4-30) and (4-31). By combining the two sets we find that

$$x_E = R \{\cos B (\cos v \cos \varphi + \sin v \sin \varphi \cos \theta) \\ + \sin B (\sin \varphi \sin \theta)\}, \quad (4\text{-}35)$$

$$y_E = R \{\cos B (-\cos v \sin \varphi + \sin v \cos \varphi \cos \theta) \\ + \sin B (\cos \varphi \sin \theta)\}, \quad (4\text{-}36)$$

$$z_E = R \{\cos B (\sin v \sin \theta) - \sin B \cos \theta\}; \quad (4\text{-}37)$$

and their cross-products can be expressed exactly in the form

$$y_E z_E = R^2 \cos^2 B \{[\sin v \cos \varphi \cos \theta - \cos v \sin \varphi] \sin v \sin \theta \\ - [\sin v \cos \varphi \cos 2\theta - \cos v \sin \varphi \cos \theta] \tan B \quad (4\text{-}38) \\ - [\sin \theta \cos \theta \cos \varphi] \tan^2 B\},$$

$$z_E y_E = R^2 \cos^2 B \{[\sin v \sin \varphi \cos \theta + \cos v \cos \varphi] \sin v \sin \theta \\ - [\sin v \sin \varphi \cos 2\theta + \cos v \cos \varphi \cos \theta] \tan B \quad (4\text{-}39) \\ - [\sin \theta \cos \theta \sin \varphi] \tan^2 B\},$$

$$x_E y_E = \tfrac{1}{2} R^2 \cos^2 B \{\cos 2\varphi \sin 2v \cos \theta - \sin 2\varphi \cos 2v$$
$$- \sin 2\varphi \sin^2 v \sin^2 \theta$$
$$+ 2 \sin \theta [\cos v \cos 2\varphi + \sin v \sin 2\varphi \cos \theta] \tan B$$
$$+ \quad [\sin^2 \theta \sin 2\varphi] \tan^2 B \}. \tag{4-40}$$

After these preliminaries let us return to the definition of our Eulerian angles $\theta$, $\psi$ and $\varphi$. If Cassini's laws as reproduced in the preceding section were exact, the statements embodied in them assert that, at all times, we should have

$$\theta = I, \quad \psi = \Omega, \quad \varphi = 180° + l_{\mathfrak{c}} - \Omega, \tag{4-41}$$

where $l_{\mathfrak{c}}$ denotes the mean longitude of the Moon. Therefore, the angle of inclination $\theta \equiv I = 1°32'$ can certainly be regarded as small; and so can (to a lesser extent) the latitude $B$ which must, in accordance with (4-34), be always less than $i = 5°9'$. If, accordingly, we regard $\theta^2$ as well as $B^2$ as small quantities whose squares and cross-products can be ignored, the foregoing expressions for the products of the coordinates $x_E$, $y_E$, $z_E$ simplify into

$$\left. \begin{aligned} y_E z_E &= R^2 \cos^2 B \{\sin (v - \varphi) - \mu \sin v \cos \varphi\} \frac{\sin (\theta + i)}{\cos i} \sin v \\ &= R^2 \sin B \cos B \{\mu \sin v \cos \varphi - \sin (v - \varphi)\} \frac{\sin (\theta + i)}{\sin i}, \end{aligned} \right\} \tag{4-42}$$

$$\left. \begin{aligned} z_E x_E &= R^2 \cos^2 B \{\cos (v - \varphi) - \mu \sin v \sin \varphi\} \frac{\sin (\theta + i)}{\cos i} \sin v \\ &= R^2 \sin B \cos B \{\mu \sin v \sin \varphi - \cos (v - \varphi)\} \frac{\sin (\theta + i)}{\sin i}, \end{aligned} \right\} \tag{4-43}$$

and

$$\left. \begin{aligned} x_E y_E &= \tfrac{1}{2} R^2 \cos^2 B \{\sin 2 (v - \varphi) - 2\mu \sin v \cos (v - 2\varphi)\} \\ &= \tfrac{1}{2} R^2 \cos^2 B \{(1 - \mu) \sin 2 (v - \varphi) - \mu \sin 2\varphi\}, \end{aligned} \right\} \tag{4-44}$$

where we have abbreviated

$$\mu = \tfrac{1}{2} [(\theta + i)^2 - i^2]. \tag{4-45}$$

If, furthermore, we regard both $\theta$ and $B$ so small that their squares or cross-products can be ignored, these expressions reduce further to

$$y_E z_E = R^2 [\theta \sin v - B] \sin (v - \varphi), \tag{4-46}$$

$$z_E x_E = R^2 [\theta \sin v - B] \cos (v - \varphi), \tag{4-47}$$

$$x_E y_E = R^2 \sin (v - \varphi) \cos (v - \varphi). \tag{4-48}$$

These latter expressions were used as a basis for the treatment of our problem by TISSERAND (1890, Chapter XXVIII) – an approximation to which we shall also adhere through the first part of this section. We wish, however, to note in this connection that an approximation intermediate between (4-42)–(4-44) and (4-46)–(4-48) – in which quantities of the order of $\theta^3$ or $\theta^2 B$ were regarded as ignorable – was developed by HAYN (1902–23) and used subsequently by KOZIEL (1948–49).* As, however, its form is not any simpler than that of our present equations (4-42)–(4-44) in which such terms are retained and only $\theta^4$ or $\theta^2 B^2$ ignored (the forms of the products $x_E y_E$ being, in fact, identical in both cases), Hayn's approximation has little to commend itself for practical work and should be superseded by the present equations (4-42)–(4-44) whenever the use of (4-46)–(4-48) may lead to too large an error.

With this preparatory ground duly covered, let us return now to the differential equations (4-4)–(4-6) governing the motion of the Moon about its center of gravity. If Cassini's laws were exact, the integrals of motion would, in fact, be represented by equations (4-41). In order to investigate the extent to which this can be true, let us generalize (4-41) by setting

$$\theta = I + \rho, \tag{4-49}$$

$$\psi = \Omega + \sigma, \tag{4-50}$$

$$\varphi = 180° + l_{(} - \psi + \tau, \tag{4-51}$$

where $\rho$, $\sigma$, $\tau$ represent the *perturbations* of $\theta$, $\psi$, $\varphi$; and as such will constitute the *physical librations* of the Moon in latitude, longitude, and node.

In view of the approximate validity of Cassini's laws as embodied in Equations (4-41), we may anticipate that $\rho$, $\sigma$ and $\tau$ will be small quantities whose squares and cross-products can be ignored. The same will be true of the angles $\theta = I$ and (to a lesser extent) of $i$, rendering the ratio $z_E/R$ similarly small (while $x_E/R$ and $y_E/R$ are of zero-order). Similarly, the angular velocities $\omega_x$ and $\omega_y$, as defined by Equations (4-7) and (4-8), should then be expected to be small (consisting as they do of terms factored by the time-derivatives $\theta$ or $\psi$ of slowly varying functions) in comparison with $\omega_z$ (which is essentially equal to the time-derivative of $\varphi$ or, approximately, the mean angular rate of the Moon's revolution in its orbit).

If we disregard the squares or cross-products of all small quantities, our fundamental equations (4-4)–(4-6) of motion become *linear* in the dependent variables and, by use of (4-7)–(4-9) and (4-46)–(4-48) assume the simplified forms

$$\ddot{\theta} \cos \varphi + \alpha \omega_z \theta \sin \varphi = - 3Gm_\oplus R^{-3}\alpha \left[\theta \sin v - B\right] \sin (v - \varphi), \tag{4-52}$$

$$\ddot{\theta} \sin \varphi + \beta \omega_z \theta \cos \varphi = - 3Gm_\oplus R^{-3}\beta \left[\theta \sin v - B\right] \cos (v - \varphi), \tag{4-53}$$

$$\ddot{\psi} + \ddot{\varphi} \qquad\qquad = + 3Gm_\oplus R^{-3}\gamma \sin (v - \varphi) \cos (v - \varphi), \tag{4-54}$$

where, as before, the dots on the left-hand side denote differentiation with respect to the time.

An inspection of the foregoing equations (4-52)–(4-54) reveals that the first two do not contain any explicit reference to $\psi$ and can, therefore, be treated separately

---

* In particular, the auxiliary quantity $\mu = 1 - \cos I + \tan i \sin I$ introduced by Hayn proves to be identical (to the order of accuracy we are working) with our $\mu$ as defined by Equation (4-45) when $\theta$ is identified with $I$.

from the first. In order to do so, let us introduce (following Lagrange) two new dependent variables

$$p = \sin \theta \cos \varphi \simeq \theta \cos \varphi, \left.\right\}$$
$$q = \sin \theta \sin \varphi \simeq \theta \sin \varphi. \left.\right\}$$
(4-55)

Within the scheme of this approximation,

$$\omega_x = - \dot{p} - q\omega_z, \left.\right\}$$
$$\omega_y = + \dot{q} - p\omega_z; \left.\right\}$$
(4-56)

so that

$$\dot{\omega}_x = - \ddot{p} - \omega_z \dot{q} - q\dot{\omega}_z, \left.\right\}$$
$$\dot{\omega}_y = + \ddot{\eta} - \omega_z \dot{p} - p\dot{\omega}_z. \left.\right\}$$
(4-57)

Suppose that the angular velocity $\omega_z$ is hereafter regarded as constant (i.e., $\dot{\omega}_z = 0$). If so, Equations (4-52)–(4-53) obviously can be rewritten in the alternative form

$$\ddot{p} + (1 - \alpha)\omega_z\dot{q} + \alpha\omega_z^2 p = 3Gm_\oplus\alpha R^{-3}(B - \theta \sin v) \sin (v - \varphi), \quad (4\text{-}58)$$

$$\ddot{q} - (1 - \beta)\omega_z\dot{p} + \beta\omega_z^2 q = 3Gm_\oplus\beta R^{-3}(B - \theta \sin v) \cos (v - \varphi). \quad (4\text{-}59)$$

Let us first explore the solutions of this system obtaining when the right-hand sides of these equations are set equal to zero. These will be purely periodic if, on insertion of an anticipated solution of the form

$$p = Fe^{\omega_z st}, \quad q = Ge^{\omega_z st}, \quad (4\text{-}60)$$

with arbitrary coefficients $F$ and $G$, the constant $s$ proves to be imaginary. Inserting (4-60) in (4-58)–(4-59) we find the latter two equations to reduce to

$$Fs^2 + (1 - \alpha)Gs + \alpha F = 0 \left.\right\}$$
$$Gs^2 - (1 - \beta)Fs + \beta G = 0 \left.\right\}$$
(4-61)

a homogeneous system which will admit of a nontrivial solution for $F$ and $G$ only if its determinant vanishes – i.e., if $s^2$ is a root of the quadratic equation

$$s^4 + (1 + \alpha\beta)s^2 + \alpha\beta = 0, \quad (4\text{-}62)$$

which are equal to $s^2 = -1$ or $-\alpha\beta$. Both will lead to purely imaginary roots for $s$ provided that $\alpha\beta > 0$ – i.e., that (cf. 4-10),

$$(C - A)(C - B) > 0. \quad (4\text{-}63)$$

In order that this be true, the moment $C$ must be either greater or smaller than both $A$ and $B$; hence, the axis of rotation with the angular velocity $\omega_z$ must be either the longest or the shortest axis of our configuration.

The results just obtained define the stability and periods of *free* oscillations about the center of gravity of a configuration undisturbed by external forces; and as such they verify the frequency $\omega_z \sqrt{\alpha\beta}$ deduced previously (p. 26) on the same premises. What interests us more now is, however, to investigate the forced oscillations in $p$ and $q$ (or, what is equivalent, in $\theta$ and $\varphi$) arising from the presence of the appropriate forcing terms on the right-hand sides of Equations (4-58)–(4-59). In order to do so, let us hereafter assume, for the sake of simplicity, that the radius-vector $R$ of the relative orbit of the Moon is constant; for if so, Kepler's second law

$$\frac{G(m_\oplus + m_\mathbb{C})}{R^3} = n^2 \tag{4-64}$$

permits us to write*

$$\frac{Gm_\oplus}{R^3} = \frac{n^2}{1+\mu}, \tag{4-65}$$

where $\mu$, the mass-ratio of the Moon to the Earth, will also be hereafter disregarded in comparison with unity on account of its smallness (0.01 229). Next, in accordance with Cassini's first law, let us identify the period of axial rotation of the Moon exactly with that of its revolution, and set

$$\omega_z = n \tag{4-66}$$

treating both as constant.

Moreover, as by (4-32) $v = L + 180° - \psi$ and, by (4-51), $\varphi = 180° + l_\mathbb{C} - \psi + \tau$, it follows that the angle

$$v - \varphi = L - l_\mathbb{C} - \tau, \tag{4-67}$$

representing as it does essentially a difference between the optical and physical libration of the Moon in longitude, can be regarded as small (its principal part $L - l_\mathbb{C}$ vanishes for a circular orbit, as the true and mean longitude of the Moon are then obviously equal). If we neglect it, $v = \varphi$; and since to the same degree of approximation Equation (4-34) reduces to $B = -i \sin v$, it follows that

$$\left.\begin{array}{l} (\theta \sin v - B) \cos (v - \varphi) = (\theta + i) \sin v = (\theta + i) \sin \varphi \\ = q + i \sin (180° + l_\mathbb{C} - \Omega) = q - i \sin (l_\mathbb{C} - \Omega) \end{array}\right\} \tag{4-68}$$

plus higher-order terms; while

$$(\theta \sin v - B) \sin (v - \varphi) = \text{second and higher-order terms only.} \tag{4-69}$$

If so, one system of Equations (4-58)–(4-59) will reduce now to

$$\ddot{p} + (1 - \alpha) n\dot{q} + \alpha n^2 p = 0, \tag{4-70}$$

$$\ddot{q} - (1 - \beta) np + 4\beta n^2 q = 3\beta n^2 i \sin (l_\mathbb{C} - \Omega). \tag{4-71}$$

The complementary function of this system – we can no longer call it a free oscillation – will be a purely periodic function of the time (with arbitrary amplitude and

---

* Strictly speaking, since the observed duration of $27^d.321\,661$ of the sideric month corresponds to $n = 2.661\,700 \times 10^{-6}$ sec$^{-1}$, and $G(m_\oplus + m_\mathbb{C}) = 4.03502 \times 10^{20}$ cm$^3$/sec$^2$, the value of $R$ computed from (4-64) comes out equal to 384 747 km = 1.0009 times the observed value of $R = 384\,400$ km – a difference due to solar perturbations of the Moon's orbit. Accordingly, the right-hand sides of Equations (4-64) or (4-65) should be multiplied by a factor $\lambda = 1.0027$ – a fact which we shall later (in Equation 4-102) indeed take into account.

phase), whose frequency $s$ will be a root of the Equation

$$s^4 + n^2(1 + 3\beta + \alpha\beta)s^2 + 4n^4\alpha\beta = 0,\tag{4-72}$$

obtained by setting the determinant of the left-hand sides of the Equations (4-70)–(4-71) equal to zero. Let us consider next the particular integral of the same system, representing forced oscillations of the form

$$p = P\cos(l_( - \Omega), \quad q = Q\sin(l_( - \Omega),\tag{4-73}$$

where $P$ and $Q$ are constants. By substituting and noting that

$$\frac{d}{dt}(l_( - \Omega) = n(1 + m),\tag{4-74}$$

where $-mn$ denotes the rate of recession of the line of the nodes, we find the system (4-70)–(4-71) of equations to assume the form

$$\{\alpha - (1 + m)^2\}P + (1 - \alpha)(1 + m)Q = 0,\tag{4-75}$$

$$\{4\beta - (1 + m)^2\}Q + (1 - \beta)(1 + m)P = 3\beta i.\tag{4-76}$$

The determinant of this system becomes equal to

$$\Delta = (1 + m)^2(2m + m^2 - 3\beta - \alpha\beta) + 4\alpha\beta;\tag{4-77}$$

and under certain conditions it may vanish or become very small: such would be the case, for instance, for prolate spheroids ($\alpha = 0$) if $3\beta = 2m + m^2$; or, in general, if $m \sim 1.5\beta$. In such a case, $\Delta$ would play the role of a small divisor – a circumstance which would render the solution of our problem quite difficult.

Suppose, however, that this is not the case (an assumption which we shall later be able to substantiate); if so, the constants $P$ and $Q$ result from

$$P = -3\beta i \left\{ \frac{(1 - \alpha)(1 + m)}{\Delta} \right\}\tag{4-78}$$

and

$$Q = +3\beta i \left\{ \frac{\alpha - (1 + m)^2}{\Delta} \right\}.\tag{4-79}$$

Since, however, from the definition (4-55) it follows that

$$p^2 + q^2 = \sin^2\theta = P^2\cos^2(l_( - \Omega) + Q^2\sin^2(l_( - \Omega)$$
$$= \frac{P^2 + Q^2}{2}\left\{1 - \frac{P^2 - Q^2}{P^2 + Q^2}\cos 2(l_( - \Omega)\right\};\tag{4-80}$$

then remembering that, in accordance with (4-49), $\theta = I + \rho$, we find that

$$I = \sqrt{\frac{P^2 + Q^2}{2}}\tag{4-81}$$

and

$$\rho = -\frac{\sin I}{2} \left\{ \frac{P^2 - Q^2}{P^2 + Q^2} \right\} \cos 2(l_{\text{\textdollar}} - \Omega) \tag{4-82}$$

as a partial solution of our problem. Moreover, as (by 4-51)

$$\left. \begin{aligned} q \cos (l_{\text{\textdollar}} - \Omega) - p \sin (l_{\text{\textdollar}} - \Omega) &= \sin \theta \sin (\varphi - l_{\text{\textdollar}} - \Omega) \\ &= \sin \theta \sin (180° + \tau - \sigma) \\ &= \sin \theta \sin (\sigma - \tau) \\ &= (Q - P) \sin (l_{\text{\textdollar}} - \Omega) \cos (l_{\text{\textdollar}} - \Omega), \end{aligned} \right\} \tag{4-83}$$

it follows that, to the same approximation,

$$I(\sigma - \tau) = \tfrac{1}{2}(Q - P) \sin 2(l_{\text{\textdollar}} - \Omega), \tag{4-84}$$

where (from 4-78 and 4-79),

$$P - Q = 3\beta i m \Delta^{-1}(1 + \alpha + m) \tag{4-85}$$

and

$$P^2 + Q^2 = (3\beta i/\Delta)^2 \{[(1 + m)^2 + \alpha^2](2 + 2m + m^2) - 4\alpha(1 + m)^2\}, \tag{4-86}$$

$$P^2 - Q^2 = -(3\beta i/\Delta)^2 m(m + 2)[(1 + m)^2 - \alpha^2]. \tag{4-87}$$

If, in particular, we combine (4-81) with (4-86) we find that, literally, the ratio of the inclination $I$ of the lunar equator to the ecliptic should bear to the angle $i$ between the lunar and terrestrial equators the ratio of

$$\frac{I}{i} = \frac{3\beta \sqrt{[(1 + m)^2 + \alpha^2][1 + m + \tfrac{1}{2}m^2] - 2\alpha(1 + m)^2}}{(1 + m)^2[2m + m^2 - 3\beta - \alpha\beta] + 4\alpha\beta} \tag{4-88}$$

which, if the quadratic terms $\alpha^2$ as well as $\alpha\beta$ are ignored within the framework of our approximation, will reduce to a simpler expression

$$\frac{I}{i} = \frac{3\beta \sqrt{1 + m + \tfrac{1}{2}m^2 - 2\alpha}}{(1 + m)(2m + m^2 - 3\beta)}. \tag{4-89}$$

If, furthermore, we ignore in this latter expression $\alpha$ or $m$ in comparison with unity or $m^2$ in comparison with $m$, we can solve it for $\beta$ to obtain

$$\beta = \tfrac{2}{3}\left(\frac{mI}{i + I}\right) \tag{4-90}$$

a relation furnishing the value of the ratio of the momenta $(C - A)/B$ in terms of the observable constants $i$, $I$, and $m$.

The reader may notice that, accordingly, $\beta$ could approach the critical value of $\tfrac{2}{3}m$ for which the determinant vanishes only if $i \ll I$; while, in actual fact, the converse is true; for Brown's value of $i = 5°8'43''$ and Koziel's value of $I = 1°32'1''$ are in the ratio

$$\frac{I}{i} = 0.2981 \pm 0.0004, \tag{4-91}$$

or just under 0.3. Moreover, the period of the regression of lunar nodes is known with great precision to be equal to 18.6133 tropical years or 248.827 sideric months. Therefore, the value of $m$ involved in Equations (4-88)–(4-90) is equal to

$$m = (248.827)^{-1} = 0.00401886. \tag{4-92}$$

If so, Equation (4-90) readily yields

$$\beta = 0.000615 \tag{4-93}$$

as the corresponding value of the ratio $(C-A)/B$ – a value which for several reasons (such as the adoption of circular orbit for the Moon, reglect of the Moon-Earth mass-ratio $\mu$, etc.) we regard still as preliminary and shall improve later on.

Within the same scheme of approximation, the corresponding periodic terms in (4-80) or (4-83) of physical libration in latitude as well as the node should oscillate in a period of one-half of the (draconic) month, and approximate amplitudes of $\frac{1}{2}mI -$ 11".1. The actual separation of the libration in longitude $\tau$ and node $\sigma$ is not possible until the third equation (4-54) of our fundamental system has similarly been solved; and to this task we now wish to direct our attention.

Within the scheme of approximation adopted so far in this section this task will fortunately prove simple: for, in view of the assumed constancy of $l_\mathfrak{c} \equiv n$,

$$\ddot{\varphi} + \ddot{\psi} = \ddot{\tau}; \tag{4-94}$$

while the angle

$$v - \varphi = L - l_\mathfrak{c} - \tau = l - \tau \tag{4-95}$$

i.e., essentially a difference of the optical and physical libration of the Moon in longitude. Both these angles are small – a fact which entitles us to rewrite Equation (4-54) approximately as

$$\ddot{\tau} = 3Gm_\oplus R^{-3}\gamma(l - \tau) = w^2(l - \tau). \tag{4-96}$$

The solution of this equation will consist of the complementary function $K \cos(wt + k_0)$ with arbitrary amplitude and phase, plus a particular integral obtained by expanding $l$ in a periodic function of the time. To this end, let us rewrite (4-96) in the form

$$\ddot{\tau} + w^2\tau = \sum_j H_j \sin(h_j t + h'_j); \tag{4-97}$$

its complete solution can then be expressed as

$$\tau = K \cos(wt + k_0) + \sum_j \frac{H_j \sin(h_j t + h'_j)}{h_j^2 - w^2}. \tag{4-98}$$

Its first term is sometimes referred to as the "free libration" of the Moon in longitude, but this is a misnomer; for although its amplitude and phase are arbitrary, its period is governed by the external force (i.e., the terrestrial attraction), and would become infinitely long (i.e., the motion would stop) if the Earth were at an infinite distance. It has recently been identified in the Moon's motion by KOZIEL (1964) to possess an amplitude $K=19''\pm5''$, and a period of 1045 days leading to the value of

$$\gamma = 0.000\,230.    \tag{4-99}$$

Of the terms due to forced libration, the largest two are associated with the elliptic inequality of the Moon's motion ($H=22\,639''.1$) with the period $2\pi/h$ of one anomalistic month; and with the annual equation ($H=-668''.9$) with the period $2\pi/h$ of one anomalistic year. Higher terms in the particular integral may become large if the divisor $h_j^2-w^2$ happens to be small (resonance between "free" and forced oscillations); but no troublesome resonance seems to occur at least among the low-order terms.

The effect of the orbital eccentricity of the Moon's libration in latitude can be ascertained from Equations (4-58)–(4-59) by an expansion of the perturbing function on their right-hand sides in ascending powers of $e$. If, following JEFFREYS (1961) we retain terms up to the quares of $e$, Equations (4-58) and (4-59) will assume the more explicit forms

$$\ddot{p} + (1-\alpha)\,n\dot{q} + \alpha n^2 p =$$
$$= -\alpha k\,\{(2e^2 - \tfrac{1}{2}\sin^2 i)\,p - (2e^2 - \tfrac{1}{2}\sin^2 \tfrac{1}{2}i)\sin i\cos(l_{\langle} - \Omega)\} - \tfrac{3}{2}(n'/n^2)\,\beta p,  \tag{4-100}$$

$$\ddot{q} - (1-\beta)\,n\dot{p} + \beta n^2 q =$$
$$= -\beta k\,\{(1 - \tfrac{1}{2}e^2 - \sin^2 i)\,q - (1 - \tfrac{1}{2}e^2 - \tfrac{3}{2}\sin^2 \tfrac{1}{2}i)\sin i\sin(l_{\langle} - \Omega)\} - \tfrac{3}{2}(n'/n^2)\,\beta q,  \tag{4-101}$$

where we have abbreviated

$$k = \frac{3\lambda}{1+\mu}    \tag{4-102}$$

and where the last terms on the right-hand sides of (4-100)–(4-101) as well as the constant $\lambda$ in (4-102) approximate the principal effects of solar attraction on the Moon's libration ($n'$ being the mean angular rate of the annual motion of the Earth around the Sun). If so, on insertion of anticipated periodic solutions from (4-73) these lead to

$$[\alpha - (1+m)^2]\,P + (1-\alpha)\,(1+m)\,Q =$$
$$= k\alpha\,[(2e^2 - \tfrac{1}{2}\sin^2 \tfrac{1}{2}i)\sin i - (2e^2 - \tfrac{1}{2}\sin^2 i)P] - \tfrac{3}{2}v^2\beta P    \tag{4-103}$$

and

$$[\beta - (1+m)^2]\,Q + (1-\beta)\,(1+m)\,P -$$
$$= k\beta\,[(1 - \tfrac{1}{2}e^2 - \tfrac{3}{2}\sin^2 \tfrac{1}{2}i)\sin i - (1 - \tfrac{1}{2}e^2 - \sin^2 i)Q] - \tfrac{3}{2}v^2\beta Q,    \tag{4-104}$$

where we have abbreviated $n'/n=v$.

The foregoing equations constitute the corresponding improvements of (4-75)–(4-76), and can be solved for $P$ and $Q$ in much the same way as we have done with the latter; but their coefficients are more complicated and the requisite algebraic work would become rather tedious. As, however, the values of most of these coefficients (except for $\alpha$ and $\beta$) are well known, the numerical evaluation of $P$ and $Q$ in terms of $\alpha$ and $\beta$ only offers no difficulty. As for $e = 0.05490$ and $i = 5°8'43''$,

$$\left.\begin{aligned}
1 - \tfrac{1}{2}e^2 - \sin^2 i &= 0.990450, \\
1 - \tfrac{1}{2}e - \tfrac{3}{2}\sin^2 \tfrac{1}{2}i &= 0.995471, \\
2e^2 - \tfrac{1}{2}\sin^2 i &= 0.002006, \\
2e^2 - \tfrac{1}{2}\sin^2 \tfrac{1}{2}i &= 0.005020,
\end{aligned}\right\} \tag{4-105}$$

and

$$m = 0.00401886, \qquad \nu = 0.0748013 \tag{4-106}$$

$$k = \frac{3 \times 1.0027}{1 + 0.01229} - 2.9716, \tag{4-107}$$

Equations (4-103) and (4-104) assume the more explicit forms

$$\begin{aligned}
(1.005961\alpha + 0.008393\beta - 1.008054)P& \\
+ 1.004019(1 - \alpha)Q &= 0.001338\alpha
\end{aligned} \tag{4-108}$$

and

$$1.004019(1 - \beta)P + (3.951614\beta - 1.008054)Q = 0.265292\beta. \tag{4-109}$$

The determinant of this system

$$\varDelta = 0.008119 - 2.983847\beta - 0.006009\alpha + 0.033166\beta^2 + 2.967116\alpha\beta, \tag{4-110}$$

and the determinants $\varDelta_P$, $\varDelta_Q$ corresponding to the two unknowns

$$\varDelta_P = -0.001338\alpha - 0.265292\beta + 0.270579\alpha\beta, \tag{4-111}$$

$$\varDelta_Q = -0.001343\alpha - 0.267429\beta + 0.268216\alpha\beta + 0.002227\beta^2; \tag{4-112}$$

which for KOZIEL's (1964) determination of

$$\alpha = f\beta = (0.633 \pm 0.011)\beta \tag{4-113}$$

simplify to

$$\varDelta = 0.008119 - 2.980043\beta + 1.911350\beta^2, \tag{4-114}$$

$$\varDelta_P = \qquad -0.266139\beta + 0.171277\beta^2, \tag{4-115}$$

$$\varDelta_Q = \qquad -0.268279\beta + 0.172008\beta^2. \tag{4-116}$$

As the last step, we square equation (4-81) obtaining

$$2I^2 = P^2 + Q^2, \tag{4-117}$$

which for $P = \Delta_P/\Delta$, $Q = \Delta_Q/\Delta$ assumes the more explicit form

$$2\Delta^2 I^2 = \Delta_P^2 + \Delta_Q^2. \tag{4-118}$$

For $I = 1°32'1''$ (Koziel, 1964), $I^2 = 0.0007164$ radians. An insertion of this value in the foregoing equation together with the expressions (4-114)–(4-116) for the respective determinants renders (4-118) an algebraic equation for $\beta$ of fourth degree, the requisite root of which renders

$$\beta = 0.000627 \pm 0.000001 \tag{4-119}$$

as the final outcome of our analysis. The uncertainty of $\pm 7''$ in Koziel's value of $I$ renders this value of $\beta$ uncertain by approximately one part in a thousand. A comparison with the preliminary value (4-93) in the derivation of which the effects of $e$ or $i^2$ were ignored, discloses that a retention of the second-order terms increased our preliminary value by 0.000015, or approximately 2% – a result which gives us hope to believe that the retention of third or higher-order terms would alter the present value of $\beta$ by less than half a unit of its third significant digit.

Accordingly, the observed motion of the moon about its center of gravity leads to the following final values for the fractional differences of the momenta of the lunar globe around different axes:

$$\left.\begin{array}{l} \alpha = 0.000397 \pm 0.000008\,, \\ \beta = 0.000627 \pm 0.000001\,, \\ \gamma = 0.000230 \pm 0.000006\,, \end{array}\right\} \tag{4-120}$$

satisfying the relation (4-11).

The reader may pardon the somewhat elaborate nature of the discussion we had to go through before establishing these final values of the differences of the momenta; but as they constitute the principal information we now possess concerning the internal structure of our satellite, it seemed worthwhile to expound in some detail the method by which they can be deduced before asking the reader to put his full faith in the results. The extent of the information contained in them will be discussed repeatedly in subsequent chapters of this volume; but we may at least note that the Eulerian period of free oscillations of the lunar globe, which we established earlier in this section to be $(\alpha\beta)^{-\frac{1}{2}}$ months, proves in the light of the above results to be equal to 2004 sideric months or 149 years and 10 months – far longer than a period of 2 years and 10 months established recently by Koziel for the complementary function of the libration in longitude – and much too long to be verified on the basis of existing observations.

In point of fact, the physical librations of the Moon have not been actually detected by the observers much before the middle of the last century; and the reason was their smallness. For, unlike the optical librations which may attain several degrees of arc, the amplitude of the physical librations of the lunar globe have turned out not to exceed two minutes; the leading terms of the expansion of the forced oscillations in latitude, node, and longitude being of the order of

$$\rho = -107'' \cos g + 36'' \cos (g + 2\omega) - 11'' \cos 2(g + \omega) + \ldots, \tag{4-121}$$

$$I\sigma = -109'' \sin g + 36'' \sin (g + 2\omega) - 11'' \sin 2(g + \omega) + \ldots, \qquad (4\text{-}122)$$

and

$$\tau = -14'' \sin g + 73'' \sin g' + 19'' \sin 2\omega + \ldots, \qquad (4\text{-}123)$$

where $g$ denotes the mean anomaly of the Moon's motion; $g'$, the mean anomaly of the Sun; and $\omega$, the longitude of the perigee of the lunar orbit. It is by such small amounts that the Cassini laws as represented by Equations (4-49)–(4-51) are inexact! The leading terms in $\rho$ and $\sigma$ arise from the Moon's elliptic inequality (which we neglected earlier by assuming the lunar orbit to be circular); the leading term in $\tau$ being due to the annual equation. The third term on the right-hand sides of Equations (4-121) and (4-122) with amplitudes of $11''$, are identical (since $g + \omega = l_( - \Omega$) with the periodic term on the right-hand sides of our previous Equations (4-80) and (4-84); although their amplitudes are quite small, they do not vanish even when the lunar orbit is treated as circular.

The relative smallness of the Moon's physical librations becomes even more apparent when we consider their effects visible from the Earth. As the radius of the lunar globe is 221 times smaller than the mean distance separating us from our satellite, a selenocentric libration by $2'$ would be seen from the Earth as an angular displacement of $2'/221$ or only $0''.54$ at the center of the apparent lunar disk (and progressively less towards the limb). No wonder that motions so small had to await their discovery long after Newton and Lagrange predicted their existence! At the beginning of the 19th century, Laplace encouraged Arago and Bouvard to search for them without avail; and it was not until Bessel took up this problem in 1839 that persistent efforts to identify the physical libration in the Moon's motion were at last crowned with success.

In a paper 'Über die Bestimmung der Libration des Mondes durch Beobachtungen' (1839) which has since become classic, BESSEL outlined the whole strategy which the investigators of this problem have followed ever since: namely, to measure the apparent position of a distinct feature near the center of the Moon's disc – and it was Bessel who selected for this purpose the crater Mösting A (see Figure 13-1) – relative to the Moon's limb, with the aid of the heliometer (a new type of astrometric telescope, designed by Bessel, possessing an objective split up in two parts, which permitted accurate measurements of relatively large angles to be performed on optical axis).

Bessel himself did not carry out any such measurements; but this was done at Königsberg by his pupils SCHLÜTER (1841–43) and WICHMANN (1844–46 and 1848), followed a generation later by HARTWIG whose three valuable series of heliometric observations (1875–79 in Strasburg, 1884–85 in Dorpat, and 1890–1922 at Bamberg) represent one of the most important contributions to the study of our subject ever made. Of other German astronomers, FRANZ at Königsberg and later Breslau initiated the use of photographic material for selenodetic purposes; while HAYN carried out important work between 1902–1923 at Leipzig (preferring, however, the use of micrometric measurements with a refractor to a heliometer), which was continued by NAUMANN until the complete destruction of the Leipzig observatory during the Second World War; and since resumed by HOPMANN and SCHRUTKA-RECHTENSTAMM in Vienna.

The other great school of selenodetic work developed at the Kazan Observatory in Russia, where a heliometer was set up for this purpose in 1891; and the early pioneer work of DUBIAGO, KRASNOV and MIKHAILOVSKY was, in more recent years, continued by YAKOVKIN (1916–1931), BELKOVICH (1931–1948), and is still being carried on by their successors NEFEDIEV and HABIBULLIN today. The Engelhardt Observatory at Kazan is the only astronomical institution of the world where a heliometer is still in practical use. And we cannot pass without appropriate mention the important contributions which have come to our subject from the Polish school of astronomers at Kraków, founded by BANACHIEWICZ, and continued at present by KOZIEL.

Accurate reductions of heliometric observations represent an arduous task, to which most observers of bygone days proved unequal during their lifetime; and the work is being completed only now with the aid of electronic computers (KOZIEL, 1964). In order to extract the maximum precision from heliometric observations, measurements referred to different parts of the Moon's limb should be corrected for local irregularities of lunar topography. Maps giving such limb corrections for different angles of libration have first been prepared by HAYN (1914); and – more recently and extensively – by WEIMER (1952), NEFEDIEV (1958) and WATTS (1963).

These maps are, in general, based on observational (photographic) evidence secured with sufficiently large telescopes of adequate resolving power. This is, however, not the case with the original heliometric measurements, made with instruments of very moderate apertures, whose resolution was, consequently, limited by diffraction to a serious extent. No heliometer used in the past (none are in use now, save for the Kazan instrument) possessed an aperture $D$ exceeding 7 inches (the majority were 5–6 inch instruments); and their Rayleigh limit of resolution in visible light was, therefore, of the order of $1.22(\lambda/D) \sim 0''.8$ – corresponding to an uncertainty of the order of 1.5 km at the mean distance of the Moon. In order to obtain, from such measures, results significant to $0''.1$ (as quoted by heliometric observers), it was necessary to take the mean of great many individual settings, on the assumption that no appreciable systematic errors were present to impair their mean – a rather questionable hypothesis.

Positional astronomers have, to be sure, long been accustomed to search for information inside optical diffraction patterns, and have done so extensively when measuring, e.g., stellar parallaxes. The success of such a process requires, however, a knowledge of the geometrical relation of the actual shape of the light source to that of its diffraction image (i.e., a point to a disk in the case of a star), in addition to a great many measures to minimize the accidental errors. In the case of selenodetic measurements the first condition cannot, unfortunately, be met; for the actual shape of lunar details on which heliometric settings are made are not known to us *a priori*; and neither is, therefore, the form of their diffraction image (which may, moreover, for the same detail very with the phase, as a result of different illumination). Under these conditions, the reader would do well to retain still certain reserve with respect to the results based on heliometric observations and keep, in particular, in mind the possibility that their actual uncertainty may be greater than claimed in the literature.

## CHAPTER 5

# PHOTOMETRIC CONSEQUENCES OF THE MOON'S MOTION: PHASES AND ECLIPSES

In the preceding chapters of this part an outline has been given of the theory of the Moon's motion in space as well as about its own center of gravity. This motion affects, of course, vitally most observational manifestations of our satellite; and the most elementary aspect of these are the conditions of illumination of the Moon by the Sun and the Earth – i.e., the *phases* of the Moon – which we shall now proceed to describe.

The fact that the Moon emits no light of its own (within limitations to be mentioned in Chapter 7), but derives most of it from the Sun, was realized already by THALES around 600 B.C.; while the illumination of the Moon by the Earth (giving rise to the "ashen light" of lunar night-time) was noted and explained as such by LEONARDO DA VINCI. For the present we shall leave aside the photometric aspects of this problem for later discussion (Chapter 19), but wish to outline the geometry of the illumination of the Moon by the Sun and the Earth in the following manner.

Let the points Sun, Earth, and Moon denote the vertices of a *plane* triangle representing at any time the relative positions of the Sun, Earth, and the Moon (cf. Figure 5-1). The inclination of the plane of this triangle to the ecliptic oscillates, of course, in the course of each month between $\pm (r/R) \sin I$, where $r$ and $R$ denote the distances $EM$ and $ES$, respectively; and an intersection of this plane with the lunar globe represents the *equator of illumination* of the Moon, which represents the axis of symmetry of its illuminated portion commonly referred to as the *phase* (cf. Figure 5-2).

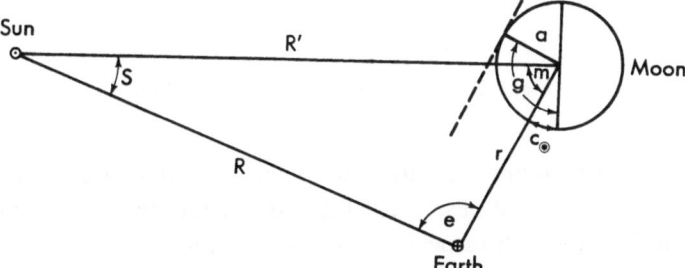

Fig. 5-1.  Illumintion of the Moon by the Sun.

This illuminated portion is, in general, limited by two arcs: the half-circle representing the limb of the apparent lunar disc (i.e., an intersection of the lunar sphere with a plane perpendicular to the line of sight of the terrestrial observer), and the *terminator* representing a locus of points at which the sun rays are tangent to the lunar globe. Strictly speaking, the center of the apparent solar disc should never be visible at the poles of illumination of the Moon because of the finite dimensions of its globe. However, the Sun's distance is so great in comparison with lunar size that the pencil of illuminating rays can be considered parallel, and the terminator regarded as a great circle perpendicular to the direction *MS*. One-half of this circle marks the zone of advancing illumination, and will hereafter be referred to as the *sunrise terminator*; while the other half, marking the zone of retreating illumination, will be called the *sunset terminator*. Needless to say, only one of these terminators can be visible from the Earth at the same time. The arcs of the sunrise and sunset terminators constitute an ellipse whose semi-major axis – the *line of cusps* – oscillates in direction in the course of each month and whose eccentricity is zero at the time of the "new" or "full" Moon (when the terminators coincide with the limb), and unity at the first or last quarter.

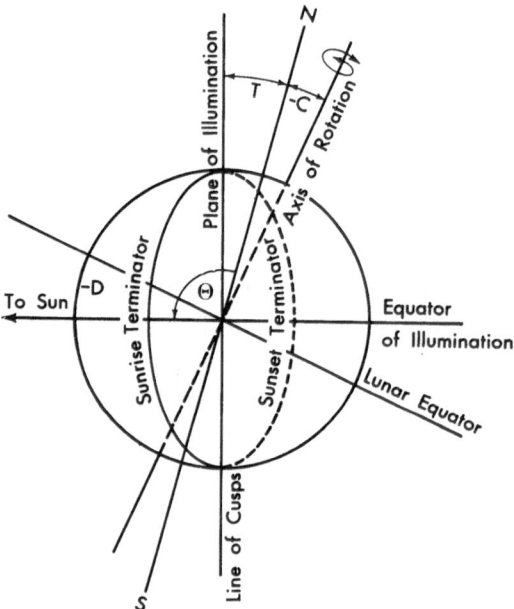

Fig. 5-2.   Illumination of the Lunar Globe.

In order to describe more accurately the shape and orientation of this terminator ellipse as seen by a terrestrial observer, let $L_\odot$, $B_\odot$ denote the selenocentric longitude and latitude of the Sun with respect to the plane of the *ecliptic*, and $A_\odot$, $D_\odot$ be the same selenocentric coordinates of the Sun referred to the *lunar equator* (both $L_\odot$ and $A_\odot$ being measured from the direction of the vernal point). Let, moreover, $l_\odot$ denote the longitude of the Sun; and $EM \equiv r$, $ES \equiv R$, and $MS \equiv R'$ be the distances Earth-Moon, Earth-Sun, and Sun-Moon, respectively. If so, it follows that

$$\left.\begin{array}{l} R' \cos B_\odot \cos L_\odot = - r \cos \beta_{\langle} \cos \lambda_{\langle} + R \cos l_\odot \\ R' \cos B_\odot \sin L_\odot = - r \cos \beta_{\langle} \sin \lambda_{\langle} + R \sin l_\odot \\ R' \sin B_\odot \quad\quad = - r \sin \beta_{\langle} \end{array}\right\} \quad\quad (5\text{-}1)$$

where $\lambda_{\mathbb{C}}$, $\beta_{\mathbb{C}}$ are the geocentric ecliptical longitude and latitude of the Moon (related with the Moon's right-ascension $\alpha_{\mathbb{C}}$ and declination $\delta_{\mathbb{C}}$ by means of the equations 3-3); while the values of $A_{\odot}$ and $D_{\odot}$ are related with $L_{\odot}$ and $B_{\odot}$ by the set of the equations

$$
\left.
\begin{aligned}
\cos D_{\odot} \cos (A_{\odot} - \Omega) &= \cos B_{\odot} \cos (L_{\odot} - \Omega), \\
\cos D_{\odot} \sin (A_{\odot} - \Omega) &= \cos B_{\odot} \sin (L_{\odot} - \Omega) \cos I - \sin B_{\odot} \sin I, \\
\sin D_{\odot} &= \cos B_{\odot} \sin (L_{\odot} - \Omega) \sin I + \sin B_{\odot} \cos I,
\end{aligned}
\right\}
\tag{5-2}
$$

where $\Omega$ denotes, as before, the longitude of the ascending node of the lunar orbit.

A solution of these equations offers no difficulty. Setting as usual, $R = \csc \pi_{\odot}$, $r = r_{\oplus} \csc \pi_{\mathbb{C}}$, ($r_{\oplus}$ denoting the mean radius of the terrestrial globe) in terms of the solar and lunar parallaxes, and $R' = R$ Equations (5-1) yield, successively,

$$
\tan (L_{\odot} - l_{\odot}) = -\frac{\sin \pi_{\odot}}{\sin \pi_{\mathbb{C}}} \cos \beta_{\mathbb{C}} \sin (\lambda_{\mathbb{C}} - l_{\odot})
\tag{5-3}
$$

and

$$
\left.
\begin{aligned}
\tan B_{\odot} &= \frac{\sin (L_{\odot} - l_{\odot})}{\sin (\lambda_{\mathbb{C}} - l_{\odot})} \tan \beta_{\mathbb{C}} \\
&= \frac{\sin \pi_{\odot}}{\sin \pi_{\mathbb{C}}} \sin \beta_{\mathbb{C}} \cos (L_{\odot} - l_{\odot}).
\end{aligned}
\right\}
\tag{5-4}
$$

Equations (5-2) can be solved in much the same way. Since, moreover, the Moon is never seen from the distance of the Sun to rise above the ecliptic by more than 50″, we can ignore in (5-2) the terms containing $\sin B_{\odot}$, and to set $\cos B_{\odot} = 1$ – in which case

$$
\tan (A_{\odot} - \Omega) = \cos I \tan (L_{\odot} - \Omega)
\tag{5-5}
$$

and

$$
\sin D_{\odot} = \sin I \sin (L_{\odot} - \Omega).
\tag{5-6}
$$

With the values of the angles $A$ and $D$ thus established, let us proceed to specify the position of the terminator ellipse; and to begin with determine the selenocentric angle $m$ between the vectors $MS$ and $EM$ at $M$ (cf. Figure 5-1). As the direction cosines of $MS$ and $EM$ are

$$
\left.
\begin{aligned}
\cos D_{\odot} \cos (A_{\odot} - \Omega), & \quad \cos B_{\mathbb{C}} \cos (L_{\mathbb{C}} - \Omega), \\
\cos D_{\odot} \sin (A_{\odot} - \Omega), & \quad \text{and} \quad \cos B_{\mathbb{C}} \sin (L_{\mathbb{C}} - \Omega), \\
\sin D_{\odot}; & \quad \sin B_{\mathbb{C}};
\end{aligned}
\right\}
\tag{5-7}
$$

respectively, it follows that

$$
\cos m = \sin B_{\mathbb{C}} \sin D_{\odot} + \cos B \cos D_{\odot} \cos (L_{\mathbb{C}} - A_{\odot}).
\tag{5-8}
$$

If we ignore the small angle $I$ and set, in accordance with (5-5) and (5-6), $A = L$ and $D = 0$, a sufficiently approximate expression for $m$ will be represented by

$$
\cos m = \cos B_{\mathbb{C}} \cos (L_{\mathbb{C}} - L_{\odot});
\tag{5-9}
$$

or, ignoring also the squares of $B$,

$$
m = L_{\mathbb{C}} - L_{\odot}.
\tag{5-10}
$$

The Sun's selenographic co-longitude $C_\odot$ (i.e., the longitude of the sunrise terminator, measured westward from the lunar prime meridian) will then be given by

$$C_\odot = L_\mathrm{\zodiacleo} - L - 90° = m - 90°;\qquad(5\text{-}11)$$

and, if we disregard the effects of orbital eccentricity, will be advancing uniformly westward at a rate of $12°.19$ per day.

In a similar manner, the inclination $T$ of the line of cusps to the direction of the terrestrial north pole (i.e., to the terrestrial meridian) is defined by the equation

$$\tan T = \frac{\cos e}{\cos \delta_\odot \sin (\alpha_\odot - \alpha_\mathrm{\zodiacleo})},\qquad(5\text{-}12)$$

where $e$ denotes the geocentric angle in the triangle $EMS$ between the vectors $R$ and $r$ (see Figure 5-1), and is given by the equations

$$\left.\begin{array}{l}\cos e = \cos \delta_\mathrm{\zodiacleo} \sin \delta_\odot - \sin \delta_\mathrm{\zodiacleo} \cos \delta_\odot \cos (\alpha_\odot - \alpha_\mathrm{\zodiacleo}) \\ = \cos \beta_\mathrm{\zodiacleo} \cos (\lambda_\mathrm{\zodiacleo} - \lambda_\odot)\end{array}\right\}\qquad(5\text{-}13)$$

in equatorial or ecliptical coordinates. Should we again disregard the inclination $I$ of the lunar orbit to the ecliptic, $T$ becomes the angle between the declination and latitude circles of the Moon in the sky, and as such may be found from the alternative formulae

$$\sin T = -\frac{\cos \alpha_\mathrm{\zodiacleo}}{\cos \beta_\mathrm{\zodiacleo}} \sin \varepsilon = -\frac{\cos \lambda_\mathrm{\zodiacleo}}{\cos \delta_\mathrm{\zodiacleo}} \sin \varepsilon,\qquad(5\text{-}14)$$

where $\varepsilon$ denotes, as before, the obliquity of the ecliptic. The inclination of the line of the cusps to the projected axis of rotation of the lunar globe is then equal to the algebraic sum $T + C$, where the angle $C$ between the axis of rotation and northern direction has already been defined by Equations (3-27) and marked on Figure 5-2.

The *phase angle* of the Moon, varying between $0°$ at the "new" Moon and $180°$ at full-Moon (cf. Figure 5-1), can be defined as

$$g = 180° - m = 90° + C_\odot,\qquad(5\text{-}15)$$

where the angles $m$ and $C_\odot$ are as given by Equations (5-9)–(5-11) before. These formulae define, strictly speaking, the *geocentric* phases of the Moon; and (5-12), the inclination of the line of the cusps as would be seen by an observer situated at the center of the Earth. The *topocentric* values of the respective quantities are, however, obtainable from the same equations, in which the geocentric coordinates $\alpha_\mathrm{\zodiacleo}, \delta_\mathrm{\zodiacleo}$ or $\lambda_\mathrm{\zodiacleo}, \beta_\mathrm{\zodiacleo}$ have been replaced by their topocentric values (for the Sun, the difference between the two is insensible). The formula (5-15) for the phase as given above ignored also the finite distance of the Moon from the Earth in assuming that the line of sight tangent to the lunar limb is parallel with the line to the center. In actual fact, the convergence of this beam will *diminish* the geocentric phase angle $g$ by $\sin^{-1}(r_\mathrm{\zodiacleo}/r) = \sin^{-1}[(r_\mathrm{\zodiacleo}/r_\oplus) \sin \pi_\mathrm{\zodiacleo}]$, where $r_\mathrm{\zodiacleo}$ denotes the radius of the lunar globe and $r_\oplus$, that of the Earth; and $\pi_\mathrm{\zodiacleo}$, the lunar horizontal parallax; or the topocentric phase $g'$ by $\sin^{-1}(r_\mathrm{\zodiacleo}/r') = \sin^{-1}[(r_\mathrm{\zodiacleo}/r_\oplus) \sin \pi_\mathrm{\zodiacleo}']$, where $r'$ stands for the lunar topocentric distance and $\pi'_\emptyset$, the topocentric parallax. Since, however, the maximum value of the ratio $r_\emptyset/r$ does not exceed 0.0048, the phase excess due to this cause is less than $0°.27$ or $16'$ and can very often be ignored.

Lastly, for the sake of completeness, we may wish to note that the heliocentric angle $s$ between the vectors $R$ and $R'$ of the triangle $EMS$, being equal to the difference $L_\odot - \lambda_\odot$ between the selenocentric and geocentric longitude of the Sun, follows at once from the exact equation (5-3) as

$$\tan s = -\frac{\sin \pi_\odot}{\sin \pi} \cos \beta_\emptyset \sin (\lambda_\emptyset - \lambda_\odot). \tag{5-16}$$

Since, moreover, $EMS$ constitutes a plane triangle, the angles $s$ and $e$ are related exactly by the equation

$$\tan s = \frac{r \sin e}{R - r \cos e}, \tag{5-17}$$

from which, to a sufficient approximation,

$$\tan s \simeq \frac{\sin \pi_\odot}{\sin \pi_\emptyset} \sin e, \tag{5-18}$$

where the ratio

$$\frac{\sin \pi_\odot}{\sin \pi_\emptyset} = \frac{8''.79415}{3422''.62} = 0.00256942. \tag{5-19}$$

From the same triangle we can also deduce the following exact equation

$$\cot m = \left\{ \frac{\sin \pi_\odot}{\sin \pi_\emptyset} - \cos e \right\} \csc e, \tag{5-20}$$

which can be used as an alternative to (5-8) for the computation of the selenocentric angle $m$ when the geocentric angle $e$ has been evaluated from likewise exact relation (5-13). Should all three angles $e$, $m$, and $s$ have been evaluated with the aid of exact equations (5-13), (5-16) or (5-17), and (5-20), the necessary requirement that

$$e + m + s = 180° \tag{5-21}$$

can serve as an algebraic check.

The geometrical circumstances of the illumination of the Moon by the Earth follow likewise readily from the foregoing analysis. It goes without saying that, for a terrestrial observer, the phase of the Earth-lit Moon is always "full" – though the intensity of illumination will vary periodically in the course of each month with the phase which sunlit Earth exhibits to the Moon. The phases of the Earth as seen from the Moon, and of the Moon as seen from the Earth, are obviously complementary: the Earth appears "full" on the Moon when the latter appears "new" to us, and vice versa. This is why the "ashen light" of lunar night hemisphere – due to the illumination of the Moon by the Earth (i.e., "secondary reflection" of sunlight) is so bright near new Moon, and diminishes with advancing phase. Lunar landscape illuminated by neither the Sun nor the Earth is not visible from the surface of our planet, and can be witnessed only from the vantage point of a space probe.

When the phase of the Moon is full or nearly so, another phenomenon may arise which invites close attention: namely, the *eclipses of the Moon*, due to its passage through the shadow cone cast by the Earth into space. This phenomenon is the resultant of a combination of geometrical circumstances connected with the relative dimensions of the Earth and the Moon, their distances from the Sun, as well as the inclination of the lunar orbit to the ecliptic and the orientation of the line of its nodes in space.

In order to investigate the circumstances under which lunar eclipses may occur, let $r_\odot$, $r_\oplus$ denote hereafter the radii of the Sun and the Earth, and let $R$ be their mutual

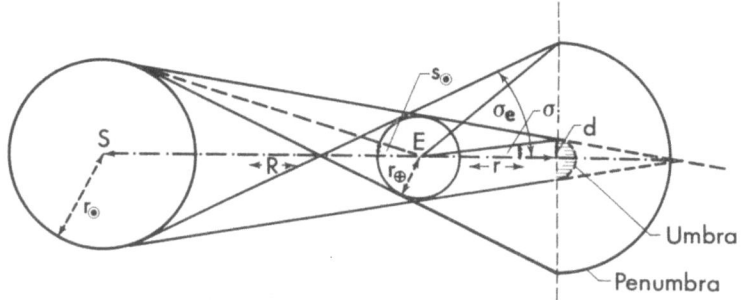

Fig. 5-3.   Geometry of Lunar Eclipses.

distance (Figure 5-3). If so, the semi-apex angle $\theta$ of the terrestrial shadow-cone will obviously be equal to

$$\theta = \sin^{-1} \frac{r_\odot - r_\oplus}{R} \qquad (5\text{-}22)$$

which, for $r_\odot = 695\,980$ km, $r_\oplus = 6371.0$ km and $R = 149\,598\,500$ km, becomes equal to $0°.26412$ or $15'51''$. The distance $d$ of the Earth's center from the vertex of this cone should then be equal to

$$\varDelta = r_\oplus \csc(s''_\odot - \pi_\odot), \qquad (5\text{-}23)$$

where $s''_\odot = 959''.6$ stands for the apparent solar semi-diameter, leading to $\varDelta = 1\,382\,000$ km. As the distance $r$ of the Moon from the Earth oscillates between $364\,400$ km and $406\,730$ km, its distance from this vertex will oscillate from $975\,000$ to $1\,018\,000$ km between perigee and apogee.

Moreover, as the angular semi-diameter $\sigma''_i$ of the Earth's shadow at the distance of the Moon is clearly given by

$$\sigma''_i = \pi_\odot + \pi_{(} - s''_\odot \qquad (5\text{-}24)$$

where the horizontal parallax $\pi$ of the Moon oscillates from $3173''.3$ to $3621''.1$ between perigee and apogee, it follows that the linear semi-diameter

$$d = r \tan \sigma''_i \qquad (5\text{-}25)$$

of the Earth's shadow at the mean distance of the Moon ($r = 384\,400$ km, $\pi_{\mathbb{C}} = 57'2''.6$) will be equal to 4601 km, oscillating by $\pm 6°.5$ per cent between perigee and apogee. This mean value of $d$ proves to be equal to 27.647 times the mean radius of the lunar globe. Therefore, the eclipses of the Moon arising from its ingress in the terrestrial shadow can be either partial or total, but never annular.

It may be of interest to point out that the simple geometrical picture of lunar eclipses unfolded so far affords, in fact, an opportunity to determine the distance between the Earth and the Moon from the observed duration of the eclipse along the following lines. Let, as before, $r_{\odot, \oplus}$ denote the radii of the Sun and the Earth, and let $r_{\mathbb{C}}$ stand for the radius of the Moon, while $d$ represents the radius of the terrestrial shadow at the Moon's distance. Let, furthermore, $R$ and $r$ denote the distance between Earth-Sun and Earth-Moon, respectively. If so, it follows from the similarity of the triangles that

$$\frac{r_{\oplus} - d}{r_{\odot} - r_{\oplus}} = \frac{r}{R} = \frac{1}{n}. \tag{5-26}$$

If, moreover (as was noted already by the ancients) the apparent semi-diameters of the Sun and the Moon as seen from the Earth are approximately equal,

$$r_{\odot} = n r_{\mathbb{C}}. \tag{5-27}$$

Lastly, the value of $d$ can be deduced from observed duration of the eclipse (i.e., fraction of a month which the Moon will take to pass through the Earth's shadow).

Thus from observations made in the 2nd century B.C., Hipparchos estimated

$$d = \tfrac{8}{3} r_{\mathbb{C}}; \tag{5-28}$$

and inserting (5-27)–(5-28) in (5-26) he found that

$$r_{\mathbb{C}} = \frac{3}{11}\left(1 + \frac{1}{n}\right) r_{\oplus}. \tag{5-29}$$

As, moreover, the angular radius of the Moon was known to be approximately $\frac{1}{4}$ of a degree, its distance $\rho$ from us should be $4 \times 180°/\pi \sim 229$ times the radius $r_{\mathbb{C}}$, and equal to

$$r = 229 \times \frac{3}{11}\left(1 + \frac{1}{n}\right) r_{\oplus} = 62\left(1 + \frac{1}{n}\right) r_{\oplus}. \tag{5-30}$$

As to the ratio $R/r = n$, Hipparchos considered its value to be between 20 and 30 (making the Sun 20–30 times as far from us as the Moon) – a hopeless underestimate; but as it entered on the right-hand side of (5-30) only through its reciprocal, it made little difference to Hipparchos's conclusion that the distance separating us from the Moon was equal to some 62 radii of the Earth (the correct value being 60.27)! As, moreover, the dimensions of the terrestrial globe were, at that time, known correctly within a few per cent of their exact value (through the preceding work by Eratosthenes), it follows that the absolute distance between the Earth and the Moon – our nearest celestial neighbour – has been known with an error of less than 2.9% since the second century before Christ – a certainly no mean achievement!

But let us return in more detail to the geometry of the Earth's shadow in space. If the Moon is to enter it in its orbit and thus give rise to the phenomenon of a lunar eclipse (whether total or partial), its geocentric angular distance $180° - e = m + s$ from the axis of the shadow must be less than

$$\sin^{-1}(\sin \pi_\odot + \sin S''_\odot) - \sin^{-1}(\sin \pi_\mathrm{C} - \sin S''_\mathrm{C}), \qquad (5\text{-}31)$$

where $e, m, s$ denote, as before the interior angles of the triangle $EMS$ (given by Equations 5-13, 5-16, and 5-20); $\pi_{\odot,\mathrm{C}}$, the parallax of the Sun or the Moon; and $S''_{\odot,\mathrm{C}}$, their apparent semi-diameters.

In more precise terms, as the axis of the Earth's shadow always lies in the plane of the ecliptic, the limits of the eclipses will be given by the angle, along the ecliptic,

$$\phi = (\sigma''_i \pm s''_\mathrm{C})(1 - 2q \cos i + q^2)^{\frac{1}{2}} \csc i, \qquad (5\text{-}32)$$

between the axis of the shadow cone and the nearest line of the nodes, where $\sigma''_i$ denotes, as before, the geocentric angular radius of the terrestrial shadow cone at the Moon's distance; $q$, the ratio of the Earth's angular velocity of revolution to that of the Moon (or, approximately, the ratio of one sideric month to a year); and $i$, the inclination of the Moon's orbit. The upper algebraic sign in the expression $(\sigma''_i \pm s''_\mathrm{C})$ refers to the occurrence of partial eclipses, and the lower sign $(-)$ to those which are total.

Inserting in (5-31) appropriate limits for $q$, $\sigma''_i$, and $s''_\mathrm{C}$ (arising from orbital eccentricity) we find that if the ecliptical angle $\phi$ happens to be less than $9°30'.5$, a *partial* eclipse of the Moon *must* occur near full phase; but it *cannot* occur if this angle exceeds $12°3'.7$; in between these two limits (obtaining if the Moon at the time of eclipse is near the perigee or apogee) partial eclipses may or may not occur.

On the other hand, a *total* eclipse *may* occur when $e$ differs from $180°$ by less than $5°34'.0$, and *must* occur if it differs by less than $4°9'.5$; in between we may have either partial or total eclipses. Therefore, if:

$$\phi < 4°\ 9'.5: \quad \text{total eclipse must occur}$$
$$4°\ 9'.5 < \phi < 5°34'.0: \quad \text{either total or partial eclipse must occur}$$
$$5°34'.0 < \phi < 9°30'.5: \quad \text{partial eclipse must occur}$$
$$9°30'.5 < \phi < 12°\ 3'.7: \quad \text{partial eclipse may occur}$$
$$12°\ 3'.7 < \phi \qquad : \quad \text{no eclipse can occur}$$

All eclipses considered so far are the *umbral* eclipses – arising when the Moon enters the cone of *full shadow* of the Earth. *Penumbral eclipses* arising when the Moon is illuminated by partly eclipsed Sun, are much more frequent. The condition for their occurence is of the same form as (5-32), provided only that $\sigma''_i$ is replaced by the angular semi-diameter

$$\sigma''_e = \pi_\odot + \pi_\mathrm{C} + S''_\odot \qquad (5\text{-}33)$$

of the terrestrial *outer* shadow cone, with a vertex *between* the Sun and the Earth. As, however, such purely penumbral eclipses are much less important (and noticeable), we do not propose to investigate their limits in the same detail; and their evaluation can be left as an exercise for the reader.

The average *duration* of a total eclipse of the Moon is close to 226 minutes (3 hours 46 min), and the average duration of totality is 103 minutes (1 hour 43 min). The *maximum* duration (in hours) of an umbral (or penumbral) eclipse can be evaluated from the formula

$$\frac{2(\pi_{\odot} + \pi_{\mathbb{C}} \mp S''_{\odot} \mp S''_{\mathbb{C}})}{m} \left\{ 1 + \frac{n \cos^2 i}{m} \right\} \cos i, \tag{5-34}$$

where $n$ and $m$ denote the hourly motions in longitude of the Sun and the Moon; and where, of the first pair of $\pm$ signs in front of $S''_{\odot}$, the upper $(-)$ refers to umbral, and lower $(+)$ to penumbral eclipses; while the second pair in front of $S''_{\mathbb{C}}$ refers to total $(-)$ and partial $(+)$ eclipses, respectively.

In contrast with the solar eclipses (or, rather, of the Earth) which may occur when the angle $e$ is small, eclipses of the Moon are objective phenomena, visible wherever the Moon happens to be above the horizon. In fact, lunar eclipses are visible over the entire night-time hemisphere; and since the Moon is rising, this will always be more than one-half of the surface of the Earth. For any particular place, therefore, lunar eclipses are bound to be more frequent than solar eclipses; though for the Earth as a whole the opposite is true.

Each year or 12 synodic months will witness either no eclipse at all, or one partial eclipse, or two eclipses of either kind (within six synodic months). Each eclipse of the Moon occurs half a month before or after the eclipse of the Sun – in particular, between two partial solar eclipses at the same node a total eclipse of the Moon must occur at the opposite node. This happened, for instance, in 1946, 1953, and will happen again in 1964 or 1971, when we shall have three eclipses during the same synodic month. Conversely, if a full Moon takes place far from the nodes, years may occur when there are no lunar eclipses at all (such as, for instance, will be true of 1966, 1969, 1980, 1984, etc).

The prediction of the eclipses of the Sun and the Moon has exercised the ingenuity of astronomers since the dawn of the civilization; and remarkable advances have been achieved even before the advent of the Greeks. As lunar eclipses occur only at night-time, they did not attract quite the same attention as total eclipses of the Sun. Nevertheless, several of them are known to have influenced historical events – such as the lunar eclipse of 413 B.C. which (according to Plutarch) caused panic in Greek Navy off the coast of Sicily, or the eclipse of 330 B.C. which enabled the historians to date the battle of Arbel, in which Alexander the Great defeated the Persian king Dareios. The lunar eclipse on 21 November of the year 167 B.C. occurred again on the eve of the battle of Pydna, in which the Roman consul Paulus Aemilius destroyed the might of Perseus, the king of Macedonia; and (as we learn from Titus Livius) the Roman tribune Sulpicius Gallus interpreted the event as a favourable omen. Or who would forget the memorable lunar eclipse of 21 February 1504 A.D., visible in the West

Indies, whose prediction (albeit with the help of the tables of Regiomontanus) saved the lives of Christopher Columbus and all mates of his fourth expedition from the Indians?

It must be owned that remarkable feats of prediction of the eclipses in ancient times were largely due to empirical discovery of certain periodicities in their occurence, which were noted long before they were properly understood. As we explained already, the occurrence of eclipses depends critically on the relative position of the lunar nodes on the ecliptic. If, therefore, the draconic month (i.e., a time interval between two successive nodal passages) and synodic month (time-interval between successive identical phases) were identical, then each new or full Moon would occur in the same relative position with respect to the nodes; and we should either have each new Moon a solar eclipse (and each full Moon a lunar one), or none at all. Observations disclose that this is not the case – a fact which alone proves the *motion* of the nodes. However, it happens that 223 synodic months are almost equal to 242 draconic months (the difference between the two multiples being only 51 min 41.2 sec); so that the positions of the Moon relative to the nodes should be almost the same every 6585 days or 18 years and 10–11 days (depending on whether this interval contains 4 or 5 leap years). In consequence, should an eclipse of the Sun or of the Moon occur at a given time, it should recur at the same place after a time interval of 6585 days – a period already known to the Chaldeans 24 centuries ago, and described by the Greeks as Saros. Even closer is the coincidence of 716 synodic and 777 draconic months, leaving a residue of only 9 min. 46.1 sec. Therefore, the eclipses should recur more

TABLE 5-1

Lunar eclipses 1965–1975

| Date | Time of mid-eclipse | Semi-Duration: | |
|---|---|---|---|
| | | partial phase | totality |
| 1965 June 14 | 1$^h$ 51$^m$ UT | 50 min | – |
| 1967 Apr. 24 | 12  7 | 107 | 41 |
| Oct. 18 | 10  16 | 103 | 28 |
| 1968 Apr. 13 | 4  49 | 103 | 28 |
| Oct.  6 | 11  41 | 104 | 31 |
| 1970 Feb. 21 | 8  31 | 26 | – |
| Aug. 17 | 3  25 | 71 | – |
| 1971 Feb. 10 | 7  42 | 107 | 39 |
| Aug.  6 | 19  44 | 112 | 51 |
| 1972 Jan. 30 | 10  53 | 102 | 21 |
| Jul. 26 | 7  18 | 80 | – |
| 1973 Dec. 10 | 1  48 | 36 | – |
| 1974 Jun.  4 | 22  14 | 93 | – |
| Nov. 29 | 15  16 | 106 | 38 |
| 1975 May 25 | 5  46 | 109 | 45 |
| Nov. 18 | 22  24 | 102 | 23 |

closely after an interval of 21 144 days or just under 58 years; and periods still longer exist after which this is even more accurately the case.

In modern times, the prediction of an eclipse presents no problem; all the astronomer has to do is to reach for OPPOLZER's *Canon der Finsternisse* (1887) containing detailed description of the circumstances of all eclipses between 1206 B.C. to 2163 A.D. From this source we find that, in the decade between 1965–1975, a total of 16 lunar eclipses will occur, as characterized in Table 5-1.

In conclusion of the present section, one retrospective remark concerning the definition of lunar phases may possibly be of use. In the current astronomical terminology it is customary to refer to the smallest and largest phase each month as "new Moon" and "full Moon", respectively; but (although this is already implied in our analysis) it should be stressed that *not every full-Moon is equally "full", nor is every new Moon equally "new"*. Strictly speaking, the Moon is "new" or "full" only when the centers of the Sun, the Earth, and the Moon lie exactly on the same line; and we know that this can be only during an eclipse – a central annular eclipse of the Sun for the theoretical "new" Moon, and a central eclipse of the Moon when the latter is "full". No sunlit full-Moon is, therefore, really full, but possesses a finite *phase defect* equal to the minimum value of the angle $180°-e$ during that particular lunation. Since, moreover, this angle exceeds $\sin^{-1} [r_{\mathbb{C}}/r] = \sin^{-1} [(r_{\mathbb{C}}/r_{\oplus}) \sin \pi_{\mathbb{C}}]$ which is usually around 16', the limb of the lunar disc visible at that time will depart from a circle by perceptible amount.

## CHAPTER 6

# DYNAMICS OF THE EARTH-MOON SYSTEM

The dynamical symbiosis between our Earth and its only natural satellite has been at the basis of all topics discussed so far in this chapter; for the motion of the Moon in space, and even more that about its center of gravity, are largely controlled by the Earth – with distant, but powerful, cooperation of the Sun. The aim of the present section will, however, be to focus specific attention on certain dynamical aspects of the Earth-Moon symbiosis which were so far not discussed (or only briefly referred to) in the text, and which are concerned with the gravitational interaction of the two bodies, or properties of their motion over long intervals of time.

In order to approach this subject, let us return to one particular property of the problem of three bodies, the existence of which was briefly mentioned in Chapter 2: namely, the Jacobi integral (2-11) of the equations of motion (2-6)–(2-8). Suppose that these bodies are identified with the Sun, Earth, and Moon, respectively. If we ignore (for the moment) the fact that the distance $X'$ between the Sun and the Earth is not strictly constant (varying somewhat as it does in the course of a year around its mean value of $1.49599 \times 10^8$ km, on account of the small eccentricity $e = 0.01675$ of the terrestrial orbit), ignore also the small inclination $i = 5°8'33''$ of the lunar orbit to the ecliptic, and regard the latter as a circle of mean radius $r = 3.844 \times 10^5$ km, then with the values of

$$\left. \begin{array}{l} Gm_\odot = 1.327149 \times 10^{11} \\ Gm_\oplus = 3.98601 \ \times 10^5 \\ Gm_\text{☾} \ = 4.9026 \ \ \ \times 10^3 \end{array} \right\} \frac{\text{km}^3}{\text{sec}^2}, \tag{6-1}$$

corresponding to the mean velocity $V = 1.047$ km/sec of the Moon in its relative orbit, and to the mean angular velocity $n = 1.990987 \times 10^{-7}$ sec$^{-1}$ of the terrestrial orbit, we find from (2-11) that the value of the constant $C$ of the Jacobian integral in our case should be equal to

$$\left. \begin{array}{llll} C = & 1774.275 & & 2Gm_\odot/X' \\ & + & 2.099 & 2G(m_\oplus + m_\text{☾})/r \\ & + & 0.006 & n^2 r^2 \\ & - & 1.096 & V^2 \\ & = & 1775.284 & \text{km}^2/\text{sec}^2. \end{array} \right\} \tag{6-2}$$

Consider now the surface of zero velocity as defined by the equation

$$2U = C. \tag{6-3}$$

With the foregoing values of different constants occurring in (2-10), this surface proves to be an oval *closed* around the Earth and differing but a little from a sphere of radius

$$r = 8.046 \times 10^5 \text{ km},\qquad(6\text{-}4)$$

which *exceeds* the present mean radius $r = 3.844 \times 10^5$ km of lunar orbit by a factor of 2.14. Inside this oval the square of the velocity $V$ is by (2-10) positive; and a mass particle can move freely within it as it pleases in accordance with the given equations of motion. Outside the oval, however, $V^2$ becomes negative and the velocity itself imaginary. This signifies that no particle moving inside the oval can ever traverse the surface of zero velocity defined by Equation (6-3); and as the value of the Jacobian constant $C$ for the Moon is such as to place its orbit inside such a surface closed around the Earth, it would appear that the Moon could never recede from the Earth at a distance much more than twice its present value. This celebrated argument, advanced in the last century by DARWIN (1880) and HILL (1878), would, therefore, render the symbiosis between the Moon and the Earth a cosmically permanent affair. It was not until quite recently that its validity was disproved by KOPAL and LYTTLETON (1963); and in view of the cosmological importance of this point this argument will be given in some detail.

In order to do so, consider quite generally two finite masses $m_{1,2}$ situated at points of rectangular coordinates $(X_1, 0, 0)$ and $(X_2, 0, 0)$, with the origin at the center of mass – so that $m_1 X_1 + m_2 X_2 = 0$. Let, moreover, these two finite masses (i.e., the Sun and the Earth in our case) revolve around their common center of gravity in an ellipse, situated in the plane $Z = 0$ (i.e., the ecliptic) in accordance with the well-known integrals of the problem of two bodies, expressible as

$$l = R(1 + e \cos v),\qquad(6\text{-}5)$$

where $l$ stands for the parameter $a(1 - e^2)$, and

$$R^2 \, dv = \sqrt{G(m_1 + m_2)} \, dt,\qquad(6\text{-}6)$$

where $v$ denotes the true anomaly in the two-body orbit.

If a particle of infinitesimal mass $m$ (i.e., the Moon) occupies a position of coordinates $X_1 Y_1 Z_1$ at a distance $R_1 \equiv r$ and $R_2 \equiv \Delta$ from the attracting masses $m_{1,2}$, respectively, then its equations of motion are readily found to be

$$\ddot{X} - 2\dot{v}\dot{Y} + \dot{v}^2 X - v Y = -\frac{Gm_1}{R_1^3}(X - X_1) - \frac{Gm_2}{R_2^3}(X - X_2),\qquad(6\text{-}7)$$

$$\ddot{Y} + 2\dot{v}\dot{X} - \dot{v}^2 Y + vX = -\frac{Gm_1}{R_1^3} Y \qquad - \frac{Gm_2}{R_2^3} Y,\qquad(6\text{-}8)$$

$$\ddot{Z} \qquad\qquad = -\frac{Gm_1}{R_1^3} Z \qquad - \frac{Gm_2}{R_2^3} Z,\qquad(6\text{-}9)$$

where dots denote differentiation with respect to the time. If now $dt$ is eliminated in favour of d$v$ by means of Equation (6-6); and if also we replace $X_1 Y_1 Z$; $R_{1,2}$ by new coordinates $\xi$, $\eta$, $\zeta$; $S_{1,2}$ defined by

$$X = R\xi, Y = R\eta, Z = R\zeta \left. \atop R_{1,2} = RS_{1,2} \right\} \tag{6-10}$$

so that the (variable) radius-vector $R$ becomes our unit of length, we find that

$$S_1^2 = (\xi + \mu)^2 + \eta^2 + \zeta^2, \left. \atop S_2^2 = (\xi + \mu - 1) + \eta^2 + \zeta^2, \right\} \tag{6-11}$$

where

$$\frac{\mu}{1-\mu} = \frac{m_2}{m_1}, \qquad \frac{1}{1-\mu} = 1 - \frac{m_2}{m_1}. \tag{6-12}$$

If so, the Equations (6-7)–(6-9) of the *elliptic* case of the restricted three-body problem assume the interesting and concise form

$$\xi'' - 2\eta' \quad = \frac{R}{l} \frac{\partial U}{\partial \xi}, \tag{6-13}$$

$$\eta'' + 2\xi' \quad = \frac{R}{l} \frac{\partial U}{\partial \eta}, \tag{6-14}$$

$$\zeta'' \qquad + \zeta = \frac{R}{l} \frac{\partial U}{\partial \zeta}, \tag{6-15}$$

deduced first by NECHVÍLE (1917, 1926), where primes denote now differentiation with respect to $v$, and

$$U = (1 - \mu)(S_1^{-1} + \tfrac{1}{2}S_1^2) + \mu(S_2^{-1} + \tfrac{1}{2}S_2^2). \tag{6-16}$$

It may be noted that this potential $U$ does not depend explicitly on the eccentricity $e$ of the orbit of the two finite masses, which enters only through the ratio $R/l$ in (6-13) to (6-15), nor is it dependent explicitly on the time.

With these modified variables let us proceed now to construct the vis-viva integral of Equations (6-13)–(6-15) in the customary manner, by multiplying these equations successively by the angular velocity components $\xi'$, $\eta'$, $\zeta'$ and forming their sum. The result can be expressed as

$$\frac{1}{2} \frac{d}{dv}(\xi'^2 + \eta'^2 + \zeta'^2) + \frac{1}{2}\frac{d\xi^2}{dv} \left. \atop = \frac{R}{l}\left\{\frac{\partial U}{\partial \xi}\frac{\partial \xi}{\partial v} + \frac{\partial U}{\partial \eta}\frac{\partial \eta}{\partial v} + \frac{\partial U}{\partial \zeta}\frac{\partial \zeta}{\partial v}\right\} = \frac{R}{l}\frac{dU}{dv}, \right\} \tag{6-17}$$

since $U$ does not contain the time explicitly. Therefore, the foregoing equation can be formally integrated as written as

$$\xi'^2 + \eta'^2 + \zeta'^2 + \zeta^2 = 2 \int \frac{dU}{1 + e \cos v}. \tag{6-18}$$

For $e=0$, this latter equation integrates at once explicitly to

$$\begin{aligned}
\xi'^2 + \eta'^2 + \zeta'^2 &= -\zeta^2 + 2U + C \\
&= \zeta^2 + \eta^2 + 2\left\{\frac{1-\mu}{S_1} + \frac{\mu}{S_2}\right\} + C,
\end{aligned}\right\} \tag{6-19}$$

which is identical with the Jacobi integral (2-11) of the "circular" problem in the scaled variables. However, when $e>0$, it is *not* possible to integrate (6-18) explicitly to any corresponding form; for (6-18) represents simply an identity which would be satisfied by any solution of Equations (6-13)–(6-15). The merit of the Jacobi integral rests on the fact that, if $e=0$ (i.e., $R=l$), each side of equation (6-17) is a perfect differential, so that on integration the values of the various terms depend only on the end-point position $\xi$, $\eta$, $\zeta$ of the particle $m$. But this does not hold for (6-18) if $e \neq 0$; and to evaluate the integral therein we should need to know the value of $dU$ for each value of $t$ (or $v$) all along any selected path. To solve (6-13)–(6-15) to this end it would be necessary to adopt six initial values corresponding to the initial position and velocity of $m$; and if integration in (6-18) is carried out, the result would be a function of the time involving six arbitrary constants. Apart from the isolated term $-\zeta^2$, the coordinates of $m$ would not appear explicitly, and (6-18) would merely give the value of $\xi'^2 + \eta'^2 + \zeta'^2$ appropriate to any selected starting values. In other words, the integral in (6-18) does not depend solely on the coordinates involved in the expression (6-16) for $U$, and in no way does the right-hand side of (6-19) equated to zero correspond to a surface.

We conclude, therefore, that *when e is not equal to zero* (i.e., excluding the "circular" problem of three bodies), *no vis-viva integral or associated surfaces corresponding to constant (or zero) velocity exist, with constant C independent of the time.* Inasmuch as the eccentricity of the terrestrial orbit around the Sun is finite (and has without doubt been so since the formation of our planet), the closed nature of the surface of zero velocity surrounding the present lunar orbit around the Earth based on the *present* value of $C$ does *not*, therefore, guarantee that the Moon may not recede from the Earth to arbitrarily large distance in the future, nor that it could not have approached us from an indefinite distance in the past. Thus the celebrated argument which Darwin and Hill advanced in the last century to prove that the Moon must have been a permanent satellite of our Earth since the time of its formation is clearly insufficient to prove their point.

Although, therefore, for $e>0$ (which will generally be the case) the quantity $C$ in Equations (2-11) or (6-19) can no longer be regarded as constant, actual integrations of the underlying equations of motion reveal its variation with the time to be so slow that, for time-intervals of the order of decades or even centuries, it can be effectively treated as a constant; and in what follows we wish to take advantage of this particular property of the restricted three-body problem to refine our acquaintance with the concept of the *velocity of escape* from the gravitational field of the Moon.

In order to do so, consider now a three-body problem in which $m_{1,2}$ stand for the Earth and the Moon, and let $m$ represent now a mass-particle moving in the field of this gravitational dipole and referred again to a system of coordinates, with origin at the center of mass of the Earth-Moon system, and rotating in space with the latter about the $Z$-axis. If so, the corresponding equation of the surface of zero velocity then assumes the form

$$C = \frac{2(1 - \mu)}{\Delta} + \frac{2\mu}{r} + X^2 + Y^2 - V^2, \qquad (6\text{-}20)$$

where*

$$\mu = \frac{m_{\mathrm{C}}}{m_{\oplus} + m_{\mathrm{C}}} = 0.01215. \qquad (6\text{-}21)$$

and $\Delta$, $r$ denote the distances of $m$ from $m_{\oplus}$ and $m_{\mathrm{C}}$, respectively. If the mean distance $R = 384400$ km separating $m_{\oplus}$ and $m_{\mathrm{C}}$ is hereafter adopted as our unit of length, and the sum $m_{\oplus} + m_{\mathrm{C}}$ as that of mass, the corresponding units of time $t$ and velocity $V$ become $[G(m_{\oplus} + m_{\mathrm{C}})R^{-3}]^{-\frac{1}{2}} = 4.3423 \ldots$ days and $[G(m_{\oplus} + m_{\mathrm{C}})R^{-1}]^{\frac{1}{2}} = 1.0246 \ldots$ km/sec, respectively.

For a particle situated on the lunar surface of mean radius $a = 1738$ km

$$r = \frac{a}{R} = \frac{1738}{384400} = 0.004521, \qquad (6\text{-}22)$$

while $X$ varies between $1 - \mu \pm (a/R)$ – so that

$$0.9834 \leqslant X \leqslant 0.9924,$$

and

$$Y \leqslant \tan(a/R) = 0.00452.$$

Accordingly,

$$X^2 + Y^2 \simeq 0.9759 \qquad (6\text{-}23)$$

and

$$\Delta \simeq \sqrt{1 - (a/R)^2} = 0.9999898; \qquad (6\text{-}24)$$

in consequence of which Equation (6-20) assumes the particular form

$$C = 8.3261 - V^2. \qquad (6\text{-}25)$$

Now a geometry of the surfaces defined by Equation (6-20) reveals (cf., e.g., KOPAL, 1959; Chapter III) that, for $\mu = 0.01215$, the following five values of $C$

$$\left. \begin{array}{l} C_1 = 3.1876 \\ C_2 = 3.1707 \\ C_3 = 3.0122 \\ C_{4,5} = 2.9878 \end{array} \right\} \qquad (6\text{-}26)$$

are of critical significance. For $C \geqslant C_1$ Equation (6-20) represents two oval surfaces closed around the two finite mass centers. For $C = C_1$ these two ovals come in contact

---

* The reader should note that (on account of tradition) this definition of $\mu$ differs from that employed in Chapter 1.

at the inner Lagrangian collinear points $L_1$. For $C < C_1$ the two ovals coalesce and engulf both $m_\oplus$ as well as $m_\mathbb{C}$ in a common dumb-bell like surface which remain closed as long as $C > C_2$, but opens up at the outer Lagrangian collinear points $L_{2,3}$ behind the less massive, and more massive component, respectively, as $C = C_{2,3}$. Finally, when $C = C_{4,5}$, the surfaces defined by Equation (6-20) disappear completely from the real domain at the Lagrangian equilateral points $L_{4,5}$.

Now, for the Earth-Moon system, these critical values of $C$ correspond to (normalized velocities as defined by Equation (6-25)). When we evaluate them and convert into absolute units, we find these critical velocities to be

$$\left.\begin{array}{ll} V_1 & = 2.322 \\ V_2 & = 2.326 \\ V_3 & = 2.362 \\ V_{4,5} & = 2.367 \end{array}\right\} \text{km/sec.} \qquad (6\text{-}27)$$

These results reveal that mass particles can actually escape from the Moon through the inner Lagrangian point at $L_1$ if their velocity exceeds 2.322 km/sec; and for $V_2 = 2.326$ km/sec they can escape altogether from the Earth-Moon system! Compare this with parabolic velocity of escape

$$V = \sqrt{2Gm_\mathbb{C}/a} = 2.375 \text{ km/sec} \qquad (6\text{-}28)$$

which would be valid in the absence of the Earth. We see that the gravitational action of the latter will facilitate escape by lowering the requisite velocity by 112 m/sec.

And, we may add, the presence of the Sun (of which no account is taken in Equation (6-20) will further lower the escape velocity by 250 m/sec if acting in the same direction as the Earth – i.e., at the time of the full Moon – and increase it by the same amount half a month later. Therefore, under most favorable circumstances, lunar surface particles can leave the Moon to fall towards the Earth if endowed with a velocity as low as 2.071 km/sec; and only a slightly larger velocity would enable them to leave us altogether and to be captured directly by the Sun.

Any particles ejected from the lunar surface with velocities less than 2 km/sec but more than approximately $\sqrt{Gm_\mathbb{C}/a} = 1.68$ km should orbit around the Moon within the space limited by the closed oval surrounding the Moon for $C = C_1$ – the mean radius of which is approximately 38 000 km (i.e., more than twenty times the radius of the Moon itself). Only if $V < \sqrt{Gm_\mathbb{C}/a}$ will the particle so ejected fall back on the lunar surface, along trajectories which will be discussed in Chapter 17.

So far in this section we have explored certain dynamical properties of the Earth-Moon system by regarding both these bodies as mass-points. In what follows we wish, however, to consider more intimate effects of their interaction which stem from the *mutual distortion* of their form. It is well known that if a body of finite size is subjected to attraction of an external mass (be it a point or another configuration of finite size) at a finite distance, the forces exerted by this comparison will, in general*, raise *tides*

---

* I.e., unless the body in question possesses infinite rigidity.

on the central body which will distort its form and thus influence its external potential; and, conversely, tides are raised by the primary on the comparison of finite size. The manner in which this depends on the relative dimensions, mass-ratio, and internal structure of both components will be quantitatively discussed in the next Chapter 7; it is the dynamical effects exerted by such tides on the relative motion of such a pair that we wish to outline in this place.

In order to do so, suppose first that both the Earth and the Moon possess perfect elasticity in their solid parts, or perfect fluidity if they were liquid. Under these conditions the equilibrium theory of tides reveals that the maximum height of the tides occurs in the direction of the attracting body, and also (for even-harmonic tides) in the opposite direction. Because of this symmetry, it can be shown that the lunar tides on the Earth (or the terrestrial tides on the Moon) would then perturb neither the rotation nor the motion of either body – save for the secular advance of the apsidal line of their relative elliptic orbit.

If, however, the bodies in question are not perfectly elastic (or, if fluid, of finite viscosity), the high tides raised mutually by them no longer point in the direction of the attracting body (or diametrically opposite to it), but *lag* behind this direction (or advance ahead of it) depending on the difference between the velocity of axial rotation and orbital revolution. Thus if tidal deformation of the Earth by the Moon is accompanied by friction (due to viscosity, or imperfect elasticity), the rotation of the Earth will carry the lagging tidal bulge *forward* (see Figure 6-1). Moreover, the gravitational attraction on this bulge is *asymmetric* with respect to the radius-vector joining the

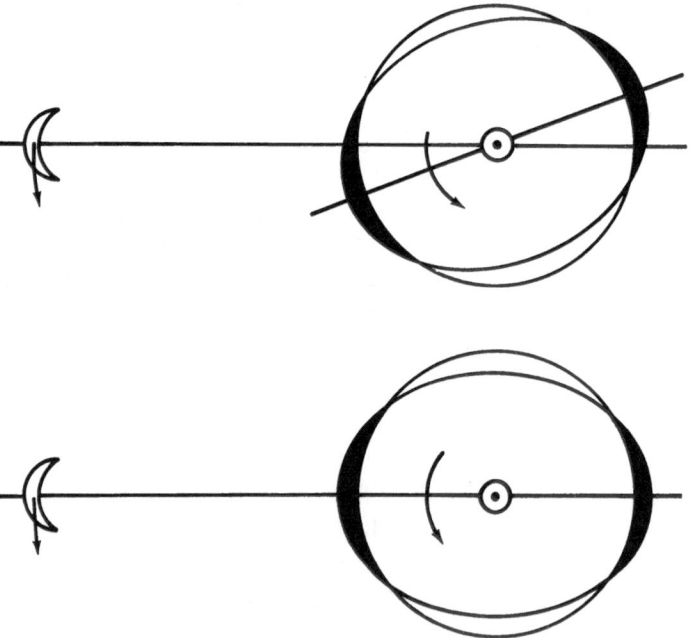

Fig. 6-1.   Tidal Bulge raised by the Moon on the Earth.

centers of the two bodies; this asymmetry gives rise to a *torque* tending to decelerate the Earth; and an equal and opposite torque tending to accelerate the Moon. Because of the friction this transfer is, moreover, accompanied by a loss of the mechanical energy of the Earth-Moon system (and its conversion into heat). In addition, as the Moon does not remain in the equatorial plane of the Earth, so that the tidal bulge is asymmetric with respect to the equator. As a result, one component of the torque acts constantly on the position of the orbital plane of the Moon; and an equal and opposite component tends to change the direction of the terrestrial axis of rotation.

Similarly, if the Moon's angular velocity of axial rotation were greater than the component of the Earth's rotation perpendicular to the orbital plane, the suface of the Moon would move ahead of the lagging tidal bulge, and angular momentum would be transferred from the Moon's orbital motion into the Earth's rotation. Should the component of the Earth's angular velocity perpendicular to the orbital plane equal the angular velocity of the Moon, frictional coupling between rotation and revolution would disappear; though (in eccentric orbits) radial component of the tides would still continue to convert mechanical energy into heat at the expense of motion (for fuller details of such a process cf. the forthcoming Chapter 10).

The most important characteristic of this tidal friction is, however, the fact that its effectiveness for transferring angular momentum between rotation and revolution depends so strongly on the distance between the attracting masses. In particular, when the Earth and the Moon are separated by the distance $r$, the height of $j$-th harmonic tides (of which that corresponding to $j=2$ is the most important) raised by one on the other proportional to $r^{-j-1}$ (cf. Equation (7-67)) and, consequently, the corresponding tidal torque varies as $r^{-2(j+1)}$ for $j=2,3,4\ldots$. Clearly, the secular changes in the orbital elements of the Earth-Moon system will be rapid when $r$ is small, and become negligible as $r \to \infty$.

What are their changes expected as a result of tidal friction? Let, in what follows, $a$ denote the semi-major axis of the relative Earth-Moon orbit (regarded as an ellipse); $e$ and $\tilde{\omega}$, its eccentricity and longitude of perigee; $i$, inclination of the orbit to the ecliptic; and $\Omega$, the longitude of the ascending node, then (cf., e.g., MOULTON, 1914, pp. 404–405) the variational equations for the respective elements assume the forms

$$\frac{da}{dt} = \frac{2e \sin v}{n\sqrt{1-e^2}} R + \frac{2a\sqrt{1-e^2}}{nr} T, \tag{6-29}$$

$$\frac{de}{dt} = \frac{\sqrt{1-e^2} \sin v}{na} R + \frac{\sqrt{1-e^2}}{na^2 e} \left\{ \frac{a^2(1-e^2)}{r} - r \right\} T, \tag{6-30}$$

$$\frac{d\tilde{\omega}}{dt} = -\frac{\sqrt{1-e^2} \cos v}{nae} R + \frac{\sqrt{1-e^2} \sin v}{nae} \left\{ 1 + \frac{r}{a(1-e^2)} \right\} T$$
$$- \frac{r \sin(\omega+v) \cot i}{na^2 \sqrt{1-e^2}} W, \tag{6-31}$$

$$\frac{di}{dt} = \frac{r \cos (\omega + v)}{na^2 \sqrt{1 - e^2}} W,$$  (6-32)

and

$$\frac{d\Omega}{dt} = \frac{r \sin (\omega + v)}{na^2 \sqrt{1 - e^2} \sin i} W,$$  (6-33)

where

$$n^2 = \frac{G(m_\oplus + m_\mathbb{C})}{a^3}$$  (6-34)

and $v$ is the true anomaly in the relative orbit, while

$$R, T, W$$

represent the three rectangular components of the tidal torque defined so that $R$ acts in the direction of the radius-vector $r$; and $T$, in the tangential direction to it in the orbital plane.

The explicit formulation of these components for both the Earth and the Moon is a matter of some complexity, which we do not wish to reproduce in this place (for fuller details cf. DARWIN, 1880; GERSTENKORN, 1955; MACDONALD, 1964); but it can be done in terms of the appropriate expansions of elliptic motion; and once these have been ascertained and inserted in Equations (6-29)–(6-33), the latter can be integrated simultaneously to yield the corresponding secular variation of the respective elements of the Moon's orbit in the course of the time.

Such integrations aiming to unravel the distant past of the Earth-Moon system were first undertaken by G. H. DARWIN (1879, 1880) on the assumption that the Earth behaved as a globe of viscous liquid (of viscosity $\eta$), and the Moon was regarded as a mass-point raising tides which lagged by an angle

$$\delta = \frac{19}{2r_\mathbb{C}g} \left(\frac{\eta}{\rho_\oplus}\right) (\omega - n)$$  (6-35)

where $r_\mathbb{C}$ denotes the mean radius of the Earth; $g$ the gravitational acceleration on Earth; $\rho_\oplus$, the mean density of the terrestrial globe; and $\omega$, its angular speed of rotation. More recently, Darwin's computations were extended (by a somewhat different method) to longer intervals of time by GERSTENKORN (1955); and still more recently MACDONALD (1964) repeated them for an elastic Earth (with such departures from perfect elasticity as are indicated by geophysical observations) as well as the Moon (for which, in the absence of other data, the same anelasticity as shown by the Earth was assumed).

The results of all these studies converge to certain interesting conclusions which we wish briefly to report in this place. First, if Equations (6-29)–(6-33) are integrated from initial condition obtaining at the present time, when the axial rotation of the Moon is completely synchronised with its revolution while the Earth rotates 24.32 times faster, the secular effects of the tangential component $T$ on the right-hand side of (6-29) will be to make an increase in the future (and diminish in the past); but the radial component $R$ arising from the eccentricity $e$ of the Moon's orbit will have an opposite effect (as it removes energy from the orbit without altering its angular momentum, it would tend to reduce $a$ secularly). At the present time, the effects of tangential tides dominate those of the radial tides on the Earth; and, as a result, the Moon must be slowly moving away from us. However, as any increase in the semi-major axis $a$ entails a growth of the orbital angular momentum

$$h^2 = \frac{Gm_{\oplus}m_{\mathbb{C}}}{m_{\oplus} + m_{\mathbb{C}}} a(1 - e^2),\qquad (6\text{-}36)$$

of the Earth-Moon system, while its total angular momentum

$$H = m_{\oplus}k_{\oplus}^2\omega_{\oplus} + m_{\mathbb{C}}k_{\mathbb{C}}^2\omega_{\mathbb{C}} + h \qquad (6\text{-}37)$$

(where $k$, and $\omega$, denote the radii of gyration of the two bodies and their angular velocities of axial rotation, respectively) must remain conserved in the course of time, it follows that *any increase in a must correspondingly slow down the rate of axial rotation of the Earth* (and vice versa). It, furthermore, appears that the resulting increase in length of the day (and the month) must continue, until both equal some 55 of our present days. At that time – probably many times as long as the present age of the Moon (at least since it became our closest celestial neighbour) – the synchronism between rotation and revolution of all three bodies should lead to "stationary" tides (cf. Figure 6-1, top) producing no torque; and in the absence of any dissipation the system should then be stable against further changes (if we disregard much weaker solar influences).

This view of the dynamical evolution of the Earth-Moon system is supported empirically by the observed secular deceleration of the Earth's rotation and of the Moon's motion. According to a recent review of the relevant data by MUNK and MacDONALD (1960) the astronomical evidence points to a tidal lengthening of the day by $1.8 \times 10^{-3}$ seconds per century, or a fractional change of $2.1 \times 10^{-10}$ per annum. This figure has recently been tentatively verified in a remarkable way by WELLS (1963), who found that certain ridges in the skeletal structure of corals might be evidence of a diurnal growth cycle. Recent corals show about 360 such ridges per year, while fossil corals from the Middle Devonian (age 400 million years; cf. HOLMES, 1959) have about 400 ridges per year – corresponding to a fractional lengthening of the day by 40 days in $4 \times 10^8$ years or $2.8 \times 10^{-10}$ per annum.

Accordingly, the semi-major axis $a$ of the Moon's orbit should presently be increasing at a rate close to 3.2 cm/year; its inclination $i$ diminishing by $1''.9 \times 10^{-6}$ per annum; and the eccentricity $e$ increasing by $1.2 \times 10^{-10}$ per annum. It is, of course, possible also to integrate our equations (6-29)–(6-33) backward in time, in order to find out what the orbit of the Moon around the Earth may have looked like in the more remote past, beyond the ages of known geological epochs. It is evident from the above estimates of the present orbital changes that, in more distant past, these could have been considerably greater (particularly for small values of $a$, when the effects of tidal torque increased as $a^{-6}$ for the second-harmonic tides). The actual results of such computations should be regarded with some reserve, as they necessarily entail extrapolation of the present internal structure of the Earth as well as of the Moon back to the times when these could have been considerably different (for example, the Earth may then not have had its present metallic heavy core).

Nevertheless, already DARWIN showed in 1880 that, when we go back in time, the duration of both the day and the month shortens, until they become synchronous when the Moon's distance is reduced to only a few radii of the Earth. This circumstance suggested to Darwin that the Moon may have originated as a part of the Earth, perhaps by fissional break-up induced by a resonant solar tide.

Gerstenkorn's work resulted in a number of significant changes in this picture. Since the tidal torque tends to increase the dimensions of the lunar orbit, the latter must have been progressively smaller in the past; its minimum value being equal to 2.89 mean radii of the Earth at a time when our day was but 4.8 hours long, and the month (i.e., period of revolution of the Moon) amounted to 6.9 hours of our time. These figures differ but little from subsequent results by MacDonald, who (on the assumption of a constant tidal lag of $\delta = 2°.16$ and constant moments of inertia of the Earth throughout the ages) arrived at a minimum distance for the Moon of 2.78 Earth radii, 5.0-hour day and 6.5-hour month some 1.79 thousand million years ago. At that time the lunar tides were some 8000 times higher than at present – attaining amplitudes comparable with the present mean depth of the oceans. Moreover, the angle of inclination between the lunar orbit and the terrestrial equator has increased from its present value of 24° to approximately 46°. This minimum approach of the Moon to the Earth occurred at a time when the component of rotational angular velocity of the Earth along the Moon's orbital axis became equal to the Moon's orbital angular velocity, so that the sub-lunar point did not move East-West on the Earth's surface. Under these conditions no torques were exerted; and for inclinations higher than 46° the torque would be reversed, transferring angular momentum from the Moon to the Earth.

Darwin did not continue his calculations beyond this points, as he expected the events then to become catastrophic and the Moon to be born from the womb of the Earth, which some of this followers (Fisher) identified with the "scar" of the Pacific ocean. However, this view had eventually to be abandoned; for with a 4.8-hour period of axial rotation the terrestrial globe would still have been far from rotational instability; and any build-up of resonant tides should have been effectively damped out by viscosity (JEFFREYS, 1930). The first investigator who set out to unravel the more distant past of the Earth-Moon system beyond the time of their closest approach was Gerstenkorn, who found that under the reversed torque the Moon receded again away from the Earth and the inclination of its orbit kept increasing rapidly until a retrograde polar orbit was achieved at a distance of 4.7 Earth radii; and since tidal effects in a retrograde orbit tend to decrease the eccentricity $e$ of the orbit, Gerstenkorn found that this eccentricity approached one when the perigee distance was equal to approximately 25 Earth radii, and inclination of the orbit to the terrestrial equator about

149°. As for $e \geqslant 1$ the Moon would no longer be gravitationally attached to the Earth, such a nearly parabolic orbit – Gerstenkorn conjectured – may have been one in which the Moon may have been originally captured by the Earth a billion years or less before the time of the closest approach, and some three billion years before our time.

Gerstenkorn did note one additional interesting fact: namely, that the minimum distance of approach between the Moon and the Earth comes remarkably close to the terrestrial "Roche Limit" $R_L$, given (cf., e.g., JEANS, 1918) approximately by the formula

$$\frac{R_L}{R_\oplus} = 2.44 \left(\frac{\rho_\oplus}{\rho_{\text{C}}}\right)^{\frac{1}{3}}, \tag{6-38}$$

where $R_\oplus$ denotes the mean radius of the Earth and $\rho_\oplus$, $\rho_{\text{C}}$, the mean densities of the Earth and the Moon, respectively. For $\rho_\oplus = 5.52$ g/cm$^3$ and $\rho_{\text{C}} = 3.34$ g/cm$^3$, we find from the above equation that $R_L = 2.88 R_\oplus$ – which is practically identical with the distance of minimum approach according to the calculations of either Gerstenkorn or MacDonald.

Now, as is well known, the Roche limit signifies the distance at which the tidal forces exerted by the central planet on a fluid satellite of infinitesimal mass become equal to its own self-attraction; and at a closer distance this satellite would be torn apart by the merciless planet into small pieces. The Moon is not fluid, of course, but possesses a global rigidity of the order of $10^{12}$ dynes/cm$^3$ (cf. Chapter 8, p. 116); in addition, with its mass of 0.0123 of that of the Earth it does not represent exactly a mass particle in the theoretical sense. However, it is interesting to speculate what would happen if the Moon did enter the Roche danger zone and suffered physical harm in consequence. How gruesome would be this experience? This thought was recently developed by ALFVÉN (1963), who accepted the qualitative features of Gerstenkorn's orbit calculations for the time preceding the closest approach – i.e., capture in a retrograde orbit with subsequent tidal transfer of angular momentum from the Moon's orbit to the Earth's rotation. But, Alfvén argues, the initial angular momentum of the Moon might actually have been considerably less than Gerstenkorn calculated, so that the minimum distance would have been well inside the Roche limit. If so, at least the surface of the Moon would have been virtually torn off into pieces and its fragments thrown out into space. Some of them may have fallen down on the Earth (directly, or after performing one or more orbits in the field of the Earth-Moon gravitational dipole), thus speeding up its axial rotation and forming, perhaps, an essential ingredient of its outer crust; moreover, this process could have continued until the Earth acquired the same angular velocity (projected on the Moon's orbital axis) as the Moon. At that point the tidal couple would vanish; the Moon would cease to approach the Earth and begin to recede from it. In the meantime, other fragments torn apart from the Moon (which were not captured by the Earth) may eventually have fallen back on its surface; and, hence, as the Moon was receding from the Earth, it may have been subject to intense bombardment by its own former fragments. It is possible that many impact craters (or maria) on the Moon (cf. Chapter 17) may have been produced in this way and at that time. Further evolution, following this brutal rendezvous with the Earth, could again take place in accordance with the Gerstenkorn-MacDonald calculations up to the present time.

Whether or not the dramatic surface break-up postulated by Alfvén did actually take place remains as yet uncertain, and may have to await the verdict of the geologist's hammer on the spot in the (hopefully) near future. However, an assumption that the Moon originated outside the Earth's orbit, was captured by our planet at some time in the past (of the order of $10^9$ years ago), and eventually brought to its present orbit by tidal evolution, seems consistent with all known dynamical features of the problem; and may be regarded as the likeliest hypothesis of the way in which our Earth acquired its only natural satellite.

# BIBLIOGRAPHICAL NOTES

## Chapter 1

Modern determinations of the mass of the Moon from accelerations of man-made spacecraft moving in its gravitational field and observed by means of highly accurate Doppler tracking have added at least an order of magnitude to their precision over its previous determinations by purely astronomical methods. For a determination of the lunar mass from the observed perturbations exerted by it on the motion of Mariner 2, cf. ANDERSON and NULL (1963); for results based on Rangers 6 and 7, cf. SCHURMEIER *et al.* (1964); while those deduced from the trajectories of Rangers 8 an 9, or of Mariner 4 are still unpublished. For a determination of the lunar distance cf. also O'KEEFE and ANDERSON (1952) or FISCHER (1962).

## Chapter 2

The subject of this section is severely classical, and has been extensively treated in the existing literature. E. W. BROWN's *Lunar Theory* (1896) or F. TISSERAND's *Mécanique céleste* (Volume III, 1894) are classics available to the modern reader in paperback editions of recent date. A shorter but adequate account of the lunar theory is contained in Chapters 12–14 of the *Methods of Celestial Mechanics* by BROUWER and CLEMENCE (1961); while for still more concise survey of the field cf. BROUWER and HORI's Chapter 1 in the *Physics and Astronomy of the Moon* (ed. by Z. KOPAL, 1962).

For a recent discussion of the coupling between the motion of the Moon in space and about its center of gravity cf. KONDURAR (1963) or ECKERT (1965).

## Chapter 3

The discovery of the existence of lunar optical librations goes back to Galileo GALILEI (1932), though the correct nature of the phenomenon was not recognised till by RICCIOLI and HEVELIUS between 1638–41. Their explanation was placed on a solid mathematical footing by NEWTON (1687), who (in Volume 3 of his *Principia*) predicted also the existence of the physical librations.

The empirical foundations of the theory of the rotation of the Moon were laid down by CASSINI in 1693; but almost a century had to elapse before LAGRANGE developed between 1764 and 1780 an adequate mathematical theory of the phenomenon, which has remained a cornerstone of all subsequent work.

Most part of the geometrical formulae developed in this section are classical. For their previous compilations cf., e.g., GRAFF (1901), recently made available in the English translation edited by KOPAL and MILLS (1965).

## Chapter 4

The possibility that, apart from optical librations, the Moon may librate physically around its center of gravity was first mentioned by NEWTON in the third volume of his *Principia* which appeared in 1687; but such a motion remained undetectable from the observations for almost 150 years. Throughout the 18th century the observations by MAYER and LALANDE failed to detect any significant displacements of the apparent positions on the Moon which could be due to this cause; and the same was true of subsequent efforts by BOUVARD and ARAGO or NICOLET, undertaken in Paris (at the encouragement of LAPLACE) in the first two decades of the 19th century, or somewhat later by KREIL in Milan.

The success in this age-long quest for the detection of physical librations of the Moon was finally scored by BESSEL (1839), the founder of modern astrometry, who devised for this purpose a new type of telescope (i.e., the helimeter) which remained in use for visual observations of this kind for almost a century. BESSEL (1839, 1941) worked out also in full detail the basic procedures for the reductions of the observations which were continued after his death by his pupil SCHLÜTER. Death claimed Schlüter only two years after the passing of his master; and their work fell into the hands of WICHMANN, another pupil of Bessel, who in 1846–1847 published the combined results of the three (WICHMANN, 1846, 1847). In subsequent decades, heliometric studies of the lunar librations in Germany became associated with the lifelong efforts of Ernst HARTWIG, the founder and first director of the Bamberg Observatory. Hartwig's interest in lunar librations has apparently been evoked by WINNECKE, under whose encouragement Hartwig completed at Strasbourg between 1877–1879 his first series of 42 heliometric observations of the positions of Mösting A (known as the Hartwig Strasbourg Series), eventually published in his doctoral thesis (HARTWIG, 1880). In 1884 Hartwig moved from Strasbourg to the Dorpat (now Tartu) Observatory in Russia, where between 1884–1885 he carried out his second (Dorpat) series of 36 heliometric observations of the Moon. In 1885 Hartwig returned to Germany to become director of the newly founded observatory at Bamberg, where between 1885 and 1922 he carried out his last and longest (Bamberg) series of such observations. On the whole, over an interval of 42 years between March 1880 and May 1922 Hartwig carried out 266 sets of heliometric observations of the Moon (and, between 1841 through 1922, three German observers – Schlüter, Wichmann and Hartwig – performed a total of 552 such observations) representing an important contribution to our science.

Since Hartwig's death in 1923 heliometric observations of the Moon in Germany have come to an end; for no successor was ready to follow in his footsteps. However, the earlier series of observations made in the 19th century were re-discussed by subsequent investigators several times. Thus Hartwig's Strasbourg series was later re-discussed by FRANZ (1887) who re-discussed later also the observations by Schlüter (FRANZ, 1899). Schlüter's measurements were re-discussed once more by STRATTON (1909); while Hartwig's Bamberg series was re-reduced later by NAUMANN (1939); and, still more recently, Hartwig's Dorpat series was thoroughly re-discussed by KOZIEL (1948, 1949). Of other work in this field carried out at Leipzig in preceding years cf. HAYN (1902–23); but is has since come to the end owing to a complete destruction of the Leipzig Observatory during the second world war.

Apart from the German contributions to the study of the physical librations of the Moon which over a century were on a massive scale, the second most important school of this subject has arisen in Russia and has been centered around the Engelhardt Observatory in Kazan. Moreover, unlike the German school (going back to Bessel) which has now become extinct, the Russian school has remained active to this date and continues to produce important results in this field. Following the Hartwig Dorpat series of 1884–1885, heliometric observations of the relative positions of Mösting A were commenced at Kazan by A. V. KRASNOV between 1895 and 1898, consisting of 112 sets of observations; while between 1900 and 1905 A. A. MIKHAILOVSKY completed the second Kazan series consisting of 54 sets of observations. This latter series of observations was subsequently reduced by VÖLKEL (1908) and BELKOVICH (1936). A third Kazan series of 130 such observations was carried out between 1910–1915 by Th. BANACHIEWICZ, whose observations were reduced by YAKOVKIN (1928). Yakovkin's contributions to the study of the figure and motion of the Moon over the past forty years have been on a massive scale. Between 1916–1931 he performed 251 sets of heliometric measurements of Mösting A which he reduced himself in two separate publications (YAKOVKIN, 1939, 1945). Since 1931 the Kazan heliometric observations were continued by I. V. BELKOVICH, who between 1931–1948 secured further 247 sets of positional measurements of Mösting A, as well as reduced jointly all Kazan and Hartwig series (BELKOVICH, 1949). In addition, since 1938 heliometric observations of the Moon have been carried out at Kazan by A. A. NEFEDIEV and Sh. T. HABIBULLIN. NEFEDIEV (1955) reduced also the old lunar heliometric observations by Krasnov; while HABIBULLIN (1958) is the author of the only comprehensive monograph on physical librations of the Moon which has appeared so far in any language.

Photographic determinations of the physical librations have been attempted early in the history of lunar photography by PRITCHARD and WARREN DE LA RUE (cf. PRITCHARD, 1879, 1882); and, more recently, by PUISEUX (1916, 1925), CHANDON (1941) and HABIBULLIN (1958). With the exception of the work reported in this last reference, the results of photographic work compared less than favourably with visual heliometric observations (cf. WEIMER, 1954). This is not, however, due to any basic unsuitability of the photographic technique to this type of work, but rather to the fact that (at least until the work of Habibullin) photographic techniques in selenodesy have not yet been properly developed – in contrast to the care with which the reductions of visual observations have been worked out.

Of other more theoretical, investigations of the rotation of the Moon we may refer to JÖNSSON (1917), HAYN (1923), YAKOVKIN (1923, 1934, 1939, 1945, 1950, 1952), BELKOVICH (1948, 1949), KOZIEL (1948, 1949, 1957, 1962), NEFEDIEV (1951, 1954, 1957), FRIDLAND (1959, 1961), GORYNIA (1960), JEFFREYS (1957, 1961), DROFA (1962) or ECKHARDT (1965). This latter paper develops a new method for treatment of the perturbing function, due to the terrestrial attraction, which is well amenable to solution by automatic computing machines; and such represents a significant advance in further development of our subject.

Another significant advance was initiated by MAKOVER (1962) and MIETELSKI and MASLOWSKI (1963) in treating Equation (4-54) for forced librations in longitude as one of Mathieu type, in place of a nonhomogeneous Equation (4-97) with constant coefficients.

For a more expository treatment of the problem of physical librations of the lunar globe, cf., e.g., TISSERAND (1891), Chapter 28; PLUMMER (1918), Chapter 23; or DANBY (1962), Chapter 14.

## Chapter 5

The subject matter discussed in this section is largely classical. For the geometrical formulae governing the illumination of the Moon cf., e.g., GRAFF (1901). For the theory of lunar eclipses and their predictions standard source remains still Th. v. OPPOLZER's *Canon der Finsternisse* (1887), which recently became available in an English translation by O. Gingerich (Dover Publ., New York, 1962).

For a more detailed account of the mathematical methods by which the Greek scholars of the Alexandrian period attempted to determine the distance of the Moon, cf., e.g., DREYER (1906), pp. 183–185.

## Chapter 6

For the anisotropy of the escape velocity from the Moon with the direction, cf., e.g., GOLD (1960).

The present treatment of the elliptically restricted problem of three bodies goes back to ι transformation of coordinates introduced originally by NECHVÍLE (1917, 1926) and followed up later by OVENDEN and ROY (1961), KOPAL and LYTTLETON (1963), SZEBEHELY and GIACAGLIA (1964), DANBY (1964), BENNETT (1965), CONTOPOULOS (1965) and others. The presentation of the subject as given in this chapter follows largely that by KOPAL and LYTTLETON (1963).

Investigations of the past evolution of the Earth-Moon system due to the action of tidal friction have been opened up with two monumental papers by G. H. DARWIN (1879, 1880) and followed up in more recent times by GERSTENKORN (1955), ÖPIK (1961), FIELD (1963), ALFVÉN (1963, 1964), MacDONALD (1964a, b) and KAULA (1964). The essential results presented in this section are due to Gerstenkorn; subsequent and more extensive computations by MacDonald filled in many details. All these results are based on the assumed extrapolability of tidal friction at its present rate in the past. Different views critical of this concept have been raised by the Russian school of investigators; cf. RUSKOL (1960, 1962, 1963a, b), SOROKIN (1965) *et al.*

For recent studies of the secular changes of specific orbital elements of the Earth-Moon system caused by tidal friction, cf., GROVES (1960, 1962), JEFFREYS (1961), GOLDREICH (1963) and others. The meaning and magnitude of the Roche limit for self-gravitating elastic solid bodies has been established by JEFFREYS (1947).

PART TWO

# INTERNAL CONSTITUTION OF
# THE LUNAR GLOBE

# INTRODUCTION

An investigation of the internal structure of any celestial body – be it star or a planet – possess many features in common. It consists, in effect, of the construction of a model based on differential equations safeguarding the conservation of the mass, momentum and energy, for a given type of boundary condition – such as the total mass, radius, rotational period and (for the stars) luminosity of the respective configuration. In point of fact, the stars and planets represent two classes of self-gravitating astronomical bodies differing mainly in mass (by several orders of magnitude); and all different external manifestations (such as the size, or luminosity) stem directly from this source.

Some forty years ago, in the early stage of development of the study of stellar structure, Russell and Vogt discovered independently a theorem (bearing their name) which asserts that the structure of a star should be completely determined by its mass and chemical composition; and the validity of this theorem transfers readily also to astronomical bodies of the mass of the Moon or the planets. The aim of the present part of this book will, in fact, be to demonstrate the extent to which this is the case.

In order to do so, in Chapter 7 which follows this brief introduction, we shall consider the problem of the *hydrostatic equilibrium* of the Moon, and point out the extent of the information one can deduce from the departures from its validity as exhibited by the lunar globe. In the next section, we shall take up the question of the *thermal equilibrium* of the lunar interior, its nuclear heat sources and the consequences of the conservation of energy as embodied in the equation of heat conduction. Chapter 9 will then be devoted to a *stress history* of the lunar globe considered as an elastic body, due to stresses arising from self-compression and secular heating of the interior; together with an analysis of free oscillation which an elastic lunar globe is likely to exhibit. Chapter 10 will then be concerned with the *possibility of fluid flow* in molten parts of the lunar interior; Chapter 11, with a brief survey of what we know of the probable *chemical composition* of the Moon; while the concluding Chapter 12 will give an account of an *exosphere* (partly exuded from the interior, and partly accreted from the interplanetary medium) which may be surrounding the lunar globe.

In dealing with these subjects we shall, of course, be concerned primarily with our Moon; but the treatment will be obviously extensible to other "cold" bodies of the solar system of comparable mass. In this connection, the writer would like to recall the following classical statement found in Professor Titchmarsh's preface to his well-known book *Introduction to the Theory of Fourier Integral* (1937); "I have retained, as having a certain picturesqueness, some references to "heat", "radiation" and so forth; but our interest is purely analytical; and the reader need not know whether such things exist". In a somewhat similar sense, the term "Moon" (or possibly, Mercury or Mars) will frequently be used in this part of the book as a descriptive term for a certain class of astronomical bodies because they possess perhaps a certain picturesqueness; but the theories underlying our treatment are more general in nature; and the reader interested mainly in essentials may forget on occasion that such bodies actually exist.

## CHAPTER 7

# HYDROSTATIC EQUILIBRIUM

The most important characteristic of any celestial body is, indeed, its mass; and in the case of the Moon this is known (cf. Chapter 1) to be $7.35 \times 10^{25}$ g or 1.2 per cent of that of the Earth. This mass is, moreover, contained inside a globe of $r_{\mathbb{C}} = 1738$ km mean radius, rendering its mean density $\rho_m = 3.34$ g/cm$^3$. What is the corresponding *pressure* inside such a configuration? If the latter were in hydrostatic equilibrium – an assumption which will have to be tested on its merits – the pressure $P(r)$ and the density $\rho(r)$ at a distance $r$ from the Moon's center should, to a high degree of approximation* be related by the well-known equation

$$\frac{dP}{dr} = -g\rho \tag{7-1}$$

of hydrostatic equilibrium where the gravitational acceleration $g$ inside a spherically-symmetrical configuration will be given by

$$g = G \frac{m(r)}{r^2}, \tag{7-2}$$

where $G$ is the constant of gravitation and $m(r)$, the mass interior to a volume of radius $r$, follows from the equation

$$\frac{dm}{dr} = 4\pi\rho r^2. \tag{7-3}$$

as a consequence of the conservation of mass.

In order to solve these equations let us assume, to begin with, that the density $\rho$ is constant and equal to the mean density $\rho_m$. If so,

$$m(r) = \tfrac{4}{3}\pi\rho_m r^3; \qquad g(r) = \tfrac{4}{3}\pi G\rho_m r; \tag{7-4}$$

and Equation (7-1) can be likewise immediately integrated yielding

$$P = \tfrac{2}{3}\pi G\rho_m^2 (r_{\mathbb{C}}^2 - r^2), \tag{7-5}$$

where the integration constant has been adjusted so as to make $P$ vanish on the surface. The central pressure

$$P_c = \tfrac{2}{3}\pi G\rho_m^2 r_{\mathbb{C}}^2 = 4.71 \times 10^{10} \text{ dynes/cm}^2 \tag{7-6}$$

---

* Only the effects of the small centrifugal force due to the Moon's axial rotation being ignored.

or 46 500 atmospheres – a pressure easily attained in terrestial laboratories and exceeded in the Earth's crust at a depth of mere 150 km below the surface.

This hydrostatic model of the Moon can be improved upon somewhat along the following lines. The mean density of the lunar globe is well within the limits of the densities (2.8–3.6 g/cm³) encountered in the Earth's outer shell; and it can be reasonably surmised that the equations of state in the two regimes way possess similar characteristics. Now from his studies of the propagation of elastic earthquake waves in the terrestrial mantle BULLEN (1947) found that, very approximately,

$$\rho \frac{dP}{d\rho} = a + bP, \tag{7-7}$$

where the empirical values of the constants $a$ and $b$ characterizing the outer shell of the Earth prove to be

$$\left. \begin{array}{l} a = 1.167 \times 10^{12} \ \text{dyn/cm}^2 \\ b = 3.5. \end{array} \right\} \tag{7-8}$$

With this in mind let us return to the equations (7-1)–(7-3): eliminating $g$ and $m$ between them we find that $P$ should be defined as a solution of the second-order differential equation

$$\frac{d}{dr}\left(\frac{r^2}{\rho}\frac{dP}{dr}\right) + 4\pi G \rho r^2 = 0. \tag{7-9}$$

On the other hand, the empirical relation (7-7) admits of an easy separation of $P$ and $\rho$ in the form

$$\frac{dP}{a+bP} = \frac{d\rho}{\rho}, \tag{7-10}$$

and can be integrated to yield

$$\rho^b = c(a+bP) \tag{7-11}$$

or, conversely,

$$P = \frac{\rho^b - ab}{bc}, \tag{7-12}$$

where the integration constant $c$ is obviously equal to $\rho_1/a$; $\rho_1$ being the surface density of our satellite. Let us eliminate now the pressure $P$ between Equations (7-9) and (7-12), retaining $\rho$ as our sole dependent variable. If so and if we set, for the sake of normalization,

$$\rho = \rho_c y^n \tag{7-13}$$

and

$$r = \left\{\frac{n\rho_c^{b-1}}{4\pi Gc}\right\}^{\frac{1}{2}} x, \tag{7-14}$$

where we have abbreviated

$$n = \frac{1}{b-1} = \frac{2}{5}, \tag{7-15}$$

the resulting equation will assume the neat form

$$\frac{d^2y}{dx^2} + \frac{2}{x}\frac{dy}{dx} + y^n = 0 \tag{7-16}$$

and can be easily identified with Emden's differential equation for polytropic gas spheres with an index $n = 0.4$.

The particular solution of this equation for which $y(0) = 0$ (implying that $\rho(0) = \rho_c$) and $y'(0) = 0$ (implying the density to be maximum at the center) can be shown to be of the form of a rapidly converging series

$$y = 1 - \frac{x^2}{6} + \frac{x^4}{300} + \frac{x^6}{21\,000} + \dots; \tag{7-17}$$

and the density concentration of the respective configuration follows (cf., e.g., CHANDRASEKHAR, 1939) from

$$\frac{\rho_m}{\rho_c} = -\left(\frac{3}{x}\frac{dy}{dx}\right)_1 \tag{7-18}$$

where the subscript 1 refers to the surface values of the respective quantities; while the density anywhere else becomes given by

$$\rho = \rho_c y^{0.4} = \frac{\rho_m y^{0.4}}{\left(-\frac{3}{x}\frac{dy}{dx}\right)_1}. \tag{7-19}$$

The observations furnish the values of $\rho_m$ and $r_{(}$, so that, from (7-14)

$$x_1 = \left\{\frac{4\pi G\rho_1}{0.4a\rho_c^{2.5}}\right\}^{\frac{1}{4}} r_{(}. \tag{7-20}$$

Moreover, the ratio of the surface density $\rho_1$ to the mean density $\rho_m$ follows readily from (7-19) as

$$\frac{\rho_m}{\rho_1} = \left\{-\frac{3}{xy^{0.4}}\frac{dy}{dx}\right\}_1; \tag{7-21}$$

while the ratio of the central to mean density $\rho_c/\rho_m$ has already been given by (7-18). With the adopted values of $\rho_m$, $r_{(}$ and of the constant $a$, Equations (7-18)–(7-21) can obviously be solved for the values of $x_1$, $\rho_1$, and $\rho_c$. In doing so we obtain

$$x_1 = 0.778, \tag{7-22}$$

and

$$\left.\begin{array}{l} \rho_1 = 3.28 \text{ g/cm}^3 \\ \rho_c = 3.43 \text{ g/cm}^3, \end{array}\right\} \tag{7-23}$$

which, together with the mean value of $\rho_m = 3.34$ g/cm$^3$, specify an improved model of the interior of our satellite. Its internal density distribution should, accordingly, vary essentially as

$$\rho_1 = \rho_c \left\{ 1 - \frac{0.4}{6} x^2 + \ldots \right\}$$

$$\simeq \rho_c \left\{ 1 - \frac{(0.778)^2}{15} \left( \frac{r}{r_{\mathbb{C}}} \right)^2 \right\};$$

(7-24)

and the pressure $P$ obtained by integrating the equation (7-1) of hydrostatic equilibrium with this non-constant density distribution results eventually as

$$P = \tfrac{2}{3}\pi G \rho_c^2 (r_{\mathbb{C}}^2 - r^2) - \frac{k\pi G \rho_c^2}{5 r_{\mathbb{C}}^2} (r_{\mathbb{C}}^4 - r^4),$$

(7-25)

where $k \equiv (0.778)^2/15 = 0.040$. The resulting value of $P_c = 4.91 \times 10^{10}$ dynes/cm$^2$ turns out to be by 4% larger than the preliminary value (7-6) deduced on the basis of a homogeneous model.

In a subsequent section of this report we shall show that such variation in density as represented by Equation (7-24) based on Bullen's law (7-7) is, in fact, to be expected merely as a result of self-compression of a lunar globe whose density at zero pressure would be everywhere the same (cf. Chapter 9); so that no evidence for real heterogeneity of the body of our satellite can so far be drawn.

After this preliminary outline of the probable distribution of density and pressure in lunar interior, let us proceed now to evaluate the *momenta* of the lunar globe, on the assumption of hydrostatic equilibrium, along different axes, as defined by Equations (4-1)–(4-3). In order to do so, let $\phi$, $\theta$ be the spherical polar coordinates (i.e., the azimuth and polar distance) of an arbitrary point in the interior, referred to the Moon's principal axes of inertia, such that

$$\left. \begin{array}{l} x = r \cos \phi \sin \theta = r\lambda, \\ y = r \sin \phi \sin \theta = r\mu, \\ z = r \cos \theta \qquad = r\nu, \end{array} \right\}$$

(7-26)

i.e., the $xy$-plane coinciding with the lunar equator and the $z$-axis with its axis of rotation. Let, moreover (consistent with the assumption of hydrostatic equilibrium) the surfaces of equal density be equipotentials defined by the equation

$$r = a \left\{ 1 + \sum_{i,j} Y_j^i (a, \theta, \phi) \right\},$$

(7-27)

where $a$ denotes the mean radius of such an equipotential, and the $Y$'s are tesseral harmonics arising from the departure of the Moon's globe from spherical form – whatever their cause (rotation, tides).

As is well known (cf., e.g., KOPAL, 1959, 1960) these tesseral harmonics are constrained to satisfy Clairaut's equation

$$Y_j^i \int_0^a \rho a^2 \, da - \frac{1}{(2_j + 1) a^j} \int_0^a \rho \frac{\partial}{\partial a} (a^{j+3} Y_j^i) \, da$$

$$- \frac{a^{j+1}}{2j + 1} \int_a^{a_,} \rho \frac{\partial}{\partial a} (a^{2-j} Y_j^i) \, da = \frac{c_{i,j}}{4\pi G} a^{j+1} P_j^i(\theta, \phi),$$

(7-28)

where $a$ denotes the mean external radius of the respective configuration, and the $c_{i,j}$'s are constants in the expansion of the disturbing forces

$$W = \sum_{i,j} c_{i,j} r^j P_j^i(\theta, \phi)$$

(7-29)

which, in the case of the Moon, may be due to the centrifugal force of its axial rotation, or to the tidal pull of the Earth.

If we divide now Clairaut's Equation (7-28) by $a^j$, differentiate with respect to $a$, and divide by $a^{2j}$, we find that

$$\left\{ \frac{jY_j^i}{a^{j+1}} + \frac{1}{a^j} \frac{\partial Y_j^i}{\partial a} \right\} \int_0^a \rho a^2 \, da - \int_a^{a_,} \rho \frac{\partial}{\partial a} \left( \frac{Y_j^i}{a^{2-j}} \right) da$$

$$= \frac{c_{i,j}}{4\pi G} (2j + 1) P_j^i(\theta, \phi);$$

(7-30)

while if we divide it by $a^{j+1}$, differentiate and multiply by $a^{2j+2}$, we obtain

$$\int_0^a \rho \frac{\partial}{\partial a} (a^{j+3} Y_j^i) \, da = a^j \left\{ (j + 1) Y_j^i - a \frac{\partial Y_j^i}{\partial a} \right\} \int_0^a \rho a^2 \, da.$$

(7-31)

At $a = a_1$, Equation (7-30) can be reduced to

$$Y_j^i(a_1) = c_{i,j} \Delta_j \frac{a^{j+1}}{Gm} P_j^i(\theta, \phi),$$

(7-32)

where

$$4\pi \int_0^a \rho a^2 \, da = m$$

(7-33)

denotes the mass of the whole configuration, and

$$\Delta_j = \frac{2j + 1}{j + \eta_j(a_1)}, \qquad \eta_j(a) = \frac{a}{Y_j} \left( \frac{\partial Y_j}{\partial a} \right)$$

(7-34)

denotes a non-dimensional constant. Similarly,

$$\int\limits_0^{a_1} \rho \frac{\partial}{\partial a}(a^{j+3}Y_j^i)\,\mathrm{d}a = a_1^j Y_j^i(a_1)\{j+1-\eta_j(a_1)\}\frac{m}{4\pi} \tag{7-35}$$

$$= \frac{2j+1}{4\pi G}(\varDelta_j - 1)\,a_1^{2j+1}c_{i,j}P_j^i(\theta,\phi)$$

by a combination with (7-32).

The actual values of the constants $\varDelta_j$ depends on the internal structure of the respective configuration. In order to determine them, let us differentiate (7-31) once more with respect to $a$: the result will be

$$a^2 \frac{\partial^2 Y}{\partial a^2} + 6\frac{\rho}{\bar{\rho}}\left(a\frac{\partial Y}{\partial a} + Y\right) = j(j+1)\,Y, \tag{7-36}$$

where

$$\bar{\rho} = \frac{3}{a^3}\int\limits_0^a \rho a^2\,\mathrm{d}a \tag{7-37}$$

denotes the mean density interior to $a$. Equations (7-34) reveal, however, that the constants $\varDelta_j$ depend on the internal structure, not through $Y_j^i(a)$ directly, but only through its logarithmic derivative $\eta_j(a)$, which satisfies the first-order differential equation

$$a\frac{\partial\eta_j}{\partial a} + 6\frac{\rho}{\bar{\rho}}(\eta_j + 1) + \eta_j(\eta_j - 1) = j(j+1), \tag{7-38}$$

subject to the initial condition

$$\eta_j(0) = j - 2. \tag{7-39}$$

For a homogeneous configuration $(\rho = \bar{\rho})$ this equation can be solved immediately to yield

$$\eta_j(a) = j - 2, \tag{7-40}$$

leading to

$$\varDelta_j = \frac{2j+1}{2(j-1)}; \tag{7-41}$$

while for a nearly homogeneous model such as characterized by the density distribution (7-24), the Radau approximation (cf. e.g., KOPAL, 1959, p. 35)

$$\sqrt{1+\eta_2(a)} = \frac{5}{\bar{\rho}a^5}\int\limits_0^a \bar{\rho}a^4\,\mathrm{d}a \tag{7-42}$$

with

$$\bar{\rho} = \rho_{\mathrm{c}}(1 - \tfrac{3}{5}kx^2) \tag{7-43}$$

yields

$$\eta_2(a_1) = \tfrac{12}{35}k = 0.0138 \tag{7-44}$$

leading, by (7-34), to the value of

$$\Delta_2 = 2.483 \tag{7-45}$$

as an improvement of the round value of 2.5 following from (7-41).

After these preliminaries, let us return to our main task of the computation of lunar momenta about different axes. In accordance with equations (4-1)–(4-3),

$$A = \int_0^{r(} \int_0^{2\pi} \int_0^{\pi} \rho r^4 (\sin^2 \phi \sin^2 \theta + \cos^2 \theta) \sin \theta \, dr \, d\phi \, d\theta \tag{7-46}$$

$$= 2 \int_0^{r(} \int_{-1}^{1} \int_{-\sqrt{1-\mu^2}}^{\sqrt{1-\mu^2}} \rho r^4 \frac{1-\lambda^2}{\lambda} \, dr \, d\mu \, dv;$$

and, similarly,

$$B = 2 \int_0^{r(} \int_{-1}^{1} \int_{-\sqrt{1-\mu^2}}^{\sqrt{1-\mu^2}} \rho r^4 \frac{1-\mu^2}{\lambda} \, dr \, d\mu \, dv \tag{7-47}$$

$$C = 2 \int_0^{r(} \int_{-1}^{1} \int_{-\sqrt{1-\mu^2}}^{\sqrt{1-\mu^2}} \rho r^4 \frac{1-v^2}{\lambda} \, dr \, d\mu \, dv. \tag{7-48}$$

where, of course, $\lambda^2 = 1 - \mu^2 - v^2$ by (7-26).

As the next step in our evaluation, let us change over from $r$ to $a$ as the variable of integration by means of the relation (7-27), leading to

$$A = 2 \int_0^{a_1} \int_{-1}^{1} \int_{-\sqrt{1-\mu^2}}^{\sqrt{1-\mu^2}} \rho r^4 \frac{\partial r}{\partial a} \frac{1-\lambda^2}{\lambda} \, da \, d\mu \, dv, \tag{7-49}$$

and quite analogously for $B$ and $C$. Moreover, the use of equation (7-27) reveals that, correctly to the first powers of the tesseral harmonics $Y_j^i$,

$$Y^4 \frac{\partial r}{\partial a} = a^4 + \frac{\partial}{\partial a} \{a^5 \sum_{i,j} Y_j^i\}; \tag{7-50}$$

and, in particular, for $j=2$,

$$\rho r^4 \frac{\partial r}{\partial a} = \rho a^4 + \rho \frac{\partial}{\partial a} (a^5 Y_2). \tag{7-51}$$

If so, however, then according to Equation (7-35)

$$\int_0^{a_1} \rho \frac{\partial}{\partial a} (a^5 Y_2) \, da = \frac{5 c_2 a_1^5}{4\pi G} (\Delta_2 - 1) P_2(\theta, \phi); \tag{7-52}$$

so that

$$A = 2 \int_{0}^{r} \int_{-1}^{1} \int_{-\sqrt{1-\mu^2}}^{\sqrt{1-\mu^2}} \rho a^4 \frac{1-\lambda^2}{\lambda} \, da \, d\mu \, dv +$$

$$+ \frac{5(\Delta_2 - 1)c_2 a_1^5}{2\pi G} \int_{-1}^{1} \int_{-\sqrt{1-\mu^2}}^{\sqrt{1-\mu^2}} \frac{1-\lambda^2}{\lambda} P_2(\theta, \phi) \, d\mu \, dv. \tag{7-53}$$

An evaluation of the angular integrals in the first expression on the right hand side of (7-53) offers no difficulty; and in doing so we find that

$$\int_{0}^{a_1} \int_{-1}^{1} \int_{-\sqrt{1-\mu^2}}^{\sqrt{1-\mu^2}} \rho a^4 \frac{1-\lambda^2}{\lambda} \, da \, d\mu \, dv = \tfrac{4}{3}\pi \int_{0}^{a_1} \rho a^4 \, da = \tfrac{1}{2}I, \tag{7-54}$$

where $I$ stands for the moment of inertia of the respective configuration about its center of gravity; and a replacement of the binomial $1-\lambda^2$ by $1-\mu^2$ or $1-v^2$ (appropriate for the evaluation of $B$ or $C$) would lead to the same result. A further evaluation of the second expression on the right-hand side of (7-53) necessitates, however, a specification of the forces which are responsible for the departure of our configuration from a spherical form. If, in particular, this is caused by a *rotation* about the $z$-axis with a constant angular velocity $\omega$, the corresponding centrifugal potential will be of the form

$$W_{\text{rot}} = -\tfrac{1}{3}\omega^2 a^2 P_2(\cos \theta), \tag{7-55}$$

leading (in accordance with 7-29) to

$$c_2 = -\tfrac{1}{3}\omega^2 \quad \text{and} \quad P_2(\theta, \phi) \equiv P_2(v). \tag{7-56}$$

If so, the entire expression on the right-hand side of equation can be explicitly evaluated, yielding

$$A = I - \frac{(\Delta_2 - 1)\omega^2 a_1^5}{9G}; \tag{7-57}$$

whereas the corresponding expressions for $B$ and $C$ similarly assume the forms

$$B = I - \frac{(\Delta_2 - 1)\omega^2 a_1^5}{9G} \tag{7-58}$$

and

$$C = I + \frac{2(\Delta_2 - 1)\omega^2 a_1^5}{9G}. \tag{7-59}$$

Accordingly,

$$A = B \tag{7-60}$$

while

$$C - A = C - B = \tfrac{1}{3}(\Delta_2 - 1)\frac{\omega^2 a_1^5}{G}. \tag{7-61}$$

If, on the other hand, our configuration becomes distorted as a result of the *tides* raised on it by an external mass $m_2$ situated at a distance $R$ from the center of gravity of the mass $m_1$, the corresponding tide-generating potential will be of the form

$$W_{\text{tidal}} = \frac{Gm_2}{\sqrt{R^2 - 2a_1 \cos \Theta + a_1^2}}, \tag{7-62}$$

where $\Theta$ denotes the angle between the direction of $R$ and the $x$-axis of our system of coordinates. In order to specify this angle, let us define the direction cosines of $R$ be given by

$$\left. \begin{aligned}
\lambda' &= \cos u \cos \Omega - \sin u \sin \Omega \cos i, \\
\mu' &= \cos u \sin \Omega + \sin u \cos \Omega \cos i, \\
\nu' &= \sin u \sin i,
\end{aligned} \right\} \tag{7-63}$$

where $u$ stands for the true anomaly of the disturbing body in its orbital plane; $\Omega$, the longitude of the intersection of this plane with the $xy$-plane; and $i$, the inclination between these two planes. Letting

$$\left. \begin{aligned}
p &= \Omega + \tan^{-1} (\cos i \tan u) \\
q &= \qquad \cos^{-1} (\sin i \tan u)
\end{aligned} \right\} \tag{7-64}$$

the primed direction cosines can be expressed as

$$\left. \begin{aligned}
\lambda' &= \cos p \sin q \\
\mu' &= \sin p \sin q \\
\nu' &= \cos q
\end{aligned} \right\} \tag{7-65}$$

and, accordingly,

$$\left. \begin{aligned}
\cos \Theta &= \lambda\lambda' + \mu\mu' + \nu\nu' \\
&= \cos q \cos \theta + \sin q \sin \theta \cos (p - \phi).
\end{aligned} \right\} \tag{7-66}$$

If, for the sake of simplicity, we make $i=0°$ (i.e., $q=90°$) and $p=0$, the foregoing equation (7-66) reduces to $\cos \Theta = \lambda$; and if so, an expansion of the right-hand side of Equation (7-62) in spherical harmonics leads to

$$c_j = \frac{Gm_2}{R^{j+1}} \quad \text{and} \quad P_j(\theta, \phi) \equiv P_j(\lambda). \tag{7-67}$$

The moments of inertia of so distorted a configuration then result as

$$\left. \begin{aligned}
A' &= I - \tfrac{2}{3}(\Delta_2 - 1) m_2 \frac{a_1^5}{R^3} \\[4pt]
B' &= I + \tfrac{1}{3}(\Delta_2 - 1) m_2 \frac{a_1^5}{R^3} \\[4pt]
C' &= I + \tfrac{1}{3}(\Delta_2 - 1) m_2 \frac{a_1^5}{R^3}
\end{aligned} \right\} \tag{7-68}$$

so that now

$$B' = C' \tag{7-69}$$

while

$$B' - A' = C' - A' = (\Delta_2 - 1) m_2 \frac{a_1^5}{R^3}. \tag{7-70}$$

Moreover, a combination of Equations (7-27) and (7-32) with (7-56) and (7-67) results in the superficial distortion of the respective configuration to be described by

$$r = a_1 \left\{ 1 - \frac{\omega_1^2}{4\pi G \rho_m} \Delta_2 P_2 (\cos \theta) \right.$$
$$\left. + \sum_j \Delta_j \frac{m_2}{m_1} \left( \frac{a_1}{R} \right)^{j+1} P_j (\cos \phi \sin \theta) \right\} \tag{7-71}$$

With these preliminaries safely established, let us proceed now to apply their results to the Moon. First, Equation (7-54) yields for the density distribution as represented by (7-24) the momentum of the Moon about its center of mass the value of

$$\left. \begin{aligned} I &= \frac{8\pi}{15} \rho_c \left\{ 1 - \tfrac{5}{7} \times 0.004 \right\} r_{\mathbb{C}}^5 \\ &= 8.828 \times 10^{41} \text{ g cm}^2 = 0.398 \, m_{\mathbb{C}} r_{\mathbb{C}}^2 ; \end{aligned} \right\} \tag{7-72}$$

and, for the angular velocity

$$\omega_{\mathbb{C}} = 2.6617 \times 10^{-1} \text{ sec}^{-1} \tag{7-73}$$

of lunar axial rotation, the difference $C - A$ follows from Equation (7-67) to be

$$\left. \begin{aligned} C - A &= 8.33 \times 10^{36} \text{ g cm}^2 \\ &= 0.00000944 \, I ; \end{aligned} \right\} \tag{7-74}$$

while the difference $C' - A'$ due to the tidal action of the Earth of mass $m_2 = 5.98 \times 10^{27}$ g and at the mean distance of $R = 384400$ km results from (7-70) as

$$\left. \begin{aligned} C' - A' &= 2.48 \times 10^{37} \text{ g cm}^2 \\ &= 0.0000280 \, I. \end{aligned} \right\} \tag{7-75}$$

The Equation (7-1) defining the form of the lunar surface which corresponds to these, difference assumes then the explicit form

$$\begin{aligned} r = 1738 \{ 1 &- 0.0000062 \, P_2 (\cos \theta) \\ &+ 0.0000187 \, P_2 (\cos \phi \sin \theta) + ... \} \text{ km} ; \end{aligned} \tag{7-76}$$

the first term in the curly brackets being due to the Moon's own rotation; the second, to the tidal action of the Earth.

How closely does the present state of the lunar globe approximate the requirements of hydrostatic equilibrium under the prevailing centrifugal and tidal forces? The simplest and most directly observable fact bearing on this situation is, of course, the nearly *spherical form* of our satellite. Extensive and accurate observations carried out by many competent investigators (for their closer discussion, cf. the forthcoming Chapter 13) attest the Moon to be a sphere of mean radius $1738 \pm 1$ km; departures from this spherical surface are very difficult to ascertain at the distance of the Earth and will be discussed in Chapter 13.

This observational result is in complete agreement with consequences of hydrostatic theory; for even if the material constituting the lunar globe reacts to temporary impulses as an elastic solid, no rock is known which could withstand indefinitely a pressure of the order of $10^4$ atmospheres prevailing in most part of the lunar interior. Given a sufficiently long time (probably short in comparison with the age of our satellite) the material which on a short time-scale may behave as a solid is bound to get crushed under its own weight to settle to a form of minimum potential energy – which is a sphere. This is why not only gaseous stars like the Sun, but also "solid" bodies like major or terrestrial planets, are bound to become spherical in time, whatever their initial state may have been; a retention of non-spherical shape being the prerogative of those self-gravitating bodies (asteroids, meteorites) whose mass is too small to give rise to internal pressures capable of overcoming the molecular cohesion of solid state. If small bodies were initially aspheric, they could remain so permanently and defy any efforts of self-attraction to establish hydrostatic equilibrium – as is, for instance, attested by observed light variations of the asteroids. The masses of even the largest of them – Ceres, Juno, Pallas, Vesta – are apparently too small for hydrostatic equilibrium to establish spherical shape; but the Moon, the first four Jovian satellites, or Mercury are safely beyond this limit.

The departures of the lunar globe from spherical form permissible for a figure of equilibrium in the prevailing field of force are indeed small. Equation (7-79) disclosed that the axial rotation of the Moon in the period of one month should be expected to give rise to an ellipticity $e$ of its meridional cross-section close to $e^{-1} = 106000$ – as compared with $e^{-1} = 297$ for the Earth – corresponding to a difference of only 16.3 meters between equatorial and polar semi-axes. Similarly, the equilibrium tides raised on the Moon by the Earth should produce a second-harmonic distortion of the lunar equator, corresponding (at the mean distance of the Moon) to a difference 47.5 m between the two equatorial semi-axes. Needless to say, the available observational data are utterly insufficient to verify so small a flattening at the poles, or so insignificant a tidal bulge which would be virtually concealed by foreshortening. The observations are, however, sufficient to rule out the existence of a substantially larger deformation; and to this extent the theory of hydrostatic equilibrium and the observations are in complete agreement.

However, this agreement becomes noticeably *worse* when we come to compare the *momenta* of the lunar globe, as deduced from its physical librations, with their equilibrium values evaluated in the present section. In Chapter 4 we found that the most probable observed values of the ratios $\alpha$, $\beta$, and $\gamma$ are as given by Equation (4-120); while their "hydrostatic" values due to the combined rotational and tidal distortion should clearly be given by

$$\alpha = \frac{C - B}{A} + \frac{C' - B'}{A'} = 0.0000094 , \tag{7-77}$$

$$\beta = \frac{C - A}{B} + \frac{C' - A'}{B'} = 0.0000374 , \tag{7-78}$$

$$\gamma = \frac{B - A}{C} + \frac{B' - A'}{C'} = 0.0000280 ; \tag{7-79}$$

and these, regrettably, do not seem to bear any relation to the observed values. In particular, the observed value of $\beta = 0.000627 \pm 0.000001$ turns out to be 17 times as large as the hydrostatic one; while for $\alpha$ and $\gamma$ the corresponding excess factors are 42 and 8, respectively; and this gross disparity could be only marginally improved by assuming the Moon to be completely homogeneous.

Could, perchance, the discrepancy so encountered be ascribed to the fact that the present dynamical characteristics of the Moon reflect a petrified state of hydrostatic equilibrium at a time when our satellite was much nearer to us than it is now? We may note that, in particular, the observed value of $\beta$, being 17 times as large as required by the hydrostatic theory at the present time, would indeed be equal to the hydrostatic one if the Moon were $(17)^{\frac{1}{3}} = 2.57$ times closer to the Earth than it is now; and its tidal bulge facing the Earth would vary as $0.88 \, P_2 \, (\cos \phi \sin \theta)$ km.

However, at this reduced distance the Keplerian angular velocity of the Moon's orbit would have been $(17)^{\frac{1}{2}} = 4.12$ times larger than today (i.e., one month lasting $6\frac{2}{3}$ days); and in the case of synchronism between rotation and revolution the corresponding polar flattening would have been 0.28 km. Neither this flattening, nor the corresponding tidal bulge seem to be borne out by the observations; but the evidence is not yet conclusive.

However, any such avenue of escape becomes effectively ruled out when we come to consider the ratio $f \equiv \alpha/\beta$ of the respective differences of the momenta. For configurations in hydrostatic equilibrium, to the first order in small quantities, Equations (7-61) and (7-70) reveal that

$$f = \frac{\frac{1}{3}(\Delta_2 - 1)\dfrac{\omega^2 r_{\text{\Moon}}^5}{G}}{\frac{1}{3}(\Delta_2 - 1)\dfrac{\omega^2 r_{\text{\Moon}}^5}{G} + (\Delta_2 - 1)\, m_{\oplus}\dfrac{r_{\text{\Moon}}^5}{R^3}} \tag{7-80}$$

which, for

$$\omega^2 = \frac{G(m_{\oplus} + m_{\text{\Moon}})}{R^3} \tag{7-81}$$

reduces to

$$f = \frac{m_{\oplus} + m_{\text{\Moon}}}{4m_{\oplus} + m_{\text{\Moon}}} \tag{7-82}$$

regardless of the Moon's internal structure. As, moreover, $m_{\oplus}/m_{\mathbb{C}}=81.303$ (p. 7) the "hydrostatic" value of the ratio $f$ should be 0.2523 or approximately one-quarter; whereas the observed $f=0.633\pm0.011$ (KOZIEL, 1964). This latter value is so much at variance with the requirements of hydrostatic equilibrium that the discrepancy must be considered as real.

Therefore, in spite of the approximate prevalence of hydrostatic equilibrium in lunar interior, as attested by the nearly spherical form of its globe, the Moon's motion around its center of gravity reveals the presence of small but unmistakable *departures* from this equilibrium, to the extent discussed in the preceding paragraphs. What can support such deviations from hydrostatic equilibrium over long periods of time – whether a finite strength of the lunar material, or some dynamical processes, is not yet clear. However, the following analogy between the Earth and the Moon may possibly prove of assistance to understand this result. Geodetic measurements supplemented by more recent satellite work have revealed that while the rotational distortion of our Earth is essentially hydrostatic (at least to 99 per cent of its observed amount), the bodily tides raised on the Earth by lunar attraction amount to only about 24 per cent of their theoretical equilibrium values (cf., e.g. TOMASHEK, 1957) – a disparity reflecting probably the fact that the centrifugal force which causes the Earth to be flattened at the poles is acting continuously, while the tides constitute an oscillatory phenomenon.

The terrestrial bodily tides are raised mainly in the mantle (because the height of a $j$-th harmonic tide is known to vary as $r^{j+1}$), which by its density should not be dissimilar in composition to our Moon. It may, therefore, be reasonable to anticipate that the same deficiency $q$ in tides as compared with the rotational distortion may also exist on the Moon; and if so, Equation (7-82) should be replaced by

$$ f = \frac{m_{\oplus} + m_{\mathbb{C}}}{(1 + 3q)\,m_{\oplus} + m_{\mathbb{C}}}, \tag{7-83} $$

which for the terrestrial value of $q=0.24$ yields $f=0.584$ – a result much closer to the observed value of this ratio. Moreover, an agreement between theory and observations could be made complete by adopting for the Moon a slightly lower value of $q=0.20$ which would indicate the lunar globe to be somewhat more rigid that the Earth's mantle – a perhaps not unreasonable expectation; though so far only a surmise lacking independent confirmation.

One last remark which may be added in this connection: namely, one concerning the sum total of the influences which partial superficial distortions varying as surface harmonics $Y_j^i(\lambda, \beta)$ of order $j$ and index $i$, can contribute towards the observe values of the ratios of the differences of momenta $\alpha$, $\beta$ or $\gamma$. It is often tacitly assumed that, because of the orthogonality properties of spherical harmonics, only the second-harmonic distortion (i.e., terms characterized by $j=2$, and $0 \leqslant i \leqslant 2$) can contribute to their magnitude. In actual fact, this is true only as far as the *first* powers of the respective harmonics $Y_j^i$ are concerned. Since, however, the integrand in the expression (7-52) to (7-54) for $A$, $B$ and $C$ contains $r^4$, powers higher than the first will be involved in a complete expression for it; and these will all contribute towards the total amount of the respective momentum, regardless of whether $j$ is even or odd.

In this connection, the most interesting term should be the *first harmonic*, whose presence in the expansion (7-27) for the radius-vector $r$ of the lunar globe would imply a finite displacement between the centre of gravity and centre of symmetry of our configuration. For bodies in hydrostatic equilibrium such a distortion would, of course, be impossible; but if (as many hold now) our Moon originated as a "cold" body by an agglomeration of solid particles, there is indeed no certainty that the location of the two centres must be coincident. This has already been pointed out by UREY, ELSÄSSER and ROCHESTER (1959). Without wishing to enter into a full discussion of such a hypothesis in this place, let us merely mention in passing the effect which such displacements would be bound to exert on the differences of the momenta – and, in particular, on the best-determined constant $\beta$.

If, for the sake of argument, the Moon did constitute a sphere, but its centre of mass were located at a point $(x_0, z_0)$ relative to the origin of coordinates at the centre of our sphere, it can be shown that (for a nearly homogeneous configuration) the second-order effect of the first harmonic should approximately result in a value of

$$\beta \sim \tfrac{3}{2} \frac{x_0^2 - z_0^2}{r_{\mathbb{C}}^2}. \tag{7-84}$$

Let us, furthermore, assume that the centre of gravity is situated on the $x$-axis (i.e., that $z_0 = 0$). If so, the observed value of $\beta = 0.000627 \pm 0.000001$ would require a displacement of $(x_0/r_{\mathbb{C}}) = 0.0205$ – i.e., $\pm 36$ km from the geometrical centre – for the complete explanation; or $\pm 26$ m to alter $\beta$ by an amount equal to its present uncertainty. It is not our wish to advocate that the second-order effect of the first harmonic does account for the entire amount of the observed value of $\beta$, or indeed any appreciable part of it; but rather to point out that (at least until more is known on the origin of the lunar globe) such complicating circumstances cannot also be completely ruled out.

## CHAPTER 8

# THERMAL HISTORY OF THE MOON

Having considered the bearing of the theory of hydrostatic equilibrium on internal structure of the Moon, and compared its consequences with available observations, let us turn now to investigate the conclusions which can be drawn from an application, to the Moon's interior, of the principle of the conservation of energy in so far as the internal *temperature* in lunar interior is concerned.

What is the absolute temperature prevailing inside the Moon now, or at any time of its past? First, let it be re-iterated that the Moon, in common with most other bodies of the solar system, originated probably by an agglomeration of solid particles of relatively short time.

Once the body of the Moon thus grew up by coalescence of solid particles at low temperatures, additional heat must have been generated continuously throughout the Moon's interior by spontaneous decay of such traces of naturally radioactive elements (e.g., potassium $K^{40}$, thorium $Th^{232}$, or the two isotopes of uranium $U^{235}$ and $U^{238}$) as are likely to present in lunar material. Therefore – as in the stars – the energy produced now inside the Moon is also due to nuclear transformations. But whereas, inside the stars, the reactions concerned are essentially of the fusion type, and the prevailing conditions are sufficiently extreme for the rate of these reactions to be influenced by local density and temperature, all exothermic nuclear reactions occurring now in the Moon are limited to spontaneous disintegration of heavy nuclei, the origin of which must be sought in the primordial state of lunar matter before its body was formed.

Let us, following UREY (1952) and subsequent investigators, assume that the principal elements responsible for internal heating of the Moon are $K^{40}$, $Th^{232}$, $U^{235}$ and $U^{238}$; and let us set

$$\varepsilon = \varepsilon_j e^{-\lambda_j t}, \tag{8-1}$$

where $\varepsilon_j$ denotes the amount of energy produced by a spontaneous decay of the $j$-th element of $\lambda_j$ half-life. The heat so dispersed in microscopic amounts through the entire mass of the Moon then flows outward towards the surface; but its escape is impeded by all kinds of physical obstacles (low thermal conductivity; high opacity or viscosity of lunar rocks) which together render the secular cooling of the lunar globe an exceedingly slow process.

In more specific terms, the energy equation which controls this rate of heat flow can be expressed in the form

$$\rho C_v \left( \frac{DT}{Dt} \right) = \text{div} \left( \kappa \, \text{grad} \, T \right) + \rho \varepsilon - P\varDelta + \mu \varPhi, \tag{8-2}$$

$$\text{I} \qquad \qquad \text{II} \quad \text{III} \quad \text{IV}$$

where $T$ denotes the local temperature; $t$, the time, $C_v$, the specific heat of the material at constant volume; $\kappa$, the coefficient of heat conduction which, in the case of radiative transfer, should be identified (cf., e.g., KOPAL, 1963) with

$$\kappa_r = \frac{4}{3} \frac{acT^3}{k\rho}, \tag{8-3}$$

where $a$ denotes the Stefan constant; $c$, the velocity of light; $\rho$, the density; and $k$, the absorption coefficient per unit mass. The term $\varepsilon$ represents the rate (per unit mass) of spontaneous energy liberation; $\varDelta \equiv \mathrm{div}\ \mathbf{u}$ is the divergence of the velocity vector of fluid flow (if any); and $\varPhi$ is the function representing the dissipation of kinetic energy of viscous flow into heat (which is a known quadratic function of the velocity components). Moreover, the symbol $D/Dt$ on the left-hand side of the preceding equation (8-1) of energy balance stand for the total (Lagrangian) time derivative (following the motion, if any).

If the interior of the Moon can be regarded as rigid – an assumption which we shall provisionally adopt – the terms III and IV on the right-hand side of the energy equation would be identically zero; and the only terms balancing the left-hand side would be I + II; the latter consisting of a linear combination of decreasing time-exponentials which characterize the respective radioactive decay. If, moreover, for the sake of simplicity we consider the coefficient $\kappa$ of heat conduction as constant and assume spherical symmetry, Equation (8-1) can be reduced to the ordinary equation of heat conduction

$$\frac{\partial T}{\partial t} = K \left\{ \frac{\partial^2 T}{\partial r^2} + \frac{2}{r} \frac{\partial T}{\partial r} \right\} + \sum_j \phi_j e^{-\lambda_j t}, \tag{8-4}$$

where

$$K = \frac{\kappa}{\rho C_v} \tag{8-5}$$

denotes the coefficient of thermal diffusivity (in $\mathrm{cm}^2/\mathrm{sec}$); while the terms

$$\phi_j = \frac{\bar{\varepsilon}_j}{C_v} \tag{8-6}$$

represent the rise in temperature (in degrees) of a unit mass, per unit time, due to the action of the $j$-th radioactive element in its appropriate abundance.

In what follows we shall, following UREY (1962), adopt for the $\bar{\varepsilon}_j$'s the same values with which the respective elements are found to occur in chondritic meteorites, and use for $C_v$ a value of $7 \times 10^{-6}$ erg/g deg (appropriate for an average of common silicate rocks; cf., e.g., BIRCH, 1952). The values of $\varepsilon_j$ and $\lambda_j$ are accurately measurable in the laboratory for each respective decay. Accepting their best values now available we arrive at the data listed in the accompanying Table 8-1, which we shall hereafter adopt.

TABLE 8-1

Radioactive Elements

| $j$ | Nuclide | $\lambda_j$ (in $10^{-9}$ years) | $\phi_j$ (in $10^{-6}$ deg/year) |
|---|---|---|---|
| 1 | $K^{40}$ | 0.546 | 0.873 |
| 2 | $Th^{232}$ | 0.0499 | 0.0328 |
| 3 | $U^{235}$ | 0.972 | 0.0942 |
| 4 | $U^{238}$ | 0.1537 | 0.0533 |

The boundary conditions of our heat conduction problem require that, at the time $t=0$,

$$T(r,0) = f(r) \tag{8-7}$$

be an arbitrary (pre-assigned) function of $r$, describing the initial distribution of temperature inside the new-born Moon; while, on the boundary $(r=r_{(})$, the fact that the Moon can lose heat only by radiation into the surrounding empty space will be expressed by the requirement that, at all times,

$$\kappa\left(\frac{\partial T}{\partial r}\right)_{r_{(}} + \sigma T^4(r_{(},t) = 0, \tag{8-8}$$

where (for a black-body radiator) $\sigma = ac/4$ denotes the Stefan-Boltzmann constant.

Unlike (8-7), the exterior boundary condition (8-8) is highly nonlinear in $T$ (due to the underlying nonlinearity of the Stefan-Boltzmann radiation law). However, the magnitude of the first term on the left-hand side of (8-8), approximately equal to that of $\kappa T/r_{(}$, is smaller than that of $\sigma T^4$ by some eight orders – which means that the condition (8-8) will effectively be fulfilled if we set

$$T(r_{(},t) = 0 \quad \text{for} \quad t > 0. \tag{8-9}$$

Although this simplification does not represent the most general way of satisfying the physical requirement (8-8), it should constitute a close approximation; and for this reason we shall, in what follows, explore its consequences in some detail.

The so simplified a boundary-value problem, consisting of the equation (8-4) subject to the conditions (8-7) and (8-9) was solved by LOWAN (1933) for an arbitrary internal distribution $\phi_j(r, t)$ of the radioactive heat sources. Adapting his solution to the present case, let us set

$$T(r,t) = T_1(r,t) + \sum_j T_2^{(j)}(r,t) = T_1(r,t) + T_2(r,t), \tag{8-10}$$

where the complementary function

$$T_1(r,t) = \frac{2}{x} \sum_{n=1}^{\infty} e^{-K(n\pi/r_{()})^2 t} \sin(n\pi x) \int_0^1 x f(x) \sin(n\pi x)\,dx \tag{8-11}$$

with $x \equiv r/r_{\langle}$ denoting the fractional distance from the center; while the particular integral $T_2(r, t)$ constitutes a sum of partial integrals

$$T_2^{(j)}(r, t) = \frac{2}{x} \sum_{n=1}^{\infty} e^{-K(n\pi/r_0)^2 t} \sin (n\pi x)$$

$$\times \int_0^1 x \sin (n\pi x) \left\{ \int_0^t e^{K(n\pi/r_0)^2 t - \lambda_j t} \phi_j(x, t) \, dt \right\} dt \tag{8-12}$$

for each $j$-th radioactive element separately.

The complementary function $T_1(r, t)$ represents, in effect, the secular change in internal temperature from its initial distribution $f(x)$, due to an outward heat flow to the surface whose temperature for $t > 0$ was set equal to zero (and which, therefore, represents a permanent sink of heat, lost into space in a way which equation (8-9) does not specify). If, in particular, the initial temperature distribution can be approximated by an expansion of the form

$$f(x) = \tau_0 + (1 - x^2) \tau_2 + \dots, \tag{8-13}$$

the corresponding expression for the complementary function (8-12) becomes

$$T_1(r, t) = \sum_{n=1}^{\infty} (-1)^{n+1} \frac{\sin n\pi x}{n\pi x} \left\{ \tau_0 + \frac{6\tau_2}{(n\pi)^2} + \dots \right\} e^{-K(n\pi/r_0)^2 t} \tag{8-14}$$

The particular integrals (8-12) for $T_2^{(j)}(r, t)$, representing an increase in temperature due to the gradual accumulation of radiogenic heat released by each particular $j$-th element, can be evaluated with equal case. If the internal distribution of radiogenic heat sources obeys a law of the form

$$\phi_j(x) = \phi_{j,0} e^{-\alpha_j(1-x)}, \tag{8-15}$$

where the $\phi_{j,0}$'s as well as $\alpha_j$'s are constant, an evaluation of the integral on the right-hand side of (8-12) yields

$$T_2^{(j)}(r, t) = 2\pi n\phi_{j,0} \sum_{n=1}^{\infty} \frac{\sin n\pi x}{x} \left\{ \frac{e^{-\lambda_j t} - e^{-K(n\pi/r_0)^2 t}}{K(n\pi/r_0)^2 - \lambda_j} \right\}$$

$$\times \left\{ \frac{(-1)^{n+1}}{(n\pi)^2 + \alpha_j^2} - 2\alpha \frac{(-1)^{n+1} + e^{-\alpha_j}}{[(n\pi)^2 + \alpha_j^2]^2} \right\} \tag{8-16}$$

which, for uniform distribution of radioactive elements reduces to

$$T_2^{(j)}(r, t) = 2\phi_{j,0} \sum_{n=1}^{\infty} (-1)^{n+1} \left\{ \frac{e^{-\lambda_j t} - e^{-K(n\pi/r_0)^2 t}}{K(n\pi/r_0)^2 - \lambda_j} \right\} \frac{\sin n\pi x}{n\pi x}. \tag{8-17}$$

If, however, the uniform distribution of heat sources were confined to an outer shell comprised between the fractional radii $\xi \leqslant x < 1$ – so that

$$\phi_j = 0 \quad \text{for} \quad 0 \leqslant x < \xi \atop \phi_j = \phi_{j,0} \quad \text{for} \quad \xi \leqslant x < 1, \Bigg\}$$ (8-18)

ALLAN (1956) has shown that the corresponding particular integral of the present problem assumes the form

$$T_2^{(j)}(r,t) = \frac{\phi_{j,0}}{\lambda_j} \left\{ \frac{\sin \beta j x}{x \sin \beta_j} - 1 \right\} e^{-\lambda_j t}$$

$$+ \frac{\phi_{j,0}}{\lambda_j} \frac{\sin \beta_j (1-x)}{x \sin \beta_j} \left\{ \xi \cos \beta_j \xi - \frac{\sin \beta_j \xi}{\beta_j} \right\} e^{-\lambda_j t}$$ (8-19)

$$+ \frac{2 r_{\text{(}}^2 \phi_{j,0}}{K} \sum_{n=1}^{\infty} \frac{e^{-K(n\pi/r_{\text{(}})^2 t}}{(n\pi)^2 - \beta_j^2} \left\{ (-1)^n + \frac{\sin n\pi\xi - n\pi\xi \cos n\pi\xi}{n\pi} \right\} \frac{\sin n\pi x}{n\pi},$$

where we have abbreviated $\beta_j^2 = \lambda_j r_{\text{(}}^2 / K$. When $\xi = 0$, the foregoing expression (8-19) reduces indeed to (8-17), as can be verified with the aid of the Fourier expansion

$$\frac{\sin \beta_j x}{x \sin \beta_j} = 1 + 2 \sum_{n=1}^{\infty} \frac{(-1)^{n+1} \beta_j^2}{(n\pi)^2 - \beta_j^2} \left( \frac{\sin n\pi x}{n\pi x} \right).$$ (8-20)

Extensive numerical computations of the rate of cooling of the lunar globe from a given assumed initial state have been carried out by UREY (1962) or KOPAL (1962a). Figure 8-1 illustrates the progress of lunar fractional cooling from an initially

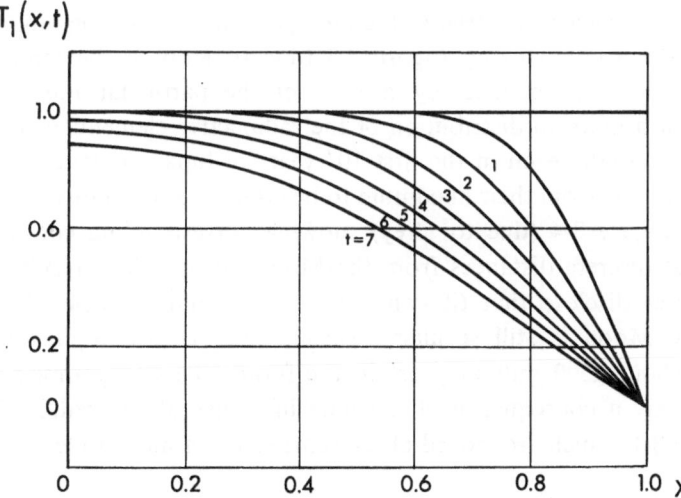

Fig. 8-1. Conductive cooling of the lunar globe from unit initial temperature in the course of the time. The individual curves correspond to the internal temperature profiles, in units of $10^9$ years, for an assumed coefficient of thermal diffusivity $K = 0.01$ cm²/sec (after KOPAL, 1962a).

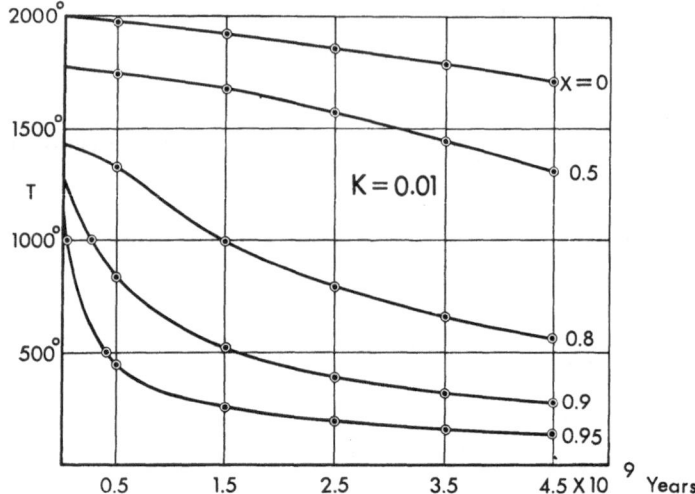

Fig. 8-2. The secular cooling of the lunar interior from an initial temperature distribution $T(x, 0) = 1100° + 900° (1 - x^2)$ for an assumed constant value of $K = 0.01$ cm²/sec (after UREY, 1962).

isothermal state for the first 7000 million years of its existence (Kopal); while Figure 8-2 shows the actual temperature variation in lunar interior, at different distances from the center, starting from an initial distribution $f(x) = 1100° + 900°(1 - x^2)$ for somewhat shorter interval of time (Urey). The reader may observed that, for plausible values of the coefficient $K$ of thermal diffusivity, the secular cooling of the lunar globe represents a long process, which did not have time to progress really very far in 4 thousand million years of the Moon's existence.

The same is, however, not true of the radiogenic heating of the lunar interior. As is shown on the accompanying Figure 8-3 (due to Kopal), the temperatures due to radiogenic heat (as defined by a sum of the particular integrals $T_2^{(j)}(r, t)$, appropriate for a uniform distribution of the radioactive elements as listed in Table 8-1) rise rather rapidly even in the first $10^9$ years of lunar existence, and appear at present to be not far from their maximum to be attained a few thousand million years hence; while Figure 8-4 (due to Urey) shows a corresponding continuous rise in temperature at discrete distances from the Moon's center. The results summarized on these figures disclose that (if conduction is the only vehicle of internal heat transport) the Moon is still secularly warming up, to attain maximum internal temperature about 2500 million years in the future. This long time-scale of lunar heating is merely a consequence of the long life-times of the radioactive elements listed in Table 8-1, which are indeed of the same order of magnitude as the age of the Moon.

The outcome of all work carried out in this field in recent years by UREY (1962), LEVIN (1960), MACDONALD (1959) or KOPAL (1962) reveals that (unless, perchance, the proportion of radioactive elements in lunar matter have been badly estimated by

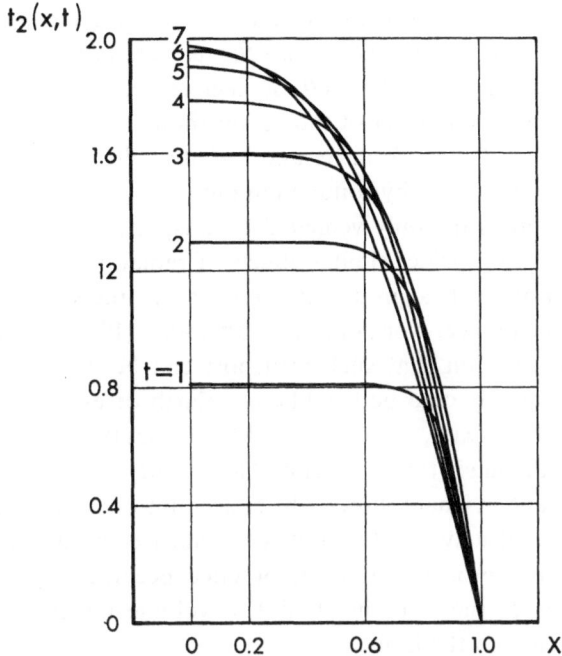

Fig. 8-3. The temperatures in the lunar interior attained by secular heating of the material by radioactive heat sources at $t = 0(1)7 \times 10^9$ years after the solidification of the lunar globe, computed for the abundances of uniformly distributed radioactive elements as listed in Table 8-1, and an assumed constant value of $K = 0.01$ cm²/sec (after KOPAL, 1962a). *Abscissae*: the temperatures $T_2(x, t)$ expressed in 1000°. *Ordinates*: fractional distance $x$ from the center of the lunar globe.

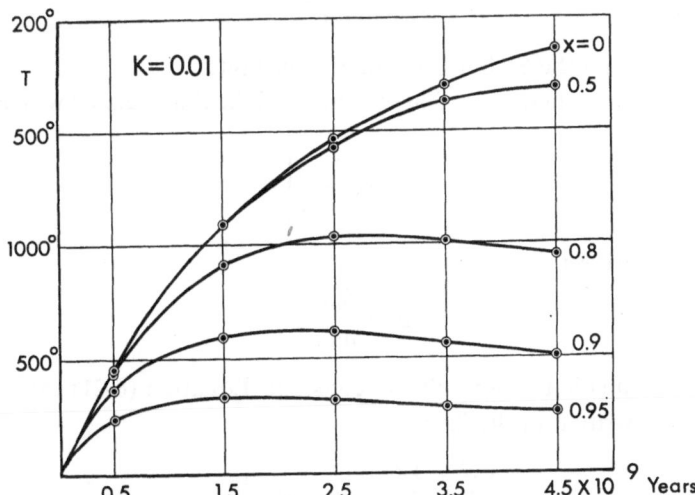

Fig. 8-4. The temperatures in the lunar interior attained by the secular heating of its material by radioactive heat sources in the first $4.5 \times 10^9$ years of the Moon's age, computed for the abundances of the radioactive elements as listed in Table 8-1, and an assumed value of $K = 0.01$ cm²/sec (after UREY, 1962). *Abscissae*: the absolute temperatures T expressed in units of 1000°. *Ordinates*: the time elapsed since the solidification of the Moon (in $10^9$ years).

analogy with chondritic meteorites) that *radiogenic heating alone of even an initially cold Moon would have been sufficient to raise the temperature of the bulk of its mass in excess of 1000 °K, and up to 1500–2000 °K near the center.* If we could only look sufficiently deep inside, the interior of the Moon should, accordingly, glow much like the inside of a terrestrial blast furnace, and radiate light which should facilitate an outward transport of heat – not by conduction only, but also radiation. So far we have considered conduction as the only vehicle for transporting heat in the interior; but now we wish to turn to consider similar effects of radiation.

Radiative transport of heat in stellar interiors has indeed been an old story to astronomers since the pioneer work of SCHWARZSCHILD (1906) at the beginning of this century; but the realization that such transport can be of importance also in the interiors of such relatively cold bodies like the Earth or even the Moon is of more recent date (cf., e.g., CLARK, 1956, 1957; MACDONALD, 1959; LEVIN, 1962). The theory of the effects arising in this connection has recently been thoroughly investigated by KOPAL (1964a); and in what follows a brief account of such a theory will be given.

In order to do so we may recall that, if the mean free path of the energy-carrying photons is allowed to approach zero (or, in practice become very small in comparison with the dimensions of the radiating body), a volume integral of the equation of radiative transfer tends to the limit

$$\frac{1}{c}\frac{\partial B}{\partial t} = \frac{1}{3r^2}\frac{\partial}{\partial r}\left(\frac{r^2}{k\rho}\frac{\partial B}{\partial r}\right) + \frac{\rho\bar{\varepsilon}_j}{4\pi}, \tag{8-21}$$

where $B$ denotes the emissivity of the respective material, related (in thermodynamic equilibrium) with the local temperature $T$ by the relation

$$\pi B = \sigma T^4, \tag{8-22}$$

with $\sigma = ac/4$ denoting the Stefan-Boltzmann constant.

If $x, \tau$ denote the normalized radial distance and the time, related with $r, t$ by means of the equations

$$\left.\begin{array}{l} r = r_{\mathfrak{c}}x, \\ t = \dfrac{3k\rho r_{\mathfrak{c}}^2}{2c}\tau \end{array}\right\} \tag{8-23}$$

and if, moreover, we set

$$B = \frac{B_0\Theta}{4\pi x}, \tag{8-24}$$

where $B_0$ denotes a suitable normalizing constant, Equation (8-21) can (for constant value of $k\rho$) be rewritten in the form

$$\frac{\partial\Theta}{\partial\tau} = \frac{1}{2}\frac{\partial^2\Theta}{\partial x^2} + f_j x e^{-\bar{\lambda}_j\tau}, \tag{8-25}$$

where we have abbreviated

$$f_j = \frac{3}{2}\frac{\overline{(\rho\bar{\varepsilon}_j)}\,\overline{k\rho}r_{\mathfrak{c}}^2}{B_0} \tag{8-26}$$

and

$$\bar{\lambda}_j = \frac{3k\rho r_{\mathbb{C}}^2}{2c} \lambda_j. \tag{8-27}$$

As to the boundary conditions of our problem, the finiteness of the temperature at the center implies, by (8-24), that

$$\Theta(0, \tau) = 0. \tag{8-28}$$

The radiative outer boundary condition (8-8) for $\kappa \equiv \kappa_R$ as defined by Equation (8-3) implies that, for $r = r_{\mathbb{C}}$,

$$\frac{4}{3} \frac{acT^3}{k\rho} \left( \frac{\partial T}{\partial r} \right) + \frac{ac}{4} T^4 = 0 \tag{8-29}$$

which in terms of our normalized variables, assumes the *linear* form

$$\left( \frac{\partial \Theta}{\partial x} \right) + (h - 1) \Theta(1, \tau) = 0, \tag{8-30}$$

where we have abbreviated

$$h = \tfrac{3}{4} k\rho r_{\mathbb{C}}. \tag{8-31}$$

The latter nondimensional quantity represents, in effect, the radius of the Moon expressed in terms of the mean free path of the energy-carrying photons and, as such, is likely to be so large in comparison with unity that the factor $h - 1$ on the left-hand side of (8-30) can be replaced simply by $h$.

The boundary-value problem arising in this connection is, therefore, linear and assumes the explicit form

$$\left. \begin{aligned} \frac{\partial \Theta}{\partial \tau} &= \frac{1}{2} \frac{\partial^2 \Theta}{\partial x^2} + f_j x e^{-\bar{\lambda}_j \tau} \\ \Theta(0, \tau) &= 0, \\ \left( \frac{\partial \Theta}{\partial x} \right), &+ h\Theta(1, \tau) = 0. \end{aligned} \right\} \tag{8-32}$$

In order to solve it, let us introduce the finite Fourier transform

$$\bar{\Theta}(\tau) = \int_0^1 \Theta(x, \tau) \sin px \, dx, \tag{8-33}$$

where $p$ is a real positive number yet to be specified. Now multiply Equation (8-25) by $\sin px$ and integrate over the interval $(0, 1)$. The, evidently,

$$\int_0^1 \frac{\partial \Theta}{\partial \tau} \sin px \, dx = \frac{\partial \bar{\Theta}}{\partial \tau}, \tag{8-34}$$

while repeated partial integration gives

$$\int_0^1 \frac{\partial^2 \Theta}{\partial x^2} \sin px \, dx = \left[\frac{\partial \Theta}{\partial x} \sin px - p\Theta \cos px\right]_0^1 - p^2 \int_0^1 \Theta \sin px \, dx . \qquad (8\text{-}35)$$

The integrated term on the right-hand side of the foregoing equation vanishes for $x=0$ by virtue of the boundary condition (8-28); while for $x=1$ it can be rewritten as

$$\left[\frac{\partial \Theta}{\partial x} - p\Theta \cot p\right]_1 \sin p .$$

If, however, we define now $p$ to be a positive root of the equation

$$p \cot p + h = 0 , \qquad (8\text{-}36)$$

the foregoing term assumes the form

$$\left[\frac{\partial \Theta}{\partial x} + h\Theta\right]_1$$

and vanishes again by virtue of the boundary condition (8-30); so that Equation (8-35) reduces to

$$\int_0^1 \frac{\partial^2 \Theta}{\partial x^2} \sin px \, dx = -p^2 \overline{\Theta} . \qquad (8\text{-}37)$$

Since, moreover,

$$\int_0^1 x \sin px \, dx = p^{-2}(\sin p - p \cos p) = p^{-2}(1+h) \sin p \qquad (8\text{-}38)$$

by virtue of (8-36), the ordinary differential equation for the Fourier transform $\Theta$ assumes the form

$$\frac{d\overline{\Theta}}{d\tau} + \frac{p^2}{2}\overline{\Theta} = \Pi_p e^{-\overline{\lambda}_j \tau} , \qquad (8\text{-}39)$$

where

$$\Pi_p = (f_j/p^2)(1+h) \sin p , \qquad (8\text{-}40)$$

subject to the initial condition requiring that, at $\tau=0$,

$$\overline{\Theta}(0) = \int_0^1 \Theta(x,0) \sin px \, dx \equiv Q_p . \qquad (8\text{-}41)$$

The solution of Equation (8-39) satisfying this condition readily assumes the form

$$\overline{\Theta}(\tau) = Q_p e^{-\frac{1}{2}p^2 \tau} + \Pi_p \left\{\frac{e^{-\overline{\lambda}_j \tau} - e^{-\frac{1}{2}p^2 \tau}}{\frac{1}{2}p^2 - \overline{\lambda}_j}\right\} . \qquad (8\text{-}42)$$

In order to invert this transform we may note that if $p$ and $q$ are any two roots of Equation (8-36),

$$\int_0^1 \sin px \sin qx\,dx = \tfrac{1}{2} + \frac{h}{2(h^2 + p^2)} \tag{8-43}$$

if $p = q$, and zero otherwise. Therefore if, as in the ordinary Fourier series, we write

$$f(x) = \sum a_p \sin px, \tag{8-44}$$

the summation being extended over all positive roots of (8-36), the coefficients

$$a_p = \frac{2(h^2 + p^2)}{h^2 + p^2 + h} \int_0^1 f(x) \sin px\,dx. \tag{8-45}$$

Hence, an inversion of the Fourier transform as defined by Equation (8-33) yields

$$\Theta(x, \tau) = 2 \sum_p \frac{(h^2 + p^2) \sin px}{h^2 + p^2 + h} \overline{\Theta}(\tau)$$

$$= 2 \sum_p \frac{(h^2 + p^2) Qp}{h^2 + p^2 + h} e^{-\frac{1}{2}p^2\tau} \sin px \tag{8-46}$$

$$+ 2 f_j \sum_p \frac{h^2 + p^2}{p^2} \left\{ \frac{(1 + h) \sin p}{h^2 + p^2 + h} \right\} \frac{e^{-\lambda_j\tau} - e^{-\frac{1}{2}p^2\tau}}{\frac{1}{2}p^2 - \lambda_j} \sin px$$

by insertion from (8-42); and, finally, by reverting to the emissivity $B(x, \tau)$ as related with $\Theta(x, \tau)$ by (8-24) we find that

$$B(x, \tau) = \frac{2}{x} \sum_p \frac{(h^2 + p^2) e^{-\frac{1}{2}p^2\tau}}{h^2 + p^2 + h} \sin px \int_0^1 B(x, 0) x \sin px\,dx$$

$$+ \rho \bar{\varepsilon}_j r_\zeta \sum_p \frac{h(1 + h)(h^2 + p^2)}{h^2 + p^2 + h} \left(\frac{\sin p}{\pi p}\right) \left(\frac{\sin px}{px}\right) \frac{e^{-\lambda_j\tau} - e^{-\frac{1}{2}p^2\tau}}{\frac{1}{2}p^2 - \lambda_j} \tag{8-47}$$

Now we mentioned earlier in this section that, in the astronomical applications which we have in mind, the constant $h$ as defined by Equation (8-31) is likely to be very large; and if so, equation (8-36) admits of roots which are asymptotically expressible as

$$p = n\pi(1 - h^{-1} + \ldots) \tag{8-48}$$

for $n = 1, 2, 3, \ldots$ . Moreover, it follows then from equation (8-36) that

$$\frac{\sin p}{p} = \frac{(-1)^{n+1}}{\sqrt{h^2 + p^2}}. \tag{8-49}$$

If $h$ is indeed so large in comparison with unity (and $h^2$ in comparison with $h$) that small factors can be ignored, our foregoing equation (8-46) will reduce to

$$
\begin{aligned}
B(x,\tau) = \frac{2}{x} \sum_{n=1}^{\infty} & e^{-\frac{1}{2}(n\pi)^2\tau} \sin n\pi x \int_0^1 xB(x,0) \sin n\pi x\, dx \\
& + \frac{\rho\bar{\varepsilon}_j r_{\text{c}} h}{\pi} \sum_{n=1}^{\infty} (-1)^{n+1} \frac{\sin n\pi x}{n\pi x} \left\{ \frac{e^{-\bar{\lambda}_j\tau} - e^{-\frac{1}{2}(n\pi)^2\tau}}{\frac{1}{2}(n\pi)^2 - \bar{\lambda}_j} \right\}
\end{aligned}
\tag{8-50}
$$

which, by use of (8-22) clearly becomes identical with (8-11) and (8-17) when we remember that, for photon gas, the specific heat

$$
C_v = \frac{4aT^3}{\rho}
\tag{8-51}
$$

which, together with (8-3), renders its thermal diffusivity to be

$$
K = \frac{\kappa_r}{\rho C_v} = \frac{c}{3k\rho}.
$$

Equation (8-47) represents an exact solution of the problem of radiative transfer of heat, in spherically-symmetrical configuration, with radiative loss at the boundary; while (8-50) represents an approximation obtaining if this leakage is very small. Unlike the latter (which reduces to zero at $x=1$), the exact value of $B(1, \tau)$ as defined by Equation (8-47) remains finite at the outer boundary, and of the order of $h^{-1}$.

How large is the magnitude of this important parameter? Equation (8-31) reveals it to be proportional to the quantity $(k\rho)r_{\text{c}}$, in which $r_{\text{c}} = 1.738 \times 10^8$ cm. The observed transparency of silicate rocks in the infrared part of the spectrum (transmitting the bulk of the radiant energy) is such that the value of the product $k\rho$ is likely to be of the order of $10^2 - 10^3$ cm$^{-1}$ (implying that the intensity of the transmitted beam will be reduced to $1/e$ by a passage through a rock layer 10–100 microns in thickness). If so, $h$ should be a quantity of the order of $10^{10}$, amply justifying the neglect of its inverse powers in (8-48).

What will, under these conditions, be the surface temperatures of the lunar globe due to radiative transfer and loss of radiogenic heat into space? These have been recently evaluated by KOPAL (1963) for assumed abundances of radioactive elements summarized in Table 8-1; and the outcome discloses that, if the central radiogenic temperatures $T_0$ (taken from KOPAL, 1962) rise in the course of time as listed in the second column of the accompanying Table 8-2, the continuing radiative heat loss should maintain the lunar surface at a temperature listed in column (3); while the last column of the same table gives the corresponding heat flux.

A glance at these results reveals that the temperature attained on the surface as a result of gradual radiative leakage of radiogenic heat into space amounts to only a few degrees Kelvin; and the corresponding heat flux, to less than 0.1 erg/cm$^2$ sec. In reality, of course, these figures constitute only a partial answer; for additional heat

TABLE 8-2

Radiative Surface Temperatures of the Moon

| Time (in $10^9$ years) | $T_0$ (in °K) | $T_1$ (in °K) | $F$ (in erg/cm² sec) |
|---|---|---|---|
| 0 | 0° | 0° | 0 |
| 2 | 1300° | 6°.3 | 0.029 |
| 4 | 1790° | 7°.6 | 0.061 |
| 6 | 2010° | 7°.8 | 0.067 |
| 8 | 2090° | 8°.1 | 0.079 |

will be transported to the surface by conduction, and also a part of the internal heat of the Moon may be primordial (i.e., independent of, and supplementary to, that generated by radioactivity). The actual amount of this latter heat store represented by the first term on the right-hand side of (8-47) is, however, still subject to uncertainty as to the proper form of $B(x, 0)$; and, on the other hand, the mathematical problem of heat conduction inside a sphere radiating through the surface into space in accordance with Stefan's law does not admit of an analytical solution leading to a finite surface temperature, because of the nonlinearity of its outer boundary condition. The amount of heat conducted to the surface and radiated away from it can, so far, be only estimated by semiqualitative arguments; and the estimates for flux $F$ range around 10 ergs/cm² sec, corresponding to absolute temperature $T_1$ about three times as high as those listed in column (3) of Table 8-2.

Direct observations of the magnitude of flux in the infrared domain of the spectrum have revealed, however, that the actual temperatures of the lunar surface are very much higher – varying as they do (cf. Chapter 20) between approximately 100 °K and 400 °K from night to day on the exposed surface, and stabilizing around 240 °K about a meter below the surface, where the temperature remains constant day and night. These temperatures cannot, therefore, have practically anything to do with the internal heat of the Moon, and are due to the illumination of its surface by the Sun. As the latter contributes thus an external source of heat (supplementary to that produced by lunar radioactivity) which clearly dominates the heat balance of the outer layers of the lunar globe our discussion of the thermal history of the Moon would be incomplete if we did not take its effects into account. In the rest of the present chapter we shall, therefore, outline a theory of the underlying problem of the conduction of external heat in the interior of a spherical configuration, and apply the results to the illumination of the Moon by the Sun.

In doing so, we shall forego until further treatment (Chapter 20) the diurnal variations of temperature in the outermost surface layer, and confine our attention to the secular inward flow of heat. The problem at issue then consists of constructing a solution of the homogeneous Equation (8-2) – with the terms II–IV absent and the coefficient $K$ of diffusivity constant – such that the surface temperature $T_1$ satisfies prescribed conditions. To approach our task in its simplest possible form, suppose that $T_1 =$ constant (i.e., the surface of a sphere being maintained everywhere at the same temperature). The solution of so simplified a problem is well known (cf., e.g., CARSLAW and JAEGER, 1959; § 9.3); and can be expressed in the form

$$T(r,t) = T_1 \left\{ 1 + 2 \sum_{n=1}^{\infty} (-1)^n \frac{\sin n\pi x}{n\pi x} e^{-K(n\pi/r_{\zeta})^2 t} \right\} \tag{8-52}$$

where, as before, $x \equiv r/r_{\zeta}$ denotes the fractional distance from the center.

The temperature $T_0$ prevailing at the center of our configuration follows as the limit of the right-hand side of (8-52) as $x \rightarrow 0$; and is found to bear to $T_1$ the ratio

$$\frac{T_0}{T_1} = 1 + 2 \sum_{n=1}^{\infty} (-1)^n e^{-K(n\pi/r_{\zeta})^2 t}. \tag{8-53}$$

The steady state is approached as $t \rightarrow \infty$; but the actual evaluation of the sum on the right-hand side of Equation (8-53) reveals that the ratio $T_0/T_1$ will approach unity within less than 1% when $Kt/r_{\zeta}^2 \sim 1$ (cf. CARSLAW and JAEGER, op. cit., Figure 29 on page 234). The time in which this will be attained will, therefore, be of the order of $r_{\zeta}^2/K$ seconds or (with $r_{\zeta} = 1.738 \times 10^8$ cm and $K \sim 0.01$ cm$^2$/sec) $10^{11}$ years. The estimated age of the Moon being less than $5 \times 10^9$ years, the value of $Kt/r_{\zeta}^2$ corresponding to it is only about 0.05, for which $T_0$ is found to equal roughly 5% of $T_1$. In other words, the inward seepage of solar heat in a stony globe of lunar dimensions constitutes so long-drawn out a process that the steady state would not get established till after some 100 billion years; and since the time the Moon was formed the progress towards it has barely begun.

As the leakage of solar heat in lunar interior is still so incomplete, in studying it we find it necessary to take account of the fact that not all parts of the lunar surface are receiving equal amount of it. Quite apart from large diurnal variations which rapidly subside with increasing depth, the temperature of the sub-crustal layer must vitally depend on lunar latitude (i.e., on the average height of the Sun above the horizon), causing dilution due to foreshortening. This latter factor will cause the temperature distribution to depart from radial symmetry; and to result as a solution of the differential equation

$$\frac{\partial T}{\partial t} = K \nabla^2 T, \tag{8-54}$$

subject to the boundary conditions requiring that

$$T(r,0) = 0 \quad \text{for} \quad r < r_{\zeta}; \tag{8-55}$$

but that, for $t > 0$, the temperature at the surface be constant and give by the equation

$$\sigma T_1^4 = (1 - A) \pi S \sin \theta, \tag{8-56}$$

where $\pi S$ denotes the flux of the solar radiation per unit area and time (i.e., the solar constant); $A$, the mean albedo (see Chapter 19) of the lunar surface; and $\theta$, the angular distance from the line of the cusps which, (if we ignore the small inclination $I = 1° 32'$ of the lunar equator to the ecliptic) becomes identical with the colatitude. Doing so, we may expect that

$$T_1 = k \sin^{\frac{1}{4}} \theta ,\tag{8-57}$$

where $k$ denotes the mean monthly temperature of the lunar equator, at a depth below the surface at which the diurnal variation of temperature no longer makes itself felt.

Let us assume now the solution of Equation (8-54) to be expansible in the form

$$T(r, \theta, \phi, t) = \sum_{j=0}^{\infty} \sum_{i=0}^{j} \tau_j(r, t) S_j^i(\theta, \phi),\tag{8-58}$$

where the surface harmonics $S_j^i(\theta, \phi)$ satisfy the partial differential equation

$$\frac{1}{\sin^2 \theta} \frac{\partial^2 S}{\partial \phi^2} + \frac{1}{\sin \theta} \frac{\partial}{\partial \theta} \left( \sin \theta \frac{\partial S}{\partial \theta} \right) + j(j + 1) S = 0 .\tag{8-59}$$

If so, the differential equation for $\tau_j$ then assumes the form

$$\frac{\partial \tau}{\partial t} = \frac{K}{r^2} \left\{ \frac{\partial}{\partial r} \left( r^2 \frac{\partial \tau}{\partial r} \right) - j(j + 1) \tau \right\};\tag{8-60}$$

and by setting

$$\tau(r, t) = r^j \tilde{\tau}(r, t)\tag{8-61}$$

reduces further to

$$\frac{\partial \tilde{\tau}}{\partial t} = K \left\{ \frac{\partial^2 \tilde{\tau}}{\partial r^2} + \frac{2(j + 1)}{r} \frac{\partial \tilde{\tau}}{\partial r} \right\}.\tag{8-62}$$

In order to separate the variables, let us set

$$\tilde{\tau}(r, t) = \alpha(r) \beta(t);\tag{8-63}$$

its insertion in (8-62) then yields

$$\frac{1}{K\beta} \frac{\partial \beta}{\partial t} = \frac{1}{\alpha} \left\{ \frac{\partial^2 \alpha}{\partial r^2} + \frac{2(j + 1)}{r} \frac{\partial \alpha}{\partial r} \right\} = -m^2 ,\tag{8-64}$$

where $m$ is a constant. The ensuing ordinary differential equations

$$\frac{d\beta}{dt} + Km^2\beta = 0\tag{8-65}$$

and

$$\frac{d^2\alpha}{dr^2} + \frac{2(j + 1)}{r} \frac{d\alpha}{dr} + m^2\alpha = 0\tag{8-66}$$

possess complete primitives of the form

$$\beta(t) = \beta_0 e^{-Km^2 t}\tag{8-67}$$

and

$$\alpha(r) = r^{-(j+\frac{1}{2})} \{\alpha_0 J_{j+\frac{1}{2}}(mr) + \alpha_1 J_{-(j+\frac{1}{2})}(mr)\} \qquad (8\text{-}68)$$

if $m \neq 0$, and

$$\alpha(r) = \alpha_2 \qquad (8\text{-}69)$$

for $m=0$; where $\beta_0$ and $\alpha_0$, $\alpha$, or $\alpha_2$ are constants, and the the $J_n$'s denote the Bessel functions of the respective argument and indices.

The requisite regularity of $\tau$ at the origin renders at once $\alpha_2 = 0$; so that an insertion from (8-36–8-69) in (8-61) readily furnishes $\tau$ as a summation of the form

$$\begin{aligned}
\tau_j(r,t) &= r^j \alpha(r) \beta(t) \\
&= a_0 r^j + \sum_{n=1}^{\infty} a_n \left(\frac{r_{\mathbb{C}}}{r}\right)^{\frac{1}{2}} J_{j+\frac{1}{2}}(m_n r) e^{-Km^2_n t},
\end{aligned} \qquad (8\text{-}70)$$

where the values of the constants $\alpha_n$ as well as $m_n$ are to be determined from the requisite boundary conditions. In our present case, two such conditions (8-55)–(8-56) must be satisfied: namely, the constancy of the temperature at the boundary $r=r_{\mathbb{C}}$; and the vanishing of the temperature at $t=0$ everywhere for $r<r_{\mathbb{C}}$.

In order to specify the former, let us expand the surface temperature, as given by Equation (8-58), in a series of surface harmonics $S_j^i(\theta, \phi)$. However, if in conformity with (8-58) $T_1$ is a function of $\theta$ only, this expansion will reduce to a series of zonal harmonics

$$T_1 = \sum_{j=0}^{\infty} A_j P_j(\cos \theta), \qquad (8\text{-}71)$$

where

$$2\pi A_j = (2_j + 1) k \int_0^{\pi} \sin^{\frac{1}{2}} \theta \, P_j(\cos \theta) \sin \theta \, d\theta. \qquad (8\text{-}72)$$

It follows by a comparison of (8-58) with (8-71) that

$$T_j(r_{\mathbb{C}}, t) = A_j; \qquad (8\text{-}73)$$

and this can be true only if, in (8-70),

$$a_0 r_{\mathbb{C}}^j = A_j, \qquad (8\text{-}74)$$

while

$$\sum_{n=1}^{\infty} a_n J_{j+\frac{1}{2}}(m_n r_{\mathbb{C}}) e^{-Km^2_n t} = 0 \qquad (8\text{-}75)$$

for unrestricted $a_n$'s. This latter condition can, in turn, be fulfilled at any time only if

$$m_n r_{\mathbb{C}} = N_n, \qquad (8\text{-}76)$$

where $N_n$ stands for the $n$-th root of the Bessel function

$$J_{j+\frac{1}{2}}(N_n) = 0. \qquad (8\text{-}77)$$

Next, the vanishing of the temperature anywhere for $t=0$ requires that, for $x = r/r_{\mathfrak{c}}$

$$A_j x^{j+\frac{1}{2}} + \sum_{n=1}^{\infty} a_n J_{j+\frac{1}{2}}(N_n x) = 0 \tag{8-78}$$

i.e., that the quantities

$$a_n \equiv A_j q_n \tag{8-79}$$

be coefficients of the Fourier-Bessel expansion of

$$x^{j+\frac{1}{2}} = -\sum_{n=1}^{\infty} q_n J_{j+\frac{1}{2}}(N_n x). \tag{8-80}$$

Such coefficient are, in turn, known (cf., e.g., WATSON, 1958; page 576) to be given by the integral expression

$$q_n = -\frac{2}{J_{j+\frac{3}{2}}^2(N_n)} \int_0^1 x^{j+\frac{3}{2}} J_{j+\frac{1}{2}}(N_n x) \, dx. \tag{8-81}$$

In order to evaluate this integral, let us set

$$j + \tfrac{3}{2} = v \quad \text{and} \quad N_n x = y. \tag{8-82}$$

If so,

$$\int_0^1 x^{j+\frac{3}{2}} J_{j+\frac{1}{2}}(N_n x) \, dx = \frac{1}{N_n^{v+1}} \int_0^{N_n} y^v J_{v-1}(y) \, dy. \tag{8-83}$$

But, by the known differential recursion formulae valid for Bessel functions,

$$y^v J_{v-1}(y) = \frac{d}{dy}\{y^v J_v(y)\}; \tag{8-84}$$

so that

$$\int_0^{Nn} y^v J_{v-1}(y) \, dy = N_n^{j+\frac{3}{2}} J_{j+\frac{3}{2}}(N_n) \tag{8-85}$$

and, eventually, from (8-79) and (8-81)–(8-82), the coefficients

$$a_n = A_j q_n = -\frac{2 A_j}{N_n J_{j+\frac{3}{2}}(N_n)}. \tag{8-86}$$

If so, however, the desired particular solution of our problem assumes the explicit form

$$T(r, \theta, t) = \sum_{j=0}^{\infty} \tau_j(x, t) P_j(\cos \theta), \tag{8-87}$$

where

$$\tau_j(x, t) = A_j \left\{ x^j - \frac{2}{\sqrt{x}} \sum_{n=0}^{\infty} \frac{J_{j+\frac{1}{2}}(N_n x)}{N_n J_{j+\frac{3}{2}}(N_n)} e^{-K(n\pi/r_{()})^2 t} \right\}. \tag{8-88}$$

In the case of spherical symmetry (i.e., $j=0$) the reader can easily verify that Equation (8-77) reduces to $\sin N_n = 0$, yielding $N_n = n\pi$ as its roots; and the corresponding Bessel functions

$$
\left.\begin{aligned}
J_{\frac{1}{2}}(n\pi x) &= \sqrt{\frac{2}{n\pi^2 x}}\, \sin\,(n\pi x), \\
J_{\frac{1}{2}}(n\pi) &= \frac{(-1)^{n+1}}{\pi}\,\sqrt{\frac{2}{n}},
\end{aligned}\right\}
\tag{8-89}
$$

leading, by (8-86), to

$$
q_n = (-1)^n \sqrt{\frac{2}{n}};
\tag{8-90}
$$

in which case Equation (8-87) becomes, as it should, identical with (8-52). Again, at $t \to \infty$, Equation (8-87) reduces to

$$
T(r,\theta) = \sum_{j=0}^{\infty} A_j x^j P_j(\cos\theta)
\tag{8-91}
$$

satisfying indeed the Laplace equation

$$
\nabla^2 T = 0
\tag{8-92}
$$

to which Equation (8-54) reduces in steady state.

In the non-steady state, however, the requisite particular solution of (8-54) satisfying the boundary conditions (8-55) and (8-56) proves to be of the form (8-77); and contains the quantitative answer to our inquiry about the secular penetration of solar heat in the lunar globe; while the actual values of the first three nonvanishing constants $A_j$ of the expansion of $T(r, \theta, t)$ on the right-hand side of (8-87) are obtained from (8-72) as

$$
A_0 = \frac{k}{2\pi} \int_0^{\pi} \sin^{\frac{3}{2}}\theta\, d\theta = -\frac{k\Gamma(\frac{1}{4})}{5\{2^{\frac{1}{4}}\Gamma(\frac{5}{8})\}^2}
\tag{8-93}
$$

$$
A_2 = \frac{5k}{2\pi} \int_0^{\pi} (1 - \tfrac{3}{2}\sin^2\theta)\sin^{\frac{3}{2}}\theta\, d\theta = \frac{k\Gamma(\frac{1}{4})}{26\{2^{\frac{1}{4}}\Gamma(\frac{5}{8})\}^2}
\tag{8-94}
$$

$$
A_4 = \frac{9k}{2\pi} \int_0^{\pi} (1 - 5\sin^2\theta + \tfrac{35}{8}\sin^4\theta)\sin^{\frac{3}{2}}\theta\, d\theta = -\frac{9k}{416}\frac{\Gamma(\frac{1}{4})}{\{2^{\frac{1}{4}}\Gamma(\frac{5}{8})\}^2}
\tag{8-95}
$$

etc., leading to

$$
\begin{aligned}
T(r,\theta,t) &= \frac{k\Gamma(\frac{1}{4})}{5\{2^{\frac{1}{4}}\Gamma(\frac{5}{8})\}^2}\{\tau_0(x,t) - \tfrac{5}{26}\tau_2(x,\tau)P_2(\cos\theta) \\
&\qquad\qquad - \tfrac{45}{416}\tau_4(x,\tau)P_4(\cos\theta) + \ldots\} \\
&= k\{0.297\,\tau_0(x,t) - 0.0471\,\tau_2(x,t)P_2(\cos\theta) \\
&\qquad\qquad - 0.0321\,\tau_4(x,t)P_4(\cos\theta) \\
&\qquad\qquad + \ldots\}
\end{aligned}
\tag{8-96}
$$

From these results it is evident that the amount of solar heat which could have found its way into the lunar interior could have raised the local temperature (especially in the sub-crustal layers) by dozens, but not hundreds, of degrees; and cannot, therefore, compare in amount with the internal radiogenic heat. The main point of interest of this mechanism attaches to the angular (i.e., nonradial) distribution of temperature to which it gives rise; and the consequences of this fact will be taken up in the next section, in which shall consider the stresses in an elastic solid arising from this source.

## CHAPTER 9

# STRESS HISTORY OF THE MOON

In the preceding chapter a survey has been given of the principal sources of heat inside an astronomical body whose mass is of the same order of magnitude as that of the Moon; and of the way in which this heat is likely to be distributed in the course of time. The time-dependence of the sources of radiogenic heat, together with the gradual cooling of lunar globe by escape of its thermal radiation into space, is bound to render the distribution of temperature inside the Moon a function of the time and, as such, will give rise to *thermal stresses* in its interior considered as an elastic solid, together with *gravitational stresses* produced by the self-attraction of the lunar mass. It is true that, in view of the relatively high temperatures likely to be prevalent in lunar interior (as described in the preceding chapter), its mass need not necessarily behave as a perfectly elastic solid; and the possibility of departures from such a state will be discussed in the next chapter; the subject matter of the present chapter being reserved for a discussion of the behaviour of the lunar globe to gravitational, thermal, or other stresses to which the body of the Moon responds as an elastic solid.

In order to set up the system of fundamental equations which govern the displacement within an elastic self-gravitating body in spherical polar coordinates $r$, $\theta$, $\phi$, let the well-known nine *stress components* $\sigma_{ik}$ causing displacement be expressed in the form

$$\left. \begin{aligned} \sigma_{rr} &= \lambda \varDelta + 2\mu\varepsilon_{rr} - (3\lambda + 2\mu)\delta T, \\ \sigma_{\theta\theta} &= \lambda \varDelta + 2\mu\varepsilon_{\theta\theta} - (3\lambda + 2\mu)\delta T, \\ \sigma_{\phi\phi} &= \lambda \varDelta + 2\mu\varepsilon_{\phi\phi} - (3\lambda + 2\mu)\delta T, \end{aligned} \right\} \tag{9-1}$$

$$\left. \begin{aligned} \sigma_{r\theta} &= 2\mu\varepsilon_{r\theta} = \sigma_{\theta r} \\ \sigma_{r\phi} &= 2\mu\varepsilon_{r\phi} = \sigma_{\phi r} \\ \sigma_{\theta\phi} &= 2\mu\varepsilon_{\theta\phi} = \sigma_{\phi\theta} \end{aligned} \right\} \tag{9-2}$$

where $\lambda$ and $\mu$ denote the two Lamé parameters characterizing the elastic behaviour of the solid, and $\delta$ is the coefficient of its linear thermal expansion (none of which is necessarily constant). The corresponding *strain components* $\varepsilon_{ik}$ are then expressible as

$$\varepsilon_{rr} = \frac{\partial u_r}{\partial r},$$

$$\varepsilon_{\theta\theta} = \frac{1}{r}\frac{\partial u_\theta}{\partial \theta} + \frac{u_r}{r}, \tag{9-3}$$

$$\varepsilon_{\phi\phi} = \frac{1}{r \sin\theta}\frac{\partial u_\phi}{\partial \phi} + \frac{u_r}{r} + \frac{u_\theta \cot\theta}{r},$$

and

$$2\varepsilon_{r\theta} = \frac{1}{r}\frac{\partial u_r}{\partial \theta} + \frac{\partial u_\theta}{\partial r} - \frac{u_\theta}{r} = 2\varepsilon_{\theta r},$$

$$2\varepsilon_{\theta\phi} = \frac{1}{r}\frac{\partial u_\phi}{\partial \theta} + \frac{1}{r\sin\theta}\frac{\partial u_\theta}{\partial \phi} - \frac{u_\phi \cot\theta}{r} = 2\varepsilon_{\phi\theta},$$

$$2\varepsilon_{\phi r} = \frac{1}{r\sin\theta}\frac{\partial u_r}{\partial \phi} + \frac{\partial u_\phi}{\partial r} - \frac{u_\phi}{r} = 2\varepsilon_{r\phi},$$

(9-4)

in terms of the three components

$$u_r \quad u_\theta \quad u_\phi$$

of the displacement vector **u**. Lastly, the symbol $\Delta$ stands for the divergence of **u** and, as such, is given by

$$\Delta = \frac{1}{r^2}\frac{\partial}{\partial r}(r^2 u_r) + \frac{1}{r\sin\theta}\frac{\partial}{\partial r}(u_\theta \sin\theta) + \frac{1}{r\sin\theta}\frac{\partial u_\phi}{\partial \phi}. \tag{9-5}$$

In terms of the quantities introduced by the preceding relations, the equations of motion (cf., e.g., LOVE, 1927, p. 91) governing the elastic displacements assume the form

$$\frac{\partial \sigma_{rr}}{\partial r} + \frac{1}{r}\frac{\partial \sigma_{r\theta}}{\partial \theta} + \frac{1}{r\sin\theta}\frac{\partial \sigma_{r\phi}}{\partial \phi} + \frac{2\sigma_{rr} - \sigma_{\theta\theta} - \sigma_{\phi\phi} + \sigma_{r\theta}\cot\theta}{r}$$

$$= \rho\left\{\frac{\partial^2 u_r}{\partial t^2} - \frac{\partial V}{\partial r}\right\}, \tag{9-6}$$

$$\frac{\partial \sigma_{r\theta}}{\partial r} + \frac{1}{r}\frac{\partial \sigma_{\theta\theta}}{\partial \theta} + \frac{1}{r\sin\theta}\frac{\partial \sigma_{\phi\phi}}{\partial \phi} + \frac{(\sigma_{\theta\theta} - \sigma_{\phi\phi})\cot\theta + 3\sigma_{r\theta}}{r}$$

$$= \rho\left\{\frac{\partial^2 u_\theta}{\partial t^2} - \frac{1}{r}\frac{\partial V}{\partial \theta}\right\}, \tag{9-7}$$

$$\frac{\partial \sigma_{r\phi}}{\partial r} + \frac{1}{r}\frac{\partial \sigma_{\theta\phi}}{\partial \theta} + \frac{1}{r\sin\theta}\frac{\partial \sigma_{\phi\phi}}{\partial \phi} + \frac{3\sigma_{r\phi} + 2\sigma_{\theta\phi}\cot\theta}{r}$$

$$= \rho\left\{\frac{\partial^2 u_\phi}{\partial t^2} - \frac{1}{r\sin\theta}\frac{\partial V}{\partial \phi}\right\}, \tag{9-8}$$

where $\rho$ denotes the (actual) density of the solid and $V$, the total potential of forces acting upon any element of its mass.

The boundary conditions which the solutions of the foregoing equations should satisfy require, first, that at the center of our configuration ($r=0$) the displacement must be zero – i.e., that

$$u_r(0) = u_\theta(0) = u_\phi(0) = 0; \tag{9-9}$$

and, secondly, that at the outer boundary $r=a$ the radial components of the stress tensor

$$\sigma_{rr}(a) = \sigma_{r\theta}(a) = \sigma_{r\phi}(a) = 0. \tag{9-10}$$

Let us assume now that, when no displacements are present, our configuration is *spherically-symmetrical*; and that, moreover, the displacements we wish to study are characterized by a *spheroidal symmetry*, requiring that the individual displacement components be of the form

$$\left.\begin{array}{l} u_r = u S^i_j, \\[2mm] u_\theta = v \dfrac{\partial S^i_j}{\partial \theta}, \\[2mm] u_\phi = \dfrac{v}{\sin \theta} \dfrac{\partial S^i_j}{\partial \phi}, \end{array}\right\} \tag{9-11}$$

where $u(r, t)$ as well as $v(r, t)$ are functions of $r$ and $t$ only; and the $S^i_j(\theta, \phi)$'s are surface harmonics satisfying equation (8-59). If we then abbreviate

$$\frac{1}{r^2} \frac{\partial}{\partial r} (r^2 u) - j(j+1) \frac{v}{r} = y \tag{9-12}$$

and

$$\frac{1}{r} \left( \frac{\partial}{\partial r} (rv) - u \right) = z, \tag{9-13}$$

it follows immediately from (9-5) that

$$\Delta = y S^i_j(\theta, \phi). \tag{9-14}$$

Assume, moreover, that the potential $V$ of the strained body can also be expressed as

$$V = \sum_{i,j} R_j(r, t) S^i_j(\theta, \phi) \tag{9-15}$$

while, in conformity with (8-58),

$$T = \sum_{i,j} \tau_j(r, t) S^i_j(\theta, \phi). \tag{9-16}$$

If so, Equation (9-7) becomes identical with (9-8) and the system (9-6)–(9-8) reduces to the following two differential equations

$$(\lambda + 2\mu) \frac{\partial y}{\partial r} + \left\{ \frac{d\lambda}{dr} + g\rho \right\} y + 2 \left( \frac{d\mu}{dr} \right) \frac{\partial u}{\partial r} + j(j+1) \mu \frac{z}{r}$$
$$= \rho \frac{\partial^2 u}{\partial t^2} + \tfrac{4}{3}\pi G \rho a^2 \frac{\partial R}{\partial r} + \frac{\partial}{\partial r} \{ (3\lambda + 2\mu) \delta \tau \} \tag{9-17}$$

and

$$(\lambda + 2\mu) y + \frac{\partial}{\partial r} (\mu rz) - 2(u + v) \frac{d\mu}{dr} = \rho r \frac{\partial^2 v}{\partial t^2} + \tfrac{4}{3}\pi G \rho a^2 R + (3\lambda + 2\mu) \delta \tau, \tag{9-18}$$

where the Lamé parameters $\lambda$ and $\mu$ as well as the thermal expansion coefficient $\delta$ have been allowed to vary with $r$ (but not with $\theta$ or $\phi$); $\rho$ denotes the (undisturbed) density – there should be no confusion in dropping the zero subscript – $g$, the (undisturbed) gravitational acceleration; $a$, the external radius of our configuration; and

$$R = \frac{3u}{a^2 r^2} \int_0^r \rho r^2 \, dr + \frac{3}{(2j+1)a^2} \left\{ \frac{1}{r^{j+1}} \int_0^r \rho \left[ yr^{j+2} - \frac{\partial}{\partial r}(ur^{j+2}) \right] dr \right.$$

$$\left. + r^j \int_r^a \rho \left[ yr^{1-j} - \frac{\partial}{\partial r}(ur^{1-j}) \right] dr \right\}.$$

(9-19)

The transformations by which we have arrived at the foregoing equations (9-17)–(9-19) are not simple; in particular, the derivation of the expression (9-19) for $R$ is rather involved, but need not be reproduced in this place. The reader desirous to follow it in more detail should consult KOPAL (1960b).

The boundary conditions which the solutions of (9-17) and (9-18) should satisfy follow from (9-9) and (9-11) as

$$u = v = 0 \qquad\qquad\qquad (9\text{-}20)$$

at $r=0$, and

$$\left. \begin{array}{l} \lambda\Delta + 2\mu\varepsilon_{rr} = (3\lambda + 2\mu)\,\delta\tau \\[4pt] 2\mu\varepsilon_{r\theta} = 0 \\[4pt] 2\mu\varepsilon_{r\phi} = 0 \end{array} \right\} \qquad (9\text{-}21)$$

at $r=a$. Since, however, from (9-3)–(9-4) and (9-11) it transpires now that

$$\left. \begin{array}{l} \varepsilon_{rr} = \dfrac{\partial u}{\partial r} S_j^i, \\[10pt] \varepsilon_{r\theta} = \left( \dfrac{\partial v}{\partial r} - \dfrac{v}{r} + \dfrac{u}{r} \right) \dfrac{\partial S_j^i}{\partial \theta}, \\[10pt] \varepsilon_{r\phi} = \left( \dfrac{\partial v}{\partial r} - \dfrac{v}{r} + \dfrac{u}{r} \right) \dfrac{1}{\sin\theta} \dfrac{\partial S_j^i}{\partial \phi}, \end{array} \right\} \qquad (9\text{-}22)$$

the conditions (9-10) will evidently be satisfied at $r=a$ provided that

$$\lambda y + 2\mu \frac{\partial u}{\partial r} = (3\lambda + 2\mu)\,\delta\tau \qquad\qquad (9\text{-}23)$$

and

$$2\mu \left( \frac{\partial v}{\partial r} - \frac{v}{r} + \frac{u}{r} \right) = 0; \qquad\qquad (9\text{-}24)$$

where, it may be noted, the right-hand side of (9-23) likewise vanishes if the temperature $\tau(a, t)$ on the free surface is kept equal to zero.

Having established the explicit form of the differential equations (9-17)–(9-18) governing spheroidal displacements of self-gravitating elastic bodies, together with their requisite boundary conditions (9-20) and (9-23)–(9-24), let us use them now to evaluate the effects, on the lunar globe regarded as an elastic solid, of slow internal heating of the kind discussed in the preceding Chapter 8. This heating represents so long-drawn out a process as to cause the magnitude of the second time-derivatives $\partial^2 u/\partial t^2$ and $\partial^2 v/\partial t^2$ on the right-hand sides of Equation (9-6)–(9-8) to be completely negligible in comparison with all other terms; so that we can, in effect, treat our thermo-elastic problem as one of steady state at an arbitrary time $t$.

Let us also assume, for the sake of simplicity, that the distribution of radiogenic heat sources inside the Moon (and, consequently, the distribution of internal temperature) is radially-symmetrical. If, however, $j=0$, $v=0$; $u$ being the sole non-vanishing component of radial displacement; in which case Equations (9-12) and (9-13) reduce to

$$y = \frac{1}{r^2} \frac{\partial}{\partial r}(r^2 u) \quad \text{and} \quad z = -\frac{u}{r}, \tag{9-25}$$

respectively; and, moreover, Equation (9-19) gives

$$R = \frac{3}{a^2}\left\{\frac{u}{r^2}\int_0^r \rho r^2\, dr + \int_r^a \rho u\, dr\right\}. \tag{9-26}$$

If so, the sole surviving Equation (9-17) of motion will reduce to the form

$$(\lambda + 2\mu)\frac{\partial}{\partial r}\left\{\frac{1}{r^2}\frac{\partial}{\partial r}(r^2 u)\right\} + \left\{\frac{\partial\lambda}{\partial r} + g\rho\right\}\frac{1}{r^2}\frac{\partial}{\partial r}(r^2 u)$$
$$+ 2\frac{\partial\mu}{\partial r}\frac{\partial u}{\partial r} = g\rho\left\{\frac{\partial u}{\partial r} - \frac{2u}{r}\right\} + \frac{\partial}{\partial r}\left\{\delta(3\lambda + 2\mu)\,T\right\}, \tag{9-27}$$

which for constant values of $\lambda$, $\mu$, and $\delta$ reduces further to

$$(\lambda + 2\mu)\frac{\partial}{\partial r}\left\{\frac{1}{r^2}\frac{\partial}{\partial r}(r^2 u)\right\} + \frac{g\rho}{r^2}\frac{\partial}{\partial r}(r^2 u) = \tfrac{4}{3}\pi G\rho a^2 \frac{\partial R}{\partial r} + (3\lambda + 2\mu)\delta\frac{\partial T}{\partial r}. \tag{9-28}$$

Since, moreover,

$$\tfrac{4}{3}\pi G\rho a^2 \frac{\partial R}{\partial r} - \frac{g\rho}{r^2}\frac{\partial}{\partial r}(r^2 u) = -4g\rho\frac{u}{r}, \tag{9-29}$$

our desired equation assumes the final form

$$\frac{\partial}{\partial r}\left\{\frac{1}{r^2}\frac{\partial}{\partial r}(r^2 u)\right\} = \left(\frac{3\lambda + 2\mu}{\lambda + 2\mu}\right)\delta\frac{\partial T}{\partial r} - \frac{4g\rho}{\lambda + 2\mu}\left(\frac{u}{r}\right), \tag{9-30}$$

where, of the two terms on the right-hand side, the first represents the effects of thermal stresses; and the second, the change in self-attraction due to displacement.

In order to assess the magnitudes of these two terms, let us recall that

$$\frac{3\lambda + 2\mu}{\lambda + 2\mu} = \frac{1 + \sigma}{1 - \sigma}, \tag{9-31}$$

where $\sigma$ denotes Poisson's ratio (i.e., the ratio of transverse compression to longitudinal expansion), which for all known substances is constrained to obey the inequality $0 < \sigma < 0.5$, and whose value can be determined for a given solid from the velocity of propagation of elastic waves. On the Earth, down to a depth of some 150 km in its mantle (where the entire range of pressures occurring inside the lunar globe are encountered) the values of $\sigma$ deduced from the seismological evidence appears to range between 0.26–0.27 (cf., e.g., BIRCH, 1952) – close enough to a round value of $\frac{1}{4}$ which we shall hereafter adopt.

On the other hand, it is known from the theory of elasticity that

$$\frac{4g\rho}{\lambda + 2\mu} = \frac{4g}{C_P^2}, \tag{9-32}$$

where $C_P$ stands for the velocity of propagation of the longitudinal (compression) waves. In the Earth's crust, $C_P$ is known to be very close to 8 km/sec; and if the same is true of the bulk of the Moon, both sides of Equation (9-32) should be of the order of $10^{-10}$ cm$^{-1}$. If so, the last term on the right-hand side of (9-30) becomes utterly negligible in comparison with all others; and the equation itself can be truncated to

$$\frac{\partial}{\partial r}\left\{\frac{1}{r^2}\frac{\partial}{\partial r}(r^2 u)\right\} = \delta\left\{\frac{1+\sigma}{1-\sigma}\right\}\frac{\partial T}{\partial r} \tag{9-33}$$

or, more simply, to

$$\frac{\partial^2 \varphi}{\partial r^2} + 4r\frac{\partial \varphi}{\partial r} = \tfrac{5}{3}\delta\left(\frac{\partial T}{\partial r}\right) \tag{9-34}$$

if we adopt the Poisson ratio $\sigma = 0.25$, and abbreviate

$$\frac{u}{r} = \varphi. \tag{9-35}$$

In what follows we wish to construct particular solutions of this nonhomogeneous equation satisfying the boundary conditions (9-20) and (9-23). The condition (9-20) requiring that $u(0) = 0$ implies that

$$\lim_{r \to 0} \varphi(r, t) = \text{finite}; \tag{9-36}$$

whereas the outer boundary condition (9-23) reduces now, by (9-35), to

$$\left\{\frac{1-\sigma}{1+\sigma}\right\}r\frac{\partial \varphi}{\partial r} + \varphi = \delta T_1 \tag{9-37}$$

or

$$3a\left(\frac{\partial \varphi}{\partial r}\right)_a + 5\varphi(a, t) = 0 \tag{9-38}$$

for $\sigma = 0.25$ and $T_1 = 0$.

If $\delta$ or $T$ were zero, the homogeneous part of (9-34) would admit of a solution of the form $\varphi = Ar^m$, where $m$ is a root of the indicial equation $m(m+3)=0$. Consequently, the particular solution of (9-34) obeying the boundary conditions (9-36) and (9-38) is found by the variation of parameters to assume the form

$$\varphi(r,t) = \frac{\delta}{3}\left\{\frac{5}{r^3}\int_0^r T(r,t)\,r^2\,\mathrm{d}r + \frac{4}{a^3}\int_0^a T(r,t)\,r^2\,\mathrm{d}r\right\} \tag{9-39}$$

and can be evaluated by quadratures for any appropriate distribution of temperature $T(r,t)$. At the boundary of our configuration, the foregoing expression readily reduces to

$$\varphi(a,t) = \frac{3\delta}{a^3}\int_0^a T(r,t)\,r^2\,\mathrm{d}r = \delta\bar{T}(t), \tag{9-40}$$

where $\bar{T}$ denotes the mean internal temperature averaged over the whole globe; while at the center (9-36) yields

$$\varphi(0,t) = \frac{\delta}{9}\left\{5T(0,t) + 4\bar{T}(t)\right\}. \tag{9-41}$$

Let us assume now that the temperature distribution $T(r,t)$ inside the Moon can be represented by Equation (8-10), with $T_1(r,t)$ and $T_2^{(j)}(r,t)$ as given by Equations (8-14) and (8-17), respectively. If, accordingly, the total fractional dilatation $\varphi(r,t)$ can be decomposed as

$$\varphi(r,t) = \tfrac{2}{3}\delta\left\{\Psi_1(x,t) + \Psi_2(x,t)\right\}, \tag{9-42}$$

then by inserting (8-14) and (8-17) on the right-hand side of Equation (9-36) and integrating term-by-term we find (cf. KOPAL, 1962c, 1963c) that

$$\Psi_1(r,t) = \sum_{n=1}^{\infty}\left\{\left(\frac{2}{n\pi}\right) + \frac{5(-1)^{n+1}J_{\frac{3}{2}}(n\pi x)}{\pi\sqrt{2(nx)^3}}\right\}$$
$$\times\left\{\tau_0 + \frac{6\tau_2}{(n\pi)^2} + \dots\right\}e^{-K(n\pi/a)^2 t} = \tau_0\Psi_1^{(0)} + \tau_2\Psi_1^{(2)} + \dots, \tag{9-43}$$

and

$$\Psi_2(r,t) = \sum_j \phi_j \Psi_2^{(j)}(r,t)$$
$$= \sum_j \phi_j \sum_{n=1}^{\infty}\left\{\left(\frac{2}{n\pi}\right)^2 + \frac{5(-1)^{n+1}J_{\frac{3}{2}}(n\pi x)}{\pi\sqrt{2(nx)^3}}\right\}\frac{e^{-\lambda_j t} - e^{-K(n\pi/a)^2 t}}{K(n\pi/a)^2 t - \lambda_j}, \tag{9-44}$$

where, it may be remembered,

$$(n\pi x)^{\frac{3}{2}}J_{\frac{3}{2}}(n\pi x) = \sin(n\pi x) - (n\pi x)\cos(n\pi x), \tag{9-45}$$

and, as before, $x = r/a$. Graphical representations of the functions $\Psi_1^{(0)}(r, t)$ and $\Psi_2(r, t)$ are shown on the accompanying Figures 9-1 and 9-2.

Let us analyze now the meaning of these results in so far as they bear on the elastic response of our Moon to its secularly changing internal temperature. If the lunar globe was initially characterized by a constant temperature $T_0$, at any subsequent time $t$ its external radius $a$ would have shrunk in size, as a result of secular cooling, by an amount $\Delta_1 a$ given by

$$\Delta_1 a = \tfrac{2}{3} a \delta T_0 \{ \Psi_1^{(0)}(a, t) - \Psi_1^{(0)}(a, 0) \} \tag{9-46}$$

where, it may be noted from (9-40),

$$\Psi_1^{(0)}(a, 0) = \tfrac{3}{2} ; \tag{9-47}$$

while, at the same time, the radiogenic heating of lunar interior would have increased this radius by an amount

$$\Delta_2 a = \tfrac{2}{3} a \delta \Psi_2(a, t) ; \tag{9-48}$$

the net change $\Delta a$ in external radius being given by the sum

$$\Delta a = \Delta_1 a + \Delta_2 a . \tag{9-49}$$

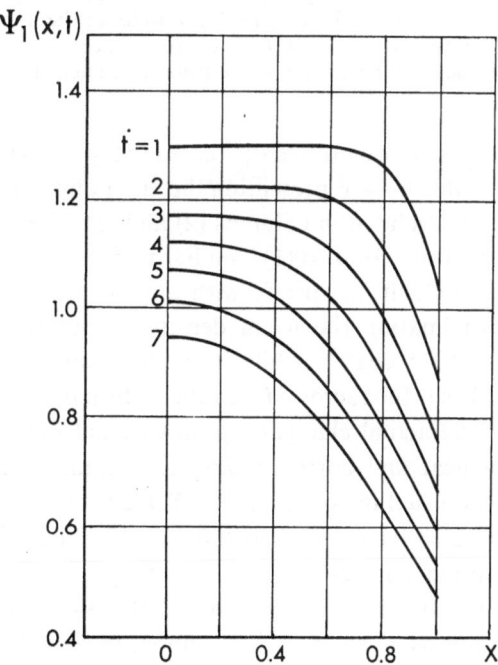

Fig. 9-1. The fractional contraction of the lunar interior due to its secular cooling from an initially constant temperature (after KOPAL, 1962a). *Abscissae*: the fractional contraction $\Psi^{(0)}(x, t)$ as defined by Equation (9-43) for a time $t = 0(1)7 \times 10^9$ years since the formation of our satellite. *Ordinates*: the fractional distance $x$ from the center of the lunar globe.

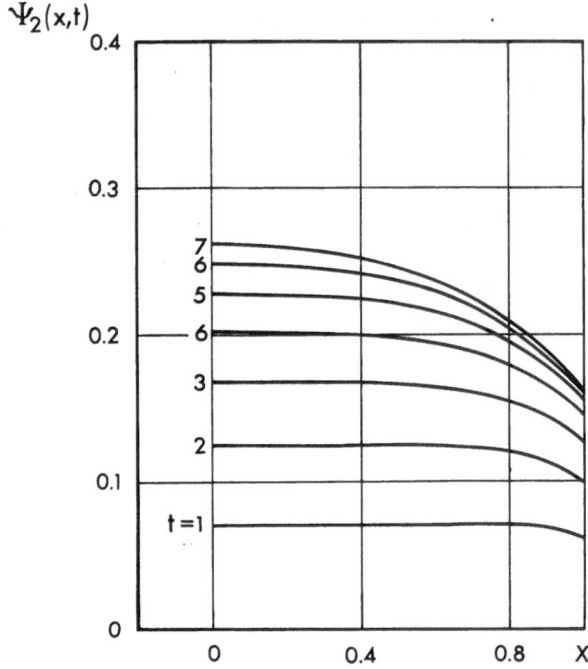

Fig. 9-2. The fractional expansion of the Moon due to the secular radiogenic heating of its interior from an initially zero temperature (after KOPAL, 1962a). *Abscissae*: the fractional expansion $\Psi_2(x, t)$ as defined by Equation (9-44) for a time $t = 0(1)7 \times 10^9$ years since the solidification of our satellite. *Ordinates*: the fractional distance $x \equiv r/a$ from the center of the lunar globe.

The numerical values of the quantities $\Psi_1^{(0)}(a, t)$ and $\Psi_2(a, t)$ – the latter for the values of $\phi_j$ and $\lambda_j$ as listed in Table 8-1 – representing the end points of the curves plotted on Figures 9-1 and 9-2, are tabulated in columns (2) and (3) of the accompanying Table 9-1 as functions of the time for $t=0(1)7$ milliard years. The actual values of the constants $\delta$ or $T_0$ are difficult to specify with any accuracy. For common silicate rocks of density approximating the mean density of our satellite, the value of $\delta$ should be close to $0.7 \times 10^{-5}$ deg$^{-1}$; while its initial temperature $T_0$ was probably between $100°–500$ °K. If so, the factor $\delta T_0$ on the right-hand side of (9-43) should be of the order of $10^{-3}$. The actual changes $\Delta_1 a$ in external radius due to the gradual leakage of primordial heat and corresponding to a round value of $\delta T_0 = 0.001$ in Equation (9-46) are then listed in column (4) of Table 9-1; while column (5) contains the values of $\Delta_2 a$ computed from (9-48) with $\delta = 7 \times 10^{-6}$ deg$^{-1}$; and, lastly, column (6) contains their algebraic sum (9-49).

An inspection of the numerical data listed in Table 9-1 and, in particular, of the resultant change in radius as given in its last column, reveals that the thermal history of our satellite as outlined in the preceding chapter had probably but little effect on its size; as the contraction due to heat loss, and expansion due to the accumulation of radiogenic heat tend very largely to cancel each other. If the Moon was warmer initially than some $140$ °K implied in our calculations of column (4), the contraction

TABLE 9-1

Thermal Expansion of the Moon

| $t$ (in $10^9$ years) | $\psi_1^{(0)}(a, t)$ | $\psi_2(a, t)$ | $\Delta_1 a$ (km) | $\Delta_2 a$ (km) | $\Delta a$ (km) |
|---|---|---|---|---|---|
| 0 | 1.500 | 0.0 | 0.00 | 0.00 | 0.00 |
| 1 | 1.035 | 62.7 | − 0.54 | 0.48 | − 0.06 |
| 2 | 0.867 | 101.3 | − 0.73 | 0.78 | + 0.05 |
| 3 | 0.756 | 127.2 | − 0.86 | 0.98 | + 0.12 |
| 4 | 0.669 | 146.3 | − 0.96 | 1.13 | + 0.17 |
| 5 | 0.593 | 156.8 | − 1.05 | 1.21 | + 0.16 |
| 6 | 0.531 | 161.7 | − 1.12 | 1.25 | + 0.13 |
| 7 | 0.475 | 162.5 | − 1.19 | 1.25 | + 0.06 |

would take the upper hand; while, conversely, if we underestimated the amount of radiogenic heat, expansion would predominate. With the data as adopted above, however, a maximum radial expansion by 0.17 km seems indicated for the present stage of the Moon's evolution, amounting to but 0.01% of its initial radius; and resulting in an increase of less than 4000 km² of its initial surface.

When we come to consider the *stresses* arising from the Moon's probable past thermal history, these likewise prove to be moderate. In the case of radial symmetry, the only non-zero components of the elastic stress-tensor reduce to

$$\sigma_{rr} = \lambda\left(\frac{\partial u}{\partial r} + \frac{2u}{r}\right) + 2\mu\frac{\partial u}{\partial r} - (3\lambda + 2\mu)\,\delta T \tag{9-50}$$

$$\sigma_{\theta\theta} = \lambda\frac{\partial u}{\partial r} + 2(\lambda + \mu)\frac{u}{r} - (3\lambda + 2\mu)\,\delta T = \sigma_{\phi\phi}; \tag{9-51}$$

and if, on the surface, $\sigma_{rr}(a, t) = 0$ by virtue of the postulated boundary condition (9-10), it follows from (9-48) that

$$\sigma_{\theta\theta}(a, t) = 2\mu\left\{\frac{1 + \sigma}{1 - \sigma}\right\}\{\varphi(a, t) - \delta T(a, t)\} = \sigma_{\phi\phi}(a, t); \tag{9-52}$$

while if, at the center, $u(0) = 0$,

$$\sigma_{rr}(0, t) = (\lambda + 2\mu)\left(\frac{\partial u}{\partial r}\right)_0 - (3\lambda + 2\mu)\,\delta T(0, t) \tag{9-53}$$

and

$$\sigma_{\theta\theta}(0, t) = \lambda\left(\frac{\partial u}{\partial r}\right)_0 - (3\lambda + 2\mu)\,\delta T(0, t) = \sigma_{\phi\phi}(0, t). \tag{9-54}$$

For the adopted value $\sigma = 0.25$ of the Poisson ratio (equivalent, by (9-31), to an identity $\lambda = \mu$), the surface value of $\varphi(a, t)$ is given by Equation (9-37); while its central value (9-38) combined with Equations (9-35) and (9-38) reveals that

$$\left(\frac{\partial u}{\partial r}\right)_0 = \frac{\delta}{9}\{5T(0, t) + 4\bar{T}(t)\} \tag{9-55}$$

Hence, it transpires from the foregoing equations that, on the surface of our configuration,

$$
\left.\begin{array}{l}
\sigma_{rr}(a, t) = 0 \\
\sigma_{\theta\theta}(a, t) = \tfrac{10}{3}\delta\mu\{\bar{T}(t) - T(a, t)\} = \sigma_{\phi\phi}(a, t);
\end{array}\right\} \qquad (9\text{-}56)
$$

while at its center

$$
\left.\begin{array}{l}
\sigma_{rr}(0, t) = \tfrac{2}{3}\delta\mu\{2\bar{T}(t) - 5T(0, t)\} \\
\sigma_{\theta\theta}(0, t) = \tfrac{4}{9}\delta\mu\{\bar{T}(t) - 10T(0, t)\} = \sigma_{\phi\phi}(0, t),
\end{array}\right\} \qquad (9\text{-}57)
$$

where $\mu$ stands for the coefficient of rigidity (shear modulus) of the respective elastic solid. If the configuration were isothermal, all stresses would vanish on the surface; while, at the center,

$$
\sigma_{rr}(0, t) = -2\delta\mu T_0 \quad \text{and} \quad \sigma_{\theta\theta}(0, t) = \sigma_{\phi\phi}(0, t) = -4\delta\mu T_0 .
$$

On the surface, the principal stress difference $|\sigma_{rr} - \sigma_{\theta\theta}| = (10/3)\delta\mu\bar{T} \sim 0.01\ \mu$ for $\delta \sim 10^{-5}$ deg$^{-1}$ and $\bar{T} \sim 10^3$ deg. Moreover, if the elastic behaviour of the Moon is similar to that of the Earth, we should expect that $\mu \sim 10^{12}$ dynes/cm$^2$ (cf., e.g., JEFFREYS, 1962; p. 222); and if so, the difference $\sigma_{rr} - \sigma_{\theta\theta}$ should be of the order of $10^{10}$ dynes/cm$^2$. This is about 10 times the crushing strength of granite at zero pressure (cf. again JEFFREYS, 1962, p. 202). Therefore, unless the adopted value of $\mu$ constitutes a gross overestimate, large-scale fractures could be expected to occur on the lunar surface as a result of thermal stresses envisaged in this chapter. Whether or not these have anything to do with the "rilles" actually observed on the Moon in great numbers (cf. Chapter 16) remains still conjectural, but the possibility should be kept in mind.

This being so, our treatment of the problem of heat conduction inside the lunar globe, as given in the preceding chapter, requires a small postscript; for the equation (8-4) then does not truly safeguard the conservation of energy, as it fails to take into account the work performed by an expanding solid against gravity. A generalization of Equation (8-4), paying due regard to the thermomechanical effects caused by a change of volume accompanying thermal expansion or contraction, was given by DUHAMEL (1837) more than a century ago, and more recently extended by LOWAN (1935) to the case of a radioactively heated sphere. In such a case, the generalized equation of heat conduction in an elastic and radially-symmetrical solid assumes the form*

$$
\frac{\partial T}{\partial t} = \frac{K}{r^2}\frac{\partial}{\partial r}\left(r^2 \frac{\partial T}{\partial r}\right) + \sum_j \phi_j e^{-\lambda_j t} - \frac{\gamma - 1}{\alpha}\frac{\partial}{\partial t}\left\{\frac{1}{r^2}\frac{\partial}{\partial r}(r^2 u)\right\}, \qquad (9\text{-}58)
$$

---

* The reader may note the close analogy between the thermo-mechanical (last) term on the right-hand side of this equation, and the term $P\Delta$ on the right-hand side of (8-2) which would arise from convective heat transport in fluids.

replacing (8-4), where

$$\alpha = -\frac{1}{\rho}\left(\frac{\partial \rho}{\partial T}\right)_P = 3\delta \qquad (9\text{-}59)$$

denotes the coefficient of *volume* thermal expansion of the respective solid (i.e., three times that of the linear thermal expansion), and

$$\gamma = \frac{C_P}{C_v} \qquad (9\text{-}60)$$

stands for the ratio of its specific heats at constant pressure and volume.

In order to solve the foregoing Equation (9-58) for $T(r, t)$, let us eliminate the div $u$ term on its right-hand side with the aid of (9-33). With due regard to the boundary conditions of our problem, Equation (9-33) can be integrated once to reveal that

$$\frac{1}{r^2}\frac{\partial}{\partial r}(r^2 u) = \frac{\alpha(1+\sigma)}{3(1-\sigma)}\left\{T(r,t) + \frac{12}{5a^3}\int_0^a T(r,t)r^2\,dr\right\}; \qquad (9\text{-}61)$$

and inserting this in (9-58) we find the latter to assume the more explicit form

$$\left\{1 + \frac{(\gamma-1)(1+\sigma)}{3(1-\sigma)}\right\}\frac{\partial T}{\partial t} + \frac{4(\gamma-1)(1+\sigma)}{5(1-\sigma)}\frac{\partial}{\partial t}\int_0^a T(r,t)r^2\,dr$$

$$= \frac{K}{r^2}\frac{\partial}{\partial r}\left(r^2\frac{\partial T}{\partial r}\right) + \sum_j \phi_j e^{-\lambda_j t}. \qquad (9\text{-}62)$$

In order to proceed further let us adopt (as before) the value of the Poisson's ratio to be $\sigma = 0.25$; if so (and provided that both $K$ and $\gamma$ can also be regarded as constants) it can be shown (cf. LOWAN, 1935) that the particular solution of (9-62) which satisfies the prescribed boundary conditions can again be expressed in the form (8-10), where the complementary function

$$T_1(r,t) = \frac{2}{x}\sum_{n=1}^{\infty}\frac{\sin\theta_n x - x\sin\theta_n}{1 - 2(\sin\theta_n/\theta_n)^2 + (\sin 2\theta_n)/2\theta_n} \times$$

$$e^{-K(\theta_n/A)^2 t}\int_0^1 xT(x,0)\sin\theta_n x\,dx, \qquad (9\text{-}63)$$

while the particular integrals

$$T_2^{(j)}(r,t) = \frac{2}{x}\sum_{n=1}^{\infty}\frac{\sin\theta_n x - x\sin\theta_n}{1 - 2(\sin^2\theta_n)/\theta_n^2 + (\sin 2\theta_n)/2\theta_n} \times$$

$$e^{-K(\theta_n/A)^2 t}\int_0^1 \sin\theta_n x\left\{\int_0^t e^{K(\theta_n/A)^2 t - \lambda_j t}\phi_j(x,t)\,dt\right\}x\,dx. \qquad (9\text{-}64)$$

In these equations the constant $A$ is defined by

$$3A = a(4 + 5\gamma)^{\frac{1}{2}} \tag{9-65}$$

and the $\theta_n$'s are positive roots of the transcendental equation

$$1 - \theta \cot \theta = 3\gamma\theta^2/4(\gamma - 1). \tag{9-66}$$

If $\gamma = 1$ (i.e., $C_p = C_v$), then $A = a$ and $\theta_n = n\pi$, in which case the foregoing equations (9-63) and (9-64) reduce to (8-11) and (8-12). To what extent is this case likely to apply? In order to answer this question, let us fall back on the definition of specific heats revealing that their difference

$$C_p - C_v = \frac{\dfrac{T_2}{\rho}\left(\dfrac{1}{\rho}\dfrac{\partial\rho}{\partial T}\right)_p^2}{\dfrac{1}{\rho}\left(\dfrac{1}{\rho}\dfrac{\partial\rho}{\partial p}\right)_T} = \frac{T}{\rho}\left(\frac{\alpha^2}{\beta}\right), \tag{9-67}$$

where $\alpha$ denotes, as before, the coefficient of volume thermal expansion, and

$$\beta = \frac{1}{\rho}\left(\frac{\partial\rho}{\partial P}\right)_T \tag{9-68}$$

stands for the coefficient of isothermal compression (i.e., reciprocal of the bulk modulus of the respective elastic solid). Therefore,

$$\frac{C_p - C_v}{C_v} = \gamma - 1 = 1 + \frac{\alpha^2 T}{\rho\beta C_v}. \tag{9-69}$$

The coefficients $\alpha$ and $\beta$ as well as the specific heats $C_v$ for several typical silicate rocks are listed in columns (3)–(5) of the accompanying Table 9-2; and its ultimate column (6) gives the values of the ratio $\alpha^2/\rho\beta C_v$. This ratio proves to be of the order of $2 \times 10^{-5}$ deg$^{-1}$, and is unlikely to vary much under conditions anticipated to exist in lunar interior. Therefore, according to (9–69), we should expect that

$$\gamma - 1 \simeq 2 \times 10^{-5} T \tag{9-70}$$

which, for anticipated temperatures between 1000°–2000 °K, becomes a quantity of the order of a few per cent – neither large, nor altogether negligible. An introduction of (9-70) in (9-62) would, of course, render the latter nonlinear in $T$; and Equation (9-58) should then be coupled with (9-33) through this nonlinear term. In such a case, the Duhamel-Lowan solution of our problem as represented by Equations (9-63)–(9-64) would, of course, no longer be exact. The fact that the coefficient of the non-linear cross-term is numerically small suggests, however an approach to the analytical solution of so generalized a thermo-mechanical problem by successive approximations: namely, to express the nonlinear term in terms of the previously determined temperature distribution which, to a first approximation, can be taken from Chapter 8 under complete neglect of the difference $\gamma - 1$. In this way, equation could be treated

TABLE 9-2

Thermal Properties of Silicates

| Mineral | $10^6\alpha$ (deg$^{-1}$) | $\rho$ (g/cm$^3$) | $10^{-2}C_v$ (erg/g deg) | $10^{13}\beta$ (cm$^2$/dyne) | $10^6\,(\alpha^2/\rho\beta C_v)$ (deg$^{-1}$) |
|---|---|---|---|---|---|
| Albite Na Al Si$_3$O$_8$ | 17 | 2.6 | 0.71 | 19 | 8 |
| Anorthite Ca Al Si$_2$O$_8$ | 15 | 2.76 | 0.7 | 11 | 11 |
| Diopside Ca Mg Si$_2$O$_6$ | 20 | 3.2 | 0.69 | 11 | 17 |
| Forsterite Mg$_2$SiO$_4$ | 25 | 3.3 | 0.79 | 8 | 30 |
| Quartz SiO$_2$ | 40 | 2.65 | 0.698 | 27 | 32 |

as linear but nonhomogeneous at each stage, and successive approximations to its solution obtained by iteration.

Before we abandon the model of a spherically-symmetrical elastic Moon for nonradial approximations, let us consider one additional problem to which our analysis as outlined so far should enable us to answer: namely so, what extent is the Moon compressed radially by self-attraction? In particular, if the Moon would consist of a material whose density at zero pressure would be everywhere the same, to what extent would this density be compelled to increase inwards by the sheer weight of overlying material?

The differential equation governing the variation of displacement due to self-attraction is (cf. KOPAL, 1963c) expressible as

$$\frac{\partial}{\partial r}\left\{\frac{1}{r^2}\frac{\partial}{\partial r}(r^2 u)\right\} = \frac{\beta(1+\sigma)}{3(1-\sigma)}g\rho, \qquad (9\text{-}71)$$

where the coefficient $\beta$ of isothermal compression continues to be defined by Equation (9-68) and $g$ denotes the local gravity. If the density $\rho$ were constant, it follows from (7-4) that

$$g = \tfrac{4}{3}\pi G\rho r, \qquad (9\text{-}72)$$

which introduced in (9-71) permits to rewrite the latter as

$$\frac{\partial}{\partial r}\left(r\frac{\partial\varphi}{\partial r} + 3\varphi\right) = \frac{4\pi G\rho^2\beta(1+\sigma)}{9(1-\sigma)}r \qquad (9\text{-}73)$$

in terms of the fractional displacement $u/r \equiv \varphi$ as defined before by Equation (9-35).

This equation can be integrated twice as it stands; and enforcing the boundary conditions (9-36) as well as (9-38) we find its particular solution relevant to our present problem to assume the form

$$\varphi(r) = -\Omega\left\{\frac{3-\sigma}{1+\sigma} - \frac{r^2}{a^2}\right\}, \qquad (9\text{-}74)$$

where

$$\Omega = \frac{2\pi G\rho^2\beta(1+\sigma)a^2}{45(1-\sigma)} \tag{9-75}$$

denotes a nondimensional constant. As for all known materials the Poisson's ratio $\sigma \leqslant 0.5$ it follows that $\varphi(r)$ as defined above will be negative throughout the entire configuration (as it should, since the gravity acts on the whole sphere to compress it); but the corresponding radial strain $e_{rr} \equiv \partial u/\partial r = \partial(r\varphi)/\partial r$ will vanish inside our configuration at a fractional distance from its center given by

$$\frac{r}{a} = \frac{3-\sigma}{3(1+\sigma)}, \tag{9-76}$$

outside of which $e_{rr}$ is positive (the radial strain being extension) and negative inside (corresponding to compression).

At the center of our self-gravitating configuration,

$$\varphi(0,t) = -\frac{3-\sigma}{1+\sigma}\Omega \tag{9-77}$$

and on the surface,

$$\varphi(a,t) = -\frac{2(1-\sigma)}{1+\sigma}\Omega; \tag{9-78}$$

whereas the average compression

$$\frac{4}{a^4}\int_0^a \varphi(r,t)r^3\,dr = -\left\{\frac{3-\sigma}{1+\sigma} - \frac{2}{3}\right\}\Omega = \varphi(a\sqrt{2/3}) \tag{9-79}$$

corresponds to that experienced at the fractional radius $x = \sqrt{2/3}$.

If, therefore, $\rho_0$ denotes the (uncompressed) density which the material of our sphere would possess in the absence of gravitation, while $\rho$ stands for the mean density of the compressed configuration as a whole, the two are obviously connected by the relation

$$\rho = \rho_0\{1 - 3\varphi(a\sqrt{2/3})\}; \tag{9-80}$$

while the actual central $(\rho_c)$ and surface $(\rho_s)$ densities in the gravitational field of the respective body should follow from

$$\begin{aligned}\rho_c &= \rho_0\{1 - 3\varphi(0)\}, \\ \rho_s &= \rho_0\{1 - 3\varphi(a)\},\end{aligned} \tag{9-81}$$

which for the Poisson's ratio $\sigma = 0.25$ assume the more explicit forms

$$\left.\begin{aligned}\rho_c &= \rho(1 + 2\ \Omega + ...) \\ \rho_s &= \rho(1 - \ \Omega + ...) \\ \rho_0 &= \rho(1 - 4.6\Omega + ...)\end{aligned}\right\} \tag{9-82}$$

where now

$$\Omega = \tfrac{2}{27}\pi G\rho^2\beta a^2. \tag{9-83}$$

For a self-gravitating elastic sphere of the mass and size of the Moon, we have $a = 1.738 \times 10^8$ cm and $\rho = 3.34$ g/cm$^3$; in addition, the gravitation constant $G = 6.668 \times 10^{-8}$ dyn cm$^2$/g$^2$. If, moreover, we assume that the compressibility $\beta$ of lunar rocks is comparable with those listed in column (5) of our Table 9-2, we should expect that $\beta \simeq 3 \times 10^{-12}$ cm$^2$/dyne; and if so, the value of the constant $\Omega$ turns out to be close to 0.0157. Accordingly

$$\rho_c = 3.34\,(1 + 2 \times 0.0157) = 3.44 \text{ g/cm}^3 \qquad (9\text{-}84)$$

and

$$\rho_s = 3.34\,(1 - 0.0157) = 3.28 \text{ g/cm}^3, \qquad (9\text{-}85)$$

while

$$\rho_o = \frac{3.34}{1 + 4.6 \times 0.0157} = 3.10 \text{ g/cm}^3. \qquad (9\text{-}86)$$

If, therefore, the Moon were not compressed by self-attraction the mean density of its material – if homogeneous – would be by almost 7% lower than the one which we actually observe; the mean radius of the Moon being increased by $1.2a\Omega = 32.7$ km to 1771 km, and its surface would be by 13430 km$^2$ larger.

The reader may notice that the values for $\rho_c$ and $\rho_s$ obtained above prove to be identical with those obtained earlier in Chapter 7 and given by Equations (7-23). This coincidence is not accidental, but due to the fact that the right-hand side of Equation (7-7) represents, indeed, a reciprocal of the bulk modulus $\beta$ as defined by Equation (9-68); and the numerical values (7-8) of the constants $a$ and $b$ in (7-7) as determined by Bullen from the elastic properties of the Earth's mantle are consistent with the value of $\beta$ adopted in this chapter.

Having finished thus with a brief survey of the elastic properties of the lunar globe regarded as a sphere, let us turn next to inquire into the thermal expansion of the Moon due to its secular illumination by the Sun. The distribution of temperature arising from this effect has already been defined by Equations (8-87)–(8-88) of the preceding chapter; and now we wish to ascertain its mechanical effects on the external form and moments of inertia of so illuminated and elastic globe.

Since the boundary condition (8-57) essential for the formulation of the underlying heat-conduction problem lacks radial symmetry we must return to the full-dress equations (9-17) and (9-18) for an appropriate formulation of the corresponding elastic problem; and if, in doing so, we regard the parameters $\lambda$, $\mu$ and $\delta$ as well as $\rho$ as constant, Equations (9-17) and (9-18) reduce to

$$\mu \left\{ \frac{\partial^2}{\partial r^2} - \frac{j(j+1)}{r^2} \right\} rz = \tfrac{4}{3}\pi G \rho^2 r y \qquad (9\text{-}87)$$

and

$$(\lambda + 2\mu) \left\{ \frac{1}{r^2} \frac{\partial}{\partial r} \left( r^2 \frac{\partial y}{\partial r} \right) - \frac{j(j+1)}{r^2} y \right\} - \tfrac{4}{3}\pi G \rho^2 \left\{ y + j(j+1)z \right\}$$
$$= \delta(3\lambda + 2\mu) \left\{ \frac{1}{r^2} \frac{\partial}{\partial r} \left( r^2 \frac{\partial \tau_j}{\partial r} \right) - \frac{j(j+1)}{r^2} \tau_j \right\} \qquad (9\text{-}88)$$

in terms of the auxiliary variables $y$ and $z$, as defined by Equations (9-12) and (9-13), for any value of $j$.

In order to estimate the relative magnitude of the individual terms in Equation (9-88), we may note that the constant

$$\tfrac{4}{3}\pi G\rho^2 a^2 = 2P_c \tag{9-89}$$

is equal to twice the hydrostatic pressure prevailing at the Moon's center. The latter is already known (cf. Equation 7-6) to be close to $4.7 \times 10^{10}$ dynes/cm$^2$ – so that the constant on the left-hand side of the foregoing equation (9-84) should be equal to $9.4 \times 10^{10}$ dynes/cm$^2$. On the other hand, the mean rigidity of the Earth is known (cf., e.g., JEFFREYS, 1962 p. 222) to be characterized by a value of $\mu = 1.5 \times 10^{12}$ dynes/cm$^3$; which for a Poisson ratio $\sigma = 0.27$ prevalent in the Earth's crust (cf. BIRCH, 1952) corresponds to $\lambda = 1.8 \times 10^{12}$ dynes/cm$^2$. But if the same were true for the Moon the sum $\lambda + 2\mu$ would become just about 100 times as large as $P_c$. On the strength of this disparity, we are led to disregard in (9-88) the second term on its left-hand side which is factored by $P_c$ in comparison with those factored by $\lambda + 2\mu$, and replace this equation by

$$\left\{\frac{\partial^2}{\partial r^2} + \frac{2}{r}\frac{\partial}{\partial r} - \frac{j(j+1)}{r^2}\right\}\{(\lambda + 2\mu)y - \delta(3\lambda + 2\mu)\tau\} = 0. \tag{9-90}$$

For the Earth or other larger terrestrial planets (Venus) the quantities $P_c$ and $\lambda + 2\mu$ are of the same order of magnitude and, consequently, the replacement of (9-88) by (9-90) would not be legitimate; but for a celestial body whose mass is as small as that of the Moon this constitutes a tolerable approximation.

If the elastic deformation which we wish to investigate is due to the heating alone the foregoing equation (9-90) will obviously be satisfied if

$$y = \delta\left\{\frac{3\lambda + 2\mu}{\lambda + 2\mu}\right\}\tau = \delta\left\{\frac{1+\sigma}{1-\sigma}\right\}\tau, \tag{9-91}$$

which inserted in (9-90) yields

$$\left\{\frac{\partial^2}{\partial x^2} - \frac{j(j+1)}{x^2}\right\}(xz) = \frac{4\pi G\rho^2 a^2}{3\mu}xy = \frac{2\delta}{\mu}\frac{1+\sigma}{1-\sigma}P_c x\tau, \tag{9-92}$$

where, as before, $x = r/a$. As consistent with our underlying assumptions both $y$ and $z$ must tend to zero with $\tau$, the complementary function of the foregoing differential equation must be identically zero; while its particular integral becomes equal to

$$xz = \frac{2\delta P_c}{2j+1}\left(\frac{1+\sigma}{1-\sigma}\right)\left\{x^{j+1}\int_0^x x^{1-j}\tau_j(x,t)\,dx - x^{-j}\int_0^x x^{j+2}\tau_j(x,t)\,dx\right\}. \tag{9-93}$$

In order to specify the elastic displacements $u$ and $v$ of spheroidal deformation as given by Equations (9-91) and (9-93) we must fall back on the definitions (9-12)–(9-13) of $y$ and $z$, revealing that

$$x \frac{\partial u}{\partial x} + 2u - j(j+1)v = \delta \left\{ \frac{1+\sigma}{1-\sigma} \right\} x\tau \qquad (9\text{-}94)$$

and

$$\frac{\partial}{\partial x}(xv) - u = xz = \frac{2\delta P_c}{(2j+1)\mu} \left\{ \frac{1+\sigma}{1-\sigma} \right\} \left\{ x^{j+1} \int_0^x x^{1-j}\tau\,dx \right.$$

$$\left. - x^{-j} \int_0^x x^{j+2}\tau\,dx \right\}, \qquad (9\text{-}95)$$

which for ignorable value of the ratio $P_c/\mu$ reduces to

$$\frac{\partial}{\partial x}(xv) = u . \qquad (9\text{-}96)$$

Multiply now Equation (9-94) by $x$, differentiate with respect to $x$, and insert in (9-96): the result will be

$$x^2 \frac{\partial^2 u}{\partial x^2} + 4x \frac{\partial u}{\partial x} + \{2 - j(j+1)\}u = \delta \frac{1+\sigma}{1-\sigma} \frac{\partial}{\partial x}(x^2\tau) \qquad (9\text{-}97)$$

for $u$, the particular integral of which assumes the form*

$$u = \frac{\delta(1+\sigma)}{(2j+1)(1-\sigma)} \left\{ jx^{j-2} \int_0^x x^{-j+1}\tau\,dx + (j+1)x^{-j-3} \int_0^x x^{j+2}\tau\,dx \right\}; \qquad (9\text{-}98)$$

while the corresponding displacement component $v$ then follows similarly from (9-96) as

$$v = \frac{\delta(1+\sigma)}{(2j+1)(1-\sigma)} \left\{ x^{j-1} \int_0^x x^{-j+1}\tau\,dx - x^{-j-2} \int_0^x x^{j+2}\tau\,dx \right\} . \qquad (9\text{-}99)$$

In order to evaluate explicitly these displacement components we should insert on the right-hand sides of the foregoing equations for $\tau_j(x, t)$ from (8-88) and integrate term-by-term – a task of some tediousness. However, in the steady state (i.e., as $t \to \infty$), Equation (8-88) reduces to

$$\tau_j(x) = A_j x^j ; \qquad (9\text{-}100)$$

---

* The reader may note that the complementary function of this equation would be identical with the solution of Clairaut's equation (7-28) defining the deformation of the respective body in its state of equilibrium (the identity becomes obvious when setting $u = xY$); but in our problem this latter deformation is taken to be identically zero.

and if so, Equation (9-98) can be readily integrated to yield

$$u_j(x) = \frac{\delta A_j(1+\sigma)}{(2j+1)(1-\sigma)} \left\{ \frac{j}{2} + \frac{j+1}{2j+3} \right\} x^{j+1}, \qquad (9\text{-}101)$$

where the $A_j$'s are constants of the expansion on the right-hand side of Equation (8-71) some of which have already been evaluated in the preceding chapter (cf. Equations 8-93 to 8-95).

Let us set out now to investigate the effect exerted by this kind of thermal dilatation on the moments of inertia of the lunar globe about the axes of inertia paralled with, and perpendicular to, the direction of incident sunlight – moments which (if we ignore the small inclination $I$ of the lunar equator to the ecliptic) become identical with $A$ and $C$ as defined by Equations (7-46) and (7-48) before. In order to evaluate the difference $C - A$ arising from this source, all we have to do is to return to the expressions (7-48) and (7-46) which define the respective moments and substract, obtaining

$$C - A = \int_0^{r_1} \int_0^{\pi} \int_0^{2\pi} \rho r^4 (\cos^2 \phi \sin^2 \theta - \cos^2 \theta)\, dr \sin \theta\, d\theta\, d\phi, \qquad (9\text{-}102)$$

where, by (9-14) and (9-101),

$$\rho = \rho_0(1 - \varDelta) = \rho_0 \{1 - y S_j^i(\theta, \phi)\} = \rho_0 \{1 - \delta \left( \frac{1+\sigma}{1-\sigma} \right) A_j x^j S_j^i(\theta, \phi)\} \qquad (9\text{-}103)$$

because of thermal expansion of material whose density in isothermal state was equal to $\rho_0$; and, likewise,

$$\begin{aligned} r &= ax + u_j \\ &= ax \left\{ 1 + \frac{\delta A_j(1+\sigma)}{(2j+1)(1-\sigma)} \left[ \frac{j}{2} + \frac{j+1}{2j+3} \right] x^j S_j^i(\theta, \phi) \right\} \end{aligned} \qquad (9\text{-}104)$$

by (9-101).

In as much as

$$\int_0^{\pi} \int_0^{2\pi} (\cos^2 \phi \sin^2 \theta - \cos^2 \theta) S_j^i(\theta, \phi) \sin \theta\, d\theta\, d\phi \quad \begin{aligned} &= \tfrac{4}{5}\pi \quad \text{if} \quad j = 2 \\ &= 0 \quad \text{if} \quad j \neq 2, \end{aligned} \qquad (9\text{-}105)$$

only the second harmonics in (9-103) or (9-104) will contribute to the difference $C - A$. Changing over from $r$ to $x$ in (9-102) as the variable of integration (in the manner of Equation 7-50), we then readily establish that the difference

$$C - A = -\frac{4\pi}{35} \left( \frac{1+\sigma}{1-\sigma} \right) \rho_0 \delta A_2 a_1^5, \qquad (9\text{-}106)$$

leading to the ratio

$$\frac{C - A}{B} = -\frac{\alpha}{14} \left( \frac{1+\sigma}{1-\sigma} \right) A_2. \qquad (9\text{-}107)$$

The numerical values of the elastic constants which occur on the right-hand side of the preceding Equation (9-107) should be close to $\alpha = 2 \times 10^{-5}$ deg$^{-1}$ and $\sigma = 0.27$; moreover from Equation (8-91) we find that $A_2 = -0.0718\,k$, where $k$ represents the mean monthly sub-surface temperature on the lunar equator. Accordingly,

$$\frac{C - A}{B} = 1.9 \times 10^{-7}\,k = 0.000045 \qquad (9\text{-}108)$$

for the observed value of $k = 240\,°$K; and from Equation (9-101) it follows then that

$$u_2(a) = -0.099\,P_2(\cos\theta)\,\text{km}. \qquad (9\text{-}109)$$

This latter equation reveals that the difference between the equatorial and polar semi-axes of insolated Moon should amount to only 149 meters – an amount too small to be verified from existing observations with any significance; while the ratio (9-108) amounts to only about 7% of its value as deduced from lunar librations (cf. Chapter 4); and for no conceivable combination of parameters it can amount to much more.

It should also be stressed that our foregoing result is exact only in the limiting case of steady state – as $t \to \infty$ – when Equation (8-88) reduces to (9-100); and before this steady state is attained, the effect of insolation on the moments of inertia is apt to be smaller. Equation (9-108) can in effect be regarded as the upper limit of the ratio $\beta = (C - A)/B$ which is unlikely to have been attained after $4\frac{1}{2}$ milliard years; and for this reason, it is all the more difficult to see how the directional heating of the lunar surface could have deformed its globe by more than a very few per cent of the extent indicated by the observed value of $\beta$.

In conclusion of the present survey of the elastic properties of the lunar globe, let us ask ourselves the following question: should the Moon get disturbed by a sudden event – whether internal (moonquake) or external (the impact of a meteorite) – what would be the periods of *free oscillations* excited in this manner? The answer to this question is contained in the fundamental equations (9-17) and (9-18), subject to the boundary conditions (9-20) and (9-23)–(9-24), for any motion characterized by spheroidal symmetry. In the case of periodic motion, the displacement components $u(r, t)$ and $v(r, t)$ should, moreover, obey the partial differential equations

$$\frac{\partial^2 u}{\partial t^2} = -n^2 u \quad \text{and} \quad \frac{\partial^2 v}{\partial t^2} = -n^2 v, \qquad (9\text{-}110)$$

where $n = 2\pi/P$ denotes the frequency of the respective oscillation of period $P$. By an insertion of (9-110) in (9-17) and (9-18), the latter two can be treated as a simultaneous system of ordinary differential equations for $u$ and $v$, subject to the boundary conditions (9-20) and (9-23)–(9-24), which can be satisfied only for a certain discrete spectrum of characteristic frequencies $n$ or periods $P$ corresponding to a given spherical-harmonic symmetry $j$.

The principal difficulty of such a problem arises from the need to eliminate between Equations (9-17) and (9-18) the potential perturbing function $R(r)$ as defined by

Equation (9-19); for this can in general (for $j \neq 0$), be done only by a recourse to the fact that $R$ satisfies the differential equation

$$\frac{\partial^2 R}{\partial r^2} + \frac{2}{r}\frac{\partial R}{\partial r} - \frac{j(j+1)}{r^2} R = -\frac{3\rho y}{a^2} + \frac{6\rho}{a^2}\left(\frac{\partial u}{\partial r}\right)$$
$$+ \left\{ r\frac{\partial^2}{\partial r^2}\left(\frac{u}{r}\right) - \frac{j(j+1)}{r^2} u \right\}\frac{\bar{\rho} r}{a^2},$$

(9-111)

where $\bar{\rho}$ denotes again the mean density of a sphere of radius $r$ as defined by Equation (7-37), and $a$ stands for the external radius of our configuration, rendering our boundary-value problem one of *sixth* order, amenable only to numerical integration. Such integrations have recently been carried out by BOLT (1960) or TAKEUCHI, SAITO, and KOBAYASHI (1961), who assumed the Moon to be homogeneous and adopted the constant values $\lambda = \mu = 0.74 \times 10^{12}$ dynes/cm$^2$; while Bolt allowed also for the variation of density due to self-compression (for essentially the same values of the elastic constants). The actual values of free periods of spheroidal oscillations which resulted from such work are listed in the accompaning Table 9-3; these periods are proportional to the

TABLE 9-3

Periods of Free Oscillations of the Lunar Globe (in minutes)

| $j$ | Homogeneous Moon Spheroidal | | Toroidal | | | Self-Compressed Moon Spheroidal | | Toroidal | | |
|---|---|---|---|---|---|---|---|---|---|---|
| 0 | 8.73, | 6.8 | | | | 8.7 | | | | |
| 2 | 14.67, | 8.0, 4.9 | 15.482 | 5.426 | 3.683 | 15.1, | 8.4 | 16.300 | 5.677 | 3.840 |
| 3 | 9.89, | 6.0 | 10.019 | 4.585 | 3.259 | 10.2, | 6.3 | 10.573 | 4.800 | 3.400 |
| 4 | 7.73, | 4.8 | 7.601 | 3.987 | 2.931 | 8.0, | 5.1 | 8.039 | 4.176 | 3.060 |
| 5 | 6.43 | | 6.18 | | | | | | | |
| 6 | 5.51, | 3.6 | 5.23 | | | 5.8 | | 5.540 | 3.339 | 2.694 |
| 7 | 4.84 | | 4.55 | | | | | | | |
| 8 | 4.32 | | 4.02 | | | | | | | |
| 9 | 3.91 | | 3.62 | | | | | | | |
| 10 | 3.57 | | 3.28 | | | | | | | |

square-root of the rigidity, and can accordingly be adjusted for other than the adopted value of $\mu$. A glance at the data presented in Table 9-3 reveals that the fundamental (longest) periods of free oscillations of the Moon characterized by radial symmetry ($j = 0$) are approximately equal to $8\frac{3}{4}$ minutes; while the longest periods of second-harmonic oscillations ($j = 2$) are close to 15 minutes – about three times shorter than they happen to be for our Earth.

The relatively high order of the boundary-value problem defining such periods becomes substantially reduced in the case of radial symmetry ($j = 0$), by virtue of the fact that the function $R(r)$ then follows directly from (9-26). In such a case the order of our problem reduces from six to two; and the corresponding differential equation becomes identical with (9-27) provided that the term $n^2\rho u$ has been added to its left-hand side.

A similar simplification results, for any value of $j$, in the case of *toroidal symmetry*, characterized by the fact that the three components of the displacement vector **u** are represented by

$$
\left.
\begin{aligned}
u_r &= 0, \\
u_\theta &= \frac{v}{\sin\theta}\frac{\partial S_j^i}{\partial\phi}e^{int}, \\
u_\phi &= -v\frac{\partial S_j^i}{\partial\theta}e^{int},
\end{aligned}
\right\}
\tag{9-112}
$$

in place of (9-11). The corresponding torsional motion does not disturb the gravitational potential; for an insertion of (9-112) in (9-5) reveals that $\varDelta = y = 0$ which together with $u = 0$ renders $R(r)$ as defined by Equation (9-19) identically zero. Moreover, the assumed form (9-112) of the velocity components $u_r$, $u_\theta$, $u_\phi$ satisfies also the first one of the fundamental equations (9-6)–(9-8) of motion identically; while the latter two reduce to the same second-order differential equation for $v$, of the form

$$
\mu\left\{\frac{\partial^2 v}{\partial r^2}+\frac{2}{r}\frac{\partial v}{\partial r}\right\}+\frac{\partial\mu}{\partial r}\left\{\frac{\partial v}{\partial r}-\frac{v}{r}\right\}
$$
$$
+\left\{\rho n^2-\frac{j(j+1)\mu}{r^2}\right\}v=0,
\tag{9-113}
$$

subject to the boundary conditions

$$
v=0 \quad\text{at}\quad r=0
\tag{9-114}
$$

and

$$
\frac{\partial v}{\partial r}-\frac{v}{r}=0 \quad\text{at}\quad r=a;
\tag{9-115}
$$

both of which are particular cases of (9-20) and (9-24).

Numerical integrations of this latter boundary-value problem for both the homogeneous and compressed lunar models have been carried out independently by TAKEUCHI, SAITO, and KOBAYASHI (1961), and CARR and KOVACH (1962); and the torsional free periods resulting from their work are likewise listed in Table 9-3. As the reader can easily verify, a difference in periods of the spheroidal and toroidal oscillations corresponding to the same harmonic modes is generally small (i.e. the fraction of a minute). For high values of $j$, the free toroidal oscillations may be regarded as dispersive surface waves (Love waves) just as the high-frequency tail of the spheroidal oscillations degenerate into surface Rayleigh waves.

Only one postscript remains to be added in this place; and that is a reminder that the results of all investigators of this subject, as summarized in Table 9-3, are valid only for *isothermal* oscillations of the lunar globe (or a Moon consisting of material of zero thermal expansion). In fact the structure of Equations (9-6)–(9-8), with the stress-components $\sigma_{ij}$ as given by (9-1)–(9-2), makes it evident that toroidal oscillations are possible only if

$$\frac{\partial}{\partial r}(3\lambda + 2\mu)\delta T = 0;\qquad\qquad(9\text{-}116)$$

and Equation (9-112) governing them will be exact only if, in addition,

$$\frac{\partial}{\partial r}(3\lambda + 2\mu)\delta T = \frac{\partial}{\partial \phi}(3\lambda + 2\mu)\delta T = 0.\qquad\qquad(9\text{-}117)$$

The actual thermal properties of our satellite may still change the free periods of spheroidal as well as toroidal oscillations as listed in Table 9-3 somewhat; but the changes arising from this source are unlikely to be appreciable.

## CHAPTER 10

# POSSIBLE CONVECTION IN LUNAR INTERIOR

In the preceding chapter of this book we have given an outline of the phenomena to be expected in the Moon's interior if its material behaves as an elastic solid. The physical reason why the Moon should behave so at least in the outer parts of its interior are indeed overwhelming; but is this necessarily the case also in the deep interior of our satellite? The relatively high temperatures which we have reasons to expect there as a result of radiogenic heating, as discussed in Chapter 8, entail a number of interesting consequences; and one should be at least a partial *melting* of rocks exposed to them for a sufficiently long time. The problem of the occurrence of melting at pressures encountered in the lunar interior has, in recent years, been discussed in particular by UREY (1962); and indications of the temperatures at which the melting of the silicate rocks should commence (or become complete) under these conditions are shown on Figures 8-2 and 8-4. If so, however, it is reasonable to inquire as to whether such material would be susceptible of actual *hydrodynamical flow* over time intervals comparable with the age of the Moon; but before attempting to answer this question, let us inquire first about a possible cause of such a motion.

As the Moon at present must be very close to the state of mechanical equilibrium (within limitations discussed in the preceding Chapter 7) the only cause of motion could be the thermal instability of its mass; and the necessary condition for this to set in is the requirement that the local temperature gradient be super-adiabatic. Now, as is well known (cf. e.g., JEFFREYS, 1959; p. 288) the adiabatic temperature gradient should be given by the equation

$$\left(\frac{dT}{dr}\right)_{ad} = -\frac{\alpha g T}{C_P},$$ (10-1)

where $\alpha$ denotes the coefficient of thermal volume expansion; $C_P$, the specific heat at constant pressure; and $g$, the gravitational acceleration which for a nearly homogeneous body will (in accordance with Equation (7-4)) be sensibly given by $(4/3)\pi G \rho_m r$. Hence,

$$\left(\frac{dT}{dr}\right)_{ad} = -\frac{4\pi G\alpha}{3C_P}\rho_m r T.$$ (10-2)

Now, for silicate rocks, $\alpha \sim 2 \times 10^{-5}$ deg$^{-1}$; $C_P \sim 7 \times 10^6$ erg/g deg (cf., e.g., BIRCH, 1952); and $\rho_m = 3.34$ g/cm$^3$; rendering

$$\left(\frac{dT}{dr}\right)_{ad} = -(2.7 \times 10^{-8})rT$$ (10-3)

if $r$ is expressed in kms, and $T$ in degrees. At $T=1500$ °K and $r=100$ km from the center of the Moon the lunar adiabatic temperature gradient proves to be $-0.°004$/km; and at $r=1000$ km its value should increase ten fold. On the other hand the conductive temperature gradients as established by Urey and other investigators, and reflected in the slopes of the curves shown on Figures 8-1 to 8-4 are, on the average, of the order of $-1°$/km – i.e. ten to a hundred times larger.

So great a disparity between the adiabatic and conductive temperature gradients leaves but little room for doubt that if, as a result of radiogenic heating, any part of the lunar interior becomes molten, convection currents are bound to arise, which should transport heat in addition to conduction or radiation. Moreover, in order that this be true, it is not necessary for actual melting of the rock to take place; for convection can likewise arise in a visco-elastic medium of finite viscosity – no matter how large – provided only that

$$(dT/dr)_{ad} \ll (dT/dr)_{conv} \qquad (10\text{-}4)$$

represents a sufficiently strong inequality, and that the time scale of the flow is sufficiently long in comparison with the Maxwellian relaxation time

$$t_* \sim \mu\beta, \qquad (10\text{-}5)$$

where $\mu$ is now the coefficient of viscosity of the respective medium, and $\beta$, the coefficient of isothermal compression as defined by Equation (9-68). For silicate rocks of density comparable with the mean density of the Moon, $\beta \simeq 10^{-12}$ cm$^2$/dyne (BIRCH, 1952); and the maximum value of $\mu$ consistent with the geological evidence bearing on the motions in the outer crust of the Earth appears to be of the order of $10^{22}$ g/cm sec. If it is permissible to adopt, by analogy the same viscosity for the Moon, the value of the product $\mu\beta$ should not exceed $10^{10}$ sec or some $10^3$ years.

To impulses lasting for a time $t < t_*$ – such as a moonquake or the impact of a meteorite or a rocket – the Moon should, therefore, react essentially as an elastic solid. However, when subject to the action of forces lasting for $t \gg t_*$ – such as the gravitational or thermal stresses – the Moon (as well as other terrestrial planets) should behave as a visco-elastic globe; and when we consider the fact that the age of the Moon is of the order of $10^7 t_*$, the possibility is at least indicated that, over such long intervals of time the bulk of the mass of the Moon may obey the laws of hydrodynamics rather than of elasticity.

In order to estimate, in such a case, the relative importance of convection for the energy transport, let us return to our fundamental equation (8-2) and attempt to estimate the numerical magnitudes of the terms III and IV on its right-hand side in comparison with the terms I and II. In doing so, let us (again consistent with the laboratory measurements for silicate rocks) adopt for the quantities $\kappa$ and $\varepsilon$ the values of $10^5$ erg/cm sec deg and $10^{-8}$ erg/g sec respectively. The magnitude of the term I should clearly be of the order of $\kappa T/R^2$ or $10^{-8}$ erg/cm$^3$ sec for $T=10^3$ deg and a scale-length $R=1000$ km $=10^8$ cm of the order of magnitude of the dimensions of the lunar globe; and if so, the product $\rho\varepsilon$ constituting the term II is then of the same order of magnitude as I.

The term III representing the heat transport by conduction is likely to be of the order of $UP/R$ where $U$ denotes the (average) velocity of convective motion. Under which conditions are the terms I and III likely to be of comparable magnitude in lunar interior? Equating $\kappa T/R^2$ with $UP/R$ for an average pressure of $10^{10}$ dynes/cm$^2$ (cf. Equation 7-25) we find this to be the case for $U \sim \kappa T/PR$ or $10^{-10}$ cm/sec – or about a centimeter for every thousand years. A velocity so low can already render convection and conduction of comparable importance for heat transport in the lunar interior; and for higher velocities it is the convection which may predominate.

Several geological phenomena (continental drift; post-glacial uplift, etc.) on the Earth indicate crustal velocities, possibly of convective origin of the order of 1 cm/year (cf., e.g., PEKERIS, 1935). If such velocities were present in the Moon – and this remains so far a hypothesis – then not only would the term III dominate the right-hand side of Equation (8-2), being about 1000 times larger than I+II; but even the viscous dissipation term IV, of the order of $\mu(U/R)^2$, should become as important as I or II.

A hypothetical convective flow in lunar interior would be bound to affect also the density distribution inside the Moon – if alone by the effect of varying temperature on the rock density through thermal expansion. Moreover, as the up-coming material must come, down somewhere else the flow is bound to be non-radial; and as such it can affect the moments of the lunar globe along its principal axes of inertia. Since the latter – or, rather, their differences – constitute the most valuable information provided so far by observations which bears on the internal structure of our satellite; and since, in particular, the anomalously large value of the ratio $(C-A)/B$ has defied so far reasonable explanation in terms of any established physical process, we wish now to inquire into the possibility that it may arise as a consequence of slow convection currents operative in the plastic part of the lunar interior.

In order to do so, let us depart from the fundamental hydrodynamical equations of viscous flow, expressible vectorially as

$$\rho \frac{D\mathbf{u}}{Dt} = \rho\nabla V - \nabla P + \tfrac{4}{3}\nabla(\mu\nabla\cdot\mathbf{u}) + \nabla(\mathbf{u}\cdot\nabla\mu)$$
$$- \mathbf{u}\nabla^2\mu + \nabla u\times(\nabla\times\mathbf{u}) - (\nabla u)\nabla\mu - \nabla\times(\nabla\times\mu\mathbf{u}), \qquad (10\text{-}6)$$

where $\mathbf{u}$ denotes the velocity vector (with components $u$, $v$, $w$); $\rho$, the density; $P$, the pressure; $V$, the gravitational potential; $\mu$, the coefficient of viscosity; and

$$\frac{D}{Dt} = \frac{\partial}{\partial t} + \mathbf{u}\nabla \qquad (10\text{-}7)$$

stands for the Lagrangian time-derivative (following the motion). The Eulerian equations (10-6) of motion safeguard the conservation of momentum of the respective dynamical system. The conservation of mass is ensured by the equation

$$\frac{D\rho}{Dt} + \rho\nabla\cdot\mathbf{u} = 0 \qquad (10\text{-}8)$$

of continuity, which together with (10-6) and the Poisson equation

$$\nabla^2 V = - 4\pi G\rho \tag{10-9}$$

constitutes a system of five equations for six unknowns

$$u, v, w ; \quad \rho, P, V .$$

One more equation – safeguarding the conservation of energy – remains to be added to render the problem determinate; and this we propose to do at a later stage of our analysis.

In order to study the nature of the viscuous flow phenomena governed by the foregoing fundamental equations (10-6)–(10-9) in spherical polar coordinates $r$, $\theta$, $\phi$, let us assume that the density $\rho$, pressure $P$, and gravitational potential $V$ characterizing the internal structure of our configuration can be expressed as

$$\rho = \rho_0(r) + \rho'(r, \theta, \phi; t), \tag{10-10}$$

$$P = P_0(r) + P'(r, \theta, \phi; t), \tag{10-11}$$

$$V = V_0(r) + V'(r, \theta, \phi; t), \tag{10-12}$$

where $\rho_0$, $P_0$, and $V_0$ describe the respective properties of our configuration in its stationary (equilibrium) state, and $\rho'$, $P'$, $V'$ stand for their changes brought about by motion with the velocity components $u$, $v$, $w$. Moreover, let the coefficient of viscosity $\mu$ be hereafter regarded as a function of $r$ only.

The functions with zero subscript, describing as they do the equilibrium structure of the respective configuration, will satisfy the equation

$$\frac{dP_0}{dr} = \rho_0\left(\frac{\partial V_0}{\partial r}\right) = - g\rho_0 \tag{10-13}$$

where, in accordance with (7-2), the gravitational acceleration

$$g = G\frac{m(r)}{r^2} = \frac{4\pi G}{r^2}\int_0^r \rho_0 r^2\, dr . \tag{10-14}$$

If we assume next that the primed functions $\rho'$, $P'$ and $V'$ as well as the velocity components $u$, $v$, $w$ are small enough for their squares and cross-products to be negligible, the Eulerian equations (10-6) can evidently be linearized to take the forms

$$\rho\frac{\partial u}{\partial t} = \rho\frac{\partial V'}{\partial r} - \frac{\partial P'}{\partial r} - g\rho' + \frac{\mu}{r}\left\{\nabla^2(ru) - 2\varDelta + \frac{r}{3}\frac{\partial \varDelta}{\partial r}\right\} + 2\frac{\partial \mu}{\partial r}\left\{\frac{\partial u}{\partial r} - \frac{\varDelta}{3}\right\}, \tag{10-15}$$

$$\rho\frac{\partial v}{\partial t} = \frac{1}{r}\left(\rho\frac{\partial V'}{\partial \theta} - \frac{\partial P'}{\partial \theta}\right) + \mu\left\{\nabla^2 v + \frac{2}{r^2}\frac{\partial u}{\partial \theta} - \frac{v}{r^2 \sin^2\theta} - \frac{2\cos\theta}{r^2 \sin^2\theta}\frac{\partial w}{\partial \phi}\right\}$$

$$+ \frac{\mu}{3r}\frac{\partial \varDelta}{\partial \theta} + \frac{\partial \mu}{\partial r}\left\{\frac{\partial v}{\partial r} + \frac{1}{r}\frac{\partial u}{\partial \theta} - \frac{v}{r}\right\}, \tag{10-16}$$

and

$$
\rho \frac{\partial w}{\partial t} = \frac{1}{r \sin \theta} \left\{ \rho \frac{\partial V'}{\partial \phi} - \frac{\partial P'}{\partial \phi} \right\} + \mu \left\{ \nabla^2 w + \frac{2}{r^2 \sin \theta} \frac{\partial u}{\partial \phi} + \frac{2 \cos \theta}{r^2 \sin^2 \theta} \frac{\partial v}{\partial \phi} \right.
$$
$$
\left. - \frac{w}{r^2 \sin^2 \theta} \right\} + \frac{\mu}{3r \sin \theta} \frac{\partial \Delta}{\partial \phi} + \frac{\partial \mu}{\partial r} \left\{ \frac{\partial w}{\partial r} + \frac{1}{r \sin \theta} \frac{\partial u}{\partial \phi} - \frac{w}{r} \right\},
\tag{10-17}
$$

where the zero subscript of $\rho_0$ has been dropped (there should be no danger of confusion), and the abbreviation

$$
\Delta = \frac{1}{r^2} \frac{\partial}{\partial r} (r^2 u) + \frac{1}{r \sin \theta} \left\{ \frac{\partial}{\partial \theta} (v \sin \theta) + \frac{\partial w}{\partial \phi} \right\}
\tag{10-18}
$$

used for the divergence of the velocity vector. Moreover, the equation (10-8) of continuity and the Poisson's equation (10-9) can be similarly linearized to yield

$$
\frac{\partial \rho'}{\partial t} + u \frac{\partial \rho}{\partial r} + \rho \Delta = 0
\tag{10-19}
$$

and

$$
\nabla^2 V' = - 4 \pi G \rho' ,
\tag{10-20}
$$

respectively.

Now take the divergence of equations (10-15)–(10-17) – i.e., operate on them, successively, by

$$
\frac{\partial}{\partial r} + \frac{2}{r}, \quad \frac{1}{r} \frac{\partial}{\partial \theta} + \frac{\cot \theta}{r} , \quad \frac{1}{r \sin \theta} \frac{\partial}{\partial \phi}
$$

and add; on insertion from (10-19) and (10-20) the outcome reveals that

$$
\nabla^2 (\rho V' - P') = V' \nabla^2 \rho + \frac{1}{r^2} \frac{\partial}{\partial r} (r^2 g \rho') + \frac{\partial \rho}{\partial r} \frac{\partial V'}{\partial r} - \frac{\partial^2 \rho'}{\partial t^2}
$$
$$
- \frac{4}{3} \mu \nabla^2 (\Delta) - 2 \frac{\partial \mu}{\partial r} \left\{ \nabla^2 u - \frac{2\Delta}{3r} + \frac{1}{3} \frac{\partial \Delta}{\partial r} + \frac{1}{r^3} \frac{\partial}{\partial r} (r^2 u) \right\}
\tag{10-21}
$$
$$
- 2 \frac{\partial^2 \mu}{\partial r^2} \left\{ \frac{\partial u}{\partial r} - \frac{\Delta}{3} \right\}
$$

$$
\frac{\partial}{\partial r} (\rho V' - P') = g \rho' + \frac{\partial \rho}{\partial r} V' + \rho \frac{\partial u}{\partial t} - \frac{\mu}{r} \left\{ \nabla^2 (r u) - 2\Delta + \frac{r}{3} \frac{\partial \Delta}{\partial r} \right\}
$$
$$
- 2 \frac{\partial \mu}{\partial r} \left\{ \frac{\partial u}{\partial r} - \frac{\Delta}{3} \right\} .
\tag{10-22}
$$

In order to eliminate the term $\rho V' - P'$ between the foregoing equations operate on (10-21) with $r \nabla^2 (r...)$ on (10-22) with $\partial / \partial r (r^2 ...)$ and subtract: by virtue of the identity

$$
\frac{\partial}{\partial r} (r^2 \nabla^2 ...) = r \nabla^2 \left( r \frac{\partial}{\partial r} ... \right)
\tag{10-23}
$$

the term containing $\rho V' - P'$ obviously drops out, and the result reveals that

$$
r\nabla^2 (rg\rho') - \frac{\partial^2}{\partial r^2}(r^2 g\rho') = \frac{\partial}{\partial r}(r^2 V' \nabla^2 \rho) + \frac{\partial}{\partial r}\left(r^2 \frac{\partial \rho}{\partial r}\frac{\partial V'}{\partial r}\right)
$$

$$
- r\nabla^2 \left(r\frac{\partial \rho}{\partial r}\frac{\partial V'}{\partial r}\right) - \frac{\partial}{\partial r}\left(r^2 \frac{\partial^2 \rho'}{\partial t^2}\right) - r\nabla^2\left(r\rho\frac{\partial u}{\partial t}\right)
$$

$$
+ \mu r\nabla^4 (ru) - \frac{\mu}{r^2}\frac{\partial}{\partial r}\{r^4 \nabla^2 (\Delta)\} \tag{10-24}
$$

$$
+ 2\frac{\partial \mu}{\partial r}\left\{\frac{\partial}{\partial r}[r\nabla^2 (ru)] - \frac{\partial}{\partial r}\left[\frac{1}{r}\frac{\partial}{\partial r}(r^2 u)\right] - r^2 \nabla^2 (\Delta) - \frac{\partial}{\partial r}(r\Delta)\right\}
$$

$$
- r\frac{\partial^2 \mu}{\partial r^2}\left\{\nabla^2 (ru) - 2\frac{\partial}{\partial r}\left[\frac{1}{r}\frac{\partial}{\partial r}(r^2 u)\right] + r\frac{\partial \Delta}{\partial r}\right\}.
$$

By making use of Equation (10-14) governing $g$ we can, however, show that

$$
r\nabla^2 (rg\rho') - \frac{\partial^2}{\partial r^2}(r^2 g\rho') = L(g\rho') = gL(\rho'), \tag{10-25}
$$

where $L$ denotes the operator

$$
L = \frac{1}{\sin \theta}\frac{\partial}{\partial \theta}\left(\sin \theta \frac{\partial}{\partial \theta}\right) + \frac{1}{\sin^2 \theta}\frac{\partial^2}{\partial \phi^2}; \tag{10-26}
$$

in consequence of which Equation (10-24) can be reduced to

$$
L(g\rho' + \rho\phi V') = -\frac{\partial^2}{\partial t^2}\left\{\frac{\partial}{\partial r}(r^2 \rho')\right\} - r\nabla^2\left(r\rho\frac{\partial u}{\partial t}\right)
$$

$$
+ \mu r\nabla^4 (ru) + \frac{\mu}{r^2}\frac{\partial}{\partial r}\left\{r^4 \nabla^2\left[\phi u + \frac{1}{\rho}\frac{\partial \rho'}{\partial t}\right]\right\}
$$

$$
+ 2\frac{\partial \mu}{\partial r}\left\{\frac{\partial}{\partial r}\left[r\nabla^2 (ru) - r\Delta - \frac{1}{r}\frac{\partial}{\partial r}(r^2 u)\right] - r^2 \nabla^2 (\Delta)\right\} \tag{10-27}
$$

$$
- r\frac{\partial^2 \mu}{\partial r^2}\left\{\nabla^2 (ru) - 2\frac{\partial}{\partial r}\left[\frac{1}{r}\frac{\partial}{\partial r}(r^2 u)\right] + r\frac{\partial \Delta}{\partial r}\right\},
$$

where we have abbreviated

$$
\phi = \frac{1}{\rho}\left(\frac{\partial \rho}{\partial r}\right). \tag{10-28}
$$

If the flow were steady, and $\mu$ constant, Equation (10-27) would reduce to

$$
L(g\rho' + \rho_0 \phi V') = \mu r\nabla^4 (ru) + \frac{\mu}{r^2}\frac{\partial}{\partial r}\{r^4 \nabla^2 (\phi u)\}; \tag{10-29}
$$

and if, in addition, the equilibrium configuration were homogeneous (i.e., $\phi = 0$), this would further reduce to

$$gL(\rho') = \mu r \nabla^4 (ru) \qquad (10\text{-}30)$$

– an equation previously used in this connection by CHANDRASEKHAR (1952).

The foregoing partial differential equation, of fourth order, safeguards the conservation of mass and momentum; but not yet of energy. Moreover, it contains two dependent variables – $\rho'$ and $u$ – one of which must be eliminated before the solution of (10-30) can proceed. This can indeed be done by an appeal to the principle of the conservation of energy as embodied in our Equation (8-2); and if we follow WASIUTYNSKI (1946) and CHANDRASEKHAR (1952) in an assumption that the principal vehicle of energy transport in our problem be conduction, this equation reduces (for constant $\kappa$) to

$$\rho C_v \frac{DT}{Dt} = \kappa \nabla^2 T + \rho \varepsilon. \qquad (10\text{-}31)$$

In order to linearize it, let us – in the manner of equations (10-10)–(10-12) decompose the temperature $T$ as

$$T = T_0(r) + T'(r, \theta, \phi; t), \qquad (10\text{-}32)$$

where $T_0$ describes the temperature distribution of our configuration in the equilibrium state, and $T'$, the temperature change due to motion, is a small quantity of first order. If so, Equation (10-31) can again be linearized to yield

$$\rho_0 C_v \left( \frac{\partial T'}{\partial t} + u \frac{\partial T_0}{\partial r} \right) = \kappa \nabla^2 T' + \rho' \varepsilon, \qquad (10\text{-}33)$$

where $T_0$ results as a solution of the equation

$$\kappa \nabla^2 T_0 + \rho_0 \varepsilon = 0 \qquad (10\text{-}34)$$

in the form

$$T_0(r) - T_0(0) = \frac{\varepsilon}{\kappa \rho_0} \int_0^r \left\{ \frac{1}{r} \int_0^r \rho_0 r \, dr \right\} dr, \qquad (10\text{-}35)$$

which for constant $\rho_0$ and the boundary condition $T_0(a) = 0$ clearly reduces to

$$T_0(r) = \frac{\rho_0 \varepsilon}{6\kappa} (a^2 - r^2). \qquad (10\text{-}36)$$

In consequence,

$$\frac{\partial T_0}{\partial r} = -\left( \frac{\rho_0 \varepsilon}{3\kappa} \right) r \qquad (10\text{-}37)$$

which, inserted in (10-33), yields

$$\left( \frac{\rho_0 C_v \varepsilon}{3\kappa} \right)(ru) + \kappa \nabla^2 T' + \rho' \varepsilon = 0. \qquad (10\text{-}38)$$

If, moreover, the term $\rho' \varepsilon$ on the left-hand side is ignored as small, we may use (following Chandrasekhar) the preceding equation as a relation between $u$ and $T'$ in the form

$$ru = \frac{3\kappa}{\rho_0 C_v \varepsilon} \nabla^2 T'.$$ (10-39)

As the next step, let us return to Equation (10-30) which, on insertion from (7-4) for $g$ appropriate for a homogeneous configuration, assumes the form

$$L(\rho') = \frac{3\mu}{4\pi G \rho_0} \nabla^4 (ru).$$ (10-40)

If, moreover, we appeal to a piezotropic equation of state relating the changes in the density $\rho'$ and temperature $T'$ by

$$\rho' = -\rho \alpha T',$$ (10-41)

where $\alpha$ stands for the coefficient of volume thermal expansion of the respective molten material (as defined by Equation 9-59), a combination of (10-40) with (10-41) yields

$$L(T') = -\frac{3\mu}{4\pi G \alpha \rho_0^2} \nabla^4 (ru).$$ (10-42)

Let us next operate on this latter equation by $\nabla^2$. As the operators $\nabla^2$ and $L$ are clearly commutative,

$$\nabla^2 [L(T')] = L[\nabla^2 (T')] = -\frac{\rho_0^2 C_v \varepsilon}{3\kappa^2} L(ru)$$ (10-43)

by use of (10-39), an elimination of $T'$ between (10-42) and (10-43) yields, as the final result of our analysis, a sixth-order partial differential equation

$$L(ru) = \frac{9\kappa^2 \mu}{4\pi G \alpha \varepsilon \rho_0^2 C_v} \nabla^6 (ru)$$ (10-44)

for $u$, first obtained by WASIUTYNSKI (1946), containing the clue to the solution of the entire problem.

The boundary conditions which this solution must satisfy are quite analogous to those met in the preceding section in connection with our treatment of the elastic Moon. First, we require, in conformity with (9-9), that

$$u = v = w = 0$$ (10-45)

at the center; while the outer boundary conditions at $r = a$ depend on the nature of the bounding surface. If the latter were rigid and subject to no slips, the same condition (10-45) must be fulfilled (thus providing the necessary complement of six boundary conditions). If, in addition, the bounding surface is constrained to remain spherical, all angular derivatives of the velocity components must vanish there as well; and since, moreover, for a homogeneous configuration the equation (10-8) of continuity reduces to

$$\Delta = \frac{\partial u}{\partial r} + \frac{2u}{r} + \frac{1}{r}\frac{\partial v}{\partial \theta} + \frac{v \cot \theta}{r} + \frac{1}{r \sin \theta}\frac{\partial w}{\partial \phi} = 0, \tag{10-46}$$

it follows from it that, on the surface $r = a$,

$$\frac{\partial u}{\partial r} = 0 \tag{10-47}$$

or, more generally,

$$\frac{\partial}{\partial r}(ru) = 0 \tag{10-48}$$

as well.

On the other hand, if the bounding surface is free but spherical, we still have the right to require that $u(a) = 0$; but in place of the vanishing of the other velocity components on the surface we require that, instead, in conformity with (9-10) the vanishing of the viscuous stress components

$$\sigma_{r\theta}(a) = \sigma_{r\phi}(a) = 0, \tag{10-49}$$

equal to the respective elastic stress components as given by Equations (9-2), in which the coefficient $\mu$ of rigidity has been identified with that of viscosity.* Since $u$ as well as its angular derivatives vanish on a spherical bounding surface, Equations (10-49) reduce to

$$\frac{\partial v}{\partial r} - \frac{v}{r} = \frac{\partial w}{\partial r} - \frac{w}{r} = 0, \tag{10-50}$$

which inserted in (10-46) yield

$$\frac{\partial^2}{\partial r^2}(ur) = 0 \tag{10-51}$$

on a free surface, replacing (10-48) valid if the surface were rigid. This form of the boundary conditions of our problem has first been laid down by JEFFREYS and BLAND (1951); though the construction of the actual solution of Equation (10-44) – due to Wasiutynski – subject to them has not been carried out till in 1952 by Chandrasekhar.

In order to solve this partial differential equation, Chandrasekhar assumed the existence of separable solutions of the form

$$ru = \sum_{j=0}^{\infty} W_j(r) S_j^i(\theta, \phi), \tag{10-52}$$

where the $S_j^i$'s denote the surface spherical harmonics satisfying Equation (8-59), and $W_j(r)$ is a function of $r$ only. Similarly, the solution for the corresponding perturbation

---

* In doing so we should, however, note that the physical dimensions of the respective quantities so denoted are different in each case: whereas the coefficient of rigidity is of the dimension of dyne/cm² = g/cm sec², that of viscosity possesses the dimension of g/cm sec.

$T'$ in temperature was assumed to exist in the form

$$T' = \sum_{j=0}^{\infty} \Theta_j(r) S_j^i(\theta, \phi), \qquad (10\text{-}53)$$

and variational methods employed to seek the appropriate forms of $W_j(r)$ and $\Theta_j(r)$. In doing so, Chandrasekhar found that

$$\Theta_j(r) = \frac{3\mu A_j}{4\pi G \alpha \rho_0^2 a^4} \frac{J_{j+\frac{1}{2}}(m_1 x)}{j(j+1)\sqrt{x}} \qquad (10\text{-}54)$$

and

$$W_j(r) = (A_j/m_1^4)\{x^{-\frac{1}{2}}J_{j+\frac{1}{2}}(m_1 x) + q m_1 (x^j - x^{j+2})J'_{j+\frac{1}{2}}(m_1)\}, \qquad (10\text{-}55)$$

where $x \equiv r/a$, $J_{j+\frac{1}{2}}(x)$ denotes a Bessel function of fractional index ($J'_{j+\frac{1}{2}}$ being its first derivative with respect to $x$); $m_1$ is the first root of the equation

$$J_{j+\frac{1}{2}}(x) = 0, \qquad (10\text{-}56)$$

$A_j$ stands for an arbitrary scale constant; and

$$q = \tfrac{1}{2} \quad \text{or} \quad -\frac{1}{2j+1}, \qquad (10\text{-}57)$$

depending on whether the boundary at $x=1$ is rigid or free. Moreover, a resort to a two-term minimizing function revealed that the foregoing equations (10-54) and (10-55) should represent the actual solutions of our problem within errors not exceeding a few per cent.

   In the light of these results, let us come now to investigate the effect which an internal convection as described by Chandrasekhar's solution could exert on the moment of inertia of a sphere consisting of homogeneous liquid of viscosity $\mu$ radioactively heated within. The density changes caused by thermal expansion of this liquid follow from a combination of (10-41) and (10-53) with (10-54). Inserting for them in (9-102) we readily find that, for a spherical configuration,

$$\left.\begin{aligned}
C - A &= -\frac{a\mu A_2}{8\pi G \rho_0} \int_0^1 \frac{x^4 J_{5/2}(m_1 x)\,dx}{\sqrt{x}} \\
&\quad \times \int_0^\pi \int_0^{2\pi} (\cos^2 \phi \sin^2 \theta - \cos^2 \theta) P_2(\cos \theta) \sin \theta \, d\theta \, d\phi \\
&= \frac{a\mu A_2}{10 G \rho_0} \int_0^1 \frac{x^4 J_{5/2}(m_1 x)\,dx}{\sqrt{x}},
\end{aligned}\right\} \qquad (10\text{-}58)$$

since the second harmonic $S_2(\theta, \phi) \equiv P_2(\cos \theta)$ alone makes a nonvanishing contribution. Moreover, as Bessel functions of half-integral index are expressible in terms of circular functions,

$$
\int_0^1 x^{7/2} J_{5/2}(m_1 x)\,dx = \frac{1}{m_1^4}\left(\frac{2}{\pi m_1}\right)^{\frac{1}{2}} \int_0^{m_1} \{(3 - y^2) y \sin y - 3y^2 \cos y\}\,dy
$$

$$
= \frac{1}{m_1^4}\left(\frac{2}{\pi m_1}\right)^{\frac{1}{2}} \{3(5 - 2m_1^2) \sin m_1 - m_1(15 - m_1^2) \cos m_1\}
$$

(10-59)

which for $m_1 = 5.76345920\ldots$ becomes equal to $0.055020$; and, in consequence, the corresponding value of

$$
\frac{C - A}{B} = \frac{3\mu A_2}{16\pi G\rho_0^2 a^4} \times 0.05502 = (4.838 \times 10^{-29})\mu A_2
$$

(10-60)

for $G = 6.668 \times 10^{-8}$ cm$^3$/g sec$^2$, $\rho_0 = 3.34$ g/cm$^2$, and $a = 1.738 \times 10^8$ cm.

Let us, for the sake of argument, that the whole of the observed constant $(C - A)/B$ $= 0.000628$ as obtained in Chapter 4 is due to this process. If so, the foregoing equation (10-60) would then require that

$$
\mu A_2 = 1.298 \times 10^{25} \text{ erg/cm}.
$$

(10-61)

The actual value of the viscosity $\mu$ in this equation is difficult to specify with any accuracy; but a broad analogy with certain terrestrial phenomena (post-glacial uplift of Fenno-Scandia, continental drift?) which may be of visco-elastic origin, indicate that $\mu$ is of the order of $10^{22}$ g/cm sec or, if anything, less (cf., e.g., JEFFREYS, 1962, pp. 343–347 or 366). If so, however, it would follow from (10-61) that $A_2 = 1.3 \times 10^3$ cm$^2$/sec, which inserted in (10-52) together with (10-55) reveals that the radial velocity $u$ of the corresponding convective flow would be of the order of

$$
u \sim \frac{A_2}{rm_1^4} \sim 10^{-8} \text{ cm/sec}.
$$

(10-62)

Velocities of this order of magnitude would lead to displacements of the order of 1 mm per year, corresponding to a flow length of the order of $10^8$ cm or one lunar radius during the entire age of the Moon! Such velocity would be scarcely adequate for the establishment of any kind of steady state (postulated in Chandrasekhar's underlying theory) in $10^9$ years – the Moon is just too large a globe for that – and the temperature difference between the top and the bottom of the convective columns requisite for the maintenance of this motion follows from (10-53) and (10-54) to be equal to

$$
\left. \begin{aligned} \Delta T' &= \frac{\mu A_2}{16\pi G\alpha\rho_0^2 a^4} \sqrt{\frac{2m_1}{\pi}} S_2(\theta, \phi) \\ &= 36° S_2(\theta, \phi), \end{aligned} \right|
$$

(10-63)

for $\alpha = 2 \times 10^{-5}$ deg$^{-1}$, corresponding to a mean temperature gradient of $-0.002$ deg/km.

In order to increase the velocity of circulation and establish the steady state within lunar life-time, it would be necessary to increase the value of the coefficient $A_2$ by at least one, and possibly two orders of magnitude; but to make this again consistent with the observed value of $(C-A)/B$, the viscosity $\mu$ would have to be diminished in the same proportion. But in this effort we run into the following impasse. It has been shown by CHANDRASEKHAR (1952) that the solution of equation (10-44) can be reconciled with the boundary conditions (10-45) and (10-49) only for certain discrete values of a nondimensional characteristic parameter

$$C_j = \frac{4\pi G\rho_0^4 \alpha \varepsilon a^6 C_v}{9\kappa^2 \mu},$$  (10-64)

which Chandrasekhar called the "Rayleigh number" and which, for $j=2$, was found to assume the value of 5224 if the outer boundary is free, and 10403 if it is rigid.

However, when we insert in the foregoing formula the values of $\alpha = 2 \times 10^{-5}$ deg$^{-1}$, $\kappa = 2 \times 10^5$ erg/cm sec deg, $C_v = 7 \times 10^6$ erg/g sec – all of which represent fair averages for silicate materials – together with $\varepsilon = 2 \times 10^{-8}$ erg/g sec (a value consistent with the data compiled in Table 8-1) while $\mu \sim 10^{22}$ g/cm sec, equation reveals that the empirical value of

$$C_2 = 2.2 \times 10^6$$  (10-65)

is several hundred times as large as those required by the theory to render the corresponding steady flow stable. Moreover, any diminution in the estimated order of magnitude of the viscosity $\mu$ – a move which would tend to increase the velocity of the corresponding convection currents – would be bound to increase the disparity between the theoretical and empirical values of the Rayleigh number by the same factor – an increasingly embarrassing situation! So large a value of $C_2$ as indicated by the above equation (10-65) could, in the light of Chandrasekhar's theory, correspond only to convection patterns characterized by very high order of spherical-harmonic symmetry $(j>22)$, regardless of whether the bounding surface is rigid or free. The zero curves of the harmonics $S_j(\theta,\phi)$ for $j>22$ are, in turn, so closely packed as to delimit, on the lunar surface, areas comparable with those of smaller circular maria. Whether or not there is indeed a causal relation between these phenomena remains still an open problem; the older literature on this subject has been ably discussed by WASIUTYNSKI (1946); and our present numerical results cannot but strengthen the case for giving this problem further consideration. But it appears very difficult to accept that, even if the Moon were convective inside, the corresponding flow could not possess any harmonic components of low orders; and, to this extent, the anomalously large value of $(C-A)/B$ still continues to defy attempts at rational explanation.

Is there any escape from this conclusion by permissible alteration of the physical constants entering the Rayleigh number (10-64), or any other way? Of the constants involved, the values of $\alpha$ as well as $C_v$ appear to be sufficiently well known from laboratory rock studies that at least their order of magnitude should be considered as fixed. The value of $\kappa$ is perhaps less well established; and that of $\varepsilon$ still less so. However an increase of $\kappa$ by not less than two orders of magnitude, or a diminution of $\varepsilon$ by four orders would be required to harmonize the characteristic Rayleigh number $C_j$ with a viscosity of the order of $10^{22}$ (not to speak of smaller viscosities, needed for the establishment of steady state, for which still larger adjustments in the same direction would be necessary). Whether or not so large a change in $\kappa$ or $\varepsilon$ may indeed be supported by independent evidence, only future investigations can decide.

However, our failure to reconcile the consequences of Chandrasekhar's theory of convective Moon with the observed value of $(C-A)/B$ may, possibly, go back to his use of the equation (10-41) for the change of state, relating the changes in density with those of the temperature alone. In a more general case, we should expect that

$$\rho' = \left(\frac{\partial \rho}{\partial T}\right)_P T' + \left(\frac{\partial \rho}{\partial P}\right)_T P' + \ldots \tag{10-66}$$

i.e., that

$$\frac{\rho'}{\rho_0} = -(\alpha T_0)\frac{T'}{T_0} + (\beta P_0)\frac{P'}{P_0} + \ldots , \tag{10-67}$$

where the coefficients $\alpha$ and $\beta$ of volume thermal expansion and isothermal compression are as previously defined by Equations (9-59) and (9-68).

For average silicate rocks, $\alpha = 2 \times 10^{-5}$ deg$^{-1}$ and $\beta = 2 \times 10^{-12}$ cm$^2$/dyn. Therefore, at moderate pressures (such as encountered, for instance, in the terrestrial laboratory experiments), the product $\beta P_0$ is likely to be quite negligible in comparison with $\alpha T_0$, and the use of (10-41) as our equation of state should then be justified; indeed, it was in this connection that valid use of (10-41) was originally made by RAYLEIGH (1916). However, in the bulk of the lunar interior we expect that $T_0 \sim 1000$ °K, while $P_0 \sim 10^{10}$ dynes/cm$^2$ (cf. Chapter 7), in which case both products $\alpha T_0$ and $\beta P_0$ turn out to be of the same order of magnitude (i.e., 0.02). Under such conditions (obtaining in planetary bodies of mass even as small as that of the Moon), the retention of any one of the terms on the right-hand side of (10-67) and neglect of the other may constitute a rather dubious procedure, and its consequences should still be viewed with caution.

In conclusion of the present section denoted to a discussion of possible fluid properties of our satellite, we wish to mention one additional flow phenomenon which, under certain conditions, may provide a new significant source of heat for the lunar interior: namely by viscous dissipation of bodily tides raised in the lunar globe by the attraction of the Earth. In Chapter 6 we pointed out that if the Moon could be regarded as a homogeneous fluid, the Earth would raise tides on it, whose dominent second-harmonic component would attain on the surface an altitude of $33P_2(\cos \phi \sin \theta)$ meters – approximately 37 times as high as the lunar tides would be on Earth in spite of the smaller fractional size of our satellite – and even if their actual height amounts to (as on the Earth) only 24% of their theoretical equilibrium value appropriate for a fluid, the amplitude of the lunar second-harmonic tides should still be close to 7.9 meters.

If the relative orbit of the Moon around the Earth were circular, and its axial rotation synchronised with the revolution, these tides would be stationary and generate no motion relative to the Moon's center. However, the finite eccentricity $e = 0.0549$ of this orbit should immediately give rise to two kinds of motion:

(a) radial motion, due to the rising and sinking of matter as the height of the tide (proportional, for the second harmonic, to the inverse cube of the radius-vector of the lunar orbit) varies periodically in the course of each month; and

(b) lateral motion, due to the fact that the tides oscillate periodically in their position with the difference between the lunar angular velocity of rotation and revolution; and, in addition, to the lag caused by the viscosity of the lunar material.

Of these two, the first type of motion should be much larger and will alone be hereafter considered.

The amplitude of the radial oscillation due to this monthly "breathing" of the mass of the Moon administered to it by the Earth should, for a $j$-th harmonic partial tide, be equal to $(j+1) e$ times its equilibrium height, which for the second-harmonic tide in a homogeneous Moon becomes equal to $130 (r/a)$ cm, anywhere in its interior, and should vary as $\cos v P_2 (\cos \phi \sin \theta)$, where $v$ denotes the Moon's true anomaly in its relative orbit. Moreover, since the period of this gravitational breathing is equal to $27.32166$ days $= 2.36 \times 10^6$ sec, the velocity $u$ of the corresponding motion should be given by

$$\left. \begin{aligned} \frac{u}{x} &= \frac{4 \times 130}{2.36} \times 10^{-6} \cos v P_2 (\cos \phi \sin \theta) \\ &= 2.20 \times 10^{-4} \cos v P_2 (\cos \phi \sin \theta) \text{ cm/sec}, \end{aligned} \right\} \tag{10-68}$$

where, as before, $x \equiv r/a$. It may be noted that, on the surface $(x=1)$, this velocity would correspond to some 69 meters/year – which is at least by four orders of magnitude greater than the hypothetical convection velocity (10-62); and in the face of such a situation, the following question becomes of cardinal importance: Does the Moon respond to a varying strain invoked by the monthly Earth-raised tides as an elastic or plastic body? In the former case, the bodily tides would not accomplish any net work in the course of an orbital cycle; but, in the latter case, a tidal-effective viscosity would bring about a degradation of motion into heat – a process which would influence also the thermal equilibrium of lunar interior, in accordance with the consequences of the conservation of energy.

At the present time, a definite answer to this question is still lacking. An indication of it may, however, be obtained by a comparison of the monthly period of disturbing force with the Maxwellian relaxation time $t_*$ as defined by Equation (10-5). For silicate rocks (of density close to the mean density of the Moon) the value of $\beta$ occurring in it is known to be of the order of $10^{-12}$ cm$^2$/dyne; but estimates of the appropriate values of $\mu$ are still largely uncertain. The velocities of tectonic motions observed in the Earth's crust are indicative of values of $\mu$ as high as $10^{22}$ g/cm sec; while laboratory measurements of the viscosity of specific rocks cluster around $\mu \sim 10^{18}$ g/cm sec, or even less. The former value, combined with the relatively well-known quantity $\beta$, leads to a time $t_*$ of the order of $10^{10}$ second or 300 years; the latter, to $10^6$ seconds or about 10 days. If the former applies to the Moon as a whole, the response of the lunar globe to a monthly tide should be essentially elastic and entail no dissipation of energy.* On the other hand, a viscosity of the order of $10^{18}$ g/cm sec would lead to a relaxation time $t_*$ short enough to permit visco-elastic flow dissipating a part of the energy of motion into heat.

In order to ascertain the actual rate at which heat could thus be produced, let us return to the fundamental Equation (8-2) for the conservation of energy, which in the case of viscous dissipation as the heat source assumes the more specific form

$$\rho C_v \frac{DT}{Dt} = \kappa \nabla^2 T + \mu \Phi , \tag{10-69}$$

where, as is well known, the dissipation function

$$\Phi = \tfrac{1}{2}(\sigma_{rr}^2 + \sigma_{\theta\theta}^2 + \sigma_{\phi\phi}^2) + \sigma_{r\theta}^2 + \sigma_{r\phi}^2 + \sigma_{\theta\phi}^2 , \tag{10-70}$$

can be expressed as a quadratic form of the components $\sigma_{ij}$ of the viscous stress tensor (obtainable from those of the elastic stress tensor if the Lamé parameter $\lambda = 0$, and $\mu$ identified with the coefficient of viscosity). If we do so in Equations (9-1)–(9-2) and, moreover, limit ourselves to the radial velocity component (ignoring the shear stresses), we find the only nonvanishing components of the viscuous stress tensor to become

$$\sigma_{rr} = 2 \frac{\partial u}{\partial r} , \quad \sigma_{\theta\theta} = \frac{2u}{r} = \sigma_{\phi\phi} , \tag{10-71}$$

while

$$\sigma_{r\theta} = \frac{1}{r} \frac{\partial u}{\partial \theta} \quad \text{and} \quad \sigma_{r\theta} = \frac{1}{r \sin \theta} \frac{\partial u}{\partial \phi} , \tag{10-72}$$

which for $u$ as given by Equation (10-68) on insertion in (10-70) yield

$$\Phi = \tfrac{3}{2}k^2 (3 \cos^4 \phi \sin^4 \theta + 1) \cos^2 v , \tag{10-73}$$

where we have abbreviated

$$k = \frac{2.20 \times 10^{-4}}{1.738 \times 10^6} = 1.27 \times 10^{-10} \text{ sec}^{-1} . \tag{10-74}$$

Moreover, the average value of $\cos^4 \phi \sin^4 \theta$ over the whole sphere is equal to 107/315 (or, very approximately, one-third), while the average of $\cos^2 v$ over an orbital cycle is one-half. Therefore, if we wish to investigate the secular variations of global temperature in lunar interior generated by viscous dissipation of tidal motion, it should be sufficient to replace, in Equation (10-69), the function $\Phi$ by its average constant value of

$$\Phi_{av} = \tfrac{3}{2}k^2 ; \tag{10-75}$$

---

* Dissipation of tidal energy due to imperfect elasticity was recently studied by KAULA (1963) and found to be small.

and, by the same argument, replace the operator

$$\frac{D}{Dt} \equiv \frac{\partial}{\partial t} + u \frac{\partial}{\partial r} \tag{10-76}$$

on the left-hand side of (10-69) by $\partial/\partial t$, as the value of $u$ (proportional, to a first approximation to the term $e \cos v$) averages out to zero in the course of each cycle.

Equation (10-69) thus simplified assumes the explicit form

$$\rho C_v \frac{\partial T}{\partial t} = \frac{\kappa}{r^2} \left( r^2 \frac{\partial T}{\partial r} \right) + \tfrac{3}{2}\mu k^2 \tag{10-77}$$

and its particular solution, subject to the boundary conditions requiring that

$$T(r,0) = T(a,t) = 0, \tag{10-78}$$

assumes (cf. KOPAL, 1962) the form

$$T(r,t) = \frac{3k^2\mu}{\rho C_v} \sum_{n=1}^{\infty} (-1)^{n+1} \left( \frac{1 - e^{-K(n\pi/a)^2 t}}{K(n\pi/a)^2} \right) \frac{\sin n\pi x}{n\pi x} \tag{10-79}$$

where, as before, the thermal diffusivity coefficient $K$ continues to be given by Equation (8-5).

Numerical computations based on this equation (for the details of which the reader is referred to KOPAL, 1963b) revealed that, for the adopted values of $C_v = 7 \times 10^6$ erg/g deg and $\kappa = 2 \times 10^5$ erg/cm sec deg, the viscous dissipation of bodily Earth tides into heat leads to the generation of temperatures inside the lunar globe which, for $\mu = 10^{17}$ g/cm sec, are of the order of 100 °K; and for $\mu = 10^{18}$, of the order of 1000 °K (i.e., comparable with the temperatures attained by a gradual release of radiogenic heat). For $\mu < 10^{17}$, the amount of heat produced by tidal dissipation would be too small to be of cosmogonic importance. On the other hand, for $\mu > 10^{18}$, the Maxwellian relaxation time $t_*$ would become long in comparison with one month; and, in consequence, the lunar response to the Earth tides would be of the nature of a forced oscillation of an elastic body, which could produce heat only due to imperfections of elasticity; and this effect, as was recently shown by KAULA (1963), is quite small. However, if the actual value of the effective viscosity happens to be of the order of $10^{18}$ g/cm sec or around it (as is suggested indeed by terrestrial experiments with silicate rock materials), the mechanical heating of the Moon through bodily tides throughout its long past could, by itself, have raised the temperature of its interior to a level comparable with that expected from the radiogenic heating alone. Moreover, as the efficiency of this process is proportional to the orbital eccentricity $e$, and inversely proportional to the cube of the distance between the Earth and the Moon, the amount of internal heat produced by tidal dissipation could have been greater in the past if the Moon was closer to the Earth, or its orbit more eccentric; also the shearing stresses (ignored in 10-73) could have made a significant – or even dominant – contribution at a time when the rotation of the Moon was not synchronised with its revolution.

Could this heating, coupled with the accumulation of radiogenic heat, have ever melted the Moon completely, and thus have brought it unquestionably within the domain of hydrodynamical treatment? A definitive answer cannot again be given until more is known about the possible role, in the lunar past, of certain short-lived radioactive isotopes – such as iodine $I^{129}$, palladium $Pd^{107}$ or, in particular, aluminum $Al^{26}$. The half-lives of these radioactive isotopes are of the order of $10^6$–$10^7$ years; therefore, if any of these had been present in the initial mass of the Moon, they would by now be virtually extinct. However, their decay products (such as $Xe^{129}$ or $Ag^{107}$) seem to occur in meteoritic matter in abundances which raise seriously the question of former radioactivity of their mother substances (cf. e.g., FOWLER, GREENSTEIN, and HOYLE, 1962; CAMERON, 1962; MARSHALL, 1962). A reasonable abundance of them in the primordial mass of the Moon might indeed have been capable of melting its mass completely in the first few million years of its existence, before such long-lived radioactive elements as $K^{40}$, $Th^{232}$, or the two isotopes of uranium had time to exert any appreciable effect.

UREY (1960) argued against such a possibility on grounds of the fact that, if the Moon were ever completely molten, $4\frac{1}{2}$ thousand million years would not have been long enough for it to solidify to its present state. This would indeed be true if the cooling were to take place by the conduction of heat only. In actual fact, however, a Moon completely molten by the action of radiogenic heat sources would almost certainly have been rather violently convective; and convection currents would drain off internal heat much more effectively than conduction alone would have been able to accomplish. In other words, even if the Moon melted completely at some initial stage of its evolution by the heating of short-lived radioactive elements like $I^{126}$, $Pd^{107}$, or $Al^{26}$, convective cooling should have enabled to solidify again in a time-span comparable with the period of heating – so that, at any rate, the surface of our satellite must have been solid throughout the greater part of its astronomical past; while the internal temperature has been slowly rising as a result of spontaneous disintegration of the long-lived radioactive elements like $K^{40}$, $Th^{232}$, $U^{235}$ or $U^{238}$, or possibly of "gravitational heating" described earlier in this chapter.

One piece of indirect evidence bearing, possibly, on lunar convection may be mentioned in this place: and that is the unequal distribution of maria over the entire globe of the Moon. It has been noted long ago that the distribution of continental land masses and ocean floors on our Earth is distinctly antisymmetric – the land-mass of Eurasia facing the Pacific ocean – a fact which is being interpreted by VENING-MEINESZ (1959), RUNCORN (1962) and others as an indication of convective flow inside the Earth's mantle. Ever since the first photographs of the far side of the Moon were obtained by the Russian investigators (cf. MIKHAILOV et al., 1960) it has become immediately obvious that the same type of antisymmetry prevails also on the Moon to an even greater extent, with virtually all maria concentrated on one (the visible) hemisphere. Now if the terrestrial continent-ocean asymmetry is indeed due to convection in the Earth's mantle, the same is likely to be true on the Moon as well, for analogous reasons. As we mentioned already, the floors of small circular maria might, possibly, represent the tops of sub-surface convective cells once reaching deep in the interior; but much more work remains to be done before this tentative suggestion could be placed on a more secure basis.

## CHAPTER 11

# CHEMICAL COMPOSITION OF THE MOON

In conclusion of the second part of this book, concerned with the internal structure of our satellite, a few words should be added on the probable chemical composition of the lunar globe – a difficult subject to approach with any confidence at a distance, when most of its discussion has to rest on indirect arguments. Yet such arguments should enable us already at this stage to draw certain probable conclusions which should facilitate our discussion of certain topics in subsequent chapters. For this reason, as well as for the sake of demonstration of certain lines of deductive reasoning to which one must often resort in the studies of celestial objects at a distance we shall, in what follows, develop this topic as far as can be done on the eve of a space age in which direct acquaintance with the object of our study will largely supersede deductions made on theoretical grounds.

First, let it be mentioned that in the present state of astronomical research it appears probable that the Moon – in common with most other bodies of the solar system – originated by an agglomeration of solid particles of small sizes at relatively low temperatures – decidedly lower than those required for complete volatilization. It does not seem possible to envisage a workable physical process which could lead to the formation of planetary bodies – let alone of so small a mass as that of the Moon – by condensation of gas at high temperature. Therefore, the view prevalent to-day assumes that the original mass of the Moon accumulated from pre-existent solid particles some $4.6 \times 10^9$ years ago (a value currently accepted for the age of the solar system, and based on laboratory determinations of relatively abundances of certain short-lived radioactive elements and of their decay products in meteoritic debris collected by the Earth), and that the internal temperature of this original agglomeration (which may have formed, under favourable conditions, in a relatively short time) was not more than a few hundred degrees Kelvin.

Under these conditions, the global chemical composition of the Moon should be essentially that of the non-volatile component – predominantly silicates – of the primordial matter from which the solar system originated, and whose unadulterated sample accessible to quantitative chemical (spectroscopic) analysis we find at present in the solar atmosphere: unadulterated because the large mass of the Sun prevented any selective escape of elements from its gravitational field even at the relatively high temperature prevailing in the solar atmosphere; and because (unlike in the interior) this temperature is still far too low to have caused nuclear transformations on any appreciable scale. When we take a sample of the gases now found in the solar atmosphere, allow for the escape of its lighter elements by relaxing the gravitational field to that generated by the lunar mass, and condense the rest, the density of the resulting matter should be close to 3.3 g/cm³ – i.e., close indeed to the mean density of the lunar globe – but should, according to UREY (1958) contain only between 11–14% of iron by weight. This, however, is less than a half of the current estimates of the iron content of the Earth (28%) and poses an interesting dilemma: either the Moon is of approximately solar composition as regards iron while the Earth (and other terrestrial planets) possess a distinctly higher concentration of this important element – or, if its iron content is the same as for the Earth, the lunar mass must contain an anomalously high proportion of some sufficiently common low-density substance (such as 2–3% of water, or 10% of graphite). Surprises may be in store when the first samples of lunar rocks are returned by rockets to Earth; but Urey considers it probable that the Moon contains substantially less iron than the terrestrial planets.

That this is indeed so may be further indicated by the virtual absence of any lunar *magnetic field*. The results of the magnetometer experiment aboard the Russian Lunik II in September 1959 indicated that the strength of such a (dipole) field – if any – does not exceed some $30\gamma$ (i.e., 0.0003 gauss).* Although the process of the generation of planetary magnetic fields is still far from fully understood, it appears that two conditions must be fulfilled: namely, rapid axial rotation of a metallic core. The Earth, rotating in one day and containing an iron core which extends up to 0.54 of the planetary radius, possesses a magnetic field of several tenths of a gauss. The planet Venus possesses probably a similar core as the Earth, but rotates but slowly (cf. GOLDSTEIN and CARPENTER, 1963); and, hence, generates but a weak magnetic field (cf. SMITH and others, 1963). The Moon rotates slowly; and, in addition, its relatively low density precludes the existence of a metallic core of any appreciable dimensions – hence, no wonder that no magnetic field of it has been recorded, or is likely to be in any appreciable amount.

A disparity in the relative iron content of the Earth and the Moon, as indicated by the foregoing evidence, may point to a marked difference in the chemical processes which preceded the formation of the lunar and planetary globes (the same would, indeed, be equally true if they were to be excess water or carbon in the Moon), rendering our satellite a very special object of considerably different origin. It is true that earlier theories, attempting to create the Moon by disruption of a liquid Earth by solar tides, brought about by resonance between the duration of the day (i.e., the period of solar semi-diurnal tides) and that of free nonradial oscillations of the primordial terrestrial globe, ran more recently against insuperable difficulties (due mainly

---

* This measurement was made at a distance of approximately one lunar radius above its sunlit hemisphere; and as was pointed out subsequently by NEUGEBAUER (1960), this evidence is insufficient to confine the intensity of a magnetic field of the Moon below this limit on its actual surface; for if such a field would prevail there, the existence of the solar wind would confine it to a relatively thin zone (the field strength falling off to less than 10% of its surface value at an altitude of 1 km) above the sunlit side – though it could extend to much greater height above the night hemisphere of the Moon (which is shielded from the solar wind). The reason is the fact that, if a reasonably weak magnetic field (of the strength of $100\gamma$ or less) existed on the lunar surface, the charged particles of the solar wind deflected by it would generate a field tending to mask it.

to high viscosity of the terrestrial matter) and had virtually been abandoned. The main weight of contemporary opinion inclines now to regard the origin of the Earth and of the Moon as unrelated events – and possibly not even simultaneous. Indeed – Urey argues – the Moon may possibly be cosmically older than the Earth, and represent the Earth's foundling rather than a true-born child; but when and by what means our planet may have acquired its only natural satellite remains as yet largely conjectural.

But let us return to our schematic chemical model of the Moon, consisting of predominantly silicate material (similar to the rocks forming the outer mantle of our own planet), whose most abundant elements (cf. UREY, 1958) should be O, Si, Mg, and Fe, followed by S, Al, Ca, Na, Ni, and Cr in approximately that order. The secular increase in internal temperature, discussed in Chapter 8, should then entail several interesting consequences; and one should be a gradual *desiccation* and degrassing of its interior. If the Moon originated – as we believe – by an accumulation of initially cold matter of average cosmic composition, this matter was bound to contain also a finite concentration of certain volatile compounds (such as $H_2O$, $SO_2$, etc.) which could, at low temperatures, have been absorbed in the crystal lattice of many minerals, and thus be present in the new-born Moon in solid state until liberated by secularly rising temperature.

The case of *water* is particularly interesting, because of its relatively high proportion (up to 0.1 per cent) in stony meteorites and of the high cosmic abundance of its constituent elements as well as great stability of its molecule. In order to estimate the probable amount of juvenile water in the mass of the Moon, broad analogy with the Earth may serve a useful purpose. The total amount of water in all oceans of the Earth is such as to be capable of submerging our entire globe under a uniform layer of approximately 1800 meters in depth; and if all this water was released by uniform desiccation of the whole mass of the Earth (equal to $5.98 \times 10^{27}$ g), each gram of this mass would have had to contribute $1.54 \times 10^{-4}$ grams of juvenile water. If the same proportion were to hold for the Moon as well, its smaller mass of $7.35 \times 10^{25}$ g should have liberated an amount of juvenile water which, if driven out completely to the surface, would be sufficient to cover the lunar globe with an ocean of a uniform depth of some 300 m.

We know from the physical chemistry of solids that, at temperatures above 1000 °K, water should be thermally expelled from almost all solid hydrates to form superheated steam (though high pressure is known to inhibit this process). The partial pressure of steam at 1000° or even 1500 °K is, to be sure, too small a fraction of the hydrostatic pressure prevailing in the bulk of the Moon's mass, to enable the steam to make its escape through molten magma under its own power as "bubbles"; and molecular diffusion of steam through such a magma constitutes an extremely slow process (of rate comparable with that of thermal conduction). The same should, however, be true of secular desiccation of our own planet; but if the latter managed to exude its oceans – so could, presumably, the Moon, in an analogous manner. Such a process must undoubtedly have been slow; and the original steam in its outward course would have condensed into liquid and eventually freeze in the sub-surface layer – beyond the reach of the diurnal heat wave – where a constant temperature of about $-35\ °C$ prevails. Therefore, unlike on the Earth where atmospheric thermal regulation enables the ocean water to remain in liquid state, hypothetical lunar oceans should be expected to remain frozen underground (for whatever fraction of it would have reached the surface, would have rapidly evaporated at diurnal temperatures and permanently escaped from the Moon).

Another point of interest may be added in this connection. Water dissolves easily in molten silicates, and lowers their melting point markedly. Hence, thermal cracking of hydrates should stimulate the formation of lava in the interior, and thus contribute to the mobility of lunar rocks. In isolated pockets – especially closer to the surface – such processes may have provided foci for surface phenomena which might be broadly termed as "volcanic". Whether or not there is evidence for such phenomena on the lunar surface constitutes a topic which we shall touch upon in the next part of this book.

Along crustal fissures which could serve as ducts facilitating more rapid escape, the formation of ice may, in fact, have locally occurred very close to the actual surface and given rise to morphological phenomena which could suggest the presence of sub-surface lunar glaciers to the external observer. Possible evidence for this will be examined again in the next part of this book, when we shall come to inspect the visible face of our satellite and try to understand the nature of the processes which may have been shaping it in the past. Whatever the case may be, however, there seems little doubt that, as a result of secular rise in internal temperature caused by radiogenic heating, the bulk of the Moon's mass must have been largely despoiled of water in the past $4\frac{1}{2}$ milliard years, which must have enriched the outer crust of the lunar globe, or escaped into space.

Another interesting group of volatile compounds, which should be mentioned in this connection, are the *hydrocarbons*. We noted earlier that the chemical composition of lunar primordial matter was probably close to that of the present-day solar atmosphere without its light volatiles – which should bring it close to the composition of chrondritic meteorites – one class of which (the carbonaceous chondrites) are known to contain several per cent of carbon. In carbonaceous chondrites which found their way into terrestrial laboratories and were chemically analyzed, as much as half of this carbon was found to occur in the form of hydrocarbons of high molecular weight – a fact which need not cause any surprise in view of the relatively high cosmic abundance of the constituent elements.

Let us assume, for the sake of argument, that the mass of the Moon is indeed chemically akin to chondritic meteorites, of which 3–4% are carbonaceous chondrites. If so, about 0.01% of this mass – representing a total of $7 \times 10^{15}$ tons – should be in the form of hydrocarbons; and if such a mass were cracked (thermally) to liquid form with an efficiency of only 10%, it could surround the entire globe of the Moon with a uniform cover to a depth of over 20 meters. Lighter hydrocarbons would presumably escape again into space; while the remainder could possibly survive as a local skin of vacuum-reduced crude oil or asphalt. Suggestions that such deposits may form a part of the surfaces of the lunar maria has actually been made (WILSON, 1962); but their merits have yet to await closer scrutiny.

Is it completely ruled out that we possess, on the Earth, any samples of actual lunar material which could have been "spilled over" from the Moon and intercepted by our Earth in the course of this long symbiosis in the past? A considerable amount of recent discussion has centered, in this connection, around the problem of the so-called *tektites* – glass-like substances, found in certain severely limited areas of the Earth's surface (australites, phillipinites, moldavites, etc.) and exhibiting certain chemical anomalies (excessive desiccation), quite unlike any rocks found on the Earth, suggesting anomalous – possibly extra-terrestrial – origin. A hypothesis that tektites might constitute material from the lunar surface removed spasmodically by meteoritic impacts was proposed by NININGER (1943); and found, more recently, strong circumstantial support from CHAPMAN's studies (1960, 1961) of the ablation patterns exhibited by many tektites, indicating that such particles have very probably undergone two distinct heating cycles sufficiently intense for partial melting of the constituent material; and that the second period of heating occurred during a rapid passage through air with a velocity of 11–12 km/sec. As the latter is practically identical with the velocity of free fall on the Earth from the Moon, a possibility of the lunar origin of the tektites is thereby strongly indicated.

On the other hand, this hypothesis has been vigorously challenged by UREY (1957, 1958, 1960), who proposed an alternative theory which arises to account for the observed tektite characteristic by cometary impacts on Earth. The lunar origin of tektites, the possibility of which has been greatly strengthened by Chapman's work, is being questioned mainly on dynamical grounds: namely, by the need to account for severe localization of the tektide finds on Earth – which the cometary hypothesis can explain in a more natural manner. At present the argument is still undecided; and its final settlement may have to await a chemical exploration of the lunar surface in situ by spacecraft. However, it remains at least a possibility that long before the first particles of the terrestrial matter (in the form of the Russian 1959 Lunik 2) were transferred to the Moon by human action, an immigration in the opposite direction – from the Moon to the Earth – repeatedly took place by natural processes in the past. The tektites in our musea may represent, indeed, the samples of lunar matter which was thrown out of the Moon at the time of the formation of large impact craters like Copernicus or Tycho; and even though a residence of many millions of years in different parts of the Earth may have earned the tektites the right of local naturalization, the vestiges of their extra-terrestrial origin appear to be still sufficiently well marked not to deceive a careful investigator.

Whatever the origin of the tektites will eventually proved to be, that some material from the Moon has come down to us on the Earth as a result of mechanical disturbances of the lunar surface by major external impacts seems very probable. In the next part of this book (Chapters 17–18) we shall demonstrate that a very large part of the lunar formations commonly called 'craters' originate by primary impacts of extraneous bodies (meteorites, asteroids, comets); and that, as a result of such impacts a large amount of lunar rocks (which may exceed a hundred times the mass of the cosmic intruder) will be thrown out from the locus of impact in all directions with a wide range of velocities. Moreover, the velocities of escape from the gravitational field of the Moon (or the Earth-Moon system) being as low as given in Chapter 6, it is not only possible, but very probable, that some part of the material disturbed by a primary impact will be thrown out of the Moon in solid state – henceforward to revolve as separate particles around the Sun until such time as may be picked up by another body with which their respective orbits may intersect – and of these the likeliest (by virtue of its relatively large cross-section) is, of course, our Earth.

How to identify such lunar "spill-over" which must undoubtedly have come down to us on the Earth from time to time? Some years ago, UREY (1959; 1965) pointed out that such bodies may be identical with certain types of stony meteorites – in particular, the carbonaceous chondrites – characterised by low cosmic-ray ages of the order of a few million years. Such ages are quite at variance with ages of the order of $10^8-10^9$ years established for the majority of metallic meteorites, which are believed to have had their origin in gradual fragmentation of particles in the asteroidal belt between Mars and Jupiter. Now ages of the order of 1–10 million years are manifestly too short for an asteroidal particle to have spiralled down to a distance of one astronomical unit from the Sun by secular perturbations of their orbits due to cumulative planetary action; and their origin must, therefore, be sought "nearer home".

Urey pointed out – and in this reasoning many more investigators are ready to follow him – that, if we eliminate the asteroidal belt as the locus of origin of short-lived stony meteorites on grounds of their low age, then (apart from a possible import by comets) from where else such particles could have come but from the Moon?* Thus it is quite possible – nay, probable – that we already possess numerous samples of genuine lunar surface materials, sufficiently large for careful analysis, in advance of the manned exploration of the Moon; and the fact that some of these samples seem to contain traces of organic material, unlikely to be due to subsequent terrestrial contamination (of e.g., UREY, 1965 and references quoted therein; or ORO, WIKSTROM and BARGHOORN, 1965) renders their analysis a problem of supreme interest. For if organic material is present in surface layers of our satellite – and the reader has to weigh for himself the chemical evidence as presented or recorded in the above-quoted sources, it may well mean that life was brought down to the Earth originally from the Moon – and that contamination of the Earth by organic material, the gradual evolution of which eventually culminated in mankind on this planet, may have come from the outside.

* It may be recalled, in this connection, that some recent evidence provided by micrometeorite detectors aboard certain artificial Earth satellites indicates that an overspill of lunar dust may annually reach us on Earth each November, when the Earth-Moon system encounters the retrograde stream of Leonid meteorites at a relative velocity close to 70 km/sec. The havoc caused on the lunar surface, unprotected by any atmosphere from the bombardment of the energetic Leonids, seems such as to cause a significant increase in counts of secondary particles reaching the Earth with velocities close to 11 km/sec (i.e. that of free fall from the Moon to the Earth), after a time close to that of such a fall following the onset of the primary Leonids.

Where and when could this have happened? Although Earth-bound transport of the material containing traces of organic matter may still occur at times even at present, once life took hold on the surface of our planet its subsequent re-acquisition may have become unimportant. The greatest probability of acquisition would have obviously been at a time of closest approach of the Earth and the Moon – possibly during a brief entry of the Moon inside the terrestrial Roche limit – which according to the present extrapolations of the tidal evolution of the Earth-Moon system (cf. Chapter 6) should have occurred about 2000 million years ago. Moreover, we have seen in Chapter 6 that, not long before that time, the Moon could have been captured by the Earth by a "free-bound" transition from its previous role of an interplanetary (or even interstellar) tramp. If so – and the chemical reasons (in particular, the Earth-Moon disparity in iron contents pointed out by Urey and discussed earlier in this chapter) do seem to indicate its different origin – it is possible that life may have originally been brought to the Earth from great distances – even such as separate us from the nearest stars. This is, however, not the place to follow such speculations in more detail – but in view of such possibilities the results of the manned exploration of the lunar surface should be expected with all the greater anticipations.

## CHAPTER 12

## EXOSPHERE OF THE MOON

The relative smallness of the mass of the Moon and the low velocity of escape from its gravitational field, together with the probable chemical composition of its globe as discussed in the preceding chapter, entail several further important consequences; and perhaps the most important one for an understanding of lunar surface features is the well-nigh complete *absence of any atmosphere*, which would protect this surface from a direct contact with the outer space.

Why should a self-gravitating astronomical body of planetary size possess an atmosphere? While, for the terrestrial planets of masses comparable with that of the Earth, their present atmospheres may constitute mixtures of primordial gases with those liberated from their interiors by essentially thermal processes in the course of their long cosmic past, it is most unlikely that a body as small as the Moon could have permanently retained any primordial gas; so that any atmosphere which it could possess would be regenerated (or possibly accreted). In order to appreciate the reason why this should be so, let us recall that the continued existence of an atmosphere around any celestial body and its composition testifies to the extent of a stalemate between two opposing tendencies: the attraction of the central body which weighs on each gas molecule in the same way as on any macroscopic object, and will prevent the escape of all those whose velocity $v$ is less than the parabolic velocity $v_{esc} = \sqrt{2GM/a}$; while, on the other hand, the heat pumped in our gas by the Sun (as well as by the surface of the respective planet) maintains the kinetic energy of the gas particles and thus keep the atmosphere distended.

In general, the frequency distribution $N$ of molecules in a gas at a temperature $T$ is governed by the well-known Maxwellian law

$$N(v) = \left(\frac{m}{2\pi kT}\right)^{\frac{3}{2}} \exp\left(-\frac{mr^2}{2kT}\right), \tag{12-1}$$

where $m$ denotes the mass of the respective molecules, and $k = 1.3805 \times 10^{-16}$ erg/deg stands for the Boltzmann constant. Only such fraction of the total number of particles will be able to escape for which

$$v^2 \geqslant \frac{2GM}{a} = 2ga ; \tag{12-2}$$

but since the mean value of $v^2$ results from (12-1) as

$$\bar{v}^2 = \frac{3kT}{m} , \tag{12-3}$$

it follows that the balance of escape will depend on the particular values of $a$ and $g$ as well as $T$ and $m$. If, in particular, $t$ denotes the time during which the density of an (isothermal) atmosphere diminishes to $1/e = 37\%$ of its initial value is equal (cf., e.g. SPITZER, 1949) to

$$t = \frac{\sqrt{6\pi\bar{v}^2}}{3g} \frac{e^Y}{Y}, \qquad (12\text{-}4)$$

where

$$Y = \frac{3}{2}\left(\frac{V_{esc}}{\bar{v}}\right)^2. \qquad (12\text{-}5)$$

When these results are applied to the Moon, for which $g = 167$ cm/sec$^2$ and $\bar{v}_{esc} = 2.38$ km/sec, and where the temperature $T$ oscillates (cf. Chapter 20) roughly between $+400$ °K at noon-time and $+100$ °K during the night, we find that the lightest gases – hydrogen and helium – should disappear from the lunar environment almost immediately. For a hypothetical atmosphere consisting of hydrogen (atomic or molecular) the time $t$ as defined by Equation (12-4) turns out to be approximately 125 minutes in daytime, and about 215 min (3.6 hrs) during the lunar night. A helium atmosphere would similarly dissipate in some 3.6 hours of daylight, or 1.4 year at night. Atomic oxygen or water vapour already take years for dispersal at daytime, and $10^{16}$ years at continuing night-time conditions. Molecular oxygen ($O_2$) could outlast in lunar daylight millions of years; and $CO_2$ still longer. But the Moon's age of $4.5 \times 10^9$ years is so great that none of the common gases could survive exposure to sunlight and remain attached to it by more than a fraction of its age; the substantially longer lifetimes at night are not really relevant in this connection – except, perhaps, in small and isolated regions near the poles of the Moon, where the Sun may never directly illuminate the surface.

Besides, some of those just mentioned (oxygen, for instance) are so reactive that they would not stay long in free state anyway if in contact with a solid surface, but would form compounds in the surface layer of solid rocks. In all, the rate of dissipation into space (or the formation of solid compounds on the surface) of all but the heaviest (or inert) gases – which are again cosmically very scarce – is so high on the Moon that we should not expect to find any appreciable permanent atmosphere around it; and this expectation has indeed been come out by all aspects of the observational evidence available to us so far.

Perhaps the most direct observation which suggested the absence of any air on the Moon already to the early students of its surface was its general telescopic appearance and perpetual absence of any visible clouds over its face. Clouds have been observed (and photographed) on Mars; but although the Moon is more than a hundred times nearer, no instance of a visible cloud has ever been reported by any observer since the discovery of the telescope (and about the only thing which was not: for otherwise meteor impacts, volcanic outbursts, flourishing cities, or even the existence of bridges were announced from time to time). The parts of the Moon near the edge of its apparent disk (which would be seen through the greatest depth of its hypothetical atmosphere) are as clearly visible as the center, without the least obscuration or limb darkening. All shadows on the Moon appear to be perfectly dark, with no evidence of atmospheric diffusion mitigating their stark outlines – though all of them must, of course, be bordered by a penumbra, due to the fact that the solar disk (which to an observer on the Moon would appear as large as we see it on Earth) rises or sets gradually above or below the lunar horizon.

Moreover, no evidence of *refraction* of light has ever been found at the Moon's limb whenever our satellite places itself between us and any more distant object. During an eclipse of the Sun, the outline of the solar limb remains completely free from any distortion where the Moon intercepts it. Furthermore, whenever on its apparent journey through the sky the Moon happens to occult a star, the latter's light is seen to vanish instantaneously (or as suddenly as is permissible by the laws of diffraction and the angular diameter of this star), and does not fade away gradually as it should if it were dimmed (selectively) by extinction in any perceptible lunar atmosphere.

This absence of refraction constitutes by itself a sufficiently stringent test to enable us to conclude that the density of a hypothetical lunar atmosphere – if any – above the surface must be less than one part in ten thousand of the terrestrial air density at sea level. In more recent years, this upper limit has further been lowered by repeated but so far fruitless quests for an indication of *twilight phenomena*, which should be produced in a hypothetical lunar atmosphere during sunrise or sunset.

In order to appreciate more fully the power of this method, consider the light reaching us, say, from any spot near the central part of the apparent lunar disk at the time of the first quarter just before sunrise, when the ground is still immersed in darkness but the space above it already receives the first rays of the rising Sun. This light should, in general, consist of three parts:

1. the light from the lunar surface as illuminated by the Earth;

2. the light of the Sun scattered in the direction of the line of sight in a hypothetical lunar atmosphere; and

3. the moonlight scattered in the terrestrial atmosphere in the direction of the observer.

In the efforts to separate the component (2) – which alone bears on the presence of any gas around the Moon – from the parasitic (but much more intense) diffuse light arising from the sources (1) and (3), Nature lends us fortunately a helpful hand: for the light scattered on gas in a direction which is perpendicular to that of its original incidence should be noticeably polarised, while the diffuse light arising from the lunar as well as the sky background should be essentially free from polarisation. A search for the manifestation of a hypothetical lunar atmosphere reduces, therefore, to the detection of a polarised component in the diffuse light reaching us from an element of surface area of the Moon just before sunrise near the first quarter.

Actual measurements of this type were first performed with sufficient accuracy by V. G. FESSENKOV, whose results were published in 1943. The outcome was again negative; no trace of polarisation was detected; and the precision of the experiment was such as to lead Fessenkov to conclude that the density of a hypothetical lunar atmosphere – if any – must be less than one part in a million (or $10^{-6}$) of that of the terrestrial atmosphere at sea level. If, incidentally, a hypothetical lunar atmosphere were as dense as $10^{-4}$ of the terrestrial one, the twilight zone illuminated by the Sun at the time of the new Moon would endow the lunar disk with an aureola whose light would, in fact, be more intense than the earthlight on the dark side of the Moon and thus be readily detectable – which it certainly is not.

In more recent years, this limit has been further lowered by Bernard LYOT, who, by examining the light reaching us from the regions just beyond the cusps of a lunar crescent eliminated the effect of earthlight represented by (1). By making his observations from the lofty height of the Observatory on Pic-du-Midi: in the French Pyrenées, through clear atmosphere renowned for its superlative seeing, and using the coronograph (a telescope specially designed by him to minimize the amount of light scattered in the instrument itself), Lyot was in a position to detect much smaller traces of polarisation in the light of the cusps of the Moon than Fessenkov was able to do near the center of the lunar disk. However, in spite of this greatly increased instrumental precision, all Lyot's results were again purely negative: no trace of polarisation was detected; and from its absence Lyot concluded in 1949 that the density of any hypothetical lunar atmosphere must be less than $10^{-8}$ of that of the terrestrial one. Moreover, after Lyot's death in 1952 his work was continued by DOLLFUS, who by resorting to further refinements depressed the possible upper limit for the density of a hypothetical lunar atmosphere to $6 \times 10^{-10}$ of the terrestrial one (1956). If the Moon possesses any gaseous atmosphere, its density above the surface cannot, therefore, exceed some $7 \times 10^{-13}$ g/cm$^3$; and how much smaller it may be remains still conjectural.

A gas density of the order of $10^{-12}$ g/cm$^3$ would represent a pretty hard vacuum from the point of view of the terrestrial physicist – and one which is, incidentally, attained in our own atmosphere at an altitude of approximately 180 km above sea level. However, even at such great heights, the number of gas particles remains still of the order of $10^{10}$ per cm$^3$; and although the mean free path of such particles between mutual collisions is of the order of 100 meters, even so rarefied a gas could manifest itself in different observable ways. It would not, to be sure, offer any protection to the surface beneath from impinging meteorites. A hypothetical lunar atmosphere of surface density of the order of $10^{-12}$ g/cm$^3$ would not decelerate any meteoritic material – small or large – enough to cushion significantly its impact. All solid particles in space intercepted by the Moon must hit its surface essentially with their cosmic velocities, and spend themselves on the surface rather than in their passage through the atmosphere as is the case on the Earth. Needless to say, the rate of meteoritic matter (of all sizes) per unit area of the Earth as well as of the Moon must be very approximately the same. In quest of its observational verification on the Moon we should not, however, look for any luminous trails during approach; but rather for instantaneous flashes of light which should be produced by impact of a meteorite on solid rocks. For meteorites of sufficient mass such flashes should be visible from the

Earth through suitable telescopes – especially against the dark surface of the night hemisphere of the Moon – and their systematic observation may thus extend the field of meteor astronomy to our satellite.

However, returning to our own atmosphere such as exists at an altitude of 180 km above sea-level, even though it is insufficient to affect the velocity of the meteors traversing through it, it can give rise to other interesting phenomena. Between 180–200 km above sea level we would find ourselves in the midst of the auroral zone, where luminescent gas stimulated by the impact of corpuscular sunrays produces the beautiful displays of "northern lights". Are there similar aurorae on the Moon? HERZBERG pointed out in 1946 that a search for emission spectra of such displays around the bright limb of the Moon might constitute one of the most sensitive tests of the presence of a hypothetical atmosphere of our satellite. No trace of any such emission has, however, so far been detected above the surface of the Moon; though a quest for fluorescent radiation from the Moon led quite recently to some very interesting discoveries on its surface, which will be briefly described in the subsequent Chapter 22.

So far as our present knowledge goes, however, the Moon does not possess any atmosphere of density in excess of $10^{-12}$ g/cm$^3$ at its surface; and how far below this limit its actual density happens to be we can only guess. Let us add that, for the astronomers concerned with the relative abundances of the elements in cosmic matter, even the upper limit just stated is already becoming rather uncomfortably low; for, after all, there really should be more gas around the Moon than this limit seems to admit.

We explained already earlier in this chapter that, on account of the feeble lunar attraction, all light gases would escape indeed completely from its gravitational field in the course of the Moon's long astronomical past; and most of the heavier gases (such as $SO_2$, for instance) would react again with surface rocks to form solid compounds. This should, however, not apply to the so-called "inert" gases – like argon, krypton, or xenon – which for reasons of their atomic structure do not enter into compounds, and are at the same time sufficiently heavy for their rate of escape from the gravitational field of the Moon to be moderately low (the mean atomic velocities of thermal agitation of these gases at 0°C being 414 m/sec for A, 287 m/sec for Kr and 229 m/sec for Xe).

The case of argon is particularly arresting, because – quite apart from any aboriginal amount of this element which the Moon may have retained from primordial times – its supply must be continually replenished by radioactive decay of the heavy isotope of potassium ($K^{40}$). If (as is highly likely) the chemical composition of the Moon is similar to that of the outer mantle of our Earth, potassium should constitute about 0.12 per cent of it by weight. The total mass of the Moon, which is known to be $7.35 \times 10^{25}$ g, should then contain about $8.8 \times 10^{22}$ g of potassium; and of this about $9.7 \times 10^{17}$ g should be the radioactive $K^{40}$ decaying (by $\beta$-disintegration) into the common isotope of argon ($A^{40}$).

The total disintegration of all lunar $K^{40}$ should, therefore, create $9.7 \times 10^{17}$ g or $1.5 \times 10^{40}$ atoms of argon – as compared with some $10^{44}$ gas particles now constituting the terrestrial atmosphere. Just how much of this argon managed to escape to the surface from the lunar mass by gradual degassing (caused by the rising internal temperature) remains conjectural within fairly wide limits. However, on almost any reasonable guess (coupled with the known rate of escape of argon from the gravitational field of the Moon) the amount of argon in the lunar atmosphere should add up to more than the upper limit of the lunar air density consistent with the absence of perceptible twilight phenomena.

When we come to consider the remaining two heavier inert gases – krypton and xenon – the situation becomes even more embarrassing. These gases likewise do not form compounds, and their atomic weights (two and three times that of argon) are such that their gravitational escape from the Moon (by collisions) becomes effectively impossible ($10^{24}$ years for Kr and $10^{41}$ years for Xe) even if exposed to noon-time conditions for the entire age of the Moon. Moreover, even if no krypton and xenon were originally present around the Moon, more than enough of them would have been produced in the lunar crust by a spontaneous disintegration of the heavy isotope of uranium ($U^{238}$), by interaction of the lighter isotope $U^{235}$ with neutrons produced on the lunar surface by impact of cosmic rays, and also (for xenon) by spontaneous decay of the radioactive isotope of iodine ($I^{129}$).

If uranium is present in average lunar rocks of density 3 g/cm$^3$ in the same percentage as on the Earth (i.e., about 1.2 parts in $10^8$), it follows from known characteristics of uranium disintegration that $2.5 \times 10^{24}$ atoms of xenon, and $1.5 \times 10^{23}$ atoms of krypton should have been produced in the lunar surface layer 1 cm thick in the course of $4 \times 10^9$ years. The liberation of a gas so formed into the atmosphere could, moreover, occur whenever the solid rocks are mechanically disturbed by some event (such as the impact of a meteorite, or tectonic processes operating in the crust).

According to the present estimates of the frequency of meteor impacts on the Moon, during $4 \times 10^9$ years its surface should have been effectively "disturbed" by such impacts down to a depth of 1–10 km. If so, the cumulative effect of it would have been a release of $10^{34}$–$10^{35}$ atoms of krypton and $10^{32}$–$10^{33}$ atoms of xenon, constituting a permanent atmosphere of $10^{-9}$–$10^{-10}$ of the terrestrial air pressure on the ground (and consisting of approximately 93% of Kr and 7% of Xe).

Why is this gas not there? Herring and Licht pointed out, some years ago, that atoms of heavy gases can be mechanically "blown off" the Moon into space by collision with corpuscular radiation (mainly protons) emitted continuously – and, sometime, in angry energetic "puffs" – by our Sun. The existence and intensity of this "solar wind" and its occasional gusts usually associated with flares and other sudden disturbances of the solar surface, is now well known from the soundings of deep-space probes, and indirectly attested by such terrestrial phenomena as polar aurorae and magnetic storms. Herring and Licht argued that the knocking-off power of this solar wind is, in fact, sufficient to remove most of the heavy inert gases from the lunar surface, and thus despoil it even of such scanty vestiges of gas envelope which its own feeble gravitational attraction would enable it to retain.

More recently, however, Öpik and Singer pointed out another and more effective way of gas removal from the lunar surface: namely, by ionization. As long as gas remains neutral, its atoms or molecules can be removed only by collisions – be it with other molecules, or particles of the "solar wind". Should, however, the gas particles become ionized and acquire thus positive electric charge (and Öpik with Singer has shown how easily this can happen to lunar argon and heavier inert gases), the positive electrostatic charge of the sunlit hemisphere of the Moon, acquired by photoionization of the light elements in its crust (of which more will be said in Chapter 21), can remove ions by repulsion far more effectively than could be accomplished by the collisions with the protons of the solar wind. It seems, therefore, possible to explain now the well-night complete absence of free gases around the Moon in more than one way, and thus reconcile theoretical expectations with the observed facts to a complete satisfacton of all investigators.

Such gases as are left by these processes to cling to the lunar surface for a limited time do not constitute any real atmosphere – in which individual gas particles are balanced up by mutual collisions – but rather a transient *exosphere*, in which the individual atoms or ions describe essentially free-flight trajectories in the prevailing gravitational or electrostatic field. Each planetary atmosphere is bound to peter out into such an exosphere on its outer fringe bordering on interplanetary space; but on the Moon this exosphere apparently reaches down to the solid surface itself. As to its chemical composition – due partly to captures from interplanetary space and partly by degassing of the lunar surface layers, Öpik estimated recently that it contains approximately $5 \times 10^5$ particles per $cm^3$, among which are about $1.2 \times 10^4$ molecules of hydrogen, $1.4 \times 10^5$ molecules of water; a comparable amount of the molecules of carbon dioxide; and not more than $1 \times 10^4$ atoms of inert gases.

If, therefore, the Moon possesses no detectable atmosphere it cannot, of course, maintain any liquid on its surface. Near the poles, to be sure, depressions may exist which are never reached by direct sunlight (and which are illuminated, at best, by sunlight scattered from adjacent slopes). In such regions, condensed volatile substances may possibly be present in the form of some kind of a permafrost; but should they even evaporate, they are apt to be lost in a very short time. Hence, no liquid – or even solid – water can be present at any spot of the Moon which can be reached by sunlight. The surface of the Moon must, therefore, be regarded as bone-dry; none of its visible features could have been formed, or even modified, by running water; or freezing and melting water. One of the most important agents producing geological changes on the Earth must have been totally absent on the Moon, and its action cannot be invoked to explain any structural characteristics of the surface of our satellite. With both the hydrosphere as well as atmosphere effectively absent from time immemorial around the Moon as disturbing agents, the fossil record of the lunar surface must possess a *vastly greater degree of permanence* that anything known to us on Earth; and in the next part of this book we shall turn our attention to its interpretation.

# BIBLIOGRAPHICAL NOTES

## Chapter 7

The model of a self-compressed Moon based on equation of state of the form (7-7) has first been worked out by JEFFREYS (1952, pp. 147–153), with further contributions made more recently by MACDONALD (1962) and LYTTLETON (1963).

Clairaut's theory of the figures of equilibrium of celestial bodies, underlying the evaluation of the lunar moments of inertia, is classical; for its fuller account cf. Sections II 1–2 of KOPAL (1959); or, more completely, KOPAL (1960). For theoretical tides in a solid Moon cf. SUTTON, MEIDELL, and KOVACH (1963).

For recent discussion of the observational determination of the lunar moments of inertia and of the "mechanical ellipticity constant" $f$ cf., e.g., FRIDLAND (1961), HABIBULLIN (1961), GORYNIA (1962), NEFEDIEV (1963) or SHAKIROV (1963); while for theoretical investigations of the lunar moments of inertia and the associated gravity field cf., GRUSHINSKI and SAGITOV (1962), or GOUDAS (1964b,c, 1965b) and MIKHAILOV (1965). For a position of the centre of mass of the Moon cf., also BYSTROV (1962) or O'KEEFE and CAMERON (1962).

## Chapter 8

Although the mathematical problem of the cooling of a sphere radioactively heated within was solved already by LOWAN (1933, 1934, 1935) more than thirty years ago, its first application to the thermal history of the Moon we owe to UREY (1952, 1955, 1957, 1962), followed by MACDONALD (1959), LEVIN and MAYEVA (1960), KOPAL (1961, 1962a), LEVIN (1962) and others. The most extensive numerical computations carried out so far are those by KOPAL (1962a). It is, however, to Urey that we owe the realization that radiogenic heating of the Moon should alone be sufficient to keep the bulk of its mass at a temperature in excess of 1000° K – possibly approaching the melting point of the Moon's mass in some parts of the deep interior. Of subsequent literature cf. also LEVIN (1963), or ORNATSKAYA (1964) and LEVIN and MAYEVA (1964).

The relative importance of a radiative (as distinct from conductive) transport of energy in the interior of the terrestrial planets has first been pointed out by CLARK (1956, 1957), and subsequently applied to the Moon by MACDONALD (1959). The mathematical treatment of the subject as given in this chapter is, however, due to KOPAL (1963a). For the actual values of the absorption coefficients $k_\lambda$ of silicate rocks in the visible light or near infrared cf., e.g., CLARK (1957a). A treatment of the problem of the heating of the Moon by insolation, as given in the latter part of this section, is due to KOPAL (1965a).

## Chapter 9

A study of the stress history of the Moon has been opened up by MACDONALD (1960) who, however, in this paper considered only the global effects. A treatment of the subject as given in this chapter follows, however, largely that by KOPAL (1961, 1962a, 1963) in which the distribution of stresses in the interior of a thermally strained sphere is explicitly taken into account. The numerical results are taken largely from KOPAL (1962a). A coupling of the thermo-mechanical effects has first been formulated by DUHAMEL (1837) in a great memoir which has since remained classic. Of more recent work on this subject cf., e.g., JEFFREYS (1929), LOWAN (1935) or LAPWOOD (1952). For the thermo-elastic effects of insolation and their influence on the form of the Moon and its moments of inertia cf. KOPAL (1965a).

The problems of free isothermal oscillation of elastic solid globes, a discussion of which concludes this section, has in recent years been considered extensively in connection with seismological problems of the Earth (cf., e.g., ALTERMAN, JAROSCH and PEKERIS, 1959); and numerical applications to the Moon are due to BOLT (1960), TAKEUCHI, SAITO and KOBAYASHI (1961), or CARR and KOVACH (1962).

## Chapter 10

The possibility that internal temperatures attained by secular release of radiogenic heat could be sufficient to melt at least a part of the matter in the lunar interior, which would then behave dynamically as a highly viscous liquid, was implicit in Urey's early work on the thermal history of the Moon (cf. UREY, 1952, 1962). Although Urey himself has remained consistently sceptical as to the extent of such a melting (cf. UREY, 1960a), the realization that molten rocks of lunar interior would be convectively unstable (KOPAL, 1961) lent some interest to theoretical investigations of the type of a flow which could be expected to arise in the lunar interior as a result of continuous release of radiogenic heat.

A mathematical theory of convection in incompressible homogeneous spheres (or shells) of viscous liquid, developed in recent years by CHANDRASEKHAR (1952, 1953, 1961), should be closely applicable to the lunar problem; and its generalization, by KOPAL (1963a) to compressible configurations in which the density $\rho(r)$ and viscosity $\mu(r)$ can be arbitrary functions of the distance $r$ from the centre* has been largely followed in this chapter.

Qualitative applications of this theory to construct a case for lunar convection have been made by RUNCORN (1962, 1963), who attempted to account in this way for the anomalously large value of the ratio $(C - A)/B \equiv \beta$ of the lunar moments of inertia, as given previously by Equation (4-119), in terms of convection currents operative in the radioactively heated interior of the Moon – currents which might give rise to a nonradial temperature distribution; and through thermal expansion influence also the internal density distribution and thus also the moments of inertia. Simultaneous work by the present writer (KOPAL, 1962b, 1963d, 1965b) has, however, revealed the existence of considerable quantitative difficulties which have been mentioned in the text.

The basic assumption underlying Runcorn's approach to the problem – namely, that the Moon is internally hot enough for a large part of its mass to behave as a molten globe – has recently been weakened by several new facts emerging from selenodetic studies (cf. Chapter 13). For the latter Runcorn relied largely on the data by BALDWIN (1963), the reliability of which will be put in proper perspective in Chapter 13. Secondly, a selenodetic establishment of level differences amounting to several kilometres over extensive regions of the lunar surface reveals that the Moon must possess a greater degree of rigidity than would be compatible with a largely molten interior. Third, a recent confirmation by KOZIEL (1964) of significant free libration of the Moon in longitude endows the lunar globe with a property which is again characteristic of solids rather than liquids. But quite apart from all these facts, in making a case for internal convection as the cause of the observed constant $\beta$ for the Moon, Runcorn invoked two different processes which are not mutually compatible: namely, he postulated internal convection to establish nonradial temperature field, and then called on the corresponding thermal expansion to deform the Moon in such a way as to account for the observed value of $\beta$. These steps lack, however, any logical connection; for one postulates the Moon to behave like a viscous liquid; the other, like an elastic solid; and both cannot be true at the same time. In point of fact, nonradial temperature distribution of convective origin can, in principle, account for any arbitrary value of $\beta$ for a perfectly spherical outer boundary (implied in Chandrasekhar's theory) – or, for that matter, a boundary of any shape – provided only that the convection currents can be made sufficiently rapid. However, the stability of such a flow seems incompatible with the observed value of $\beta$ as given by Equation (4-119) for any reasonable choice of physical parameters characterizing the lunar globe.

The concluding part of this section, discussing possible heat production in the lunar interior by viscous dissipation of bodily tides, is largely based on previous work by KOPAL (1963b). For a corresponding problem of heat production by dissipation of bodily tides in an imperfectly elastic Moon cf., KAULA (1963).

## Chapter 11

Needless to stress, the chemical composition of the lunar globe – an object which so far can be

---

* The coefficients of $\partial\mu/\partial r$ as well as $\partial^2\mu/\partial r^2$ in KOPAL (1963a) are incomplete; their present version as given in this chapter has been established by ASHWORTH (1964).

studied only at a distance – can be investigated only by indirect methods within the framework of our general knowledge of cosmochemistry. Fundamental contributions to this field are those by UREY (1956a, 1957, 1958, 1958a, 1966); cf., also BRIGGS (1962), PALM and STROM (1962) or WARREN (1963).

The problem of a possible lunar magnetic field has received considerable attention in recent years, since the flight of the Russian Lunik III, which on September 14, 1959 performed the first magneto-metric measurements in close proximity of the lunar globe (cf., DOLGINOV, YEROSHENKO, ZHUZGOV, PUSHKOV and TYURMINA, 1960; or DOLGINOV, YEROSHENKO, ZHUZGOV and PUSHKOV, 1962). For previous discussion of a possible magnetic screening by the Moon cf., e.g., SUCKSDORF (1956); and, more recently, BIGG (1963a, b; 1964); or DODSON and HEDEMAN (1964).

For theoretical discussions of possible lunar magnetic fields, cf. KOPAL (1959b), SINGER and WENTWORTH (1959), NEUGEBAUER (1960), ÖPIK and SINGER (1960), ZHARKOV and ULINICH (1960), SINGER and WALKER (1961), PIDDINGTON (1962), GOLD (1962a), ARONOVITZ and MILFORD (1965), and others.

A possibility of the occurrence of water or ice on the Moon, and estimates of its abundance, have been discussed by WATSON, MURRAY and BROWN (1961); KOPAL (1961a, 1962c, 1963c), GOLD (1962, 1965) and more recently by GILVARRY (1964b), HAPKE and GOLDBERG (1965) and UREY (1965).

The lunar origin of tektites has first been suggested by NININGER (1943, 1947, 1952) and has since become the subject of a considerable literature. Among more recent contributions to this problem cf. UREY (1955, 1957a), VARSAVSKY (1958), BARNES (1958), BARNES, KOPAL and UREY (1958), O'KEEFE (1959, 1961, 1963), BARNES (1960), CHAPMAN (1960), LOWMAN (1962), CHAPMAN and LARSON (1963), HAWKINS (1963a), GREENLAND and LOVERING (1963), GILVARRY (1965a, b) and others. A considerable number of other papers on this subject appeared, e.g., *Geochimica and Cosmochimica Acta* 28 (1964), or the Univ. of Chicago monograph on *The Tektites* (ed. by J. A. O'KEEFE, 1963).

For a case of the lunar origin of chondritic meteorites cf. UREY (1959, 1962, 1965), but also ARNOLD (1965) or ÖPIK (1966). The possibility of an existence of indigenous organic matter on the Moon has first been mentioned by SAGAN (1961); and the case for it was strengthened by more recent work by ORO, WIKSTROM and BARGHOORN (1965).

## Chapter 12

For the modern optical tests of a possible lunar atmosphere, cf. FESSENKOV (1943), LYOT and DOLLFUS (1949), DOLLFUS (1952, 1956).

Of recent studies of the selective escape of gas from the gravitational field of the Moon, cf., e.g., ÖPIK (1955, 1962), EDWARDS and BORST (1958), FIRSOFF (1959), ÖPIK and SINGER (1960), BRANDT (1960), SINGER (1961), and others. Among theoretical studies of lunar ionosphere and the interaction of the surface of the Moon with the solar wind, cf. HERRING and LICHT (1959, 1960), NAKADA and MIHALOV (1962), WEIL and BARASH (1963), BERNSTEIN, FREDERICKS, VOGL and FOWLER (1963), HINTON and TAEUSCH (1964), MANNO, SAUERMANN and ENGELMAN (1965), etc.

PART THREE

# TOPOGRAPHY OF THE MOON

# INTRODUCTION

In the third part of this book we shall emerge from the uninviting depths of the lunar interior to the surface of our satellite, in order to discuss the problems encountered there in the form of its visible sculpture. Stripped to the essentials, a study of this surface can be physically approached in two ways: as a "boundary condition" of all internal processes which we have surveyed in the preceding chapter, and as an "impact counter" of external events which had pockmarked the lunar face since time immemorial. In no other sense can the interpretation of the lunar surface possess any meaning.

To begin with, in the next Chapter 13 we shall consider the global shape of the surface of our satellite, and survey our knowledge in this field in order to learn from it what we can on the forces which may have influenced its formation. Chapter 14 will be devoted to a definition of the lunar coordinate system and its determination. Chapter 15 will contain an outline of the human efforts to map the face of the Moon up to the present time. In Chapter 16 we shall briefly survey the principal types of formations encountered on the surface of our satellite. In Chapter 17 we shall set out to identify and explain the characteristics of such formations as are likely to have been produced by external impacts of solid bodies (of widest mass-range) encountered by the Moon on its perpetual journey through space; in the concluding Chapter 18 we shall then attempt to unravel on this basis a stratigraphic system of the lunar surface and to reconstruct its probable time scale; leaving an examination of *microscopic* structures of this surface by indirect method for the concluding fourth part of this volume.

## CHAPTER 13

# GLOBAL FORM OF THE MOON;
# DEFINITION OF LUNAR COORDINATES

The fundamental problem of lunar topography consists of determining the exact shape of the lunar surface (i.e., its deviations from a sphere) on the basis of observations which can be made from the Earth. This requires, in turn, a determination of absolute (three-dimensional) coordinates of a sufficient number of specific control points on the Moon, through which a smooth surface may be interpolated by harmonic analysis. The aim of the present chapter will be to outline the methods by which this task can be approached, and to summarize briefly the results which have so far been obtained by their application.

The problem confronting us can, in brief, be divided in three parts: the determination of absolute positions in the *marginal* (or *limb-*) *zone* of the lunar globe, which from time to time appear in projection on the celestial sphere in the course of the librations (see Chapter 3); in the *terminator zone*, where sunrays are tangent to the local surface; and the *central zone* between the two, where stereoscopic measurements alone can furnish the desired results. Each of these three zones calls for special methods of reduction (of increasing complexity) and will be treated below in turn.

The measurements of the positions of the *limb* of the Moon are simplest in theory as well as practice, and lead to the desired results in the most straightforward manner. Moreover, the maximum angles of libration 7° 57' and 6° 51' in longitude and latitude expose from time to time almost 17.7 per cent of the entire lunar globe in projection on a plane perpendicular to the line of sight; and measurements of the rectangular coordinates $x, y$ in this plane of projection (at the telescope, or on photographic plates) should enable us to determine the distance of that point from the Moon's center (defined as the center of curvature of the entire arc of the limb) in a straightforward manner.

The reader scarcely needs to be reminded that, under ordinary conditions, the observed arc of the limb will extend over only one-half of the entire circumference, as the other half of the "meridian of illumination" will be vitiated by the phase effect – even at full – moon (unless both libration angles happen to be zero at the same time); and when this occurs, the Moon is usually eclipsed. The best conditions for measuring the exact shape of the entire lunar limb occur, in fact, during *annular eclipses of the Sun*; and if, moreover, such an eclipse happens to be *central*, the phase-effect completely disappears. Under such conditions can measure only the lunar entire circumference on the same plate, but also refer the actual measurements to the adjacent solar limb (only seconds of arc apart) which represents a theoretical circle of reference.*

The measured distance of a point $P$ from the center of symmetry of the limb determines then the radial coordinate of that point of the lunar globe; but in order to describe its position uniquely on the surface two additional angular coordinates must be sought. It should obviously be of advantage to define such a coordinate system with respect to fixed properties vested in the lunar globe itself – independent of the position of the observer – and this is best done *by identifying the axes of this coordinate system with the principal axes of inertia of the Moon* – i.e., as the axes around which the lunar globe librates physically. Let, accordingly, the *lunar equator* be hereafter defined as the plane inclined by the angle $I$ (i.e., the mean of the Eulerian angle $\theta$) to the ecliptic; and the lunar *prime meridian* be a plane perpendicular to it and containing the radius-vector of our satellite when the Moon stands at the mean ascending node at perigee or apogee. The *selenographic latitude* $\beta$ is then the angle measured from the equator positively northward (i.e. towards the hemisphere containing Mare Imbrium); while the *selenographic longitude* $\lambda$ is measured along the equator positively eastward (i.e., towards Mare Crisium).

The zero point of this coordinate system – i.e., a point at which the line common to the two fundamental planes intersects the visible lunar surface – is not marked by any special feature, nor can its position be measured directly with accuracy. What astronomers have done, instead, since the days of Bessel has been to measure the position of an adjacent crater called Mösting A (marked on Figure 13-1) relative to the theoretical but nondescript zero point. According to the latest results obtained by KOZIEL (1964), the location of the central point of Mösting A in this system of coordinates is specified by the values of

$$\left. \begin{array}{l} \lambda = -\,5°9'50'' \pm 5'' \ (m.e.), \\ \beta = -\,3°10'47'' \pm 4''(m.e.), \end{array} \right\} \tag{13-1}$$

implying that the principal $x$-axis of the lunar globe intersects the lunar surface $3°\,10'\,47''$ North and $5°\,9'\,50''$ West of the center of Mösting A; and the mean error of $\pm 5''$ in the estimated position is equivalent to an uncertainly of only 42 meters in the location of this spot on the lunar surface – in close agreement with the error of $\pm 7''$ within which Koziel's value of $I$ defines the position of the lunar poles.

How to determine the coordinates $\lambda$, $\beta$ so defined of any particular point of the lunar limb from visual (micrometric) or photographic observations? In order to do so, consider Figure 13-2 representing a projection of the apparent disk of the Moon on the celestial sphere, in which the $x$-axis coincides with the (terrestrial) East-West direction; and the positive direction of the $y$-axis points towards the (terrestrial) north pole; with the origin at the centre of the mean circle representing (in arbitrary units) the

---

* I.e., the departures of the Sun from a sphere due to its axial rotation are insignificant; while atmospheric or instrumental distortion acts on both limbs alike.

Fig. 13-1.  Position of Mösting A.

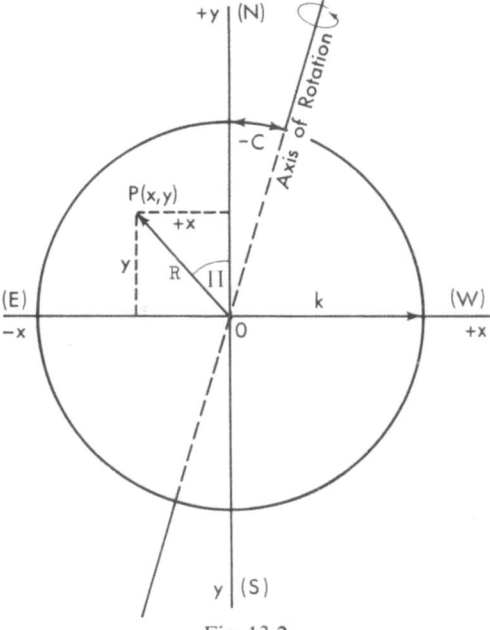

Fig. 13-2.

apparent lunar limb. Let, moreover, $x$, $y$ denote the position of any particular point $P(k, \lambda, \beta)$ of this limb. If so, then obviously

$$x^2 + y^2 = k^2;\tag{13-2}$$

while the angular selenographic coordinates $\lambda_P$ and $\beta_P$ of $P$ follow from the solution of the equations

$$\left.\begin{aligned}
\sin \beta_P &= \sin b' \cos \kappa + \cos b' \sin \kappa \cos (\Pi - C'), \\
\cos (\lambda_P - l') \cos \beta_P &= \cos b' \cos \kappa - \sin b' \sin \kappa \cos (\Pi - C'), \\
\sin (\lambda_P - l') \cos \beta_P &= \hphantom{\cos b' \cos \kappa - \sin b' \sin \kappa} + \sin \kappa \sin (\Pi - C'),
\end{aligned}\right\}\tag{13-3}$$

where

$$\Pi = \tan^{-1} \frac{x}{y}\tag{13-4}$$

denotes the position angle of $P$ in the $xy$-plane on the lunar limb (reckoned counter-clockwise from the north pole); $C'$, the (topocentric) inclination of the Moon's axis of rotation (easily obtainable from its geocentric value as given by Equations (3-27)); $b'$, $l'$ are the topocentric librations in latitude and longitude (defined in Chapter 3); $\kappa$, the selenocentric angle $90°$-$s'_{\mathbb{C}}$ of the lunar limb as seen by the terrestrial observer, with $s'_{\mathbb{C}}$ standing for the semi-diameter of the apparent lunar disk – such that where $a$

$$\sin \kappa = \sqrt{\left(1 - \frac{a^2}{r'^2}\right)} \quad \text{and} \quad \cos \kappa = \frac{a}{r'},\tag{13-5}$$

stands for the absolute radius of the Moon at the place of observation; and $r'$, for the topocentric distance of its center at that time.

Large-scale mapping of the marginal zone of the Moon was initiated by Hayn (1907) on the basis of visual heliocentric measurements[*], and greatly improved by him by means of photographic measurements between 1908 and 1912. The limb charts prepared by Hayn in this way and published in 1914 remained standard for a whole generation. In 1952, Weimer published a new atlas of 139 lunar profiles visible at different librations, based on the measurements of the Paris photographs of the Moon secured by Loewy and Puiseux. Another set of such charts based on heliometric measurements was since published by Nefediev (1957); but all these contributions were superseded in volume as well as completeness by the results of extensive measurements by Watts (1963) of the U.S. Naval Observatory, who used for this purpose a total of 867 photographic plates taken on 503 nights between 1927 and 1956; and presented his results in the form of 1800 charts for position angles spaced by $0°.2$. Measurements of the eclipse plates were initiated already by Hayn (1914); but the best results obtained in this way came out only quite recently, as a result of the measurements of the photographs of the annular eclipses of the Sun secured by the Lockheed-Manchester expeditions in 1962 and 1963 (cf. Carson et al., 1966).

---

[*] A method of star occultations by the Moon was applied by Przybyllok (1905) as an alternative to heliometer measurements.

We have already mentioned that approximately 35 per cent of the surface of the visible hemisphere of the Moon appears from time to time on the limb. This is not true of more than 70 per cent of this surface, where other methods must be applied for measurements of absolute positions. One such method can be based on the determination of the exact form of the sunrise or sunset terminator, which represents the projection on the celestial sphere of the lunar limb as seen from the Sun. If the Moon were a sphere, the terminator would appear to us in projection as the arc of an ellipse, the eccentricity of which would become zero at the time of the "full" or "new" Moon (when the terminator should coincide with the limb) and one at the first (or last) quarter. Deviations of the observed shape of the terminator from its expected elliptic form would be due to the asphericity of the Moon; and from such residuals one could then determine the absolute positions of points on the terminator. As these residuals are projected values of the actual deviations of the terminator from the mean spherical surface on a plane normal line of sight, the probable error of the results of such measurements is bound to be maximum at the time of quadratures (i.e., the first or last quarter) and minimum near the limb.

The main difficulty connected with the application of such a method is the fact that the actual positions of the terminator cannot be determined by astrometric measurements, but one must trace them photometrically. As the results of photometric measurements (which are inherently less accurate) may be vitiated by local variations in ground reflectivity, it is perhaps not surprising that all results obtained so far in this way (cf., e.g., MAINKA, 1901; RITTER, 1934) are much inferior in accuracy, and cannot compete with the outcome of other methods. It should be kept in mind that the quantities sought by measurements are close to the limits of accuracy attainable by any kind of measurements from the distance of the Earth. At its mean distance a linear distance of 1 km corresponds approximately to a geocentric angle of $0\overset{\prime\prime}{.}54$; and at a distance of $10°$ from the lunar central meridian a vertical bulge of 1 km in height would manifest itself in projection by an angular displacement of less than $0\overset{\prime\prime}{.}09$. In the focal plane of (say) the Lick refractor of focal length of 17.82 m this angle would correspond to a linear shift of only 7.9 microns – which is just about the diffraction limit of resolution in visible light attainable theoretically with a 36-inch objective. It is obvious that the measurements of a great many plates (preferably overexposed for easier identification of the terminator) would be required to obtain data of any significance; and this has not so far been done.

A variant of the foregoing technique going, in principle, back to Galileo Galilei has more recently been proposed by HOPMANN (1964). Throughout the month, all peaks and many other well-defined formations appear (and disappear again) at some distance from the terminator as bright spots illuminated by the first (or last) rays of the rising or setting Sun. If these moments are timed (visually, or by means of sequential photography), and the relative height $h$ of such peaks determined by the shadow method (as developed in the next chapter) the distance of their base from the Moon's center can be deduced from the observations.

The most general and commonly used for the determination of absolute positions on the lunar surface is the *stereoscopic* method, applicable to points distributed all over the visible hemisphere of our satellite – in the marginal as well as central zones of its surface – and truly indispensable in the latter. As was explained in Chapter 3, the librations (mainly optical) of our satellite cause the positions of all points of the lunar surface as viewed from the Earth to vary – mainly in a monthly period – with respect to the line of sights by amounts which may attain almost 18° at times when the librations in latitude and longitude are simultaneously in opposite maxima. The determination of absolute positions on the Moon by the stereoscopic method is based on the fact that features on a non-spherical surface are displaced for the terrestrial observer in a different way for different libration angles than if they were located on a spherical surface. The principle of the method is illustrated on Figure 13-3, where the arc *ABC* represents the lunar equator and *M*, its geometrical center. Also *MB* is the direction towards the terrestrial observer from which selenographic longitude $\lambda$ of surface points are measured. If *P* is such a point and $\lambda$ its longitude, then its position *D* on the lunar disc will be where the lines *DP* and *MB* intersect for the location of the observer. The orthographic projection of *P* on the lunar disc can be obtained from the position *D* on the disc after the correction for *finite distance*, defined the well-known expression

$$\frac{(MD)(PD)}{(MP)} \sin s',$$

where *s'* denotes, as before, the apparent angular radius of the Moon as seen from the

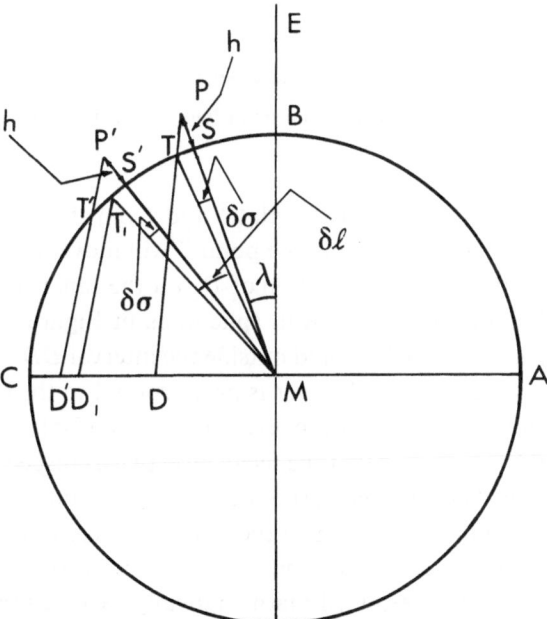

Fig. 13-3.  Application of the principle of stereoscopy for an equatorial point lying above the mean moon level.

position of the observer. The fact that the point $P$ is above the mean spherical surface whose radius is denoted by $r_0$ cannot be perceived by the observer from a single vantage point $P$, and thus the latter may lie anywhere along the line $PD$.

Let it now be assumed that the observer remains in the direction $MB$ while the Moon rotates by the small angle $\delta l$ around its axis of rotation. If the point $P$ lies on the mean sphere – i.e., is identified with $T$ – then the angle between the directions $PD$ and

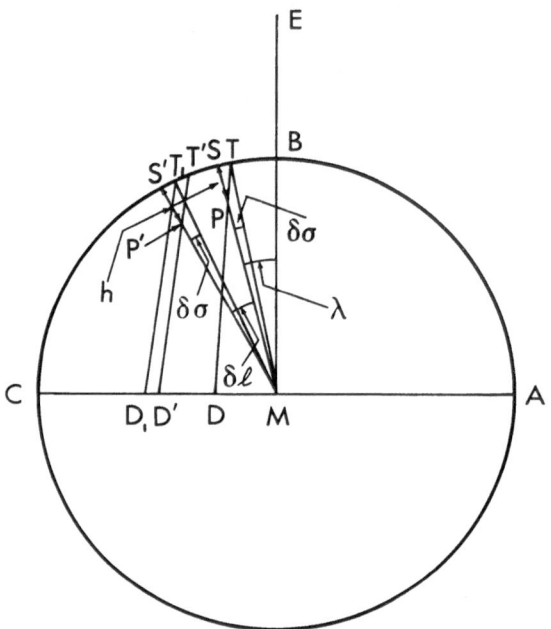

Fig. 13-4.  The case of an equatorial point lying below the mean moon-level.

$BM$ can be measured directly from the plate, and from it the angle $\lambda + \delta\sigma$ can be deduced. Then, because of the rotation, the point $T$ will move to $T_1$ with selenocentric angle $\lambda + \delta l + \delta\sigma$ and, therefore, the location of $D_1$ on the lunar disc can be predicted. If the point in question is above the mean sphere, as in Figure 13-4 then its conical projection on the lunar disc will be found outside the interval $DD_1$ and at the point $D'$. The distance $D_1 D'$ as seen by the observer is proportional to the height $SP = h$. If, on the other hand, the point $P$ is below mean level (see Figure 13-5), then the point $D'$ falls inside the interval $DD_1$. It is clear that the mean level in this respect is insignificant and can be arbitrarily fixed. The essential part of the present technique rests on the accurate measurement of the distance $D_1 D'$, where the point $D'$ is recorded on a photographic plate of the Moon as the conical projection of a small crater, rock, crater rim, etc., and $D_1$ is fictitious and will be determined mathematically. Before presenting the general three-dimensional case of the problem and its solution, we shall give some indication of the size of the quantity $D_1 D'$.

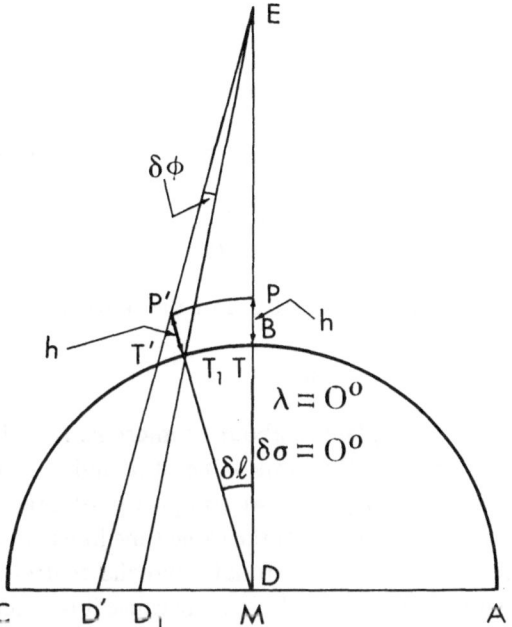

Fig. 13-5.  Representation of stereoscopic displacement of the center of the apparent lunar disc.

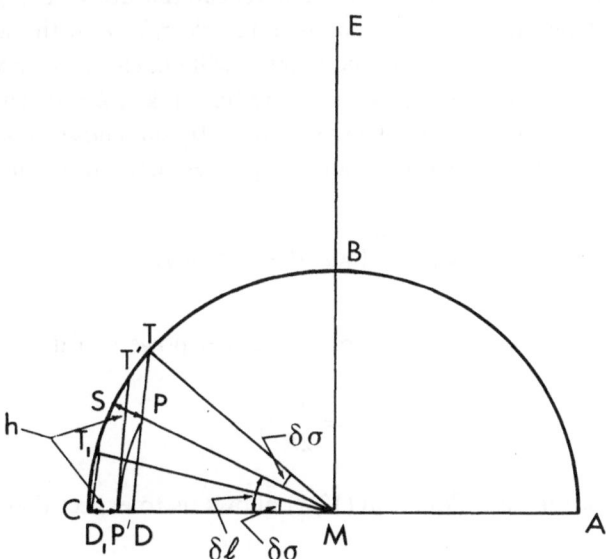

Fig. 13-6.  Stereoscopic displacement of the features inside the Moon's marginal zone.

Let us assume that $SP=1$ km, $MB=1738.0$ km, $\lambda=0$, $\delta l=15°$. Then the point $M$ coincides with the point $D$ and the point $T$ coincides with the point $B$. This case is illustrated on Figure 13-6 where $E$ denotes the Earth. The quantity to be measured is the geocentric angle $\delta\varphi$. If $MT=r_0$, $TP=1$ km and $ME=R$, then we easily establish the relation

$$R\delta\varphi = \frac{(r_0 + 1)\sin \delta l}{1 - (r_0/R + 1/R)\cos \delta l} - \frac{r_0 \sin \delta l}{1 - r_0 \cos \delta l/R}, \qquad (13\text{-}6)$$

from which by elimination of terms small or equal to $1/R^2$ one obtains

$$\delta\varphi = h\frac{\sin \delta l}{R} \qquad (13\text{-}7)$$

for a general altitude $h$ positive or negative. For the particular case examined here we have

$$\delta\varphi = 0\!''\!14; \qquad (13\text{-}8)$$

and this corresponds on a Lick plate to about 13 microns. It is clear that $\delta\varphi$ increases with $\delta l$ and, of course, $h$, but for the same value of $\delta l$ and $h$, it decreases with $\lambda$. For this reason, we can gain in accuracy when we compare photographs of the Moon taken at widely differing libration angles and at the same time limit our measurements to the central regions of the disc. It is very fortunate that the central regions can be more accurately measured with the present technique, because it is exactly these regions that are expected to depart most from the mean sphere.

In order to illustrate the case, we shall repeat the above computation on the assumption that $P$ lies in a direction making an angle $\pi/2 - \delta l$ with the $ME$ line. After the Moon is librated by an angle $\delta l$, the point $P$ will appear at the position $P'$ on the limb (see Figure 13-5 where for convenience we have taken $h < 0$). The point $D'$ coincides in this case with the point $P'$. The segment $P'D_1$ subtends a geocentric angle $\delta\varphi$, which is the quantity to be measured, and simple geometry allows us to establish the relation

$$\delta\varphi = \frac{h}{R}(\cos \delta l - \cos^2 \delta l), \qquad (13\text{-}9)$$

or, for small values of $\delta l$, ignoring third and higher powers of it,

$$\delta\varphi = \frac{h}{2R}\delta l^2 . \qquad (13\text{-}10)$$

Near the center of the disc, Equation (13-7) allows us to expect that

$$\delta\varphi = \frac{h}{R}\delta l ; \qquad (13\text{-}11)$$

and this shows that the stereoscopic technique is more effective near the lunar center than the limb because the quantity to be measured is larger by one order of magnitude at the center. From Equation (13-9) for $h = 1$ km, we find

$$\delta\varphi = 0\!''\!0035 \qquad (13\text{-}12)$$

which on a Lick plate corresponds to 0.3 of a micron, a quantity impossible to measure with any available measuring engine.

The *general case* of absolute height determinations by the stereoscopic technique can be outlined as follows: Let, in Figure 13-7, $M$ be the center of the Moon which, in this case, must be either the center of mass or the center of figure. If we define as

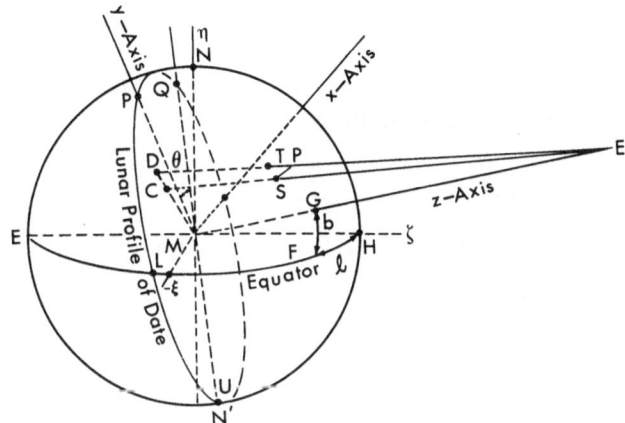

Fig. 13-7. Application of the principle of stereoscopy in a general case.

center of figure the center of a circle fitted to the lunar limb at zero total libration for a geocentric observer, then this center does not coincide with the center of mass. Their geocentric distance, according to YAKOVKIN (1962), can be as large as half a second – i.e., of the same size as the shift of lunar surface details due to the departures from a sphere. A similar estimate is given by SCHRUTKA-RECHTENSTAMM (1956). In addition the center of figure thus defined is found to vary considerably in position with the libration in latitude (YAKOVKIN, 1962). It is, therefore, essential that this variation be appropriately taken into account. The point $M$ in the theory developed by Saunder is assumed to be invariant; and since this is far from the truth, the results based on it are subject to systematic errors.

Let $P$ in Figure 13-7 be the point whose absolute coordinates are to be determined, $S$ the intersection of the line of sight of the terrestrial observer at $E$ with a mean sphere, and $D$ the projection of $P$ (not corrected for finite distance) on the disc $LPQ$. The elevation $PS$ of the point $P$ above or below the mean sphere will be denoted again by $h$. Let $QU$ be the projection of the axis of rotation of the Moon on the visible disc at that time. We shall now define a "librated" rectangular frame of reference $Mxya$ whose origin rests at the point $M$, whose $Mxy$ plane coincides with the plane of the lunar disc, and whose $Mx$-axis makes an angle $\pi/2 - \theta$ with the projection of the axis of rotation $NN'$ on the disc. This angle can be determined with the aid of the topocentric libration angles $\measuredangle HF = l$ and $\measuredangle FG = b$, after the direction of the $Mx$-axis is arbitrarily fixed on the photographic plate. Finally, the $Mz$-axis is arbitrarily fixed on the photographic plate, while the $Mz$-axis coincides with the direction $ME$ and is positive towards the observer $E$.

If now, $x, y, z$ are the coordinates of the point $T$ and $x_1, y_1, z_1$ the coordinates of point $S$ in the $Mxyz$ frame of reference, then the coordinates of the point $P$ will be

$$x_1 + \frac{x_1}{r_0} h, \quad y_1 + \frac{y_1}{r_0} h, \quad z_1 + \frac{z_1}{r_0} h.$$

The position vectors $\overrightarrow{MT}$, $\overrightarrow{MP}$, $\overrightarrow{ME}$ and $\overrightarrow{MS}$ which we shall denote by $\mathbf{p}$, $\mathbf{q}$, $\mathbf{s}$, $\mathbf{t}$, respectively, have the following projections:

$$\mathbf{p} = (x, y, z), \tag{13-13}$$

$$\mathbf{q} = [x_1(1 + h/r_0), \quad y_1(1 + h/r_0), \quad z_1(1 + h/r_0)], \tag{13-14}$$

$$\mathbf{s} = (0, 0, R), \tag{13-15}$$

$$\mathbf{t} = (x_1, y_1, z_1). \tag{13-16}$$

These vectors are related by the equation

$$\mathbf{s} = \mathbf{p} = \kappa(\mathbf{s} - \mathbf{q}), \tag{13-17}$$

where $\kappa$ is a scalar quantity. It is also true that

$$\mathbf{q} = \mathbf{p} + \overrightarrow{TS} + \overrightarrow{SP}, \tag{13-18}$$

where

$$\overrightarrow{TS} = \delta\mathbf{p} = (\delta x, \delta y, \delta z), \tag{13-19}$$

$$\overrightarrow{SP} = \delta\mathbf{q} = \left(\frac{x_1}{r_0} h, \frac{y_1}{r_0} h, \frac{z_1}{r_0} h\right).$$

In addition,

$$\delta\mathbf{q} = \frac{(\mathbf{p} + \delta\mathbf{p})}{|\mathbf{p}|} h \tag{13-20}$$

and, hence,

$$\mathbf{s}(1 - \kappa) = \mathbf{p} - \kappa\mathbf{p}(1 + h/|\mathbf{p}|) - \kappa\delta\mathbf{p}(1 + h/|\mathbf{p}|). \tag{13-21}$$

Therefore, $\delta\mathbf{p}$ will be known if the scalar $\kappa$ is determined. This can be done by employing the relation

$$|\mathbf{p}| = |\mathbf{p} + \delta\mathbf{p}| = 1, \tag{13-22}$$

which means that

$$\mathbf{p} \cdot \delta\mathbf{p} \cong 0. \tag{13-23}$$

By multiplying the vector relation (1-21) by $\mathbf{p}$, we eliminate $\delta\mathbf{p}$ from it, and the resulting relation allows us to determine $\kappa$. We find:

$$\kappa = \frac{\mathbf{p}(\mathbf{p} - \mathbf{s})}{\mathbf{p}(\mathbf{p} - \mathbf{s}) + h},\qquad (13\text{-}24)$$

and relation (13-21) yields for $\delta\mathbf{p}$ the expressions

$$\delta\mathbf{p} = \frac{h}{h + |\mathbf{p}|}\,\frac{\mathbf{s} - (\mathbf{s}\cdot\mathbf{p})\cdot\mathbf{p}}{\mathbf{p}(\mathbf{s} - \mathbf{p})}.\qquad (13\text{-}25)$$

Let us now denote by $\mathfrak{G}$ the matrix transforming the frame $Mxyz$ to the frame $M\xi\eta\zeta$. Its explicit form is

$$\mathfrak{G} = \begin{pmatrix} \cos\theta\cos l - \sin\theta\sin b\sin l & -\sin\theta\cos l - \cos\theta\sin b\sin l & \cos b\sin l \\ \sin\theta\cos b & \cos\theta\cos b & \sin b \\ -\cos\theta\sin l - \sin\theta\sin b\cos l & \sin\theta\sin l - \cos\theta\sin b\cos l & \cos b\cos l \end{pmatrix}.\qquad (13\text{-}26)$$

The components of the vector $\mathbf{p}$ in the standard frame of reference are then

$$\mathbf{p} = (\xi, \eta\,\zeta)\qquad (13\text{-}27)$$

whereas those of vector $\mathbf{t}$ are

$$\mathfrak{G}\mathbf{t} = (\xi_1, \eta_1, \zeta_1),\qquad (13\text{-}28)$$

i.e.,

$$\mathfrak{G}\delta\mathbf{p} = (\mathbf{t} - \mathbf{p}) = (\delta\xi, \delta\eta, \delta\zeta).\qquad (13\text{-}29)$$

All three coordinates of the point $T$ can thus be directly measured from the plate, and from them the coordinates of the point $S$ can be deduced after $h$ has been determined. To do this, we need another plate exposed at much different libration angles. We shall use the subscript 1 for quantities referring to this plate. Thus, let $l_1$, $b_1$ be the libration angles of this plate, where $|l - l_1|$, $|b - b_1|$ are as large as possible. Let, also, $\mathbf{p}_1$ and $\mathbf{q}_1$ be the position vectors of the points $T_1$ and $P_1$ which correspond to $T$ and $P$, respectively. If we treat $l - l_1$ and $b - b_1$ as small and call them $\delta l$ and $\delta b$, then we can write that

$$\mathbf{p} - \mathbf{p}_1 = \delta\mathbf{p}_1,\qquad \mathbf{q} - \mathbf{q}_1 = \delta\mathbf{q}_1,\qquad (13\text{-}30)$$

and

$$\delta\mathbf{p}_1 = \left(\frac{\partial\mathfrak{G}}{\partial l}\,\delta l + \frac{\partial\mathfrak{G}}{\partial b}\,\delta b\right)\mathbf{p},\qquad (13\text{-}31)$$

$$\delta\mathbf{q}_1 = \left(\frac{\partial\mathfrak{G}}{\partial l}\,\delta l + \frac{\partial\mathfrak{G}}{\partial b}\,\delta b\right)(\mathbf{p} + \delta\mathbf{p})(1 + h/|\mathbf{p}|).\qquad (13\text{-}32)$$

In the second plate, the conical projection of the point $P_1$ will not coincide with the conical projection of $T_1$ as was true (or taken by definition) in the first plate. Let us call $T'$ the intersection of the mean sphere by the line $EP_1$, and $\mathbf{p}'$ the selenocentric position vector corresponding to it. This vector can be directly obtained from plate measurements, and expressed as $\mathfrak{G}\mathbf{p}'$ in the standard coordinate system. Thus, the equation of condition will be of the form

$$\mathfrak{G}(\mathbf{p}' - \mathbf{p}) + \left(\frac{\partial\mathfrak{G}}{\partial l}\,dl + \frac{\partial\mathfrak{G}}{\partial b}\,db\right)[\mathbf{p}' - \mathbf{p}(1 + h/|\mathbf{p}|]$$

$$= (1 + h/|\mathbf{p}|)\left(\frac{\partial\mathfrak{G}}{\partial l}\,dl + \frac{\partial\mathfrak{G}}{\partial \beta}\right)\delta\mathbf{p} \tag{13-33}$$

in which $h$ and $d\mathbf{p}$ are not known. However, by post-multiplication of both sides by

$$\mathbf{p}'\cdot\left(\frac{\partial\mathfrak{G}}{\partial l}\,\partial l + \frac{\partial\mathfrak{G}}{\partial b}\,\delta b\right)^{-1}$$

we eliminate the right-hand side of Equation (13-33) and obtain

$$\mathbf{p}'\left(\frac{\partial\mathfrak{G}}{\partial l}\,\partial l + \frac{\partial\mathfrak{G}}{\partial b}\,\delta b\right)^{-1}\mathfrak{G}(\mathbf{p}' - \mathbf{p}) + \mathbf{p}'[\mathbf{p}' - \mathbf{p}(1 + h/|\mathbf{p}|)] = 0, \tag{13-34}$$

where $\mathbf{p}'$ is the transposed matrix (row vector) or $\mathbf{p}$. Equation (13-34) is scalar and can be solved for $h$; in doing so we find that

$$h = |\mathbf{p}|^{-1}\left[\mathbf{p}'\left(\frac{\partial\mathfrak{G}}{\partial l}\,\delta l + \frac{\partial\mathfrak{G}}{\partial b}\,db\right)^{-1}\mathfrak{G}(\mathbf{p}' - \mathbf{p}) + \mathbf{p}'(\mathbf{p}' - \mathbf{p})\right]. \tag{13-35}$$

Once $h$ is computed, a substitution in Equation (13-34) will give for $\delta\mathbf{p}$ the vector equation

$$\delta\mathbf{p} = \frac{1}{(1 + h/|\mathbf{p}|)}\left[\left(\frac{\partial\mathfrak{G}}{\partial l}\,\delta l + \frac{\partial\mathfrak{G}}{\partial b}\,\delta b\right)^{-1}\mathfrak{G}(p' - p) - [p' - p(1 + h/|\mathbf{p}|)]\right]. \tag{13-36}$$

The last two formulae can be generalized by dropping the assumption concerning the size of $\delta l$ and $\delta b$; and, in such a case, they become

$$h = |\mathbf{p}|^{-1}\left[\mathbf{p}'(\mathfrak{G}_1 - \mathfrak{G})^{-1}\mathfrak{G}(\mathbf{p}' - \mathbf{p}) + \mathbf{p}'(\mathbf{p}' - \mathbf{p})\right], \tag{13-37}$$

and

$$\delta\mathbf{p} = \frac{1}{(1 + h/|\mathbf{p}|)}[(\mathfrak{G}_1 - \mathfrak{G})^{-1}\mathfrak{G}(\mathbf{p}' - \mathbf{p}) - [\mathbf{p}' - \mathbf{p}(1 + h/|\mathbf{p}|)]]. \tag{13-38}$$

Thus, two plates of the Moon exposed at widely different libration angles are sufficient to calculate both $h$ and $\delta\mathbf{p}$ and from them the absolute coordinates of the point P. However, the size of the differential displacement of objects situated off the "mean sphere" is very small and, hence, repeated measurements of many photographs must be used to improve the accuracy of the results.

One essential addition to the above is the correction of the error orginating from the variation with the libration in latitude of the center $M$ of the visible disc. This correction must take the form of a translational term in transformation (13-27). If $\mathfrak{H}$ is the three-by-one correcting matrix, then Equation (13-27) should read

$$\mathfrak{G}\mathbf{p} + \mathfrak{H} = (\xi, \eta, \zeta); \tag{13-39}$$

whereas, for the second plate, the equivalent transformation must be

$$\mathfrak{G}_1\mathbf{p}_1 + \mathfrak{H}_1 = (\xi_1, \eta_1, \zeta_1), \tag{13-40}$$

where $\mathfrak{H}_1 \neq \mathfrak{H}$.

The elements of the stereoscopic method described in the preceding paragraphs go back to FRANZ (1901), SAUNDER (1905); and were subsequently developed by RITTER (1934) and especially GOUDAS (1965a) whose presentation of the subject we have largely followed in this chapter. The observations intended to put this theory to task have also been initiated by FRANZ (1901), SAUNDER (1905, 1907, 1911); and, in more recent years, carried on by SCHRUTKA-RECHTENSTAMM (1956, 1958), BALDWIN (1963) and others. SCHRUTKA (1958) measured on seven Lick plates the positions of 150 points covering only a portion of the visible lunar disk (not extending too far to the limb). BALDWIN (1963) determined similarly the positions of 696 points on the Moon on the basis of his measurements of five Lick plates; while, more recently, the U.S. Army Map Service (cf., BREECE, HARDY and MARCHANT, 1964) published a list of the positions of 256 points based on the measurements of 15 Lick plates. Still more recently, the Aeronautical Chart and Information Center (ACIC) of the U.S. Air Force published a new set of independent hypsometric data for 196 points, based on the measurements of 40 Pic-du-Midi plates; and by the use of sequential photography succeeded in eliminating better than all their predecessors the positional errors arising from the unsteadiness of seeing (MEYER and RUFFIN, 1965).

A harmonic analysis of the surfaces defined by these independent sets of data, and a systematic study of their deviations from the mean sphere was recently undertaken by GOUDAS (1963; 1964abc; 1965abc). In doing so, he assumed the actual lunar surface to be expansible in a series of tesseral harmonics of the form

$$r(a, \lambda, \beta) = a\left\{ 1 + \sum_{i=0}^{j} \sum_{j=0}^{\infty} [J_{i,j} \cos i\lambda + K_{i,j} \sin i\lambda] P_j^i(\beta) \right\}, \tag{13-41}$$

where $P_j^i(\beta)$ stands for the associated Legendre polynomials in the latitude of order $j$ and index $i$; and $J_{i,j}$, $K_{i,j}$ are numerical coefficients to be determined from the observations. By known properties of surface harmonics, $P_j^i = 0$ for $i > j$; and in order to facilitate the solution, Goudas assumed that $J_{2j-1}^i = K_j^{2i} = 0$ for $i, j = 1, 2...$, implying the geometrical similarity of the near and far side of the Moon. With the local values of $r(a, \lambda, \beta)$ given by the stereoscopic method for points of known positions $\lambda$ and $\beta$, Equation (13-41) becomes linear in the coefficients $J_j^i$, $K_j^i$, and can be solved for them by the method of least squares. As many equations of condition of the form (13-41) are available for this purpose – which means 150 equations based on the work of Schrutka-Rechtenstamm, 696 based on that of Baldwin; 256 for the Army Map Service; and 196 for ACIC.

The results of such solutions (performed with the aid of the electronic computers at Manchester and Seattle) have been given by Goudas independently for the four observers; and a partial tabulation of the corresponding values of $10^5 J_{i,j}$ and $10^5 K_{i,j}$ for the first three harmonics of orders $j=2$, 3 and 4 are given in Table 13-1. Mean errors of these quantities are listed for the results based on the measurements of Schrutka and AMS.

Thus, for instance, according to Goudas's analysis of Schrutka's data (GOUDAS, 1964c), the equation (13-41) of the lunar surface assumes the more explicit form

$$r(a, \lambda, \beta) = 1737.45 \pm 0.87$$

$$- 0.45 \sin \beta + 0.17 \cos \beta \sin \lambda$$
$$- 0.30(3 \sin^2 \beta - 1) + 0.21 \cos^2 \beta \cos 2\lambda \qquad (13\text{-}42)$$
$$- 0.18(5 \sin^2 \beta - 3) \sin \beta + 0.18(5 \sin^2 \beta - 1) \cos \beta \sin \lambda$$
$$- 0.42 \sin \beta \cos^2 \beta \cos 2\lambda + 0.04 \cos^3 \beta \sin 3\lambda$$
$$+ 0.17(35 \sin^4 \beta - 30 \sin^2 \beta + 3) + 0.27(7 \sin^2 \beta - 3) \sin 2\beta \sin \lambda$$
$$- 0.33(7 \sin^2 \beta - 1) \cos^2 \beta \cos 2\lambda$$
$$- 0.41 \sin \beta \cos^3 \beta \sin 3\lambda + 1.16 \cos^4 \beta \cos 4\lambda + \dots \text{ kms;}$$

and similarly from the results of other investigators. Hypsometric contours based on

TABLE 13-1

Values of $10^5 J_{i,j}$ and $10^5 K_{i,j}$

| Coefficient | Baldwin | Schrutka | AMS | ACIC |
|---|---|---|---|---|
| $J_{0,2}$ | 0.94 | $-0.14 \pm 0.41$ | $-0.33 \pm 0.39$ | $-0.46$ |
| $J_{1,2}$ | 0 | 0 | 0 | 0 |
| $J_{2,2}$ | $-0.03$ | $0.28 \pm 0.20$ | $0.69 \pm 0.12$ | 0.29 |
| $J_{0,3}$ | 0.14 | $-0.56 \pm 0.10$ | $0.13 \pm 0.01$ | $-0.13$ |
| $J_{1,3}$ | 0 | 0 | 0 | 0 |
| $J_{2,3}$ | $-0.15$ | $-0.05 \pm 0.01$ | $0.13 \pm 0.01$ | $-0.06$ |
| $J_{3,3}$ | 0 | 0 | 0 | 0 |
| $J_{0,4}$ | 0.29 | $1.13 \pm 0.12$ | $-0.59 \pm 0.11$ | 0.25 |
| $J_{1,4}$ | 0 | 0 | 0 | 0 |
| $J_{2,4}$ | 0.01 | $0.05 \pm 0.03$ | $-0.11 \pm 0.01$ | $-0.07$ |
| $J_{3,4}$ | 0 | 0 | 0 | 0 |
| $J_{4,4}$ | 0.006 | $0.009 \pm 0.003$ | $0.009 \pm 0.004$ | 0.009 |
| $K_{1,2}$ | 0.55 | $0.59 \pm 0.05$ | $0.49 \pm 0.02$ | 0.57 |
| $K_{2,2}$ | 0 | 0 | 0 | 0 |
| $K_{1,3}$ | $-0.08$ | $0.10 \pm 0.37$ | $0.15 \pm 0.01$ | 0.15 |
| $K_{2,3}$ | 0 | 0 | 0 | 0 |
| $K_{3,3}$ | 0.06 | $0.001 \pm 0.005$ | $0.073 \pm 0.003$ | $-0.007$ |
| $K_{1,4}$ | 0.28 | $0.33 \pm 0.04$ | $0.19 \pm 0.01$ | 0.28 |
| $K_{2,4}$ | 0 | 0 | 0 | 0 |
| $K_{3,4}$ | 0.002 | $-0.006 \pm 0.001$ | $0.002 \pm 0.001$ | 0.016 |
| $K_{4,4}$ | 0 | 0 | 0 | 0 |

the results of Goudas's harmonic analysis of the data by Schrutka-Rechtenstamm, Baldwin, the Army Map Service and ACIC are shown in projection on the actual lunar surface on the accompanying Figures 13-8, 13-9, 13-10, and 13-11 (after GOUDAS, 1964c and 1965c).

An intercomparison of these figures reveals that the similarity of the surfaces as defined by the three independent sets of the data is rather vague – if one can speak of a similarity at all. When the outcome of the work of different competent investigators

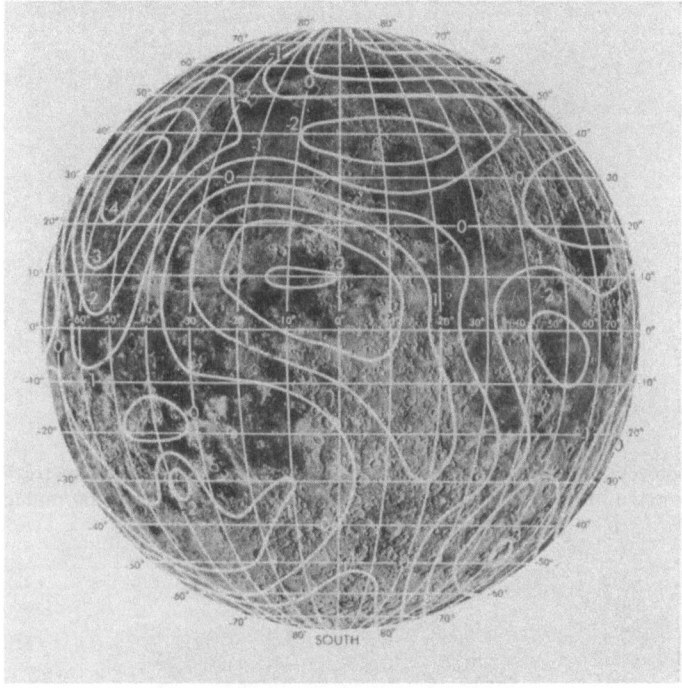

Fig. 13-8. Isolevel contour map of the Moon, deduced from the hypsometric data by SCHRUTKA-RECHTENSTAMM (1958) and projected on the LEM-1B USAF Lunar Mosaic chart. Eight surface harmonics are used in the representation (after GOUDAS, 1964c).

based on comparable observational material (i.e., the Lick plates) thus leads to results of only marginal similarity, the logical conclusion to be drawn from such a fact would be to regard the underlying data on the limit of significance, and in need of improvement in quality as well as quantity.

That they are indeed likely to be so was already anticipated by the arguments given on page 172, when it was pointed out that stereoscopic shifts expected as a result of altitude differences of 2–3 km on the Moon during the entire libration cycle are of the same order of magnitude as the optical resolving power of the telescopes employed so far for this type of work. Positional astronomers have, to be sure, long been accustomed to search for information inside the optical diffraction patterns of the light sources – and have done so extensively, e.g., in the measurements of stellar parallaxes. The

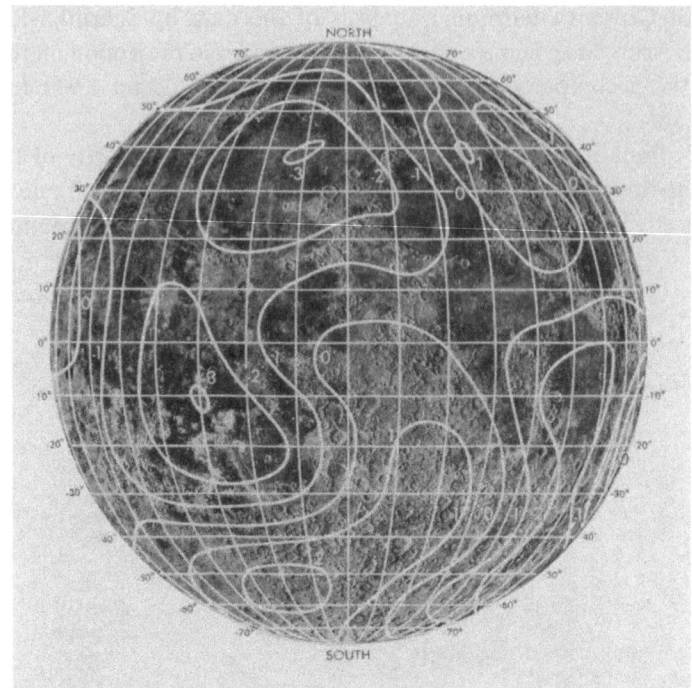

Fig. 13-9. Isolevel contour map of the Moon, as deduced from the hypsometric data by BALDWIN (1963) and projected on the LEM-1B USAF mosaic chart. Eight surface harmonics are used in the representation (after GOUDAS, 1964c).

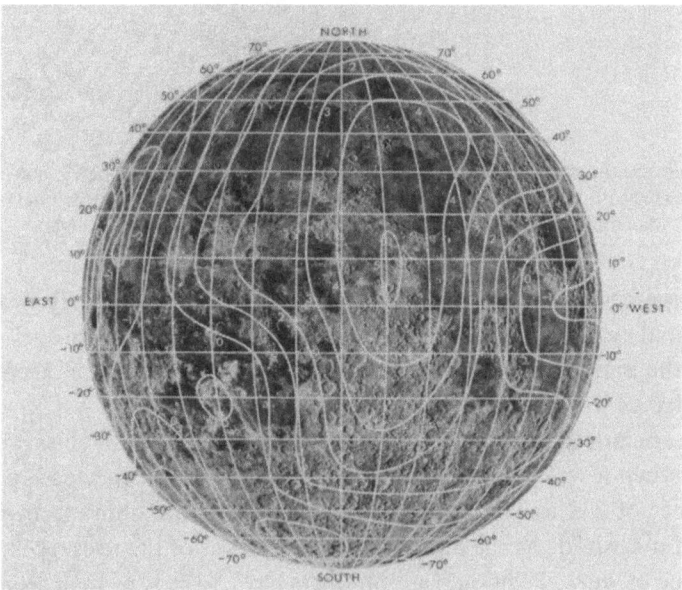

Fig. 13-10. Isolevel contour map of the Moon, as deduced from the hypsometric data by the U.S. Army Map Service and projected on the LEM-1B USAF Lunar mosaic chart. Eight surface harmonics are used in the representation (after GOUDAS, 1964c).

success of such a process requires, however, first a knowledge of the geometrical relation of the actual shape of the respective light source to its diffraction image (i.e., a light point to a central disk surrounded by a series of diffraction rings in the case of a star); and, secondly, the measurement of a great many plates to minimize the accidental errors of such a procedure.

In the case of selenodetic measurements the first condition cannot, unfortunately, be met; for the intrinsic outlines of lunar formations on which micrometric settings

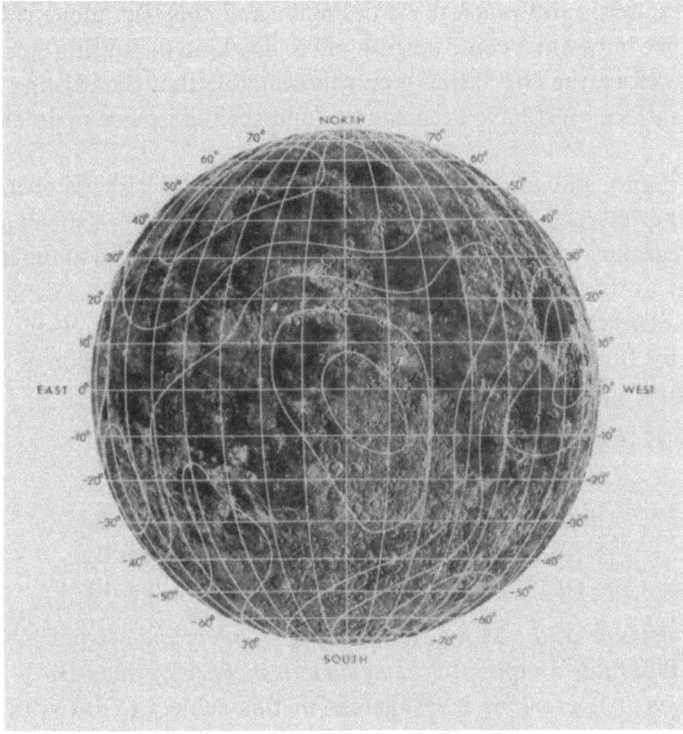

Fig. 13-11.   Isolevel contour map of the Moon, deduced from the hypsometric data by MEYER and RUFFIN (1965) and projected on the LEM-1B Lunar mosaic chart. Eight surface harmonics are used in the representation (after GOUDAS, 1965c).

are made are not known to us *a priori*; and neither is, therefore, the shape of their diffraction image (which may, moreover, vary with the phase as a result of different direction of illumination). Furthermore, a sheer weight of such work prevented Baldwin, Schrutka, or the investigators of AMS from measuring a sufficient number of plates. The five negatives measured by Baldwin, seven plates measured by Schrutka, or fifteen plates measured at the AMS are too few in comparison with 50–100 plates usually required for significant measurements of stellar parallaxes. Under these conditions, their data could scarcely have been of more than marginal accuracy for the problem in question; and, as a result, the outcome of Goudas's analysis revealing but superficial resemblance of the lunar surfaces based on such data cannot, perhaps, be regarded as unexpected.

Even within these limitations, the data on hand as analyzed by Goudas lend themselves to certain conclusions which may be pointed out in this place. First, none of the sets of available data when expanded in a harmonic surface of the form (13-41) shows a preponderance of the second zonal harmonic in the direction of the Earth, which could be interpreted as a "tidal bulge". In point of fact, the largest coefficients appear to be associated with the *fourth* harmonic (even though the consistency between different sets of the data leads still much to be desired); and this fact alone rules out a possibility that such harmonics could correspond to any kind of equilibrium tides (present or fossil). Therefore, the conclusion seems inescapable that the existing deformations of the lunar surface must have been caused (and their existence maintained) by other forces.

The uncertainty still surrounding the details of the exact shape of the lunar globe is considerably lessened in the marginal zone, where a more direct observational approach is feasible. A harmonic expansion of the apparent form of the lunar meridian perpendicular to the line of sight at the time of zero libration, as defined by the ensemble of all observational data from Hayn through Watts (and also the more recent eclipse data by Brooks and Davidson), led at the hands of GOUDAS (1965b) to an expansion of the apparent radius of the Moon of the form

$$
\begin{aligned}
r'' = 914{.}''61 &\pm 0{.}''01 \\
&- (0{.}''25 \pm 0{.}''05) \sin \beta - (0{.}''07 \pm 0{.}''08) \cos \beta \\
&+ (0{.}''46 \pm 0{.}''09) \sin 2\beta - (0{.}''24 \pm 0{.}''08) \cos 2\beta \\
&+ (0{.}''29 \pm 0{.}''05) \sin 3\beta - (0{.}''07 \pm 0{.}''04) \cos 3\beta \\
&- (0{.}''19 \pm 0{.}''08) \sin 4\beta + (0{.}''16 \pm 0{.}''05) \cos 4\beta + \ldots ,
\end{aligned}
\tag{13-43}
$$

at the mean distance of the Moon, rendering *the limb of the Moon a deformed ellipse, elongated* (rather than compressed) *along an axis inclined by about* $35° \pm 2°$ *to the lunar axis of rotation*. All previous investigators of this subject (YAKOVKIN, 1952; DOMMANGET, 1962; POTTER, 1962; WATTS, 1963; BROOKS and DAVIDSON, 1963) are in essential agreement on this fact. Yakovkin estimated independently this inclination to be 23°; Watts, to 35°; Potter and Bystrov, to 34°; while Brooks and Davidson, to 37°; and all agree that this inclination lies in the same quadrant (i.e., from SW to NE). The semi-major axis of the figure of symmetry deviates, therefore, markedly from the Moon's principal axes of inertia; and the difference between the two semi-diameters amounts to between 2–3 km in length. This fact forces us to the conclusion that *the Moon cannot be regarded as a homogeneous body*; and that appreciable changes in its mass-distribution must have occurred which deformed noticeably the surface but did not influence the positions of the principal axes; but what these forces could have been can so far be only guessed at.

## CHAPTER 14

## RELATIVE COORDINATES ON THE MOON AND
## THEIR DETERMINATION

In the foregoing chapter we have outlined several methods for the determination of absolute (three-dimensional) coordinates of individual points on the lunar surface, and pointed out that determinations of the radial coordinates (i.e., of the distance of any such point from the Moon's center) confront us with some of the most difficult and exacting problems encountered anywhere in the domain of astrometry. The outcome of such studies, and the harmonic analysis of their results, has revealed that within the limits of $\pm 3$ km (i.e., $\pm 0.2$ per cent of the lunar radius) the Moon is essentially a sphere of mean radius of 1738 km; though deviations of the actual surface from this sphere appear to be quite complicated, and their physical significance is still largely obscure.

The aim of the present chapter will be to consider a somewhat easier, and astrometrically less exacting, problem: namely, the determination of the *relative coordinates* of individual points on the Moon, consisting of the specification of their angular positions on the mean sphere of the lunar surface, and of their altitudes above the surrounding landscape. As we shall show below, this task still calls for highest-precision angular measurements, but – and this is significant – no longer of large angles, comparable with the apparent semi-diameter of the Moon; and this will make so much difference in the sheer weight of work that, not hundreds, but thousands of lunar points had their selenographic coordinates established in recent years for cartographic purposes. The data already on hand are too voluminous and extensive to be fully surveyed in a survey in this place; the space at our disposal will necessitate limiting our exposition to the essentials of the underlying methods.

In order to do so, let us consider (cf. Figure 14-1) an arbitrary point

$$P(p=a+h, \lambda, \beta)$$

anywhere on the visible hemisphere of the Moon, and turn first to a determination of its angular coordinates $\lambda_P$ and $\beta_P$. This can be accomplished by an obvious extension of the method used already in the preceding section to determine the selenographic coordinates of limb points. For let (as on Figure 13-2) $P(x, y)$ denote now any point on the apparent disc on the Moon, such that

$$x^2 + y^2 = R^2 \leqslant k^2; \tag{14-1}$$

and let the equation

$$\tan \Pi = \frac{x}{y} \tag{14-2}$$

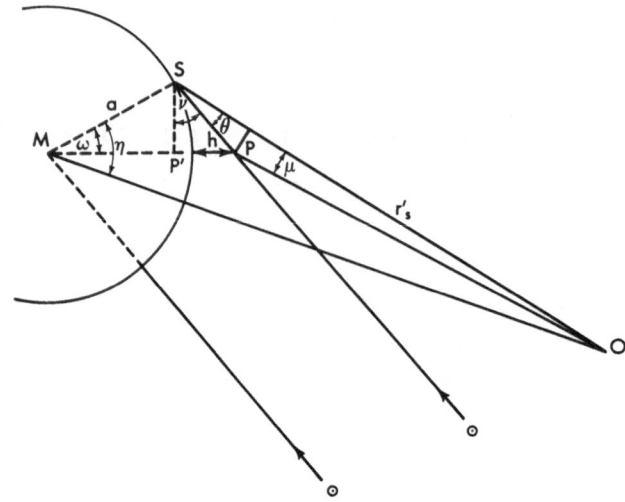

Fig. 14-1.  Geometry of the lunar shadows.

continue to define its position angle $\Pi$. If so, the selenocentric angle $\kappa$ between the radii vectors to the observer at $O$ and to $P$ will obviously be defined by the relation

$$\sin \kappa = R/p . \tag{14-3}$$

In view of the finite distance of the Moon, our unit of length $k$ in the $xy$-plane of Figure 13-2 is, however, related with its true radius $p = a + h$ by the equation

$$k = p \cos s'_{\mathbb{C}}, \tag{14-4}$$

where $s'_{\mathbb{C}}$ denotes, as before, the apparent lunar semi-diameter at the time and place of observation. If so, a combination of (13-3) and (13-4) reveals that

$$\kappa = \sin^{-1} \{(R/k) \cos s'_{\mathbb{C}}\} . \tag{14-5}$$

The reader may note that, if the point $P$ were moved to the limb (where $R = k$)

$$\sin \kappa = \cos s'_{\mathbb{C}} = \sqrt{(1 - p^2/r'^2)},$$

in agreement with Equations (13-5).

With this new generalised definition of $\kappa$, the equations defining the selenographic coordinates $\lambda_P$ and $\beta_P$ of the point $P$ continue to be of the form (13-3). Moreover, if we set

$$\tan \kappa \cos (\Pi - C') = \tan Q , \tag{14-6}$$

Equations (13-3) can be solved to yield

$$\tan (\lambda_P - l') = \frac{\tan (\Pi - C')}{\cos (Q + b')} \sin Q . \tag{14-7}$$

Conversely, if the values of $\lambda_P$, $\beta_P$ as well as $l'$, $b'$ are given and the task is to evaluate the position of the respective point on the lunar face in the $xy$-coordinates, the equations

$$
\left.
\begin{aligned}
\cos \kappa &= \sin b' \sin \beta_P + \cos b' \cos \beta_P \cos (\lambda_P - l'), \\
\cos (\Pi - C') \sin \kappa &= \cos b' \sin \beta_P - \sin b' \cos \beta_P \cos (\lambda_P - l'), \\
\sin (\Pi - C') \sin \kappa &= \qquad\qquad + \qquad \cos \beta_P \sin (\lambda_P - l')
\end{aligned}
\right\} \quad (14\text{-}8)
$$

can be used to evaluate $\kappa$ and $\Pi$, from which

$$
x = k \sec s'_{\mathbb{C}} \sin \kappa \sin \Pi \qquad\qquad (14\text{-}9)
$$

and

$$
y = k \sec s'_{\mathbb{C}} \sin \kappa \cos \Pi. \qquad\qquad (14\text{-}10)
$$

The foregoing systems of Equations (13-3) or (14-8) can be re-stated in the following more economic way. Let

$$
\left.
\begin{aligned}
\xi &= \cos \beta_P \sin \lambda_P \\
\eta &= \sin \beta_P \\
\zeta &= \cos \beta_P \cos \lambda_P
\end{aligned}
\right\} \quad (14\text{-}11)
$$

be the direction cosines of the radius from the Moon's center to the point $P$. If so, Equations (13-3) can evidently be rewritten as

$$
\left.
\begin{aligned}
\xi &= \quad x \cos l' - y \sin b' \sin l' + z \cos b' \sin l', \\
\eta &= \qquad\qquad + y \cos b' \qquad + z \sin b', \\
\xi &= -x \sin l' - y \sin b' \cos l' + z \cos b' \cos l'
\end{aligned}
\right\} \quad (14\text{-}12)
$$

where, for a spherical Moon of mean radius $a$,

$$
x^2 + y^2 + z^2 = 1; \qquad\qquad (14\text{-}13)
$$

and, by their inversion,

$$
\left.
\begin{aligned}
x &= \quad \xi \cos l' \qquad\qquad\qquad - \zeta \sin l', \\
y &= -\xi \sin l' \sin b' + \eta \cos b' - \zeta \cos l' \sin b', \\
z &= \quad \zeta \sin l' \cos b' + \eta \sin b' + \zeta \cos l' \cos b'.
\end{aligned}
\right\} \quad (14\text{-}14)
$$

The direction cosines $\xi, \eta, \zeta$ may evidently be regarded as the rectangular coordinates of a point $P$ at unit distance from the Moon's center; the $\zeta$-axis being oriented toward the Earth along the lunar "first radius" (defined as the intersection of the equator and prime meridian), and the $\eta$-axis coinciding with the Moon's axis of rotation. Their values are often used to describe the position of a point on the lunar surface in place of the angular coordinates $\lambda_P$ and $\beta_P$.

If, for some purpose, it were required to obtain the selenocentric right-ascension $\alpha_{;}$ and declination $\delta_P$ on the lunar surface, of known selenographic coordinates $\lambda_P$, $\beta_P$, this can be done with the aid of the equations

$$
\left.
\begin{aligned}
\cos \delta_P \cos (\alpha_P - \Omega') &= \cos \beta_P \cos \lambda_P, \\
\cos \delta_P \sin (\alpha_P - \Omega') &= \cos \beta_P \sin \lambda_P \cos i - \sin \beta_P \sin i, \\
\sin \delta_P &= \cos \beta_P \sin \lambda_P \sin i + \sin \beta_P \cos i,
\end{aligned}
\right\} \quad (14\text{-}15)
$$

where $i$ denotes the angle of inclination between the lunar and terrestrial equators; and $\Omega'$, the longitude of the ascending node of the lunar equator, as given by Equation (3-6). The selenocentric ecliptical coordinates $L_P$, $B_P$ would follow, similarly, from the transformation equations

$$
\left.
\begin{aligned}
\cos B_P \cos (\Omega - L_P) &= - \cos \beta_P \cos (\Omega - \lambda_P - l_\odot) \\
\cos B_P \sin (\Omega - L_P) &= - \cos \beta_P \cos I \sin (\Omega - \lambda_P - l_\odot) + \sin \beta_P \sin I, \\
\sin B_P \qquad\qquad &= + \cos \beta_P \sin I \sin (\Omega - \lambda_P - l_\odot) + \sin \beta_P \cos I,
\end{aligned}
\right\} \quad (14\text{-}16)
$$

where $\Omega$ denotes the mean longitude of the ascending node of lunar orbit on the ecliptic; $l_\odot$, the mean longitude of the Sun; and $I$, the inclination of the lunar equator to the ecliptic. The two systems $L_P$, $B_P$ and $\lambda_P$, $\beta_P$ differ only in so far as the latitude $B_P$ is measured from a plane passing through the center of the Moon and parallel with the ecliptic, while $\beta_P$ is measured from the lunar equator; and $L_P$ is measured from the vernal equinox, while $\lambda_P$, from the lunar prime meridian.

All foregoing geometrical formulae relating the angular selenographic coordinates of point $P$ with its rectangular coordinates $x$, $y$ of its projection on the celestial sphere assume the measurements of the latter to be freed from any systematic observational errors. Such errors are bound to arise from two principal sources:

(a) differential refraction, due to a difference in zenith distance between point $P$ and the center of the apparent lunar disk; and

(b) optical distortion of images in the focal plane of the telescope employed (or, rather, osculating plane to the actual focal surface), arising from the difference in scale-off axis; as well as from a finite inclination of the photographic plate from a plane perpendicular to the optical axis.

The effects of differential refraction are relatively small for observations carried out not too far from zenith, and can be amply accounted for in reductions by use of the well-known and closely approximate formulae for differential corrections

$$
\Delta\sigma = f(z)\{1 + \tan^2 z \cos^2 (\pi - M)\}\, \sigma \qquad (14\text{-}17)
$$

and

$$
\Delta\pi = - f(z)\{\tan^2 z \cos^2 (\pi - M) \tan (\pi - M) + \tan z \sin M \tan \delta'_{\mathbb{C}}\} \quad (14\text{-}18)
$$

to the measured angular separation $\sigma$ and position angle $\pi$ of any two points on the lunar surface, where the angle $z$ denotes the zenith distance of a point half-way across the arc $\sigma$; $f(z)$ is the amount of refraction at that zenith distance obtainable from standard tables; and the auxiliary angle $M$ is defined by the system of equations

$$
\cos M = \cot (N + \delta'_{\mathbb{C}}) \cot z, \qquad (14\text{-}19)
$$

$$
\tan N = \tan \delta_\odot \sec (\alpha_\odot - \alpha'_{\mathbb{C}}), \qquad (14\text{-}20)
$$

where $\alpha'$, $\delta'$ stand for the topocentric right-ascension and declination of a point half-way between the two respective points.

The errors of the category (b) arising from limitations of the optical system employed are, unfortunately, more troublesome. Only the heliometers (performing, by their nature, all measurements always on their optical axis) are reasonably free from them; but only at the price of a relatively small aperture (and, therefore, resolving power) of the optics employed. For large telescopes of more adequate resolving power, the errors arising from (b) just cannot be eliminated, and must be accounted for by independent calibration.

The best way to effect such a calibration would be by a simultaneous photograph of the adjacent star field. In practice this is, unfortunately, impossible because the optimum exposures needed to bring out lunar features are far too short to record stellar images in the neighbourhood of the lunar limb; and one cannot go after sufficiently bright stars too far a field because of the limited size of plates. In effect, the only calibration on which we can fall back to determine the requisite plate constant must be found on the lunar face itself; and the one fundamental point already mentioned – namely, the position of the position of the crater Mösting A – is obviously not enough. Tobias Mayer – the father of positional selenography – chose the central mountain of the crater Manilius as his fundamental point of reference (a choice in which he was followed by Bouvard and Nicollet); but since the days of Bessel, all more modern observers have adopted Mösting A for this purpose. Franz used, besides Mösting A, the positions of the craters Aristarchus, Byrgius A, Fabricius K, Gassendi, Macrobius A, Nicollet A, Proclus, and Sharp A as a system of nine fundamental reference points on the surface of the Moon; while Hayn used only five (i.e., Egede A, Kepler A, Messier A, Mösting A, and Tycho). Some positions of these reference points were more recently re-measured by SCHRUTKA-RECHTENSTAMM (1958).

The details just mentioned constitute a group of lunar triangulation points of first order, based (very largely) on heliometric measurements. With their aid, the positions of a much larger number of secondary points were determined, partly visually (FRANZ, 1901) but mainly photographically (SAUNDER, 1905, 1911; KÖNIG and others). Franz's list contains the positions of a total of 1446 secondary reference points on the surface of the Moon. Saunder measured 2885 of them from negatives secured by Loewy and Puiseux with the equatorial coudé at Paris (SAUNDER, 1905), and by Ritchey with the 40-inch Yerkes refractor (SAUNDER, 1911); while König's results, while more numerous, were based on measurements of plates taken with a much smaller instrument (8-inch refractor of 343 cm focal length). They were only partly reduced after König's premature death by others (Müller, Fischer) and partly remain still unpublished. Saunder's measurements were recently used, by graphical interpolation, to furnish the coordinate net of the *Orthographic Atlas of the Moon* (ARTHUR *et al.*, 1961); but the errors inherent in such a process were subsequently pointed out by HAWKINS (1963a).

So far we have been satisfied to locate the position of point $P$ on the lunar surface by assuming the latter to lie on a sphere of unit radius. Suppose, however, now that we wish to drop this assumption and set out to determine the *differences* in magnitude of the radii vectors from the Moon's center to two points $P_{1,2}$ – i.e., the *relative* elevation of one point with respect to the other. A determination of absolute magnitude of such vectors constitutes a difficult problem which has already been discussed, to some extent, in the preceding Chapter 13. What we wish to do now is to address ourselves to a simpler problem: namely, one of *relative* height determination from the *shadows* cast by a particular eminence on the surrounding terrain in the rays of the rising or setting Sun.

In order to ascertain such relative elevations of specific lunar features from the measured lengths of their shadows, let us depart from the geometry as shown on the accompanying Figure 14-1, with the aim of determining the height $h$ of a sunlit lunar eminence $P$, casting a shadow whose tip, as seen from the Earth, is situated at another point $S$. Let the slant range between $P$ and $S$ be denoted by $\delta$, and the distance between $S$ and the observing site $O$ on Earth be $r'_s$. Let, moreover, the angle $\theta$ at $S$ between the vectors $PS \equiv \mathbf{s}$ and $OS \equiv \mathbf{r}'_s$ represent the difference in elevation of the peak $P$ and the observer $O$ above the lunar horizon. If so, then by solving for $\mathbf{s}$ from the triangle $OPS$ we find that

$$\mathbf{s} = \frac{r'_s \sin \mu}{\sin (\theta + \mu)}, \tag{14-21}$$

where $\mu$ denotes the (topocentric) angular length of the shadow cast by $P$ as seen from $O$.

Furthermore, if $\omega$ denotes the selenocentric length of this shadow (i.e., the angle $PMS$; cf. again Figure 14-1) and $a$ stands for the mean radius of the Moon, it follows from the triangles involved that

$$a \sin \omega = \mathbf{s} \cos v \tag{14-22}$$

and

$$(a + h) \cos v = a \cos (v - \omega), \tag{14-23}$$

where $v$ is the angle $PSP'$. Our aim is to solve this latter equation for $h$. If we divide both sides of it by $a \cos v$, we find that

$$\frac{h}{a} = \frac{\sin (\omega - v)}{\cos v} - 1 = \sin \omega \tan v - 2 \sin^2 \tfrac{1}{2} \omega, \tag{14-24}$$

where from an elimination of $\mathbf{s}$ between (14-21) and (14-22) it readily transpires that

$$\sin \omega = \frac{r'_s \sin \mu \cos v}{a \sin (\theta + \mu)}. \tag{14-25}$$

Of the auxiliary quantities involved on the right-hand side of the foregoing equation, the topocentric angle $\mu$ can be directly measured. The topocentric distance $\mathbf{r}'_s$ of point $S$ (cf. Figure 14-2) follows from the triangles $EMO$ and $MOS$ (not necessarily co-planar) as a solution of the equations

$$r'^2_s = r'^2 + a^2 - 2 ar' \cos \eta, \tag{14-26}$$

$$r'^2 = r^2 + \rho^2 - 2 \rho r \cos \gamma, \tag{14-27}$$

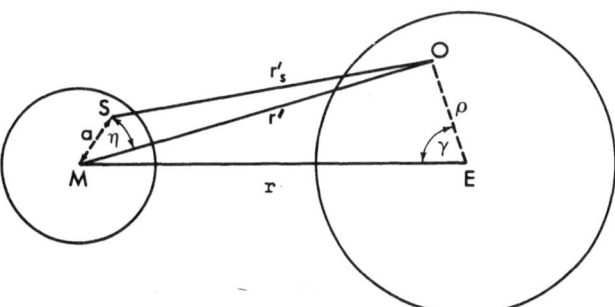

Fig. 14-2. Geometry of the Earth-Moon configuration.

where $a \equiv MS$ denotes the mean radius of the Moon; $\rho \equiv OE$, the distance between the center of the Earth and the observing place; and $\gamma$, the geocentric angle between the vectors $\overrightarrow{EO}$ and $\overrightarrow{EM}$; while $\eta$ is the selenocentric angle between $MO$ and $MS$.* An elimination of $r'$ between (14-26) and (14-27) leads to

$$
\begin{aligned}
r_s'^2 &= r^2 + \rho^2 + a^2 - 2a\rho \cos \gamma - 2a \left(r^2 + \rho^2 - 2\rho r \cos \gamma\right)^{\frac{1}{2}} \cos \eta \\
&= r^2 \left\{1 + p^2 + q^2 - 2p \cos \gamma - 2q \left(1 - p \cos \gamma + \tfrac{1}{2} p^2 + \ldots\right) \cos \eta\right\},
\end{aligned} \quad (14\text{-}28)
$$

where we have abbreviated

$$
p = \frac{\rho}{r} \quad \text{and} \quad q = \frac{a}{r}. \quad (14\text{-}29)
$$

For the mean values of $r = 384\,400$ km and $\rho = 6371$ km while $a = 1738$ km, $p = 0.01657$ and $q = 0.004521$. If we regard these latter ratios small enough for their squares and cross-products to be negligible, the square-root of Equation (14-27) can be approximated by

$$
\mathbf{r}_s' = r \left(1 - p \cos \gamma - q \cos \eta + \ldots\right) \quad (14\text{-}30)
$$

within a small fraction of one per cent – with $r$ varying between 364\,400 and 406\,730 km in the course of each month on account of the eccentricity of the lunar orbit.

In order to complete the evaluation of the angle $\omega$ as defined by (14-25) from the measured values of $\mu$, we must specify $\gamma$, $\eta$ as well as $\theta$ and $v$ in terms of observable quantities; and this can be accomplished in the following manner. The geocentric angle $\gamma$ is (from the triangle $EMO$; cf. Figure 14-2) defined clearly by

$$
\cos \gamma = \sin \delta_{\mathbb{C}} \sin \varphi + \cos \delta_{\mathbb{C}} \cos \varphi \cos (\alpha_{\mathbb{C}} - \Theta), \quad (14\text{-}31)
$$

where $\alpha_{\mathbb{C}}$, $\delta_{\mathbb{C}}$ denote the Moon's right ascension and declination; $\Theta$, the sidereal time of observation (the difference $\alpha_{\mathbb{C}} - \Theta = H$ being equal to the lunar geocentric hour

---

* The reader may note that, in terms of our preceding notations, $\kappa = \eta - \omega$.

angle); and $\varphi$, the geocentric latitude of $O$ – in terms of which the variation of the terrestrial radius vector $\rho$ with latitude can be expressed as

$$\left.\begin{aligned}
\rho &= 6378 \left(1 - 0.00235 \sin^2\varphi + \dots\right) \text{ km} \\
&= r \sin \pi_{\mathbb{C}} \left(1 - 0.00235 \sin^2\varphi + \dots\right) \text{ km}
\end{aligned}\right\} \tag{14-32}$$

where $\pi_{\mathbb{C}}$ stands for the equatorial horizontal parallax of the Moon. Similarly, from the triangle $OMS$,

$$\cos \eta = \sin b' \sin \beta_S + \cos b' \cos \beta_S \cos (l' - \lambda_S), \tag{14-33}$$

where $b'$, $l'$ are the topocentric librations of the Moon as defined in Chapter 3 (representing, in effect, the selenographic latitude $\beta_0$ and longitude $\lambda_0$ of the observer's position at $O$); and $\beta_S$, $\lambda_S$, the selenographic latitude and longitude of the point $S$ at which the tip of the shadow is cast on the lunar surface. Moreover, if $A_\odot$, $D_\odot$ denote selenographic coordinates of the Sun as seen from the Moon*, the angle $\theta$ between the direction to the Sun and the observer at $O$ will be given by

$$\cos \theta = \sin b' \sin D_\odot + \cos b' \cos D_\odot \cos (l' - A_\odot), \tag{14-34}$$

while the angle $v$ equal to the altitude of the Sun above the lunar horizon at $P$ (cf. again Figure 14-1) follows from the equation

$$\sin v = \sin \beta_P \sin D_\odot + \cos \beta_P \cos D_\odot \cos (\lambda_P - A_\odot) \tag{14-35}$$

where $\beta_P$, $\lambda_P$ denote the selenographic coordinates of the point $P$ on the lunar surface.**

With the aid of the auxiliary quantities and angles defined by the foregoing equation, the selenocentric angle $\omega$ as given by Equation (14-25) can readily be determined in terms of the measured shadow length $\mu$; and with its aid, the fractional height $h/a$ of the shadow-casting eminence $P$ evaluated from (14-24). An alternative way of determining $\omega$ would be to do so from the measured pairs of coordinates $\beta_P$, $\lambda_P$ and $\beta_S$, $\lambda_S$ in place of one of them being combined with $\mu$. If the values of both $\beta_{P,S}$ and $\lambda_{P,S}$ have been determined with the requisite precision, by the method outlined earlier in this section, it follows immediately that

$$\cos \omega = \xi_P \xi_S + \eta_P \eta_S + \zeta_P \zeta_S, \tag{14-36}$$

where $\xi_{P,S}$ etc. are direction cosines of the lunar radii-vectors to the surface points $P$, $S$, of the form (14-11); and, consequently,

$$\cos \omega = \sin \beta_P \sin \beta_S + \cos \beta_P \cos \beta_S \cos (\lambda_P - \lambda_S). \tag{14-37}$$

---

* The difference between selenocentric aud topocentric values of these coordinates on the Moon becomes immaterial because of the Moon's small size and great distance of the Sun.
** Strictly speaking, these coordinates refer to the point at which the rays from the center of the apparent disk of the Sun are tangent to the shadow-casting obstacle. Should – as may frequently be the case – the profile of this obstacle be convex, the selenographic position of the tangent point may shift somewhat during sunrise or sunset with the varying altitude of the Sun above the lunar horizon. The apparent time displacements of $P$ due to this cause are, however, likely in most cases to be too small to be significant.

Once the angle $\omega$ has thus been determined from known values of $\beta_P$, $\lambda_P$ and $\beta_S$, $\lambda_S$, and the angle $v$ from (14-35), Equation (14-24) then furnishes at once the desired ratio $h/a$. This procedure for determining $h$ may, on the face of it, seem more straightforward than the one outlined earlier, which leads to $h$ via the measured (topocentric) shadow length $\mu$. In actual fact this is, however, scarcely the case; for its complexity is stored in the absolute determination of two pairs of coordinates from the measured pair of plane coordinates $x_{P,S}$ and $y_{P,S}$, by a method outlined at the outset of this chapter; and this implies a knowledge of lunar topocentric librations $b'$, $l'$ as well as of the observer's position on Earth.

Throughout all foregoing developments we have tacitly assumed the observer $O$ to be situated on the terrestrial surface and formulated the expression for $r'_S$ in (14-25) accordingly. If, however, the actual observations are made from another vantage point – such as a spacecraft $\varDelta$ launched to the Moon by human hand – the entire procedure for shadow reductions as outlined in this chapter continues to be valid as it stands, provided that:

(1) the quantity $r'_S$ as given by (14-26) or (14-28) for an Earth-bound observer is replaced by the actual distance of the rocket from the point $S$ on the Moon at the time of observation; and that

(2) the selenographic coordinates $\beta_0$, $\lambda_0$ of the observer at $O$ are no longer identified with the topocentric libration constants $b'$ and $l'$, but replaced in Equations (14-33) and (14-34) by the actual selenographic coordinates $\beta_\varDelta$ and $\lambda_\varDelta$ of the spacecraft at the time of observation.

In order to do so, suppose that the geocentric equatorial coordinates $\alpha_\varDelta$ and $\delta_\varDelta$ of the spacecraft have been obtained from their topocentric values measured at any particular time, and that $r_\varDelta$ denotes its instantaneous distance from the Earth's center (obtained from the measured topocentric slant range $r'_\varDelta$). If so, the rectangular coordinates of the spacecraft in the geocentric equatorial system will be given by

$$\left.\begin{aligned}
x_{\oplus\varDelta} &= r_{\oplus\varDelta} \cos \delta_\varDelta \cos \alpha_\varDelta, \\
y_{\oplus\varDelta} &= r_{\oplus\varDelta} \cos \delta_\varDelta \sin \alpha_\varDelta, \\
z_{\oplus\varDelta} &= r_{\oplus\varDelta} \sin \delta_\varDelta;
\end{aligned}\right\}
\tag{14-38}$$

while the geocentric equatorial coordinates of the Moon's center in the same system are:

$$\left.\begin{aligned}
x_{\oplus\mathbb{C}} &= \mathbf{r} \cos \delta_\mathbb{C} \cos \alpha_\mathbb{C}, \\
y_{\oplus\mathbb{C}} &= \mathbf{r} \cos \delta_\mathbb{C} \sin \alpha_\mathbb{C}, \\
z_{\oplus\mathbb{C}} &= \mathbf{r} \sin \ddot{o}_\mathbb{C},
\end{aligned}\right\}
\tag{14-39}$$

where $\mathbf{r}$ denotes the radius-vector of the lunar relative orbit.

If so, the *selenocentric* rectangular coordinates, as defined at the outset of this section, will be given by

$$\begin{aligned}
x_{\mathbb{C}\varDelta} = \ &(x_{\oplus\varDelta} - x_{\oplus\mathbb{C}})(\cos \omega \cos \varOmega' - \sin \omega \sin \varOmega' \cos i) \\
&+ (y_{\oplus\varDelta} - y_{\oplus\mathbb{C}})(\cos \omega \sin \varOmega' + \sin \omega \cos \varOmega' \cos i) \\
&+ (z_{\oplus\varDelta} - z_{\oplus\mathbb{C}})(\sin \omega \sin i) \ = \ \ r_{\mathbb{C}\varDelta} \cos \beta_\varDelta \cos \lambda_\varDelta,
\end{aligned}
\tag{14-40}$$

$$y_{\mathbb{C}A} = (x_{\oplus A} - x_{\oplus\mathbb{C}})(- \sin \omega \cos \Omega' - \cos \omega \sin \Omega' \cos i)$$
$$+ (y_{\oplus A} - y_{\oplus\mathbb{C}})(- \sin \omega \sin \Omega' + \cos \omega \cos \Omega' \cos i) \qquad (14\text{-}41)$$
$$+ (z_{\oplus A} - z_{\oplus\mathbb{C}})(\cos \omega \sin i) \quad = r_{\mathbb{C}A} \cos \beta_A \sin \lambda_A,$$

and

$$z_{\mathbb{C}A} = (x_{\oplus A} - x_{\oplus\mathbb{C}})(\sin \Omega' \sin i)$$
$$+ (y_{\oplus A} - y_{\oplus\mathbb{C}})(- \cos \Omega' \sin i) \qquad (14\text{-}42)$$
$$+ (z_{\oplus A} - z_{\oplus\mathbb{C}})(\cos i) = r_{\mathbb{C}A} \sin \beta_A,$$

where

$$r_{\mathbb{C}A}^2 = x_{\mathbb{C}A}^2 + y_{\mathbb{C}A}^2 + z_{\mathbb{C}A}^2 \qquad (14\text{-}43)$$

denotes the distance of the spacecraft from the center of the Moon; and $\omega$, the angle between the lunar ascending node and prime meridian (measured along the Moons' equator) can be evaluated from the equations

$$\cos b \sin \omega = \{\cos i \cos \delta \sin (\Omega' - \alpha_{\mathbb{C}}) - \sin i \sin \delta_{\mathbb{C}}\} \cos l$$
$$+ \{\cos \delta \cos (\Omega' - \alpha_{\mathbb{C}})\} \sin l, \qquad (14\text{-}44)$$

or

$$\cos b \cos \omega = \{\cos i \cos \delta_{\mathbb{C}} \sin (\Omega' - \alpha_{\mathbb{C}}) - \sin i \sin \delta_{\mathbb{C}}\} \sin l$$
$$- \{\cos \delta \cos (\Omega' - \alpha_{\mathbb{C}})\} \cos l, \qquad (14\text{-}45)$$

where the lunar libration angles $b$, $l$ as well $\Omega'$ and $i$ possess exactly the same meaning as in Chapter 3.

Let, furthermore,

$$\left. \begin{array}{l} x_{\mathbb{C}s} = a \cos \beta_s \cos \lambda_s \\ y_{\mathbb{C}s} = a \cos \beta_s \sin \lambda_s \\ z_{\mathbb{C}s} = a \sin \beta_s \end{array} \right\} \qquad (14\text{-}46)$$

denote the selenocentric rectangular coordinates of the shadow tip $S$ on the lunar sphere of mean radius $a$ in the same system. If so, the direction cosines $l$, $m$, $n$ of the vector $\overrightarrow{OS} \equiv \overrightarrow{AS}$ joining the position of the spacecraft $A$ with the point $S$ will be given by

$$\left. \begin{array}{l} l = \dfrac{x_{\mathbb{C}A} - x_{\mathbb{C}s}}{r'_{As}}, \\[2mm] m = \dfrac{y_{\mathbb{C}A} - y_{\mathbb{C}s}}{r'_{As}}, \\[2mm] n = \dfrac{z_{\mathbb{C}A} - z_{\mathbb{C}s}}{r'_{As}} \end{array} \right\} \qquad (14\text{-}47)$$

where

$$r_{As}'^2 = (x_{\mathbb{C}A} - x_{\mathbb{C}s})^2 + (y_{\mathbb{C}A} - y_{\mathbb{C}s})^2 + (z_{\mathbb{C}A} - z_{\mathbb{C}s})^2. \qquad (14\text{-}48)$$

This equation should replace (14-27) for observations made aboard a spacecraft.

Since, moreover, the direction cosines of incident sunlight in the same system of selenographic coordinates are given by

$$l_\odot = \cos D_\odot \cos A_\odot \\ m_\odot = \cos D_\odot \sin A_\odot \\ n_\odot = \sin D_\odot \quad\quad\quad\quad\quad\quad (14\text{-}49)$$

it follows that the angle $\theta$ between the radii-vectors from $S$ to $\varDelta$ and the Sun should be given by the equation

$$\cos \theta = ll_\odot + mm_\odot + nn_\odot, \quad\quad\quad (14\text{-}50)$$

which represents a generalization of (14-34) for observations made from the vantage point at $\varDelta$ – no matter how close to the lunar surface this may happen to be. The selenographic coordinates of a spacecraft at $\varDelta$ then are given by

$$\sin \beta_\varDelta = \frac{z_{\langle\varDelta}}{r_{\langle\varDelta}} \quad \text{and} \quad \tan \lambda_\varDelta = \frac{y_{\langle\varDelta}}{x_{\langle\varDelta}}; \quad\quad (14\text{-}51)$$

and its height above the lunar surface,

$$h = r_{\langle\varDelta} - a. \quad\quad\quad\quad\quad (14\text{-}52)$$

Lastly, the selenographic components of the *velocity* of our spacecraft should follow from a time differentiation of the equations (14-40) – (14-42) for the selenographic co-ordinates $x_{\langle\varDelta}$, $y_{\langle\varDelta}$, and $z_{\langle\varDelta}$; care being merely taken to add $+ \omega_{\langle} y_{\langle\varDelta}$ to $x_{\langle\varDelta}$ and $- \omega_{\langle} x_{\langle\varDelta}$ to $y_{\langle\varDelta}$ as terms arising from the rotation of the Moon with an angular velocity $\omega_{\langle}$ about the $z$-axis (the selenographic axes rotate).

Whichever method of those outlined above we employ for a determination of the fractional difference $h$ in height between the points $P$ and $S$ by the shadow method, the underlying geometry implies that the accuracy of the results will be maximum at the center of the apparent lunar disk (where the measured shadow length $\mu$ or the coordinate differences $x_P - x_s$ etc. are subject to no foreshortening), and diminish progressively toward the limb. Near the limb, both angles $\theta$ and $\mu$ tend separately to zero; and, as a result, their ratio on the right-hand side of Equation (14-25) for sin $\omega$ becomes well-nigh indeterminate – which effectively invalidates the shadow approach to a determination of the heights of lunar mountains in the limb regions. In order to do so, another method must be sought; and this is fortunately made possible by the fact that – on account of libration – the lunar regions within peripheral regions of approximately $\pm 7°$ in width are, from time to time, seen in projection against the dark background of the sky, and their profiles can be accurately measured. Following some early exploratory work by HAYN (1914), a comprehensive atlas of lunar profiles visible at different librations has been published by WEIMER (1952) and, more recently, by NEFEDIEV (1958) and WATTS (1963). The significance of the outlines of the lunar limb as seen against the bright background of the Sun during annular eclipses has already been mentioned in Chapter 13; while similar silhouettes exhibited by the Moon during partial solar eclipses (cf., e.g., WHITWELL, 1929; FUJINAMI, 1952; FUJINAMI, INA, and KAWAI, 1954; KRISTENSON, 1954, and others) are much more limited in scope and cannot compete with the limb measurements at night.

So much for the situation encountered in the *limb* regions. On the *terminator* – i.e., at the time of sunrise (or sunset), when the peak $P$ just intercepts the first (or last) rays of the rising (or setting) Sun,

$$s_{max} = (a + h) \sin \omega \qquad (14\text{-}53)$$

and, from (14-22) and (14-23) it follows that

$$v = \omega, \qquad (14\text{-}54)$$

such that

$$\cos v = \frac{a}{a + h} = \cos \omega, \qquad (14\text{-}55)$$

where $\omega$ stands then for the selenocentric arc at which $P$ becomes visible beyond the terminator in the direction of incident sunlight. This angle is difficult to measure directly; but can be computed with the aid of the equality $\omega = v$ from known values of the position $\beta_P$, $\lambda_P$ of the peak in question, as well as of the selenocentric position $A$, $D$ of the Sun at the time of observation. This method was, in fact, first used to estimate the altitudes of the lunar mountains by Galileo Galilei. We may wish to add only that, in such a case, $h$ need not stand for the height of any isolated peak, but also for that of any plateau along the terminator. If the Moon were a perfectly smooth sphere, the sunrise (or sunset) terminator would be an ellipse. Since the terminator is defined as the locus of points at which the Sun just rises above (or sets below) the lunar horizon, any irregularities of its outline or departures from an ellipse would be indicative of the differences in level at different latitudes intersected by it.

Throughout all our discussion of the geometry of the shadow method for the determination of lunar relative altitudes we have, in fact, tacitly assumed so far that the shadow of the peak $P$ is cast on a sphere. The lunar surface is, however, by no means smooth in detail; and the shadow method – furnishing as it does information on altitude difference between points $P$ and $S$ – should enable us to ascertain not only the relative altitude of any particular eminence above the surrounding landscape, but also, in principle, any irregularities or undulation of ground on which the shadows are cast. In order to outline the steps by which this can be done, let us return to Figure 14-1, and suppose that the actual distance from the Moon's center to the point $S$ is, not $a$, but $a + \delta a$, with $\delta a$ denoting a local deviation of the mean Moon-level (such as a difference in level between the foot of a mountain and the tip of its shadow). If so, Equation (14-21) continues to hold good irrespective of surface irregularities; but Equations (14-22) and (14-23) should be replaced by

$$(a + \delta a) \sin \omega = s \cos v \qquad (14\text{-}56)$$

and

$$(a + \delta a + h) \cos v = (a + \delta a) \cos (v - \omega), \qquad (14\text{-}57)$$

leading to

$$\frac{h}{a + \delta a} = \frac{\cos (v - \omega)}{\cos v} - 1 \tag{14-58}$$

in place of (14-24), where

$$\sin \omega = \frac{r'_s \sin \mu \cos v}{(a + \delta a) \sin (\theta + \mu)} \tag{14-59}$$

exactly; with $a + \delta a$ replacing $a$ in the equation (14-26) for $r'_s$; but the definitions of $\theta$ and $v$ unchanged.

Should the shadow of $P$ be cast on a sphere, the measured angular shadow length $\mu$ should be a smooth function of the time varying as the Sun rises or sets on the Moon in a manner predictable from our theory. If, however, the landscape on which the shadow is cast is uneven, a plot of $\mu$ versus the time should show irregularities, depending on the local variation of $\delta a$. Suppose that we draw a smooth curve by free hand through such irregularities, which we shall use for a definition of the local mean Moon level. Deviations $\delta\mu$ from such a curve can then be translated into the corresponding undulations $\delta a$ of ground on which the shadows are cast with the aid of the formulae given in the preceding paragraph, in which angles $\theta$ and $v$ continue to be given by Equations (14-34) and (14-35) or (14-50).

In conclusion of the present discussion of the geometry of shadows cast by sunlit mountains, it should be stressed that we have so far assumed – for the sake of simplicity – the Sun to act as an illuminating point-source of light. In reality, of course, the apparent angular diameter of the Sun as seen from the Moon amounts to very approximately half a degree; and this fact alone is bound to provide all lunar shadows with a *penumbra* even in the complete absence of any atmosphere – covering regions a part of the apparent solar disk would be set for the observer on the ground. The intensity of illumination at any point of this penumbral zone should depend on the brightness of the visible segment of the Sun; and this will vary from full light to complete darkness over a strip whose width should depend on the altitude of the rising (or setting) Sun – becoming the greater, the lower the shadow-casting obstacle.

For this reason, dependable information on the angular length of lunar shadows cannot easily be obtained by any visual settings possible with a micrometer, but must be sought by micro-densitometric tracings – a method introduced in lunar studies and developed to its present state by the Manchester group of astronomers led by the present writer (for its partial summary cf., e.g., KOPAL et al., 1961). Illustrative examples of the application of different variants of this technique are shown on the accompanying Figures 14-3 (Theophilus) and 14-4 (Archimedes); while the results obtained from an analysis of this latter formation are diagramatically exhibited on Figure 14-5 (after TURNER, 1959).

In order to make proper use of such microdensitometric records – and, in particular, to determine the point, in the penumbral band, at which the center of the apparent solar disk (whose coordinates occur in Equations 14-34 and 14-35) just clears the lunar horizon – it is, however, necessary first to investigate the expected distribution

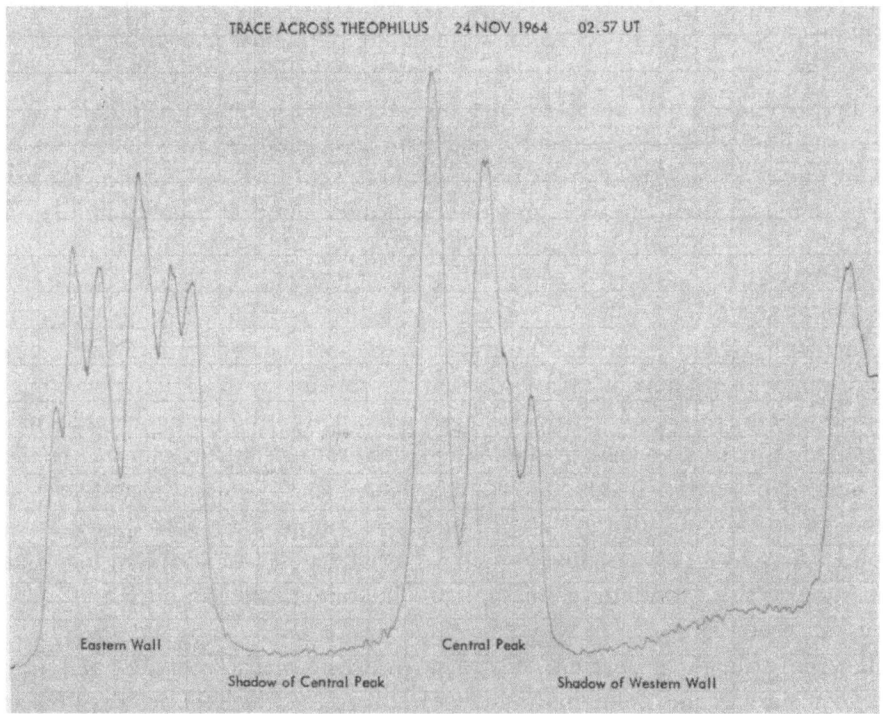

TRACE ACROSS THEOPHILUS    24 NOV 1964    02.57 UT

Eastern Wall

Shadow of Central Peak

Central Peak

Shadow of Western Wall

Fig. 14-3.

of light within the penumbral zone; and this can be approached in the following manner.

Let, as on Figure 14-1, the peak $P$ cast a shadow in the light of a light source, having the position of the Sun and an angular semi-diameter of $\rho_\odot = 15'59''.6$. The width $f_{1,2}$ of the penumbral zones on either side of $S$ will clearly be given by

$$f_{1,2} = \frac{s \sin \rho_\odot}{\sin (v \pm \rho_\odot)} \tag{14-60}$$

in the $SP'$-direction; and their projections $f_{1,2}$ on a plane tangent to the lunar surface at $S$ become

$$f'_{1,2} = \frac{s \sin \rho_\odot}{\sin (\rho_\odot - \omega \pm v)}; \tag{14-61}$$

the upper and lower sign in the denominators referring to the parts of the penumbra interior and exterior to $S$, respectively. The total width of the penumbral zone then becomes equal to

$$f'_1 + f'_2 = \frac{2 \sin \rho_\odot \cos v \sin (\rho_\odot - \omega)}{\sin (\rho_\odot - \omega + v) \sin (\rho_\odot - \omega - v)}, \tag{14-62}$$

where the angle $v$ continues to be given by (14-35) while $s$ and $\omega$ follow from (14-21) and (14-25).

Let, moreover, the position of any point within the penumbral zone of the lunar surface in the direction of incident sunlight be characterized by a single coordinate, $x$ measured positively outwards from $S$ and normalized so as to assume the values $\pm 1$ at the ends of the penumbra. If so, and if the lunar horizon acts like a straight occulting edge (or, which is more likely, the horizontal irregularities are small in comparison with the solar semi-diameter), then it can be shown (cf., e.g., KOPAL, 1959, p. 207) that the fractional illumination $I(x)$ at any point of the penumbra should vary as

$$\pi I^U (x) = \pi - \cos^{-1} x + x \sqrt{(1 - x^2)} \tag{14-63}$$

if the apparent solar disk were uniformly bright, and

$$4 I^D (x) = (1 + x)^2 (2 - x) \tag{14-64}$$

if it were completely darkened at limb.

In actual fact, the solar disk is known to be partially darkened at the limb; and in the yellow light ($\lambda = 0.56\ \mu$) is coefficient of darkening $u$ is known to be approximately equal to two-thirds; and if so, we should expect (cf. again KOPAL, 1959, p. 308) that

$$\begin{aligned} I(x) &= \frac{3(1 - u)}{3 - u} I^U(x) + \frac{2u}{3 - u} I^D(x) \\ &= \tfrac{1}{7} \{3 I^U(x) + 4 I^D(x)\}. \end{aligned} \right\} \tag{14-65}$$

Fig. 14-3. Sunset over the craters Theophilus and Cyrillus, photographed with the 24-inch refractor of the Observatoire du Pic-du-Midi (Manchester Lunar Programme). The white line across Theophilus indicates the direction of the microdensitometric tracing reproduced below.

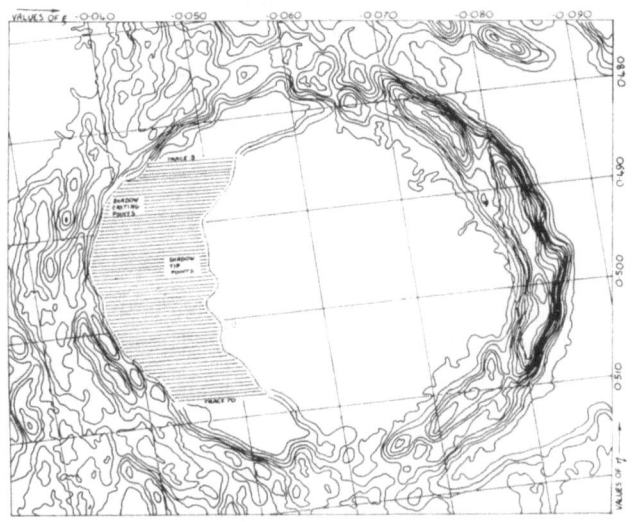

ISOPHOTES OF ARCHIMEDES: 19h O5m ON 25/10/63

CONTOUR VALUES REFER TO THE RECORDED PLATE DENSITY, THE SUPERIMPOSED LINES
ARE THOSE OF THE STANDARD ORTHOGRAPHIC MAP GRID AT INTERVALS OF 0·01 OF
A LUNAR RADIUS.

Fig. 14-4.

The essential characteristic of this expected distribution of light is the fact that both functions $I^U(x)$ and $I^D(x)$ as defined by Equations (14-63) and (14-64) – and, therefore, their weighted mean (14-65) – possess an inflection point at $x = 0$.* The variation of intensity in the penumbral shadow turns out, therefore, to be most rapid when the *center* of the apparent solar disk just rises above the lunar horizon at $S$. This fact enables us then to *define* the position of $S$ as the point at which a micro-densitometric tracing (on the intensity scale) traversing the penumbral zone exhibits inflection.

What would happen if the shadow-casting obstacle on the Moon were so low (like the lunar wrinkle ridges, for instance) that the entire width of the visible shadow were, in fact, a penumbra? The measurement of the shadow lengths $\mu$ would then of course cease to possess any meaning. However, any unevenness of even gently undulating ground can, in principle, be established from the photometric measurements of surface brightness which, at low angles of incidence, should vary as the cosine of the angle between incident sunlight and the surface normal (in accordance with Lambert's law): as the former direction is a known function of the time, photometric measures can be used to determine the latter. Such a method was indeed worked out by VAN DIGGELEN (1951) and applied by him to a determination of the profiles of lunar wrinkle-ridges in the region of the Mare Imbrium, with encouraging results.

The efficiency of the shadow method for determination of the relative altitudes of lunar mountains is at its best when the line of sunrise or sunset terminator (along which the length of the shadows magnifies the altitude difference to maximum extent) passes through the center of the apparent lunar disk (i.e., the domain of minimum foreshortening). At the limb of the Moon, the limits of precision with which one can determine the altitude of lunar mountains is given by the resolving power of the telescope employed, which for a 24-inch (60 cm) aperture in yellow light is (by Rayleigh's criterion) close to 500 meters. However, near the center of the apparent lunar disk, the vertical heights are greatly magnified in the shadows cast in the oblique rays of the Sun. Simple geometry reveals that, at a time when the whole disk of the Sun just appears above the lunar horizon, the ratio of the length of the shadow to the height of the object casting it is equal to cot $0°5 = 115$ (and exceeds 115 in the penumbral zone, illuminated by only a part of the solar disk); though when the Sun's center has risen to $2°$ above the lunar horizon, this magnification ratio has diminished to 27. If, therefore, photographs can identify on the Moon at sunrise the shadows (say) one kilometer in length, these should be indicative of the presence of vertical obstacles less than 10 meters in height; and of even smaller obstacles in the penumbral zone (which is, of course, not more than approximately 18 km wide on the Moon at any time).

* This would, moreover, continue to be true for *any* distribution of brightness over the apparent solar disk – provided only that it retains radial symmetry.

Fig. 14-4.   Direct photograph of sunrise over the crater Archimedes in the eastern part of Mare Imbrium, taken with the 24-inch refractor of the Observatoire du Pic-du-Midi on 25 October 1963 (left), and its microdensitometric transcipt (right) used to study the vertical relief of this formation by the shadow method (after JONES, 1965). Manchester Lunar Programme.

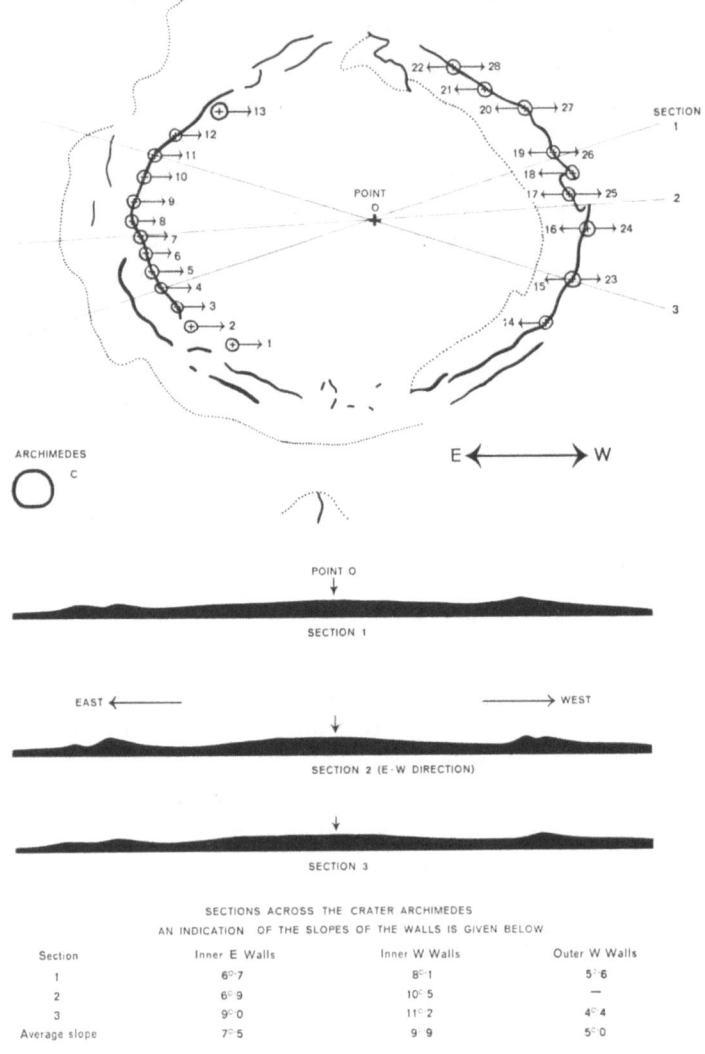

Fig. 14-5. Topography of the lunar crater Archimedes (after TURNER, 1959).

Have such objects been actually seen, or photographed, on the Moon from the distance of the Earth? Unfortunately, virtually all photographs obtained so far are ill-suited to be used for an appropriate search, as their exposure times chosen so as to bring out the entire field covered by each plate were invariably much too short to reveal any details inside the relatively narrow twilight strip. It may perhaps come to the reader as a surprise that at most – if not all – lunar photographs he may have seen, the apparent terminator marked the region where the Sun has risen already several degrees above the lunar horizon; for exposures capable of revealing details in the zone illuminated by a lower Sun would have left the rest of the field very largely burnt out.

Moreover, direct visual observations would likewise be of little avail, as the twilight (penumbral) zone would appear too dark to the eye to discern any contrast.

In order to illustrate this point, we reproduce on the accompanying Figures 14-6 to 14-7 two pairs of photographs of the same terminator region of the Moon and obtained practically at the same time – one with a "normal" exposure, the other "over-

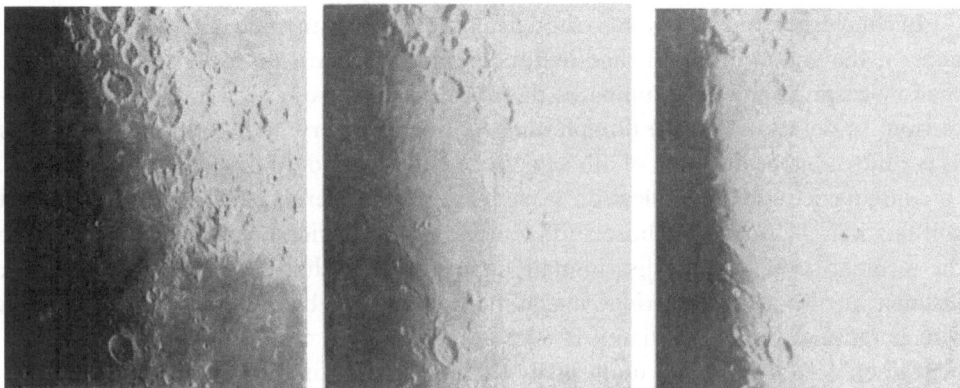

Fig. 14-6.   Sunset over Mare Fecunditatis, taken in the f/15 focus of the 43-inch reflector at Pic-du-Midi on March 2, 1964. Exposures: left, 0.025 sec; center, 0.5 sec; right, 2 sec – all within the same minute. Note increased definition of the terminator with increasing exposure time.

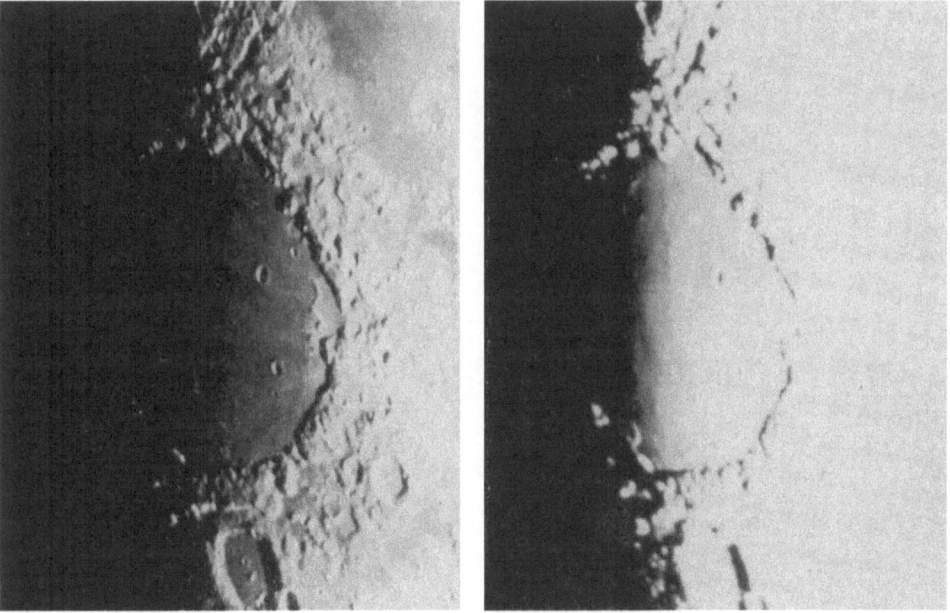

Fig. 14-7.   Sunset over Mare Crisium, photographed on low-contrast plates with the 24-inch refractor at Pic-du-Midi on January 2, 1964. Exposures: left, 1 sec; right, 50 sec – both within the same minute. Note increasing definition of the terminator with the exposure.

exposed" sufficiently to bring out details in the actual penumbral zone. The reader can judge for himself the extent to which these overexposures resulted in an increase of information in the twilight zone, much of which has not been known before. Indeed, it is quite probable that "overexposed" lunar photographs, as reproduced on Figures 14-6 and 14-7 this section, can enable us to bridge from ground – at least in certain parts of apparent lunar disk – to some extent the gap between the Earth-bound and space-born cameras recording the surface of our satellite.

In conclusion, one last point – though a hardly surprising one – should be stressed: namely, the limitation of all shadow topographic work arising from the fact that the shadow-casting sunlight illuminates the Moon always from very much the same direction. In point of fact, the illuminating sunlight can vary in direction only as much as permitted by the libration of our satellite in latitude; and this is a very small amount. In consequence, the magnification of small altitude differences by the shadow effect will facilitate detection of those surface irregularities which are situated normally to the incident sunlight, and discriminate against objects that are parallel with it. For instance, a rille or wrinkle ridge in the mare plains will be more easily detected if it runs parallel with the meridian if it were in the direction of a latitude circle. This observational selection forced upon us by the almost uni-directional nature of the illuminating light source may exert serious effects (the majority of known wrinkle ridges do indeed run more or less in the direction of the meridians), and must be kept in mind in any work on the geological interpretation of the lunar landscape.

## CHAPTER 15

# MAPPING OF THE MOON

The lunar topography constitutes a subject whose emergence and subsequent evolution can be traced with relative precision: for its sources go back to the very first days of telescopic astronomy inaugurated by Galileo Galilei in 1609. Although Galileo was not the real inventor of the telescope, he was indubitably the first to use it for observations of celestial bodies; and as he recorded a year later in his *Nuncius Sidereus*, ". . . Sed missis terrenis ad coelestium speculationes me contuli: ac Lunam prius tam ex propinquo sum intuitus, ac si vix per duas Telluris diametros abesset" *; and further (on p. 13) he continued that ". . . De facie autem Lunae, quae ad aspectum nostrum vergit primo loco dicamus; quam facillionis intelligentiae gratia in duas partes distinguo, alteram nempe clariorem, obscuriorem alteram . . . ut certo intelligamus, Lunae superficiem non perpolitam, acquabilem, exactissimaeque sphaericitatis exsistere, ut magna Philosophorum cohors de ipsa deque reliquis corporibus coelestibus opinata est, sed contra inaequalem, asperam, caritatibus tumoribus confertam, non secus ac ipsiusmet Telluris faciem, quae montium ignis vallumque profunditatibus hic inde distinguitur".**

GALILEO's *Nuncius Sidereus* (1610) contained altogether five drawings of the Moon, one of which is reproduced on the accompanying Figure 15-1. A mere glance at it will convince us that Galileo was not a great astronomical observer; or else that the excitement of so many telescopic discoveries made by him at that time had temporarily blurred his skill or critical sense; for none of the features recorded on this (and other) drawings of the Moon can be safely identified with any known markings of the lunar landscape.[†] In spite of the obvious shortcomings of these first telescopic observations of the Moon, their impact on the contemporary scientific thought was, however, profound; and promptly inaugurated the era of specifically selenographic literature and mapping, which it will be our aim to survey in this chapter.

The first drawings of the Moon following those of Galileo were made by P. Chr.

---

* "When I gave up observation of the terrestrial objects, I turned my attention to the celestial bodies: first I saw the Moon from such proximity as if it were barely two terrestrial diameters distant".
** "Further, with regard to the side of the Moon facing us, let it be said first that one part of it is noticeably brighter, the other darker . . . so that we could perceive that the surface of the Moon is neither smooth nor uniform, nor very accurately spherical, as is assumed by a great many philosophers about the Moon and other celestial bodies; but that it is uneven, rough, replete with cavities and packed with protruding eminences, in no other wise than the Earth which is also characterized by mountains and valleys".
† The ring-like configuration near the center of the apparent disk on Figure 15-1 (which Galileo compared with the central European land of Bohemia) may represent the crater Ptolemy.

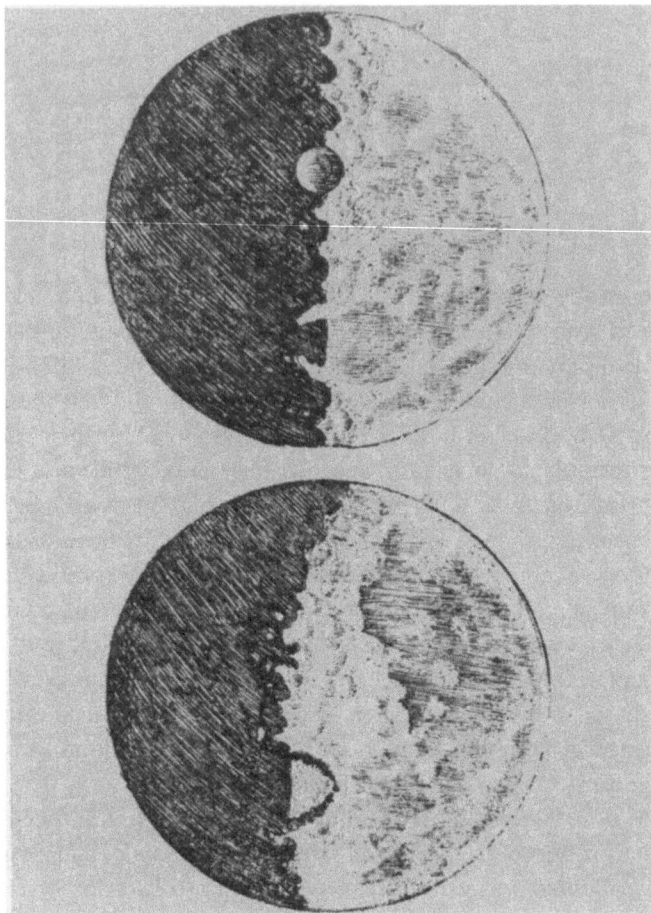

Fig. 15-1.   Drawings of two phases of the Moon by Galileo GALILEI, published in his *Nuntius Sidereus*
(Padua, 1610).

SCHEINER – his historic adversary in their subsequent dispute over the sunspots – and
can be found on page 58 of his *Disquisitiones mathematicae de controversiis et novi-
tatibus astronomicis* (Ingolstadt, 1619). The drawing, made at an unknown date, represents the Moon in proximity of the first quarter (cf. Figure 15-2), was 9.2 cm in
diameter and not much more detailed than those of Galileo.

Among subsequent early contributors to lunar topography, the name of another
Jesuit, the Belgian mathematician C. MALAPERT (1619) should be mentioned; but like
the abortive attempts of Galileo or Scheiner, his efforts remained limited to the graph-
ical representation of a single phase. The first attempts to chart the entire face of the
Moon were due to PEIRESC and GASSENDI, assisted between 1634–1635 by the painter
Claude SALVAT and, since 1636, by the engraver Claude MELLAN. Although the project

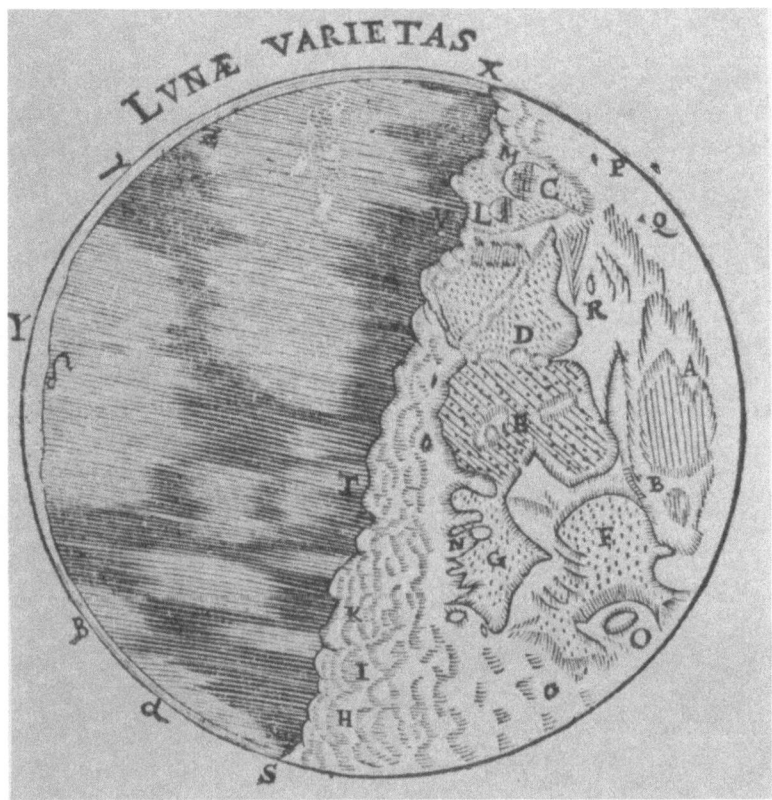

Fig. 15-2. A Map of the Moon prepared by P. Chr. SCHEINER, and reproduced on page 58 of his *Disquisitiones* (Ingolstadt, 1619).

was never completed, it led to the publication of three very fine drawings by Mellan, of which more will be said later.

Of their successors, we should mention A. ARGOLI (in his *Pandosion Sphaericum*, Padua 1644); M. F. VAN LANGREN (in his *Selenographia Langreniana*, Antwerp 1645); A. M. SCHYRLAEUS DE RHEITA (in his *Opus Theologiae, Philosophiae, et Verbi Dei Praeconibus utile et iucundum*, Antwerp 1645); F. FONTANA (*Novae Coelestium Terrestriumque Rerum Observationes*, Naples 1646); J. HEVELIUS (*Selenographia sive Lunae Descriptio*, Danzig 1648); E. DIVINI (1649); G. B. RICCIOLI (*Almagestum Novum*, Bologna 1651). Of these, the contributions by Langren and Hevelius are by far the most important; and the two (together with Riccioli) should be regarded as progenitors of the *nomenclature* of lunar craters and maria which has survived up to the present time.

Michel Florent van LANGREN (1600–1675), a mathematician and cosmographer to the King Philipp II of Spain (to whom Langren's maps of the Moon were dedicated) completed, according to Riccioli (1651), several maps of the Moon, the most notable

Fig. 15-3.    A map of the Moon prepared by M. F. van Langren, as it appeared in his *Selenographia Langreniana* (Antwerp, 1645). Reproduced from the Paris copy of the original.

of which appeared at Brussels in 1645. One specimen of this very rare map (35 cm in diameter on the original) is in the collections of the Bibliothèque Nationale de Paris (for its reproduction, see Figure 15-3); and contains markings of 322 configurations to which – for the first time in the history of selenography – proper names were assigned; but these were mainly heraldic and very few of them have survived (except for Sinus Medii, and a large peripheral crater Langrenus, which its author named after himself.) Besides this map, two others of Langren's maps have been preserved – one in Strasbourg, the other in Brussels (for its reproduction, cf. Figure 15-4). It is obviously much more primitive than the map of 1645 and probably antedates the latter – according to WISLICENUS (1902), it may have been presented to Queen Isabel of Spain as early as in 1628. On the other hand, the Strasbourg copy is probably a poor copy of the Paris map of 1645 (possibly prepared by an inferior draughtsman).

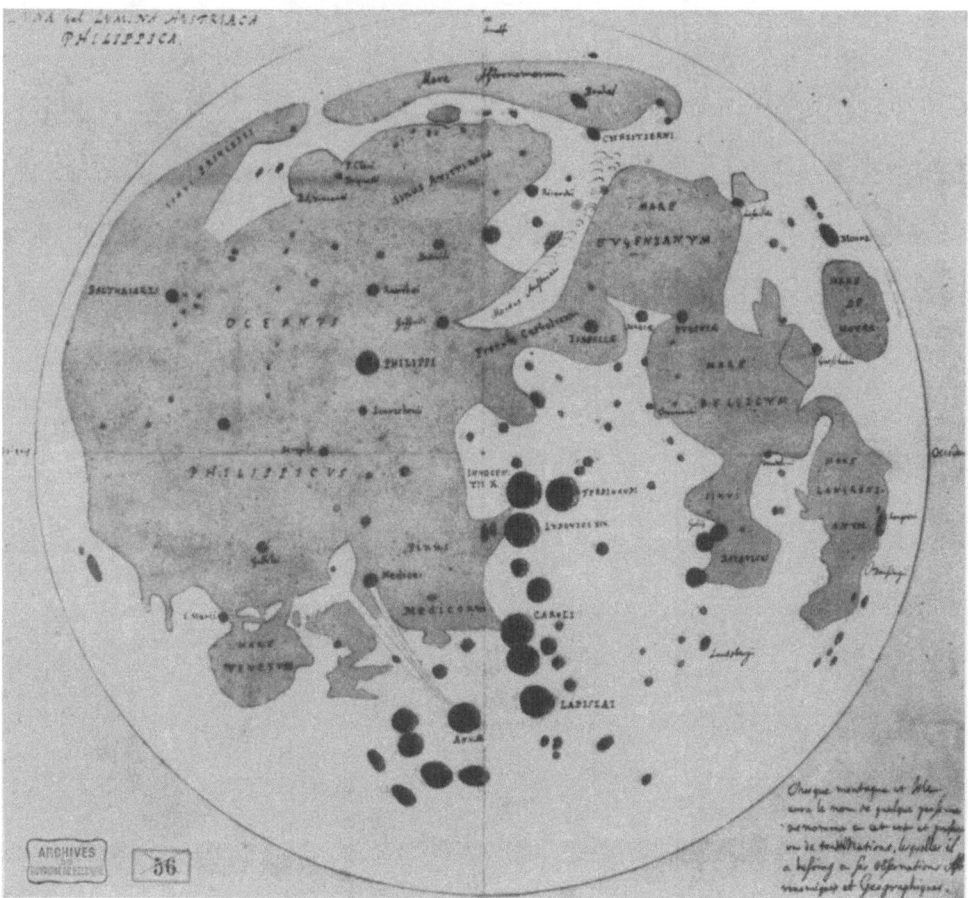

Fig. 15-4. Another copy of LANGREN'S map of the Moon, preserved in the Belgian Royal Archives in Brussels.

Fig. 15-5.   A map of the Moon by A. M. DE RHEITA (1645).

In the same year in which Langren published the first lunar map worthy of that name, another such map appeared – less valuable scientifically, but equally noteworthy. Its author was the Czech capucin P. Antonín Šírek OF REJTA (1597–1660), better known under his latinized name of Anton Maria Schyrlaeus de Rheita. A keen scholar, he invented a reverting eyepiece for the Keplerian telescope; and also conjectured – almost a century before Halley – that fixed stars possessed proper motions of their own. The map of the Moon (18 cm in diameter), which he constructed on the basis of his own observations, appeared as a part ("de facie lunae, etc.") of his large partly astronomical and partly theological work, entitled *Oculus Enoch et Eliae, Opus Theologiae, Philosophiae, et Verbi Dei Praeconibus utile et iucundum*, published in Antwerp in 1645 and dedicated to no one lesser than Jesus Christ. Rheita's map (reproduced on Figure 15-5) is rather scanty in detail, and contains no nomenclature;

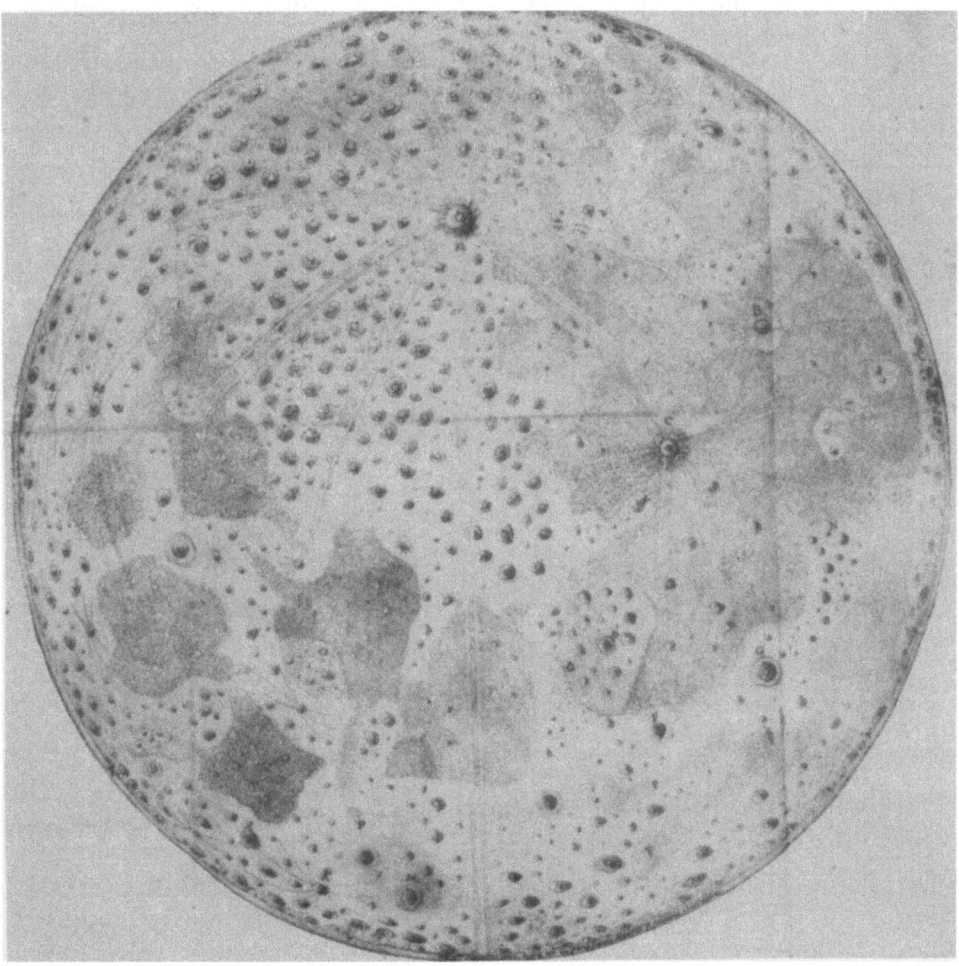

Fig. 15-6. A map of the Moon by FONTANA (1646).

but the contours of the maria as well as the ray systems of Copernicus or Tycho are quite well represented. It is of interest to note that it is the first map of the Moon oriented with the south at the top (as seen through an inverting astronomical telescope).

FONTANA's map (see Figure 15-6), published in 1646, must be considerably earlier in origin, as some drawings of the Moon included in his book go back to 1629 and were circulated prior to the appearance of his *Observationes* (as is attested by a letter to Galileo Galilei written by P. Fulgenzio Micanzio in 1638). Fontana's map is not very reliable in detail; and gives impression that its author, probably impressed by a great number of craters visible through his telescope, did not take trouble to ascertain their positions with any precision, but represented them as scattered more or less at random.

The *Selenographia* of Johannes HEVELIUS (1611–1687), which appeared in Danzig in 1647, contains three general maps (of 28.5 cm in diameter) and 40 drawings of

Fig. 15-7.   One of the three lunar maps prepared by Johannes HEVELIUS in 1645, as it appeared in his *Selenographia* (Danzig, 1648). It takes account for the first time of the phenomenon of libration.

Fig. 15-8.    Another map of the Moon by HEVELIUS, more schematic than the first, but containing
his nomenclature.

individual phases of the Moon (each 16 cm in diameter) accompanied by explanatory
text. The maps (of which two are illustrative – see Figure 15-7 – and the third, more
schematic (Figure 15-8), contains the nomenclature proposed by Hevelius) show for
the first time the lunar limb parts which become visible only during the extreme librations of our satellite (of which Hevelius was the discoverer; cf. page 65). For over
a half century did Hevelius's *Selenographia* remain a fundamental work in its field
(though little of his proposed nomenclature survived); but, unfortunately, no further
prints of his maps were made (and, after the death of their author, one of the copper
plates of the engraving was reportedly used to serve as the bottom of a tea-pot).

In 1649, shortly after the appearance of Hevelius's *Selenographia*, another map of
the Moon was published by Eustachio DIVINI (1610–1685) in Rome, reproduced on
Figure 15-9. Although its author claims to have used a micrometer in its preparation,
an examination of his map fails to bear out this contention; and the map itself pos-

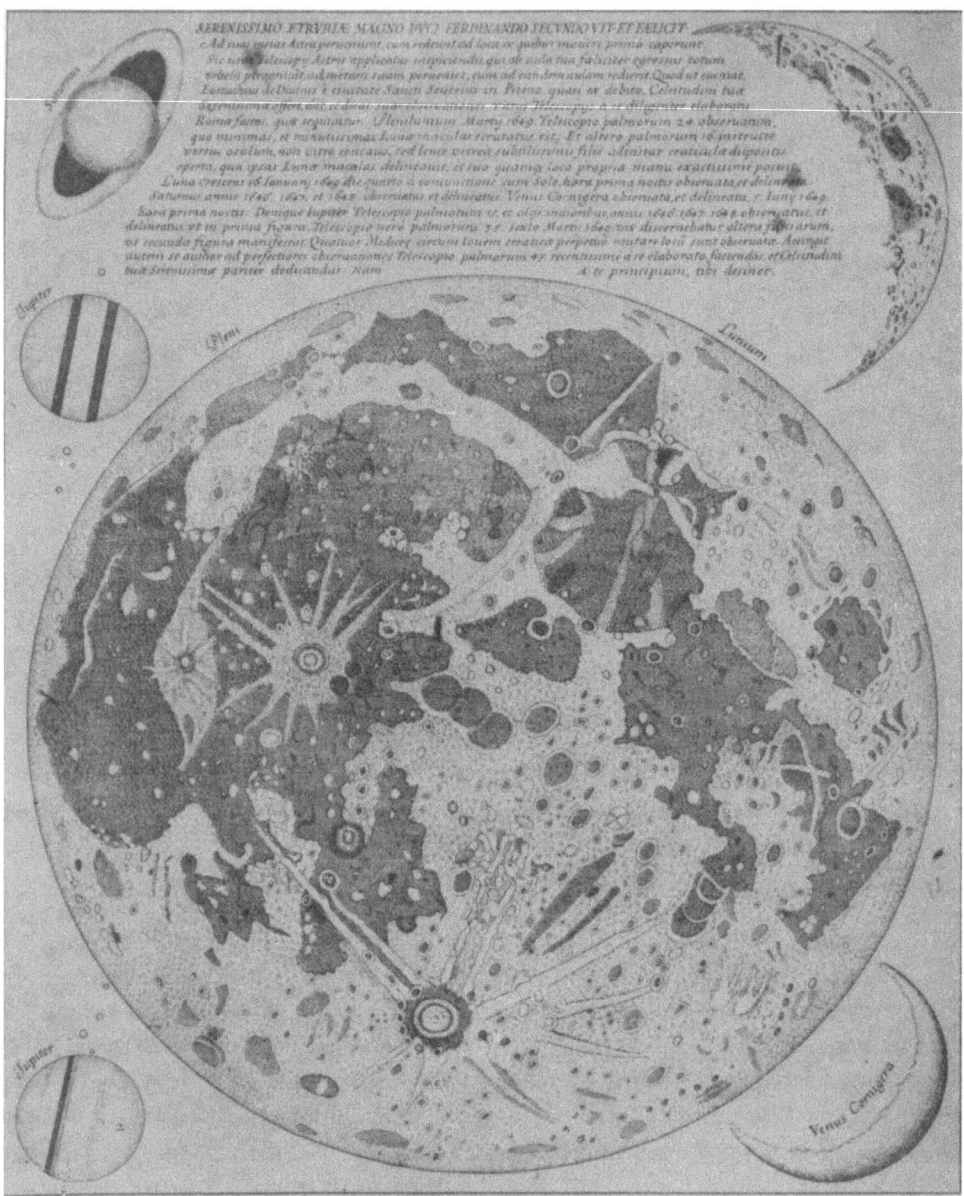

Fig. 15-9.   A map of the Moon by E. DIVINI (1649).

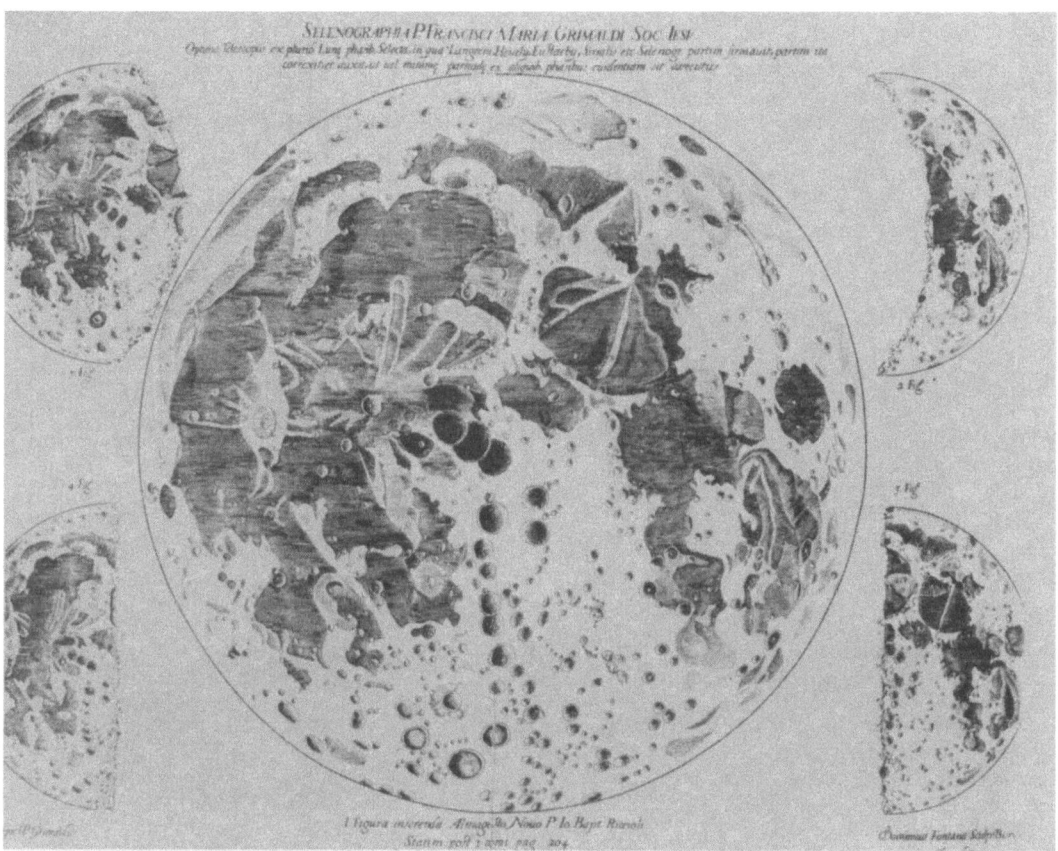

Fig. 15-10. A map of the Moon reproduced opposite page 204 of Riccioli's *Almagestum Novum* (Bologna, 1651), due to F. M. Grimaldi.

sesses such resemblance to that of Hevelius that it must have been at least influenced by it.

In 1651, a much more important map was published by Giovanni Battista Riccioli, S. J., from Ferrara, in his large work *Almagestum Novum*, the chapter of which concerning the Moon contains two maps (28 cm in diameter), – see Figures 15-10 and 15-11 – one of which (reproduced on Figure 15-10) shows the effects of librations and contains the lunar nomenclature, as introduced by Riccioli, much of which is still in use to-day. In particular, Riccioli (like Langren) referred to the lunar large dark patches as "maria", and the smaller ones as "palludes" – misnomers which have remained in use to this day – while the term "terrae", given by him to the bright regions, did not take root in general usage. However, such familiar by words as Mare Serenitatis, Oceanus Procellarum, and many others, we owe to Riccioli. He also inaugurated the custom of designating lunar craters by proper personal names, but wisely substituted those of the scholars for those of contemporary princes, and thus

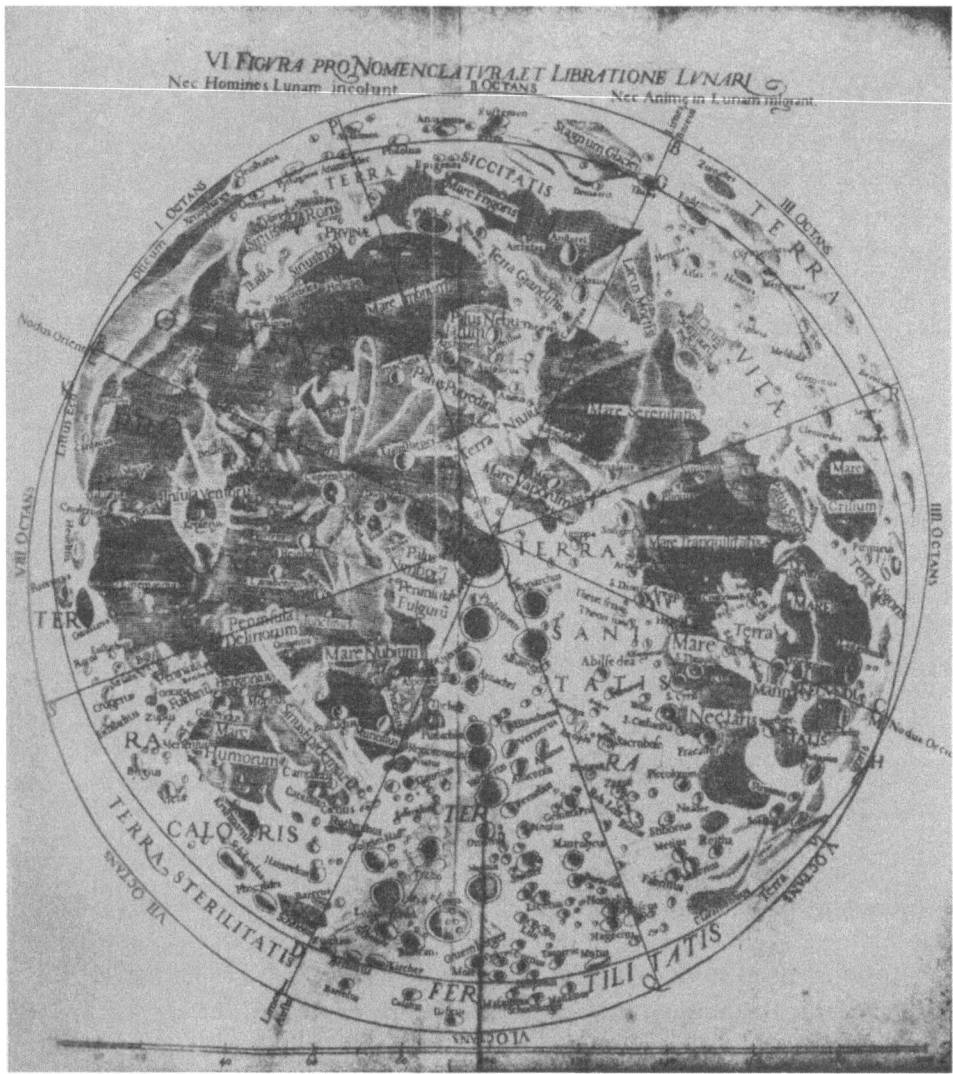

Fig. 15-11.   The second lunar map by F. M. GRIMALDI (taking account of the phenomena of libration) as it appeared in RICCIOLI's *Almagestum Novum*. Note the delightful inscription immediately below the heading "Nec Homines Lunam incolunt, nec Animae in Lunam migrant" – with a clear portent for the astronauts of the future!

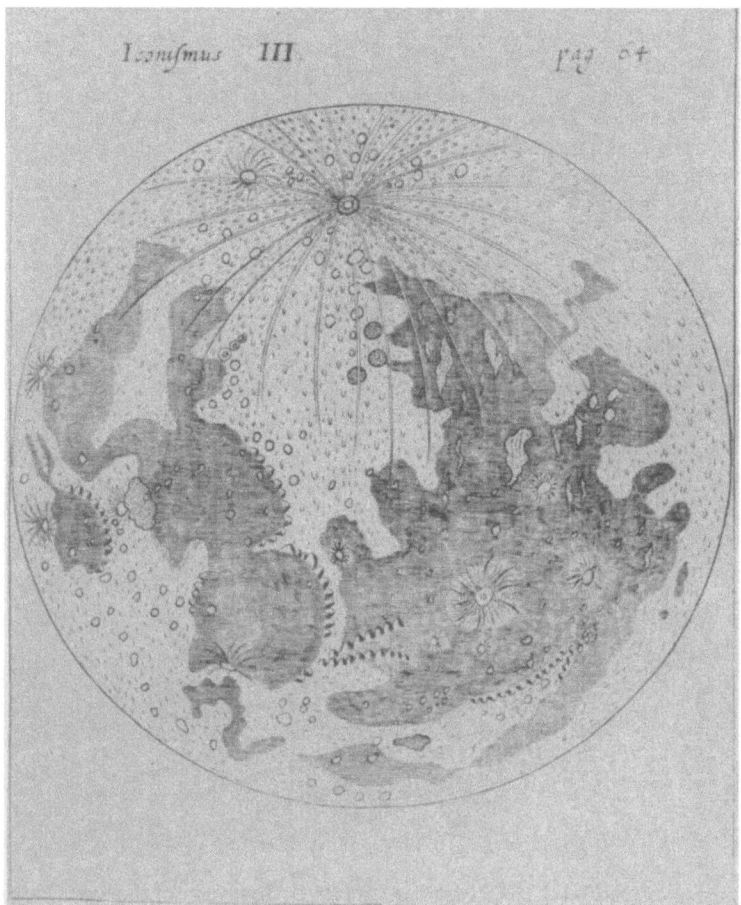

Fig. 15-12.   A map of the Moon by P. KIRCHER (1660).

divorced the face of the Moon from Langren's heraldry. In particular, Riccioli used the names of the scholars of antiquity for craters located in the northern hemisphere of the Moon (after grouping the disciples around their respective masters), and placed the names of the renaissance scholars to the north – a division which has not subsequently been too closely respected. The maps included in the *Almagestum* were not the work of Riccioli himself, but of his collaborators GRIMALDI and SIRSALIS.

After two further maps of lesser significance were published by BOREL (1650) and KIRCHER (1660) – the latter reproduced on Figure 15-12 – a new important contribution to lunar topography was made in 1660 by G. MONTANARI (1633–1687) in Modena. Little known even at the time of its publication in MALVASIA's ephemerides (1662), it was virtually forgotten for three centuries until its discovery and republication by BONACINI (1931). This map (38 cm in diameter on the original), reproduced on

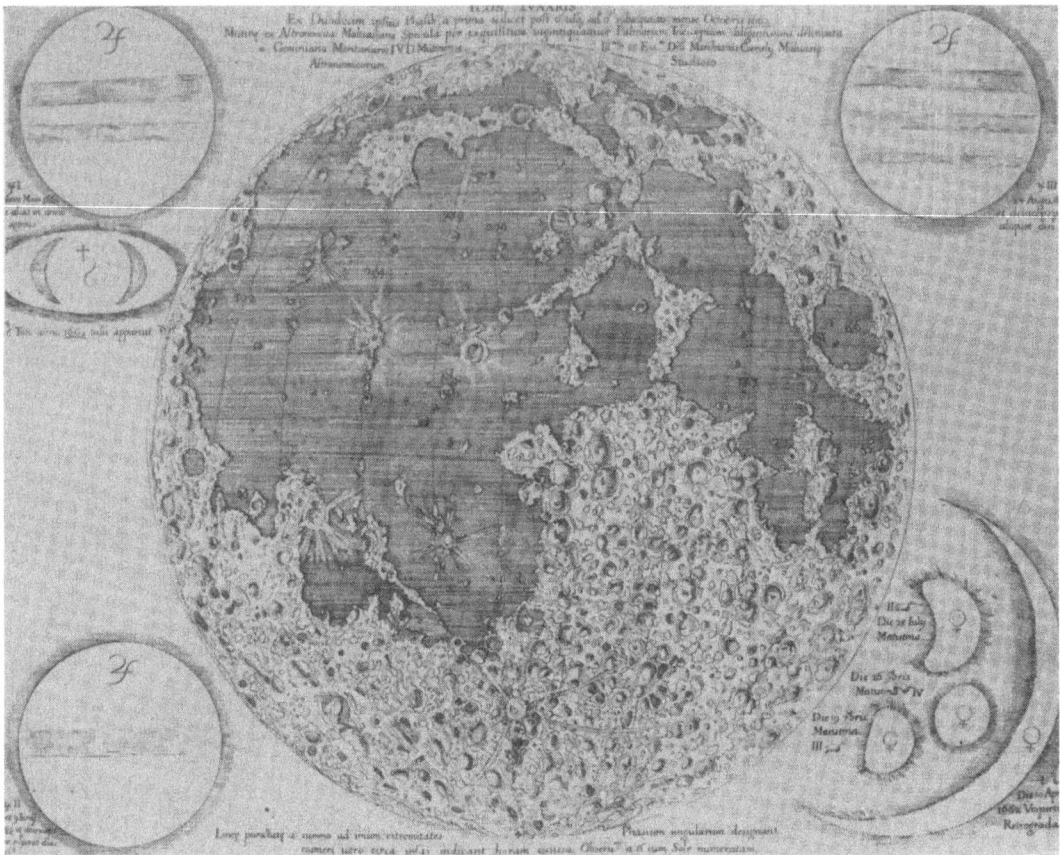

Fig. 15-13.   A map of the Moon by G. MONTANARI (1662), discoverer of the light changes of Algol.

Figure 15-13, was based on its author's observations with a telescope of 6.3 m focal distance, secured during eleven successive nights commencing October 15, 1662; and according to MALVASIA (1662), a filar micrometer was used in plotting the map.

Towards the end of the 17th century, all previous selenographic efforts were overshadowed by the great map of Gian Domenico CASSINI (1625–1712), prepared between 1671–1679 from Cassini's drawings supplemented by those of LECLERC and PATIGNY. Its reproduction as shown on the accompanying Figure 15-14 reveals several well-known formations which appeared on it for the first time in the history of lunar mapping – such as the Altai mountains, Mare Smythii, or the Rheita valley. One conspicuous formation is, however, still missing on it: namely, the Alpine valley, drawn for the first time by Francesco BIANCHINI and published in his *Hespheri et Phosphori Nova Phaenomena* (1728).

Tradition has it that Cassini's map was engraved by the same Claude Mellan, who engraved his own drawings of the Moon around 1636 (cf. Figure 15-15). By 1680

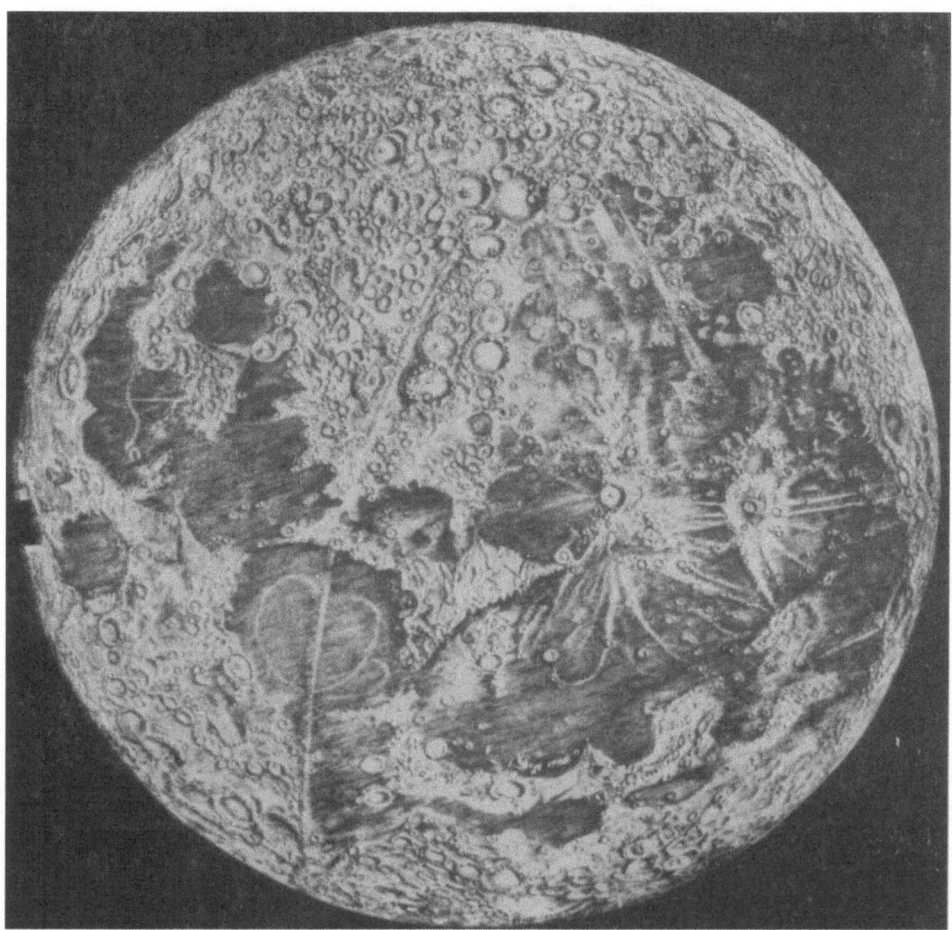

Fig. 15-14.    A map of the Moon by J. D. CASSINI from 1680, engraved by C. MELLAN.

Mellan was 82 years of age, but had still eight years to live. Of the actual engraver we have, however, no positive knowledge; an ancient inscription penciled on the verso side of the very rare print of this map in possession of the Royal Astronomical Society in London (one of the two in existence; the other being in the Observatoire National de Paris) merely mentions "engraved by an Italian artist"; but this could well apply to Mellan, who spent many years in Italy, and imported the engraving techniques from there to France.

Only a few copies of the original edition of this map were apparently ever printed. In 1785 – more than a hundred years later – the copper plate of the engraving was found at the Imprimerie Royale in Paris, and new reprint of it made; but in the turmoil of the revolutionary years no one thought of bringing the plate back to the Observatory, and a search for it made in 1828 revealed that it was sold by the Imprime-

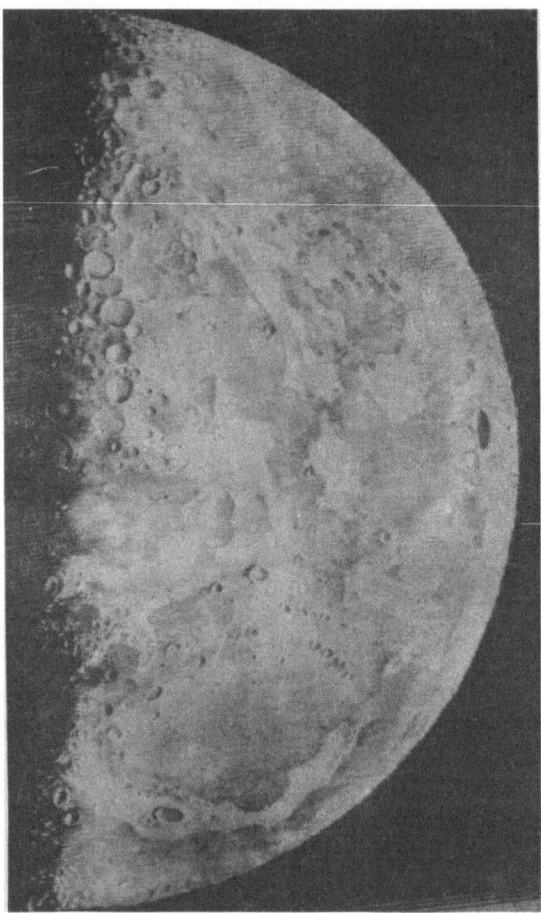

Fig. 15-15.   An engraving of the last quarter of the Moon by Claude MELLAN – one of three carried out around 1636.

rie in the meantime to a coppersmith – conceivably to serve the same profane purpose as did the copperplate with the engraving of Hevelius's map a century and a half before.

The achievements of Cassini and his contemporaries in the field of lunar topography become all the more impressive when we stop to consider the meager telescopic means at their disposal for the observations of our satellite. Throughout this period we are still in the first geological age of optical dinosaurs, characterized by small heads on huge bodies. The apertures of their simple objective lenses seldom exceeded 6–8 inches; but their focal ratios were excessively large (to lessen chromatic aberration), resulting in focal lengths unequalled by any telescopes built since. Thus the instrument with which Hevelius carried out most of his observations of the lunar surface at Danzig possessed a focal length of 49 meters (cf. Figure 15-16).

Fig. 15-16. A telescope of HEVELIUS in Danzig, used for him for observations of the Moon around the middle of the 17th century. His maps reproduced on preceding Figures 15-7 and 15-8 resulted from the observations made with such means.

Needless to say, such telescopes could be but very crudely mounted. They had no tubes, and their objectives were mostly fixed at the end of a long pole, directed to different parts of the sky by the pull of ropes. Sometimes, in desperation, the astronomer dispensed with the mounting altogether, fixed his objective on to the roof of a building, and waited for a transit of his celestial object on the ground, with an eyepiece in his hand (see Figure 15-17). This was indeed the accepted practice of telescopic work at the Paris Observatory in the days of J. D. Cassini; and most part of the details on the lunar surface as shown on his map on Figure 15-13 was apparently obtained in this way.

In the first half of the 18th century, the age of the long-necked telescopic Dinosauri of the earlier epoch gradually came to an end (particularly, since Dollond's discovery of the achromatic objective in 1759); and this fact is also fully reflected in the selenographic literature of that time – or rather a lack of it. For since about the turn of the 18th century the production of new lunar maps came gradually to a standstill. To be sure, Cassini's empirical laws of the motion of our satellite (cf. page 16) were already established with the aid of no more elaborate means; but no one did trace as yet, on any map, the actual position of the lunar equator. In fact, the concept itself of the lunar coordinates did (and could) not emerge from contemporary records until mere drawings of the Moon gave way to actual micrometric measurements of its surface features.

This stage was inaugurated by the middle of the 18th century, and the pioneer of the new technique was Tobias MAYER (1723–1762). His *Bericht von den Mondskugeln* (Nürnberg, 1750) contained an account of his method and the first quantitative measurements of the coordinates of 23 reference points on the lunar surface; but (owing to Mayer's premature death) maps based on his observations remained inaccessible till they were published by LICHTENBERG in his *Tobiae Mayeri Opera Inedita* (Göttingen, 1775); and, more than a century later, by KLINKERFUESS in 1881. A reproduction of this latter map – 35 cm in diameter on the original – is shown on the accompanying Figure 15-18. Its principal innovation is, of course, the net of equatorial coordinates, rendering this work a real chart. Mayer's critical sense was also shown by the selection of the details measured by him and recorded on his map. A comparison of it with a modern photograph reveals that no important features are missing; but all small details which could not be measured accurately with Mayer's telescope were resolutely omitted.

Mayer's work centering around 1750 represents a veritable landmark in the selenographic literature, and inaugurated the era in which selenography became at last a more exact scientific discipline. In the generations to come, its principal contributors proved to be:

J. H. LAMBERT (1728–1777), whose new lunar map based on micrometric measurements in stereographic projection was published in the years 1780–83.

J. H. SCHRÖTER (1745–1816), whose *Selenotopographische Fragmente* (Lilienthal, 1791 and Göttingen, 1802) contained not only positional measurements of numerous lunar mountains, but also determinations of their heights.

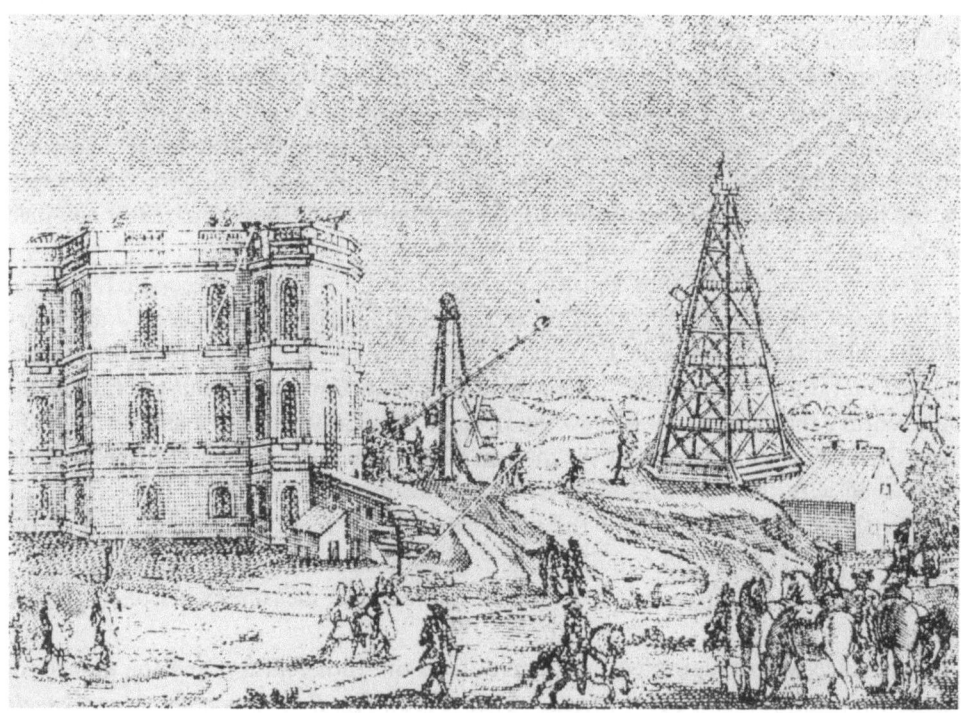

Fig. 15-17. A night scene (engraved by a contemporary artist) of observing activity at the Paris Observatory in the days of J. D. CASSINI. The data for his map of the Moon as reproduced on Figure 15-14 were obtained under such conditions.

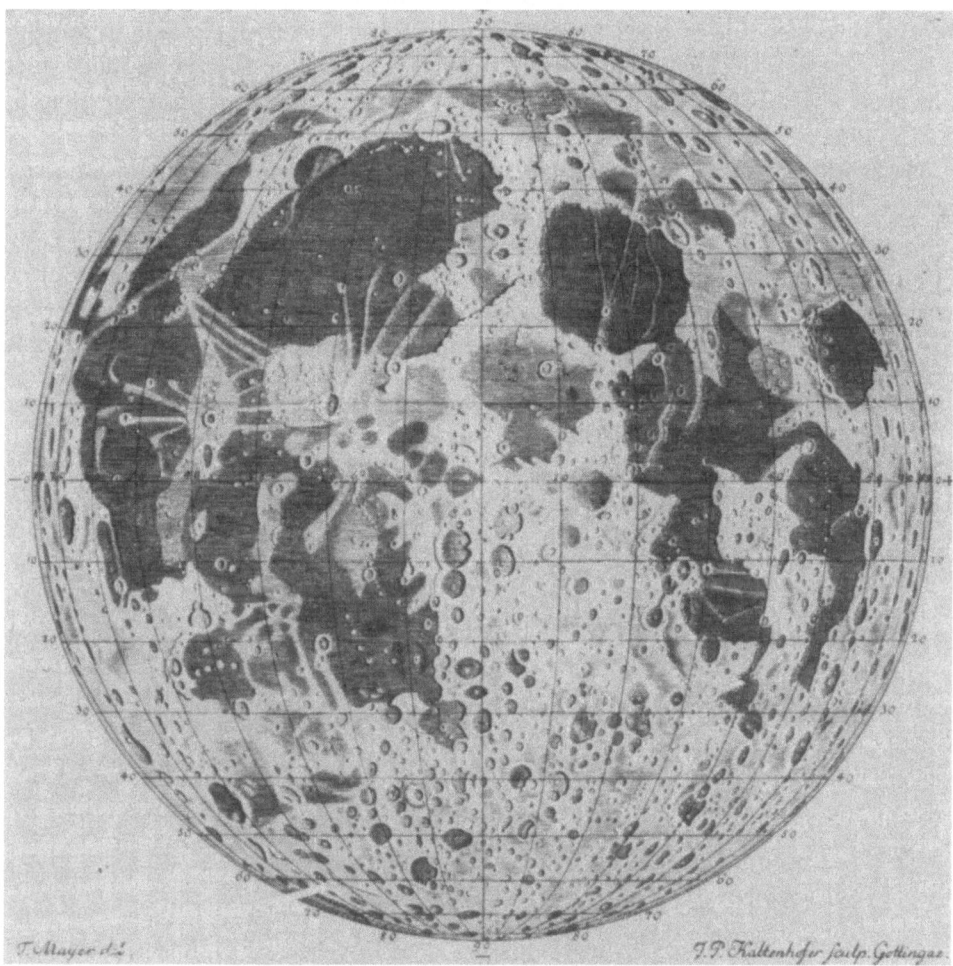

Fig. 15-18. A map of the Moon based on the observations by Tobias MAYER around 1750, and published by KLINKERFUESS in 1881 (north on top).

W. G. LOHRMANN (1797–1840), who produced two original maps of the Moon, the largest of which, published in the year of 1835 in Leipzig, was 385 mm in diameter; and left much additional material for a large map which was not published till in 1878 by J. F. J. SCHMIDT;

W. BEER (1797–1850) and J. H. MÄDLER (1794–1874) who prepared a new large map of the Moon – the first one to be divided in four quadrants – corresponding to a diameter of 97.5 cm for the apparent lunar disk, and containing practically all details that can be seen with the aid of a 4-inch refractor. The map by Beer and Mädler represents a veritable landmark in selenographic literature; and was not superseded till 1878 by the work of

J. F. J. SCHMIDT (1825–1884), whose "Charte der Gebirge des Mondes" based on observations extending between 1840–1874 and published in Berlin in 1878, represents by far the greatest selenographic work of its kind produced up to his time, and which in its class was not superseded till by the recent publication of a posthumous map by FAUTH (1964).

The work of some of the preceding selenographers deserves more than a passing mention. Thus LOHRMANN (1797–1840) can, together with Mayer, be regarded as the co-founder of scientific selenodesy. A professional geodesist of the kingdom of Saxony, at the encouragement of Gauss and Encke he set out in 1822 to secure the observations which could serve as a basis for an accurate map of the Moon with the aid of several telescopes, the largest of which (provided with a micrometer) has an objective of 12

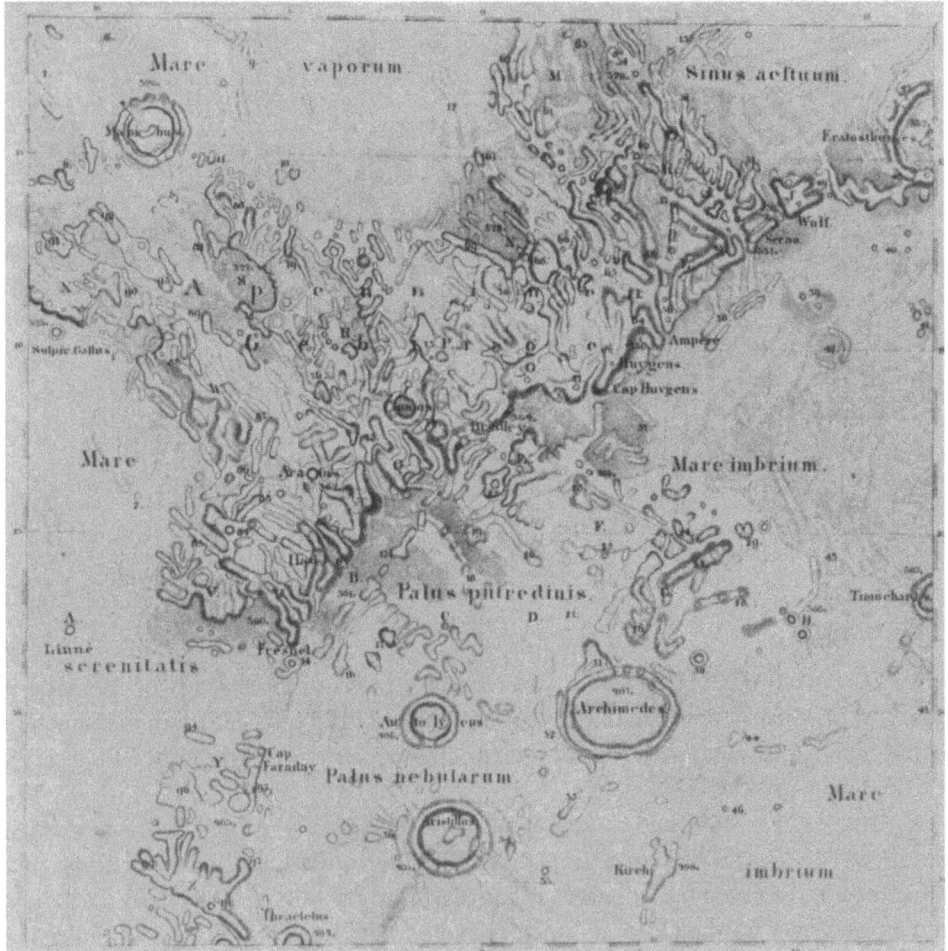

Fig. 15-19.   Region of the lunar Apennines as recorded on one section of a map by W. G. LOHRMANN (Leipzig, 1878).

cm in diameter and 1.8 m focal length. With this instrument Lohrmann measured the positions of 79 craters, recorded on a map 97 cm in diameter (divided in 25 sheets, a sample of which is reproduced on Figure 15-19). The first four were published by Lohrmann in a book entitled *Topographie der sichtbaren Mondoberfläche* (1824) which, containing as it does an accurate description of the methods employed as well as of the results obtained, can be regarded as the first modern treatise on selenography. Poor health delayed the completion of Lohmann's map until 1836; and although a smaller general map (38.5 cm in diameter) was published by him in 1838, his premature passing in 1840 (like Mayer, he barely survived the age of forty) delayed the publication of the main map for many years. It was eventually edited by Schmidt – but did not actually see the light of the day until 1878 – long after the publication of the famous map of the Moon by Beer and Mädler in 1834.

The *Mappa Selenographica* (Berlin, 1834) of the two last named authors, accompanied by their book *Der Mond* (1837) constitutes another important milestone in the development of the selenographical literature. The map itself (95 cm in diameter) was subdivided in four quadrants, and contained a remarkably faithful representation of the Moon's face as seen through its authors' Fraunhofer refractor of 9.4 cm free aperture (for a reproduction of its sample, cf. Figure 15-20). Its structure is based on the positions of 105 fundamental points measured micrometrically (related with the previous measurements by Lohrmann); moreover, the accompanying volume contains the results of micrometric measurements of the diameters of 148 craters, and of the altitudes of 830 mountains (determined by the shadow method). A second edition of this map appeared in 1877; but a reduced edition (to 32 cms) appeared already in 1837.

The maps by Beer and Mädler remained unsurpassed in wealth of information for several decades. Their supremacy ended only in 1878, with the appearance of a much larger map by Julius Schmidt, which together with its accompanying *Ergänzungsband* (Berlin, 1878) constituted a veritable mine of information concerning the surface details of our satellite, mountain heights, catalogue of rilles, etc.

Schmidt's map – a sample of which is reproduced on the accompanying Figure 15-21 – consists of twenty-five sections printed on separate sheets, and on a scale corresponding to a diameter of the apparent lunar disk of 194.9 cm. It records the positions of 32856 individual surface features, and introduces the lunar nomenclature essentially as we know it today. Schmidt did not undertake many new positional measurements (these he took mostly from Lohrmann and Mädler), but carried out determinations of the relative heights of more than 3000 lunar mountains by the shadow method (mostly while he served as director of the capitular observatory at Olomouc in Moravia), published in the *Ergänzungsband*.

In view of the vast scope of Schmidt's work it should scarcely come as a surprise that his published data contain many errors, and should not be used by a critical investigator without re-examination. In commenting on these, Miss BLAGG (1929) noted that ". . . (Schmidt) made a great many height measures (on the Moon), and in his book accompanying his lunar maps he gives them and combines them with those

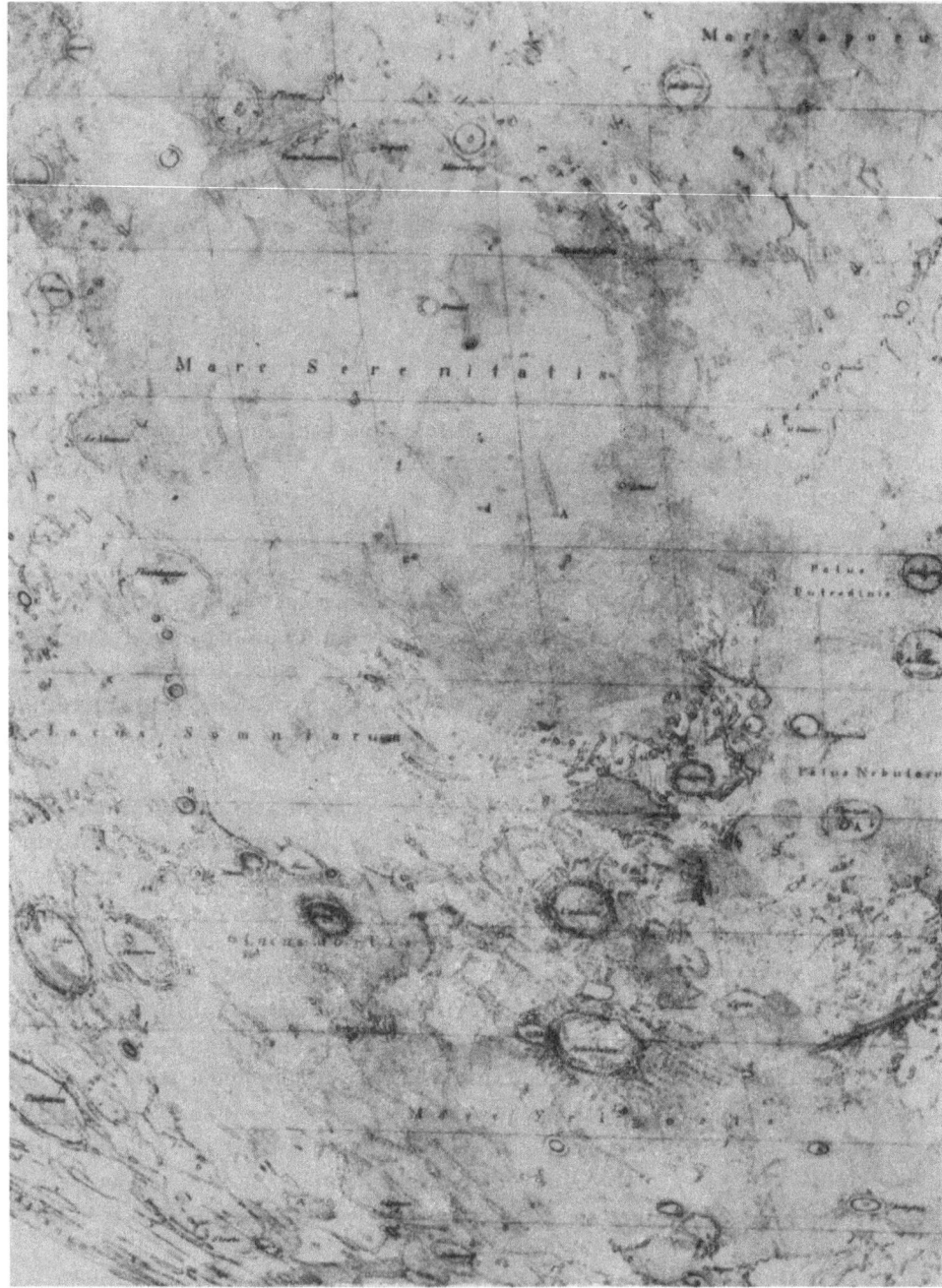

Fig. 15-20.   A section of the lunar map by BEER and MÄDLER (Berlin, 1837) of the region of
Mare Serenitatis.

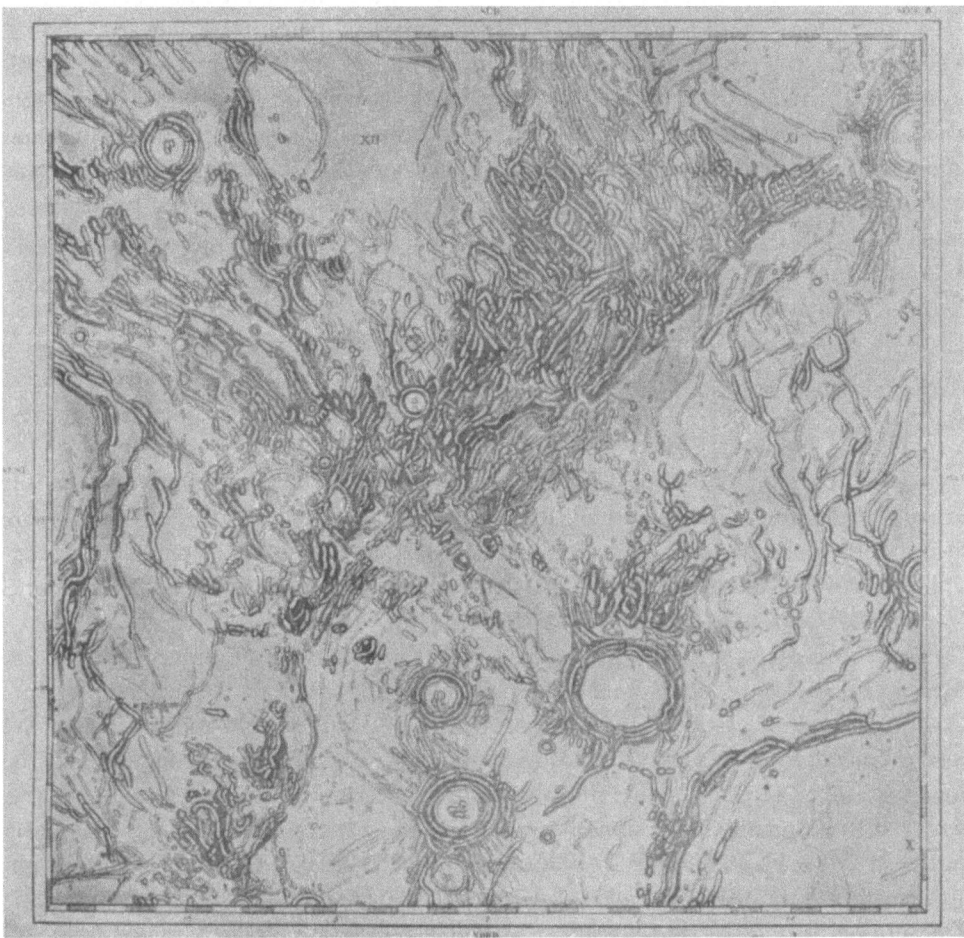

Fig. 15-21. One of the 24 sections of the lunar map by J. F. J. SCHMIDT (Berlin, 1878) representing the lunar Apennines and adjacent parts of Mare Imbrium.

of Beer and Mädler when possible. But, unfortunately, he made some mistakes of identification, and combined their measures of one height with his own of another. He also occasionally mixed up his own measures. Before, therefore", went on Miss Blagg, "any theories are seriously promulgated in connection with the comparison of one height with another, it is highly desirable that further measurements should be undertaken". It may be added that such further measurements, on a scale initiated by Schmidt, were not undertaken till in the last seven years – as a part of a large collaborative lunar mapping programme between the University of Manchester and the United States Air Force.

Abortive efforts to surpass Schmidt's map were undertaken (under the auspices of the British Association for the Advancement of Science) by BIRD between 1866–

1870; but only a small torso (four sections out of a contemplated number of 160 to constitute a lunar map 5 meters in diameter) appeared before Bird's death; and the whole project was abandoned in 1882. Among other contributions to selenographic studies forthcoming from the British Isles during the second part of the 19th century we should, however, mention three separate volumes which appeared under the same title *The Moon* – one by J. NASMYTH and J. CARPENTER (1874), the second by E. NEILSON (1876), and the third by T. G. ELGER (1895); the last accompanied by a map 45 cm in diameter. None of these works represented, however, any real advance in lunar mapping; throughout the 19th century this latter subject was still firmly in the hands of the continentals.

In the first half of the present century, efforts to construct lunar maps comparable with, or larger than, Schmidt's on the basis of visual telescopic observations (aided, in some cases, by visual transcripts of photographs) were made only three times: namely, by Goodacre, Wilkins and Fauth.

W. GOODACRE (1856–1938), for many years director of the lunar section of the British Astronomical Association, published in 1910 in London a map, in twenty-five sheets, corresponding to an apparent diameter of the Moon of 192.5 cm; but the control points of his system of selenographic co-ordinates has been based on prior photographic positional measurements by SAUNDER (1907).

More recently, other two large maps of the Moon have been published by H. P. WILKINS (1896–1960) and P. FAUTH (1867–1942). Wilkins's map, 2.5 meters in diameter (sub-divided into sixteen separate sheets) appeared first in 1946 and contains many details copied from large-scale photographs. The material plotted on this map is, however, rather heterogeneous; and the accuracy of its coordinate system as well as workmanship in detail leave something to be desired. A considerably larger map (3.5 meters in diameter, divided in 22 sections) and more reliable in detail resulted from the lifetime's work of Philip Fauth, was only recently posthumously published (1964); representing the latest – and possibly also the last – effort of visual selenography expressed in cartographic form. An example of its execution is shown on the accompanying Figure 15-22.

The reason why Fauth will probably remain the last author of a major map of the Moon based on visual observations is the fact that – as it has happened already in so many other branches of astronomical science – visual selenographic work has by now been almost completely superseded by photography. In 1839, a few years after the world was astounded by the notorious Moon hoax in the *New York Sun* on alleged discoveries of Sir John HERSCHEL (then in South Africa) of lunar inhabitants and their works; while the public chose to follow the speculations by F. P. GRUITHUISEN about the inhabitability of our satellite, a discovery was made which in time, became primarily responsible for the realization that lunar surface is dead and immutable: namely, that of the photographic process.

The history of lunar photography goes back all the way almost to the cradle of the photographic process itself. When L. J. M. Daguerre, one of the originators of this process, devised in 1839 the way of copying photographic negatives on paper, he was

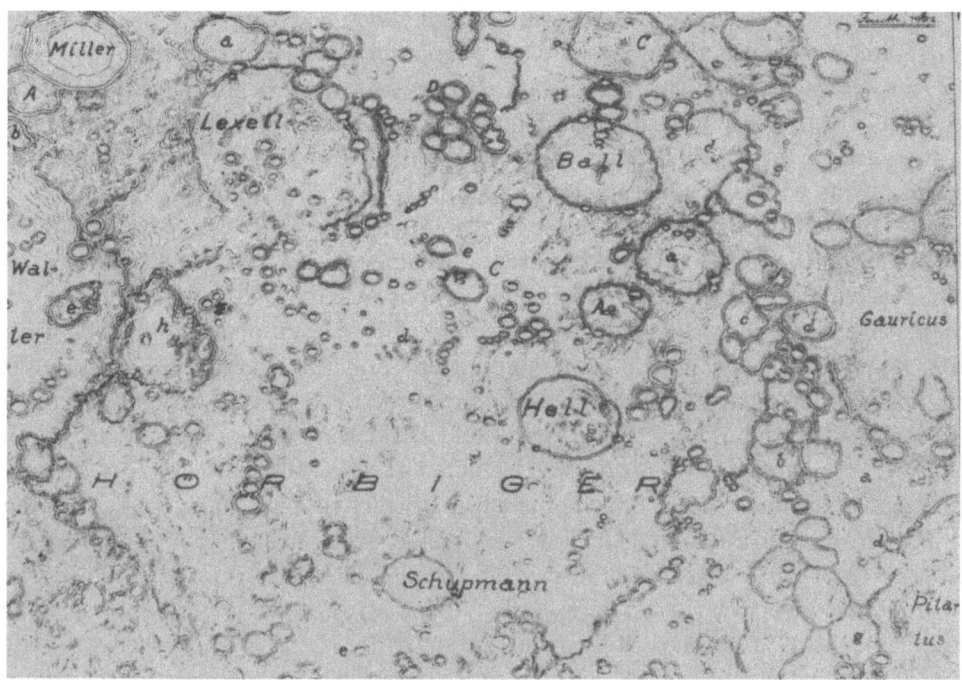

Fig. 15-22. A map of the south-central portion of the lunar disc, prepared by Ph. FAUTH in 1932, and published (posthumously) in 1964. The Hell Plain designated (privately) by Fauth on his chart as "Hörbiger" has since been renamed Deslandres (for its photograph cf. Figure 16-5); and the low-walled crater labelled "Schupmann" is more generally known as Hell B.

encouraged by D. T. J. Arago, then director of the Paris Observatory, to attempt a photograph* of the Moon, in order to discover whether or not the light of the Moon was chemically actinic. This proved indeed to be the case; though otherwise the experiment was a failure, in so far as Daguerre's plate showed no distinguishable detail.

The failure did not, however, deter J. W. DRAPER (1800–1882) from repeating promptly Daguerre's experiment with improved means. He realised that the main cause of it was the fact that Daguerre underexposed his plates; and in order to avoid this, Draper exposed, in 1840, a 25 mm image of the Moon formed in the focus of his 12-inch mirror of $f/10$ focal ratio for 20 minutes. The outcome proved to be a success; and was further improved by W. C. BOND (1789–1859), the first director of the Harvard College Observatory, who (working together with J. A. WHIPPLE) used for this purpose the 15-inch $f/20$ Merz refractor then newly installed at Harvard. By 1850, these investigators were able to obtain lunar photographs exposed in less than one minute, which were capable of enlargement and showed details of all principal features of the surface of our satellite.

* It is of interest to note that the term "photography" seems to have been first applied to the contemporary daguerrotypes by the selenographer J. H. MÄDLER.

The next forward step in lunar photography was taken by Warren DE LA RUE (1815–1889), who was the first to use the collodium plates exposed in the focus of a 12-inch reflector of 305 cm focal length. His negatives, taken in the years 1852–57, were sufficiently sharp to stand considerable enlargement, and can be regarded as true forerunners of the splendid series of photographs secured later by various large telescopes of the world. In 1854, improved photographs of the Moon were submitted at the meetings of the British Association for the Advancement of Science by Sir William CROOKES, and J. B. READE. Crookes used a refractor of 203 mm in aperture and 390 cm focal length; while Reade took his photographs with the aid of a reflector of 60 cm diameter and 23½ m focal length; the telescope itself was stationary during exposures that lasted several seconds, and the guiding was done on the plateholder. In 1857, Thomas GRUBB published many photographs of the Moon, taken with a refractor of 32 cm aperture and 610 cm focal length (guided likewise on the plateholder); and in 1858, L. M. RUTHERFURD (1816–1892) secured the first stereoscopic pairs of lunar photographs in New York. Shortly thereafter, further contributions to lunar photography were made in the United States by such well-known astronomers as Father Angelo SECCHI (1818–1878) then in Georgetown, and Benjamin Apthorp GOULD (1824–1896).

All these investigators can be regarded as fathers of lunar photography near the middle of the 19th century. With the exception of Reade, they all worked with telescopes 8–15 inches of aperture; and the entire subsequent development of the subject was inseparably connected with the construction and use of the telescopes of larger resolving power, together with a gradual improvement of the photographic material and replacement of the collodium plates by dry bromosilver emulsions.

As a culmination of this effort in the 19th century, its last decade witnessed the publication of two great photographic atlases of the Moon, based on photographs secured at the Lick and Paris Observatories. Following the erection of the 36-inch refractor of 17.34 m focal distance at Lick Observatory in 1888, Edward S. HOLDEN (1846–1914), the observatory's first director, used this excellent instrument for an extensive program of lunar photography; and a part of the results appeared under the title of the *Lick Observatory Atlas of the Moon* in 1896–7. The enlargements reproduced in this Atlas correspond to a diameter of the Moon of 97.45 cm (i.e., to a scale of 1:3547000). The actual size of the illustration is 23 × 32 cm; and 1 mm on the prints corresponds to 3567 meters on the lunar surface.

In subsequent years, Holden lent a considerable number of the Lick negatives to Ladislav WEINEK (1848–1913), director of the Prague Observatory who embarked on their systematic enlargement and printing. On the original Lick negatives, the diameter of the Moon's image oscillated between 12.4 cm at apogee to 13.9 cm at perigee. Weinek enlarged these 23.77 times, to make the diameter of the Moon correspond to 296 cm at apogee and 330 at perigee – so that 1 mm on the enlargements corresponds to 1115 meters on the lunar surface (i.e., to a scale of 1:1115000). Apart from a total of 114 Lick negatives received from Holden, Weinek received also from Paris negatives from M. LOEWY, which he enlarged 23–26 times to correspond to a lunar diameter of 396 cm (i.e. to a scale of 1:877500).

The results of this work were published by WEINEK under the title *Photographischer Mond-Atlas, wahrnehmlich auf Grund von focalen Negativen der Lick-Sternwarte in Maasstabe eines Monddurchmessers von 10 Fuss*, (Prague 1899). Originally Weinek intended to publish two volumes of his atlas, each consisting of 200 plates with the image size of 24.5 × 29.5 cm. As the Atlas is so arranged that each object is reproduced as it appears at the time of the lunar sunrise and sunset, its first 10 fascicles of 20 leaves each, which appeared between November 1897 and November 1900, recorded only 100 different sections of the lunar surface. The position and orientation of each plate was specified by the selenographic coordinates of its center, as well as by the seleno-graphic longitude of the terminator.

That the second half of Weinek's Atlas as originally planned never appeared was probably due to its high price: for although its publication was subsidized by the Vienna Academy of Sciences as well as by Baron Rothschild and Miss Catherine Wolfe Bruce, it was offered commercially for sale at 100 Gulden – a sum equivalent to at least a monthly income of the highest-paid astronomers of that day. This effect-ively prevented adequate marketing and rendered Weinek's Atlas a rarity which only well-endowed libraries could afford to possess.

Moreover, the almost 24-fold enlargement adopted by Weinek for his work did not increase proportionally the amount of information discernible on the individual prints. In order to demonstrate this, we reproduce on the accompanying Figure 15–23 a comparison of two photographs of the same region of the Moon (i.e., the crater Copernicus): Weinek's enlargement of a Lick negative (left) with a recent Pic-du-Midi photograph of the same formation and on the same scale (right). Both photographs were obtained on sites renowned for first-class seeing; but in spite of the fact that the

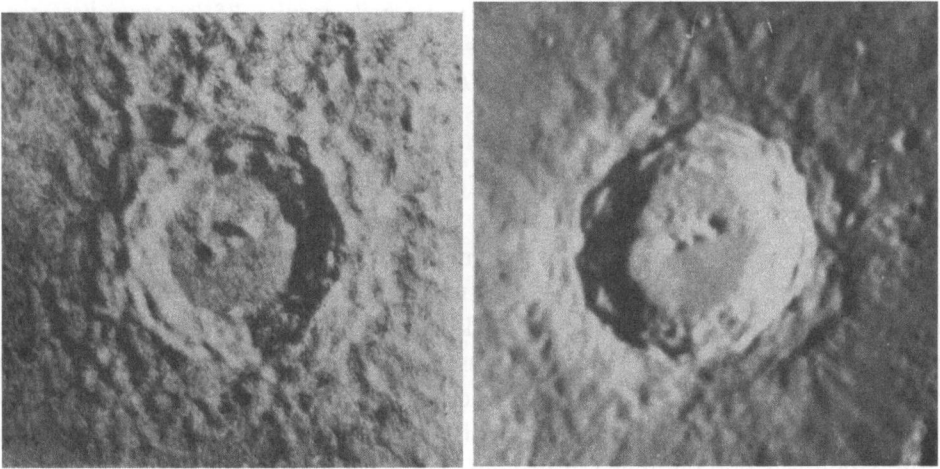

Fig. 15-23.   A comparison of a visual transcription of the Lick photographs of the crater Copernicus by L. Weinek (1899), left, with a recent Pic-du-Midi photograph of the same formation on the right. The comparison reflects a vast improvement in photographic materials since Weinek's time.

photograph on the left was taken with a 36-inch objective, while only a 24-inch was used to secure the one on the right, the greatly superior quality of the latter reflects a vast improvement in photographic material in use since the rays of Holden and Weinek. Nevertheless, such inordinate enlargements of these and other early photographs were used by other able selenographers – such as J. N. KRIEGER (1865–1901) – as a general background for detailed visual topography of individual lunar regions. His work, *J. N. Krieger's Mond-Atlas* (Wien, 1912), of which we reproduce Figure 15-24 as an illustration, was published subsequently by R. KÖNIG (1865–1927).

Apart from the Holden and Weinek Atlases based on the Lick photographs of the Moon, the last years of the 19th century witnessed the commencement of another great contribution to selenographic literature: namely, the *Atlas Photographique de la Lune*, by M. LOEWY and P. PUISEUX, Observatoire de Paris, part 1–10, plates 1–80, 1896–1909. The negatives at the basis of this Atlas were secured by its authors and their collaborators at the Observatoire National in Paris with the aid of the "grand équatoreal coudé" of 60 cm free aperture and 18 m focal length. The size of the lunar image in the focus of this refractor oscillated between 17.3 cm to 15.5 cm from perigee to apogee; and 80 enlargements of such negatives (unfortunately, not all on the same scale) constitute an atlas which remained the standard work on its subject for several decades.

As was the case with Weinek's Atlas, the physical size of the Paris Atlas made it likewise not easily accessible to every interested user; and, for this reason, several reduced editions of this standard work appeared in many countries. Thus M. C. LE MORVAN published such an abridged edition in his *Carte photographique et systématique de la Lune* (Paris 1914–1921) in two volumes (each in four parts), containing enlargements which correspond to a size of the lunar disk between 90–120 cm; the size of the individual prints being 25.5 × 32 cm. A Belgian edition of the Paris Atlas appeared in Brussels between 1899–1912 under the title of *Atlas lunaire, publié par la Société Belge d'Astronomie, reproduisant à une échelle reduite 2/5 les agrandissements photographiques de M. M. Loewy and Puiseux.* A Spanish edition by E. J. THOST, entitled *Resume del Atlas fotografico de la Luna del Observatorio Nacional de Paris*, appeared in Tarragona in 1922 (60 plates 15 × 21 cm in size); and was reproduced also in German in Stuttgart.

Apart from the Paris Atlas and its various editions which dominated the field in the first half of the present century, we may mention W. H. PICKERING's *Photographic Atlas of the Moon* which appeared as volume 51 of the *Annals of the Harvard College Observatory* in Cambridge, Mass., 1903, and contained on 88 plates reproductions (unenlarged) of photographs taken by Pickering with a 12-inch telescope of 41.5 m focal length at Mandeville, Jamaica. Because of relatively long exposure required for so large a focal ratio, the majority of Pickering's photographs lack the contrast and definition of the Paris plates; the merit of his work resting on the fact that each region of the Moon was photographed at five different elevations of the Sun.

Much more recently, a photographic atlas of the Moon on a similar scale was published by Sh. MIYAMOTO and M. MATSUI as No. 95 of the *Contributions from Kwasan Observatory*, Kyoto, in 1960. The photographs reproduced on 85 plates were

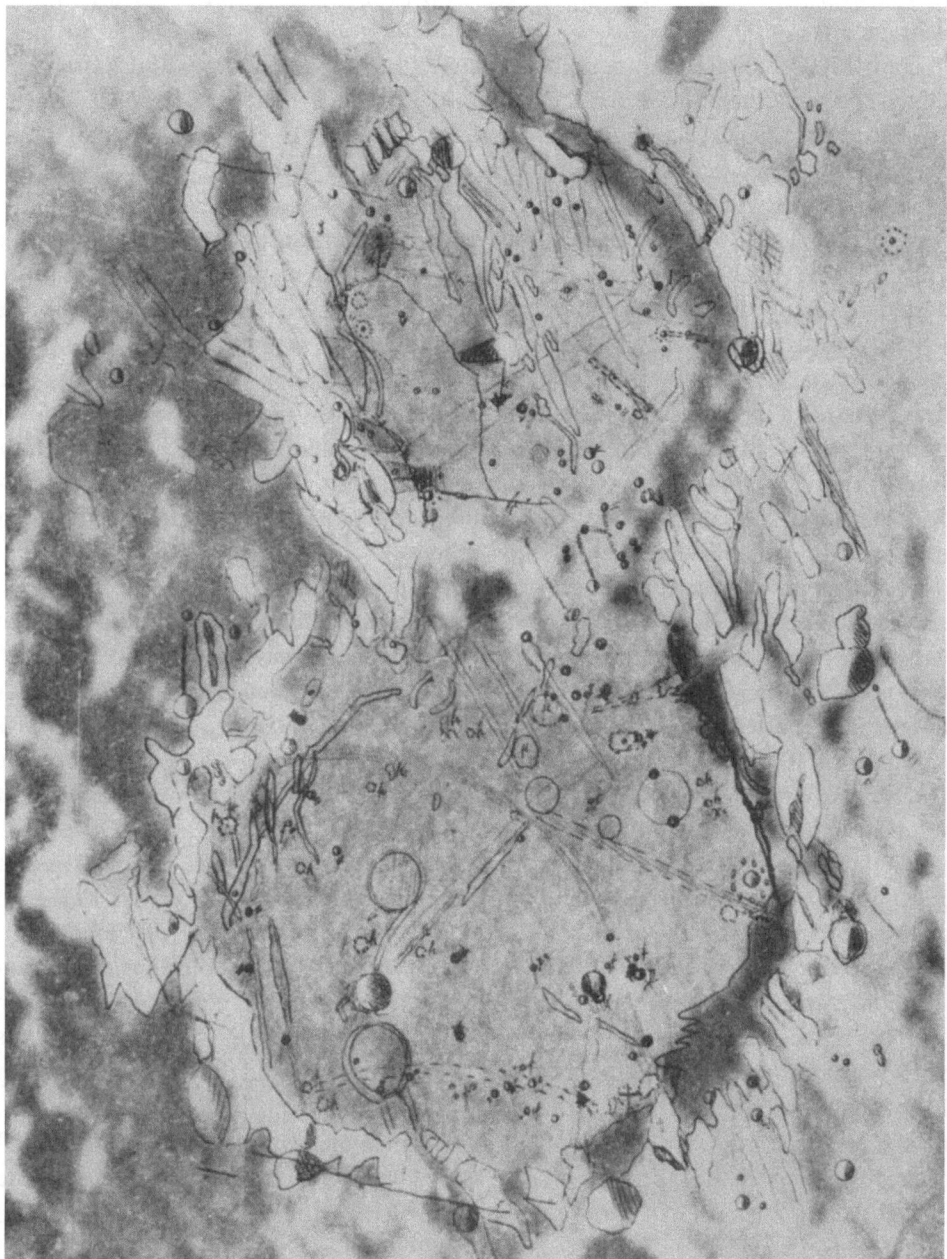

Fig. 15-24.   An example of Krieger's use of photographic enlargements as a basis for subsequent visual work in the region of the craters Ptolemaeus and Alphonsus.

secured with the aid of a 12-inch refractor, operating at an effective focal length of 20.70 meters.

Until 1960, the Paris photographic Atlas of the Moon by Loewy and Puiseux remained the most extensive work of its kind in existence. That year it was, however, superseded by a new *Photographic Lunar Atlas*, edited by G. P. KUIPER in collaboration with D. W. G. ARTHUR, E. MOORE, J. W. TAPSCOTT, and E. A. WHITAKER (Univ. of Chicago Press, 1960). This atlas contains 281 illustrations, of which 212 exhibit 44 lunar regions under 4–5 different conditions of illumination by the Sun, on the scale of 1:1 370 000. All these illustrations are based on enlargements of the photographed secured by many different astronomers at the Lick, Mt. Wilson, Pic-du-Midi, and Yerkes Observatories, and collected by the compilers. Moreover, a subsequent appendix to this atlas (*Orthographic Atlas of the Moon*, compiled by D. W. G. ARTHUR and E. A. WHITAKER in 1962) contains overprints of lunar co-ordinates superimposed over photographs of a major part of the visible lunar hemisphere; while a second appendix (in the form of *Rectified Lunar Atlas*, by E. A. WHITAKER, G. P. KUIPER, W. K. HARTMANN and L. H. SPRADLEY (1964)) contains reproductions of photographs of the lunar limb regions rectified by projection on as sphere.

The latest addition to our literature, *Photographic Atlas of the Moon* by Z. KOPAL, J. KLEPEŠTA, and T. W. RACKHAM (1965), and based wholly on photographs secured by its authors with the aid of the 24-inch refractor of the Observatoire du Pic-du-Midi* in France, between 1961–1962, differs from its predecessors in so far as only terminator photography (both sunrise and sunset) has been included in the main part of the Atlas consisting of 197 separate illustrations; and an accompanying skeleton map of the Moon by A. RÜKL contains a complete account of lunar nomenclature recognised at the present time (see the fold-out enclosed with this book).

The largest and most detailed mapping project now in existence was initiated in 1960 by the Aeronautical Chart and Information Center of the U.S. Air Force, in collaboration with the Department of Astronomy at the University of Manchester. The observational data underlying this project consist of many thousands of sequential photographs secured for this purpose with the 24-inch refractor and the 43-inch reflector of the Observatoire du Pic-du-Midi, supplemented by visual observations with a similar refractor at the Lowell Observatory in the United States. The reduction of these data has largely followed the method outlined earlier in Chapters 13 and 14; and the results continue to appear – not in the form of any atlas – but in individual sheets 50 × 60 cm in size, and on a scale of 1:1 000 000, with one millimetre on the chart equalling a length of 1 km on the surface of the Moon. Other, special purpose charts on the scales of 1:2 000 000 and 1:500 000 have been prepared for selected regions. An example of such a chart on a scale 1:2 000 000, covering the Copernicus-Kepler region of the lunar surface and the adjacent plains of the Oceanus Procellarum is reproduced on the accompanying Figure 15-25.

---

* Its objective being a twin of the lens of the grand équatoreal coudé at Paris Observatory, with which Loewy and Puiseux secured the photographs for their well-known atlas of the Moon between 1896–1909; and which in 1942 was re-mounted by LYOT at Pic-du-Midi.

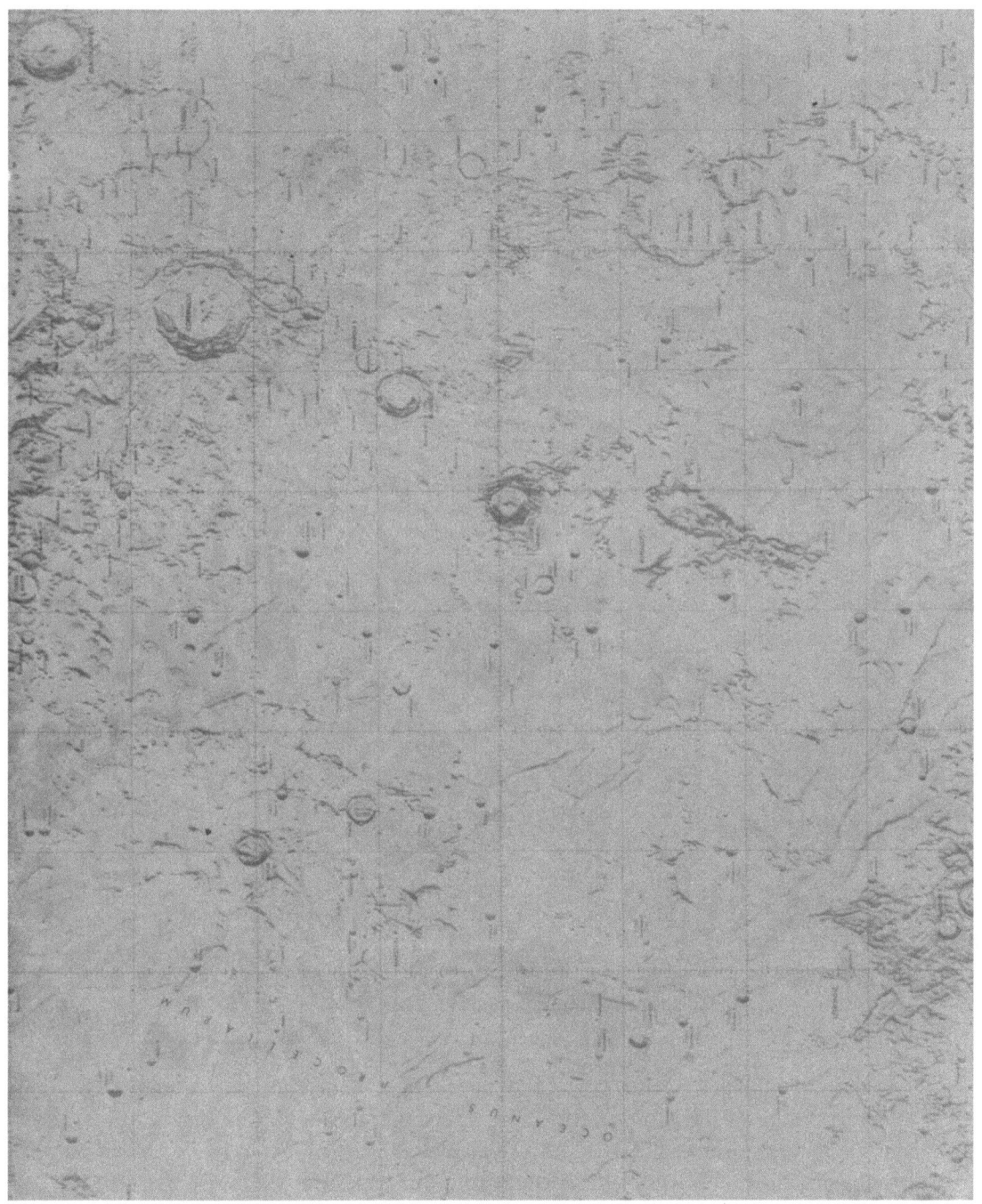

Fig. 15-25.   A section of an ACIC chart of the lunar surface of the plains of the Oceanus Procellarum, south-west of Copernicus, on a scale of 1:2000000 (1964).

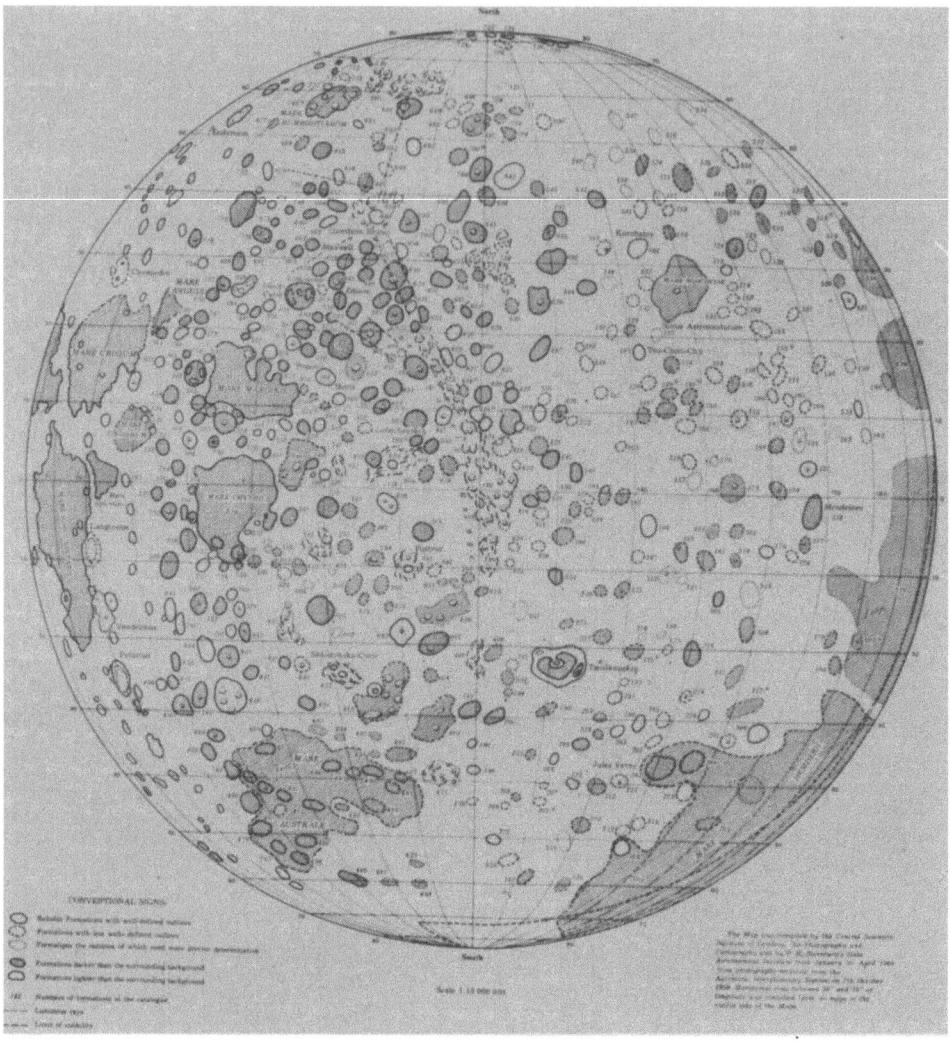

Fig. 15-26.   The first map of the far side of the Moon, prepared by the USSR Central Institute of Geodesy and the Sternberg State Astronomical Institute in Moscow (1960) on the basis of photographs received from Lunik 3 on October 7, 1959.

A total of 144 such fields (of which 24 appeared to-date) should eventually cover the entire surface of the Moon. There is no doubt that this latest and most elaborate lunar mapping project launched so far will also be the last of its kind – i.e., based on astronomical observations of our satellite from our terrestrial vantage point – to be superseded by more detailed maps constructed on the basis of information secured by space craft operating in greater proximity of the lunar surface (in particular, the Orbiter satellites), on the eve of the manned landings on the Moon in the (hopefully) near future.

All that we said on lunar mapping in this chapter has concerned the visible hemisphere of our satellite (or, more precisely, that part of the lunar surface which becomes visible from our terrestrial vantage point in the course of the lunar libration cycle. In October 1959, the memorable flight of the Russian Lunik 3 provided us with the first photographic records (albeit with a relatively low resolution) of most parts of the lunar far side. The photographic data secured during this flight have been published by the USSR Academy of Sciences in an *Atlas of the Other Side of the Moon* (Moscow, 1959; English translation London, 1961), edited by N. P. BARABASHEV, A. A. MIK-HAILOV, and Yu. N. LIPSKI; and the first map of the lunar far side constructed on the basis of these data by the Russian investigators is reproduced on the accompanying Figure 15-26.

## CHAPTER 16

# FORMATIONS ON THE LUNAR SURFACE;
## DESCRIPTIVE SURVEY

In the preceding section of this survey we got acquainted with some of the fundamentals concerning the lunar coordinates and mapping, which should provide a basis
for more detailed description of the surface of the Moon with its almost innumerable
diverse features, and thus help us to understand from them the nature of the principal
processes which have been shaping up its face since time immemorial.

What is so arresting about this face, and what can we learn from its inspection?
The Moon is a very old body – not less than 4500 million years of age – and has
probably been a close companion of our Earth since the days of its formation or not
long thereafter. A permanent and virtually complete absence of air or liquid water
on its surface makes it, moreover, certain that most part of its composite fossil record
must be of very ancient age – its oldest visible landmarks being, perhaps, not far removed in time from the days of the origin of our whole solar system. On the Earth
or other terrestrial planets, all landmarks of comparable age must have fallen prey to
the joint disturbing action of air and water aeons ago. However, as any change on the
Moon – caused by other kinds of erosion (seismic, light) – can proceed only at an exceedingly slow rate, its present wrinkled face must still bear at least traces of many
events which have taken place in the inner precincts of our solar system in the days
of its formation; and if so, their correct interpretation holds indeed a rich scientific
prize.

Even to the naked eye the Moon is a beautiful object, diversified with markings
which have been associated with numerous popular myths of many nations. If we look
at its pock-marked face through a telescope (or, more comfortably, at Figure 16-1
and many other photographs reproduced in this volume), a most cursory glance reveals
the lunar surface to consist essentially of two principal types of ground. One, rough
and articulate, is comparatively light (reflecting, in places as much as 18% of incident
sunlight) and broken up by many mountains. The other type of ground is darker
(reflecting, on the average, but 6–7% of incident sunlight), much smoother and frequently so flat as to superficially simulate a liquid surface. The first type of the ground
is commonly called the *continents*. They occupy large continuous areas – particularly
in the southern hemisphere of the Moon – and cover, on the whole, a little less than
two-thirds of the area of the visible face of our satellite; though (according to Russian
photographs of 1959) more than nine-tenths of the far side. An example of a photograph of the far side of the Moon obtained by the Russian Lunik 3 on October 7,
1959 is shown on the accompanying Figure 16-2 and 16-3; while a composite montage
of a hemisphere consisting of the visible portions of the near and far sides is shown on

Fig. 16-1.  A view of the lunar face close to the full Moon.

Figure 16-4. The flatlands – or *maria* as they were misnamed by early observers of the
Moon before the true nature of its surface was properly understood – checkered by
faults and sparingly dotted with cliffs, occupy the rest. Both types of ground are, on the
whole, remarkably uniform in reflectivity and general appearance – on a small or
large scale. In view of what has been said in chapter 13 of this part, the reader
should, however, be careful not to identify summarily the continents with the "highlands", and the maria with "lowlands", as is sometimes done in the literature of
lighter vintage; for as can be clearly seen on Figures 13-8 to 13-12, the system of
lunar isohypses does not seem to be related with that of the isophotes (marking the
distribution of the continents and the maria) in any obvious manner.

A closer look at the Moon with the aid of a telescope magnifying a few hundred
times reveals a rather bewildering array of mountains and formations, no two of
which are exactly alike. However, the dominant type of formation among them – and
by far the most numerous on any part of the Moon – appears to be ring-like walled
enclosures commonly called *craters*. This word is used here in its original sense – to
describe a cup-shaped topographic feature – as derived from the Greek root κρατερ
(meaning cup or bowl), without prejudice for the views on their origin; a too special-
ised an interpretation could easily render it as much of a misnomer as the Martian
"canals" or lunar "seas".

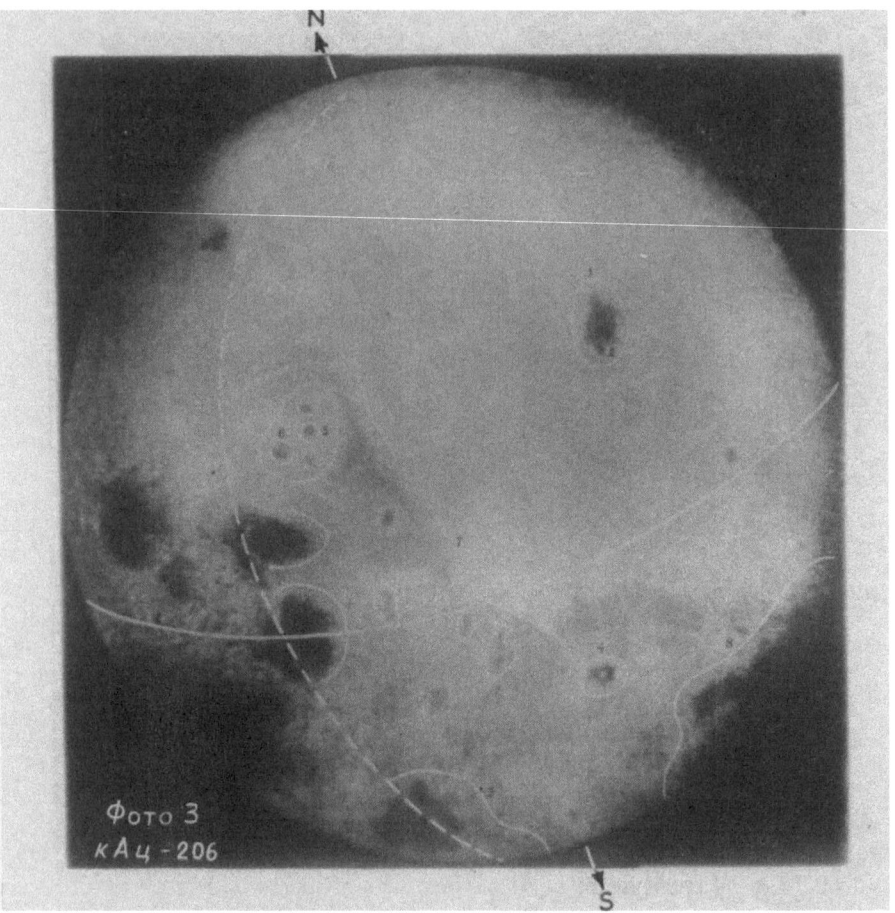

Fig. 16-2. A photograph of the far side of the Moon, secured by the Russian third space station on October 7, 1959, at a distance of 66000 km. The solid arc marks the position of the lunar equator; the dotted arc, the limit of the hemisphere visible from the earth.

The craters occur almost everywhere on the Moon – in mountainous regions as well as in the maria – in truly prodigious numbers, giving the lunar surface its characteristic appearance of a pock-marked face. The number of craters with diameters in excess of 1 km is estimated to be more than 300000 on the visible hemisphere of the Moon alone; and those smaller still are too many to be individually counted. The largest of them attain dimensions in excess of 200 km, such as the crater Clavius (Figure 16-5) whose enclosure exceeds 230 km across; or the largewalled plain on the Western shores of Mare Nubium, recently given the name of Deslandres, a sunrise over which can be well seen on Figure 16-6. There are altogether five craters on the visible side of the Moon whose dimensions exceed 200 km, and an additional 32 objects with diameters between 10 and 200 km. Smaller formations of this type are so many that their detailed description could continue almost without end.

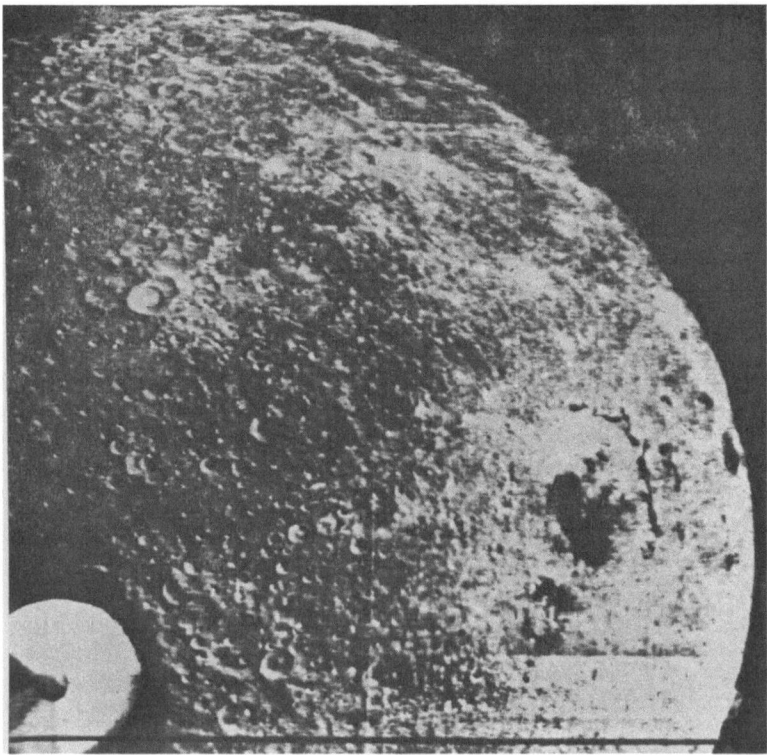

Fig. 16-3.  A more detailed view of a part of the far side of the Moon, obtained by the Russian automatic interplanetary station Zond 6 on July 20th, 1965. The large dark spot on the right is Mare Orientale (the bright disc in the lower left-hand corner is a photometric scale).

In order to facilitate references to them, the majority of more conspicuous formations on the lunar surface have been given proper names assigned to them in accordance with certain general rules. Thus most lunar plains referred to as "seas" have been given names connected with the weather, or state of the mind – hence the etymology of the Oceanus Procellarum, Mare Imbrium, Nubium, Tranquilitatis, Serenitatis or Crisium. Only certain small maria near the limb of the Moon were named after human beings (Mare Humboldtianum, Mare Smythii, etc.). By common consent, the craters are given the names of persons who are no longer living and who have entered the history of some branch of human endeavour; some 500 of such names are now recognised by the International Astronomical Union.* In earlier times, a veritable motley of personalities – saints and scholars, potentates, ancient sages and Christian ecclesiastics, but remarkably few artists or writers – have found access to the lunar Pantheon by favour of their friends. Among the more recent candidates of immortality, potentates are now definitely out of luck, and the chances of the scientists have greatly improved – witness the first group of names bestowed on the formations on

* A complete and up-to-date version of this official lunar nomenclature can be found on the Skeleton Map of the Moon, by A. Rükl, enclosed as fold-out with this volome.

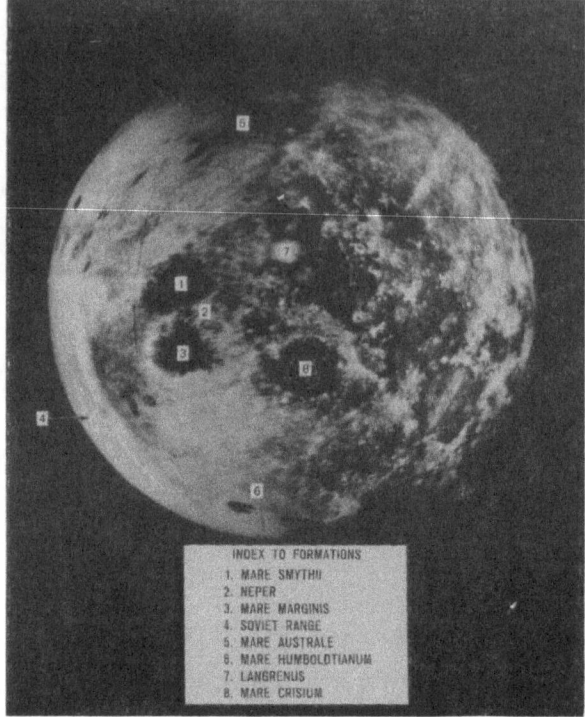

INDEX TO FORMATIONS
1. MARE SMYTHII
2. NEPER
3. MARE MARGINIS
4. SOVIET RANGE
5. MARE AUSTRALE
6. MARE HUMBOLDTIANUM
7. LANGRENUS
8. MARE CRISIUM

Fig. 16-4.   A combination of photographs of the near and far side of the Moon, projected on the same sphere.

the far side of the Moon, which included even a novelist (Jules Verne). Lastly, the lunar mountains or mountain ranges are usually named after corresponding formations on the Earth – thus on lunar maps too we encounter the Alps, Apennines, Carpathian mountains and other chains whose names are familiar to the terrestrial geographer.

No two mountains or craters on the Moon are exactly alike. However, apart from individual distinguishing features, most craters possess also many characteristics in common; and as these are of fundamental importance for all efforts to unravel the past history of the Moon – efforts which may shed much light on the history of all other terrestrial planets as well – we propose to review them, in this section, in some detail before coming to grips with the theories of their origin.

First, let us note the fact which even a cursory inspection of the lunar face bears out at a glance: namely, that *the distribution in size of lunar craters ranges continuously from the largest formations of this type* – like Clavius or Deslandres – *down to the smallest pits* discernible with our telescopes (and, of late, of the Ranger photographs secured from the close proximity of the lunar surface) *in numbers increasing rapidly with diminishing size*, which will be discussed in Chapter 18. Secondly, *the areal distribution of craters* – of all sizes – *on the visible face of the Moon appears to be essentially*

Fig. 16-5.  The lunar crater Clavius near the south pole of the Moon (photographed by the 200-inch
Hale telescope of the Palomar Observatory).

*at random.* Thirdly, *the heights of the ramparts of all craters proves*, in general, *to be
very small in comparison with their absolute dimensions.*

The measurement of the dimensions is simple enough in angular units – from
photographs or at the telescope; and their conversion into absolute unity requires
merely a knowledge of the instantaneous distance of the feature in question on the
lunar surface from the observer. A circular crater on the surface of the Moon will, in
general, appear to us in projection as an ellipse of semi-major axis $a'$ along an arc
parallel with the limb, and semi-minor axis $b' = a' \cos \kappa$, where $\kappa$ denotes (as in the
preceding Chapter 14) the selenocentric angle of the respective formation as defined
by Equations (13-5) (cf. also Figure 14-1).

Let, moreover, $s'$ denote the apparent topocentric semi-diameter of the Moon at the
time of observation (related with the geocentric semi-diameter $s$ by Equation (3-14).
If so, the absolute value of the true semi-major axis $a$ will be given by the equation

$$a = 1738 \frac{a'}{s'} \left( \frac{r'_s}{r'} \right) \text{km}, \tag{16-1}$$

where $r'_s$ denotes the topocentric distance of the respective lunar formation; and $r'$,

Fig. 16-6.   The lunar craters Deslandres, Walter, Regiomontanus, and Purbach. White arrow indicates the position of the hill-top crater Regiomontanus A (for its profile see Figure 17-9). Photograph
taken with the 24-inch reflector of the Observatoire du Pic-du-Midi (Manchester Lunar Programme).

the topocentric distance of the Moon's centre. As, however, from (14-26),

$$\frac{r'_s}{r'} = 1 - \sin s' \cos \eta + \dots \tag{16-2}$$

and, very approximately, $\eta \approx \kappa$, Equation (16-1) can to a sufficient precision be rewritten as

$$a = 1738\,(a'/s')(1 - \cos \kappa \sin s') \text{ km} \tag{16-3}$$

and

$$b = 1738\,(b'/s')(1 - \cos \kappa \sin s') \sec \kappa \text{ km}. \tag{16-4}$$

On the other hand, the relative *height* of the ramparts, or of any other mountain
on the Moon, can be determined from the length of the shadows cast on the surrounding landscape for a given altitude of the Sun above the lunar horizon, by the method
developed in some detail in the preceding Chapter 14. The length of the shadows will,
of course, vary continuously in the course of the lunar day; being greatest (and also
changing most rapidly) at the time of sunrise or sunset. A continuous photographic

Fig. 16-7.  Sunset over the crater Copernicus and the Carpathian mountains, photographed with 24-inch reflector of the Observatoire du Pic-du-Midi (Manchester Lunar Programme).

Fig. 16-8.  Sunset over the craters Theophilus, Cyrillus, and Catharina, photographed with the 24-inch refractor of the Observatoire du Pic-du-Midi (Manchester Lunar Programme).

Fig. 16-9. A photograph of the crater Tycho near the south pole of the Moon, taken with the 120-inch reflector of the Lick Observatory. Note the relatively dark rim surrounding the crater, demonstrating the extent of the region which was in the "ballistic shadow" of the ramparts during the impact.

(ciné-) record of such changes, carried out in recent years on a very large scale by astronomers from the University of Manchester working at the Observatoire du Pic-du-Midi in France in collaboration with the Aeronautical Chart and Information Center of the U.S. Air Force, has resulted in the determination of the relative elevations of thousands of points of the lunar surface (mainly in the equatorial regions of our satellite). The precision of the positional measurements on the lunar surface attainable with our ground-based telescopes is such as to enable us to measure the location of any individual point relative to others with errors of the order of 300–400 m; while the height of any mountain can be determined by the shadow method within ± 10 m – a very respectable accuracy when we consider the fact that the underlying triangulation must be carried out at a distance that never becomes less than 356 000 km.

Extensive measurements of the heights of the ramparts of many individual lunar craters carried out by the shadow method revealed that their altitudes are, in general, very moderate indeed. Thus the maximum height attained by the rims surrounding the crater Clavius (Figure 16-5) is barely 1600 m above the surrounding landscape, and not more than 4900 m above the lowest point of that crater's floor. This height is so small relative to the dimensions of the whole crater than an observer situated near the centre of Clavius would not see even the rims of its ramparts at all – they would be completely below his horizon. Photographs of three other conspicuous lunar craters, well known to any student of the surface of our satellite – Copernicus, Theo-

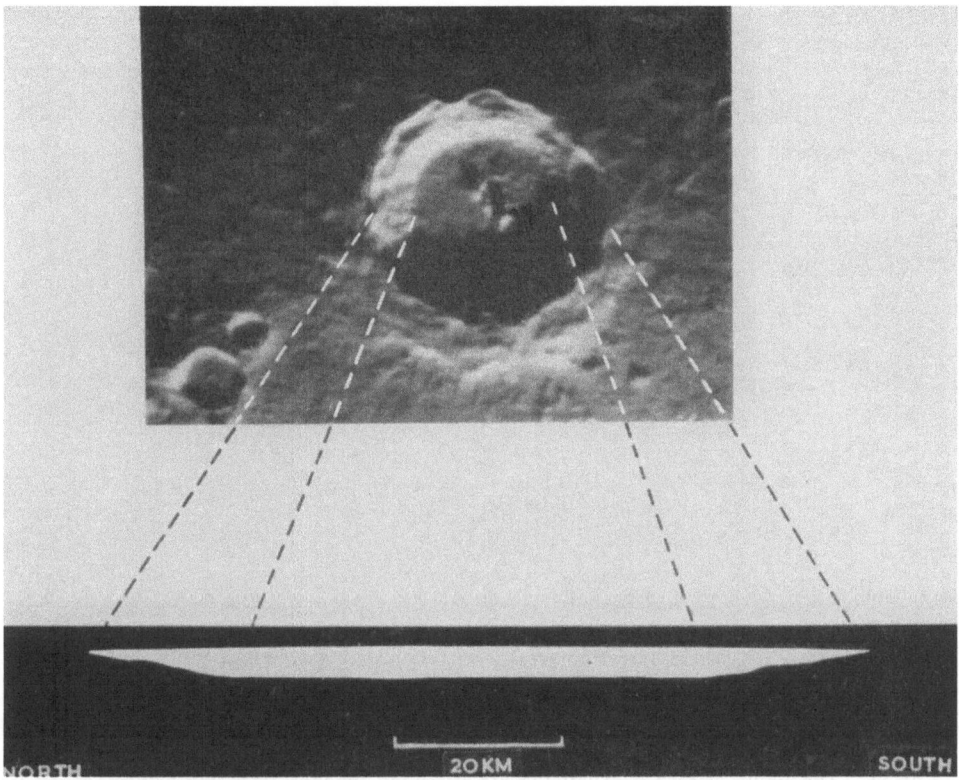

Fig. 16-10. This photograph of the sunset in Copernicus is accompanied by the crater's profile (right). Vertical and horizontal dimensions to the same scale reveal the essential symmetry of the crater. (After RACKHAM, 1962.)

philus and Tycho – are shown on the accompanying Figures 16-4 to 16-9, together with their vertical cross-sections as determined recently by RACKHAM (1959) by the shadow method; and the results demonstrated again the essential flatness of large lunar craters (Figures 16-10 to 16-12). All three of these formations are close to 100 km in diameter; but the maximum altitude of the Copernicus ramparts does not exceed 1000 m above the surrounding landscape, and 3300 m above the bottom of its floor; its central peak rises to 1200 m above its surroundings. For Theophilus, the corresponding figures are 1200 m and 4400 m for the ramparts, and 2200 m for the central peak; while for Tycho, Rackham found the latter two values to be 5400 m and 1600 m, respectively. A section of a detailed map of the lunar surface in the region of Copernicus, based on all available visual and photographic evidence, is reproduced on Figure 16-13 from the original prepared by the Aeronautical Chart and Information Center of the United States Air Force (LAC No. 58).

Reduced profiles of these and a few other large craters are shown in relation to the curvature of the lunar surface on the accompanying Figure 16-14. A glance at the

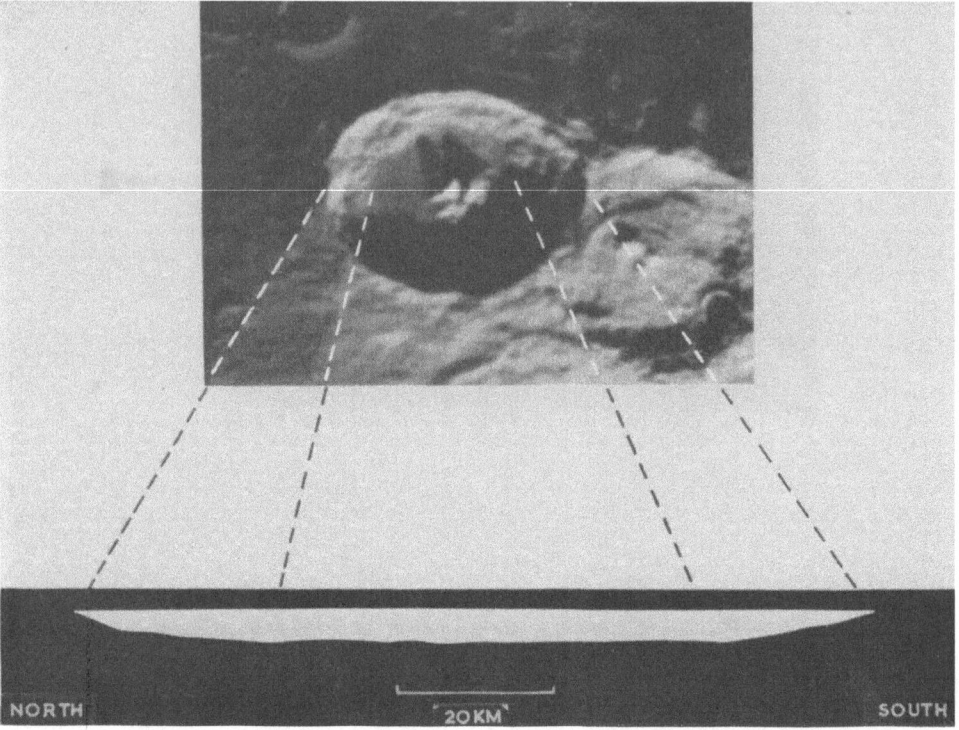

Fig. 16-11.  Theophilus is seen here with the late afternoon shadows extending halfway across the floor. The profile below is a cross section deduced from many relative height measurements of both east and west walls. (After RACKHAM, 1962.)

dimensions of these configurations plotted on the same horizontal scale reveals the markedly flat nature of such craters. Moreover, *their floors appear to all be depressed below the surrounding surface.* The extent of this depression seems statistically correlated with the diameter of the respective formation – in the sense that the larger the crater, the deeper its floor lies below the neighbouring landscape ("Ebert's rule"), and the volume of the depression seems in many cases to come close to the volume of the ramparts raised around them ("Schröter's rule") – a feature suggestive of the possibility that the material contained in the rims may have been displaced from the crust by forces which produced the entire formation – a possibility which we shall discuss in more detail in subsequent Chapters 17 and 18. However, very small craters do not appear to possess almost any ramparts to speak of, and represent mere pits sunken in the lunar surface. The craters on the Moon thus hardly deserve to be considered as real mountains; and when seen from above they resemble rather pockmarks spread profusely over the ageing lunar face, having in general no obvious terrestrial analogy.

In fact, wherever we turn on the Moon to ascertain the linear dimensions as well

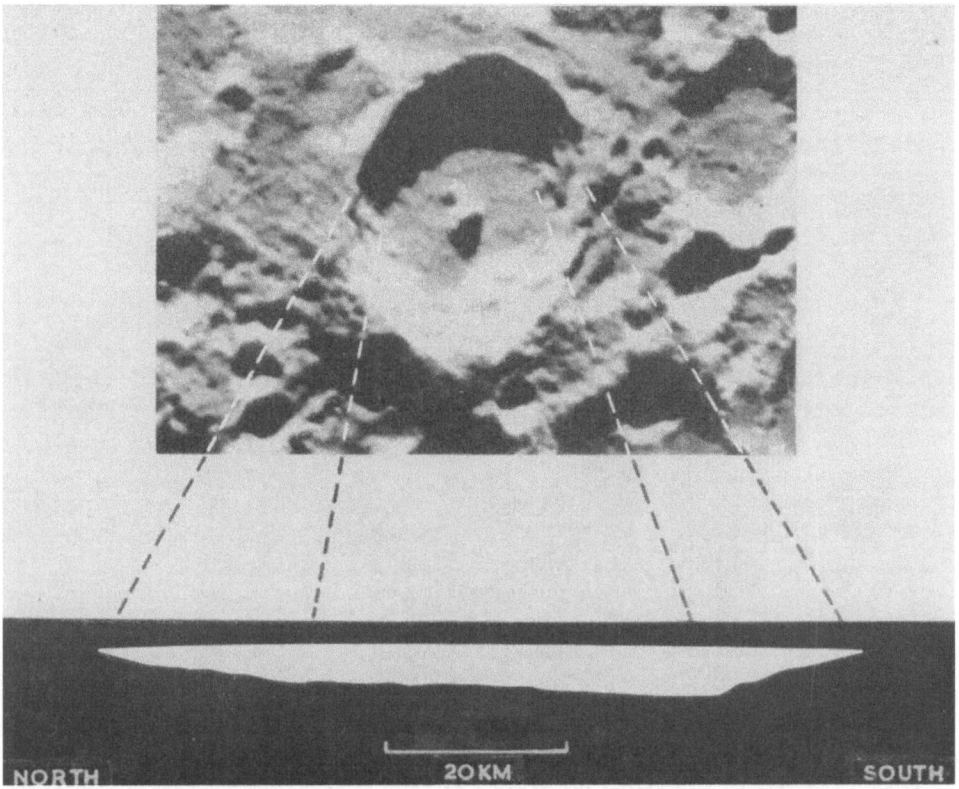

Fig. 16-12. Tycho is a deep crater and the west wall towers more than 5000 m above the floor. The curvature of the lunar surface is indicated by the smooth line in the scaled profile. (After RACKHAM, 1962.)

as the altitudes of diverse formations on the lunar surface by actual measurement, we find much the same situation. Perhaps the most surprising result of more recent work on lunar topography has been a realization – so contrary to earlier impressions based on visual inspection rather than measurement – *that there appear to be no steep slopes anywhere on the Moon of any appreciable size.* In more specific terms, an examination of those (mainly equatorial) parts of the lunar surface which have been mapped on the U.S. Air Force charts reveals that, in the regions of the maria, 63% of the ground is inclined to the horizontal direction by less than 1 degree; 90% of the ground is inclined by less than 2 degrees; while 99% of it is so inclined by less than 5 degrees. In the continental regions, 21% of the ground is inclined by less than 1 degree; 30% by less than 2 degrees; 64% less than 5 degrees; 90% less than 10 degrees, and 99% less than 12 degrees.

An impression of ruggedness, gained previously by a cursory glance at photographs on the sunrise or sunset on the Moon, largely disappears when the appropriate altitude on the Sun above the horizon (not readily apparent to mere inspection) is

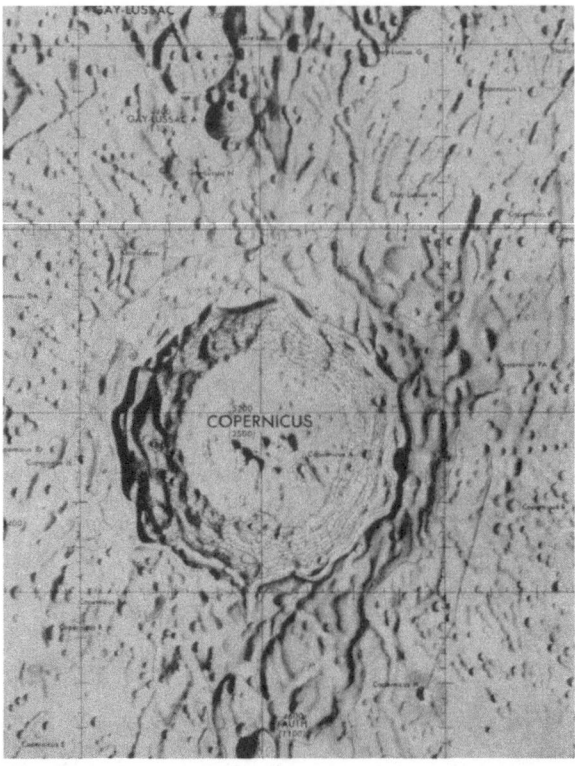

Fig. 16-13.   U.S. Air Force Chart of the Moon LAC No. 58 of the surroundings of Copernicus.

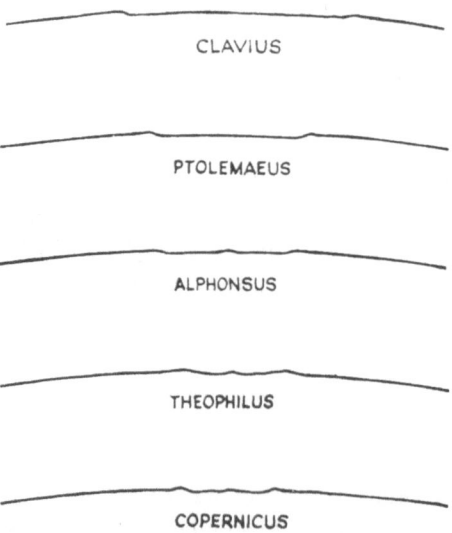

Fig. 16-14.   Vertical profiles of certain craters in relation to the curvature of the lunar surface.

duly taken in account. All slopes of large crater walls or mountains whose gradients have so far been measured are inclined by less than 15 degrees to the horizontal direction. Exceptions to this rule are the inner walls of small craters or crater pits, whose inclinations are occasionally larger; and slopes amounting locally to 30–40 degrees have indeed been noted from the relatively late hour at which such slopes become illuminated at sunrise (cf., e.g., POHN, 1963). But – and this is essential – such slopes occupy but a relatively minute fraction of the total surface area (much less than 1%) of the Moon; so that our general conclusion on the essential flatness of the lunar surface remains unaffected by them. The reader may note that the same general situation is also true of the surface of the Earth as well, where natural steep slopes are very much of an exception. The forces which keep levelling down the Earth's surface to this shape – namely, the combined action of air and water – are, however, completely absent on the Moon and have been since time immemorial; so that the reasons of the essential flatness of the lunar surface must be sought along different lines.

When, in addition, we recall the large curvature of the relatively small lunar globe, we cannot escape the conclusion that the horizontal panorama visible from most points on the Moon would be rather dull and unimpressive, with but a few landmarks on which the eye could safely rest in search of orientation. It is well to keep in mind that, on the Moon, an object 3000 m in altitude would disappear completely below the observer's horizon at a distance of 102 km (and a man of average height, at some $2\frac{1}{2}$ km). The highest mountains found on the lunar surface – the Leibnitz and Doerfel mountains in limb regions near the south pole of our satellite – attain altitudes of 6000 and 5600 metres, respectively. When one considers the fact that the Moon is but one-quarter of the Earth in size, the proportional height of lunar mountains is indeed considerably greater than on the Earth. Nevertheless, when measured in absolute terms, the Leibnitz or Doerfel mountains on the Moon are comparable in height with their terrestrial namesakes in Africa (Ruwenzori) or with Kilimanjaro, rather than those of the Himalayan giants. Real eight-thousanders are conspicuously absent on the Moon – unless, perchance, one may yet be found in the range of Soviet Mountains recently discovered on the Moon's far side (Figure 16-4).

Besides the craters and the maria, the surface of the Moon exhibits many other characteristic features of lesser magnitude possessing no obvious terrestrial homologues – four of which: namely, the domes, rilles, wrinkle ridges and bright rays deserve special mention. The lunar *domes* represent small, inconspicuous hills of approximately circular circumference and gently convex cross-section, clustering in certain regions of the lunar surface in considerable numbers. The largest formation of this type (which may, in fact, represent a conglomerate of several domes), called Rümker, is situated in Sinus Roris near the limb of the apparent disk of the Moon (Figure 16-15). A large group of individual bulges of this type occur in the midst of the vast plains of the Oceanus Procellarum, in the neighbourhood of the craters Copernicus and (Figure 16-16). Their linear dimensions attain 10–12 km, and a height of some 300–400 m; but the majority are considerably smaller. This is, in particular, true of members of another typical cluster of domes (containing at least 35 such formations on the

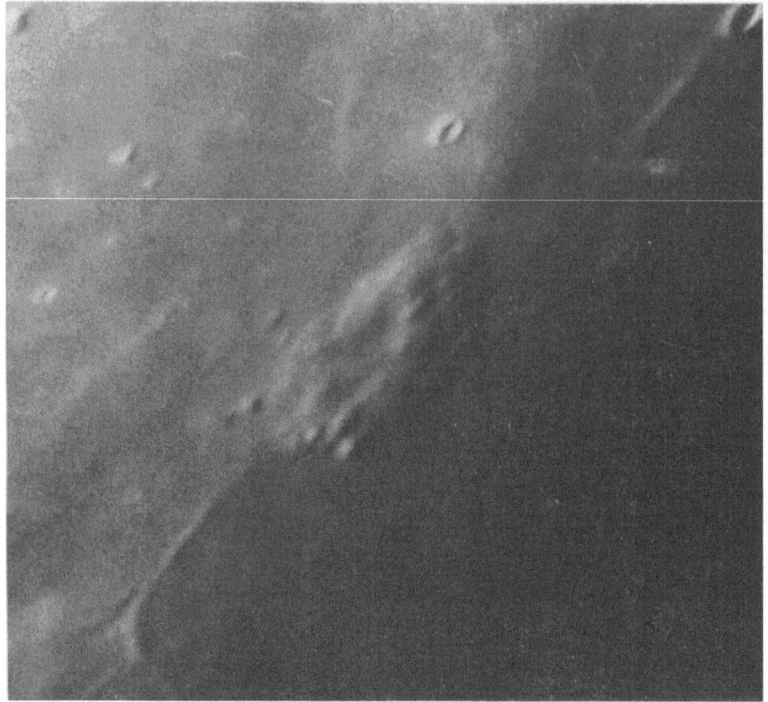

Fig. 16-15.   The Rümker formation – the largest dome on the Moon in the region of Sinus Roris. Photographed with the 24-inch refractor of the Observatoire du Pic-du-Midi (Manchester Lunar Programme).

Fig. 16-16.   Lunar domes in Oceanus Procellarum near the crater Hortensius (photographed with the 100-inch reflector of Mt. Wilson Observatory).

area of a few hundred square kilometres) which can be seen inside the crater Deslandres (Figure 16-6). The majority of its members are not more than 2–5 km across,
and barely 100–150 metres in height.

In all cases, the altitude of the lunar domes does not seem to attain more than
2–3 per cent of their diametres, indicating again that the actual slopes of these formations are very gentle (not in excess of a few degrees). The largest show some evidence crater-like central depressions or bulging – see the domes south of the crater
Cauchy on Figure 16-17 – and some (like the one north of Birt – see Figure 16-25) are
bisected by a rille. On the whole, the domes seem to occur almost exclusively in the
plains of the maria (or, like in Deslandres, in mare-like interiors of large craters) in
clusters (cf., ARTHUR, 1962; DALE, 1962); and preferentially in areas where (judging
from the presence of ghost craters and isolated peaks) the surface of the mare appears
to be shallow.

The *rilles* represent an altogether different type of lunar surface formation without
any obvious terrestrial analogy. They scarcely form a homogeneous group on the
Moon either. For instance, the conspicuous rilles in the neighbourhood of Cauchy
("Cauchy's hyperbolae", well shown on Figure 16-17) are shallow valleys with nearly

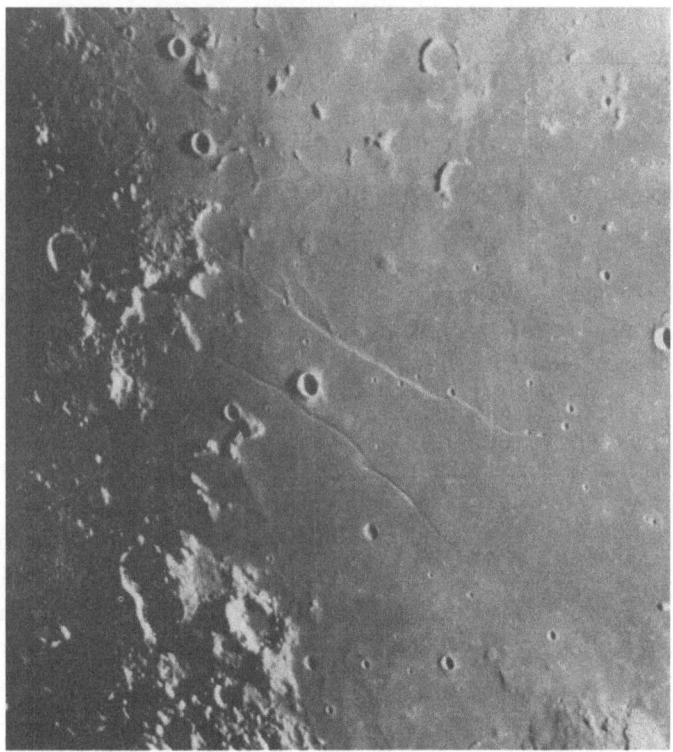

Fig. 16-17.   Eastern shores of Mare Tranquilitatis, with Cauchy's rilles and a family of domes. Photographed with the 43-inch reflector of the Observatoire Pic-du-Midi (Manchester Lunar Programme).

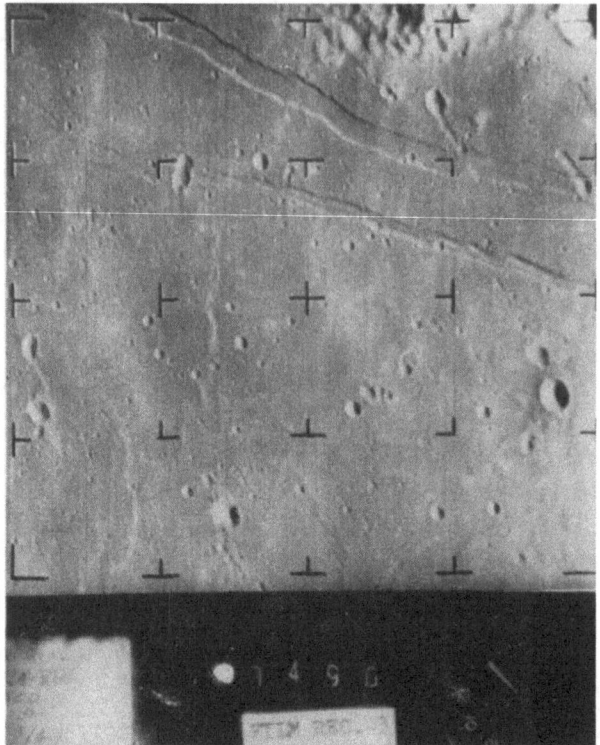

Fig. 16-18. A part of the rilles south-east of the craters Sabine and Ritter in Mare Tranquilitatis, as photographed by Ranger 8 on February 10, 1965, from an altitude of 379 km above the lunar surface. Note the extreme shallowness of such formations, similar to those seen on Figure 16-19.

level floors, depressed below the surrounding flatlands by less than 10% of their width; some even appear to possess an uplifted median ridge. Formations of this type south of the craters Sabine and Ritter have been beautifully recorded by Ranger 8 and are reproduced on the accompanying Figure 16-18. Others – like the Hyginus or Triesnecker rilles (Figure 16-19) are again very different: shallow and blurred grooves, with no well-defined distinction between walls and floor. These are often overlapped by craters (Hyginus on Figure 16-20 being a most conspicuous example), with which they may be generically related. In other instances, rilles cluster in the neighbourhood of certain craters (Aristarchus, Prinz and others) in patterns possibly reminiscent of local stress release (Figure 16-21); but they are known to occur also on crater floors where they may run roughly parallel to the crater walls; as in Alphonsus (Figure 16-22) – the long rille running along the Apennine mountains along the eastern shores of Mare Imbrium being possibly of this type – or almost normally to them (as in Lacus Mortis, shown on Figure 16-23).

Related with the rilles are *grooves* of the type shown on Figure 16-24. These are very numerous on the Moon, and are found predominantly in the continental regions

Fig. 16-19.  The Hyginus and Triesnecker rilles in Sinus Medii, photographed with the 43-inch reflector of the Observatoire du Pic-du-Midi (Manchester Lunar Programme).

Fig. 16-20.  Details of the Hyginus rille in Sinus Medii, as photographed by Bernard Lyot on March 21, 1945, with the 24-inch refractor of the Observatoire du Pic-du-Midi.

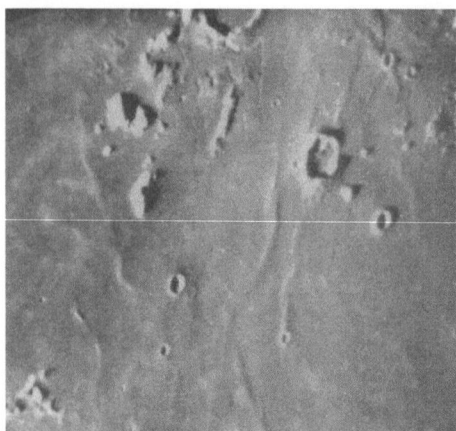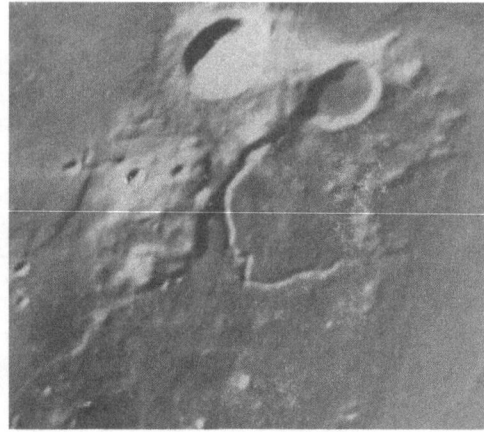

Fig. 16-21. Lunar rilles in the neighbourhood of the craters Prinz (left) and Aristarchus (right), photographed with the 24-inch refractor of the Observatoire du Pic-du-Midi (Manchester Lunar Programme).

Fig. 16-22. Rilles inside the crater Alphonsus, as photographed by Ranger 9 on March 24, 1965, from the proximity of the lunar surface. Left: photograph taken from an altitude of 413 km above the lunar surface, 2 min 50 sec before impact, recording a square field of 180 km across. Right: photograph taken from an altitude of 166 km, 1 min 9.5 sec before impact (the position of the latter marked with a white cross). Note the rugged nature of the floor of the crater, replete with innumerable pits, pointing to a great age of the ground.

of the Moon, often cutting across other mountainous formations in their way. Whether these are due to subsidence of internal origin (as seems suggested by the chains of small craters in their path), or (less probably) to ricocheting effects of solid debris ejected by major meteor impacts, is as yet uncertain.

Another type of lunar formation is the *cleft*, such as the Straight Wall in Mare Nubium, which for a distance of approximately 60 km separates two parts of this

Fig. 16-23.   Lacus Mortis east of the craters Eudoxus and Aristoteles, photographed with the 24-inch refractor of the Observatoire du Pic-du-Midi (Manchester Lunar Programme).

Fig. 16-24.   Typical lunar grooves in the neighbourhood of the crater Gutenberg on the western shores of Mare Fecunditatis, photographed with the 43-inch reflector of the Observatoire du Pic-du-Midi (Manchester Lunar Programme).

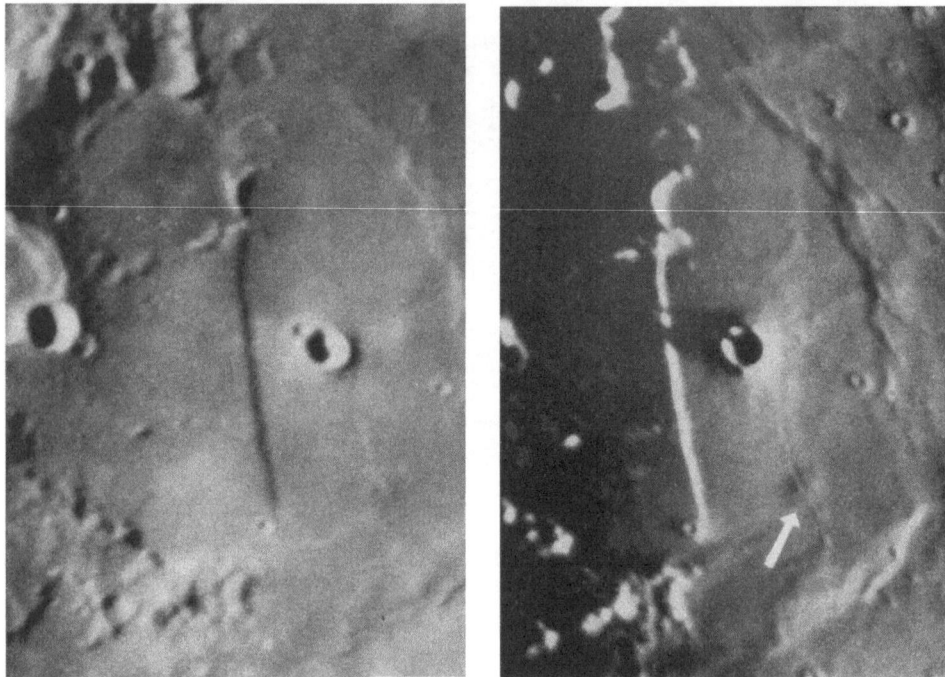

Fig. 16-25.   Sunrise (left) and sunset (right) over the "Straight Wall" in Mare Nubium, as photographed with the 24-inch refractor of the Observatoire du Pic-du-Midi (Manchester Lunar Programme). Note a shallow dome bisected by a rille north of the crater Birt.

mare displaced by about 300 metres in altitude. The slope of the dividing wall has been determined from a photometry of the sunrise and sunset over this formation (see the accompanying Figure 16-25) by the methods of Chapter 14; and found to be inclined to the horizontal by less than 11 degrees. With this figure in mind, a glance at the photographs reproduced on Figure 16-25 reveals eloquently the exaggerating effects of low sunlight on the impression of lunar slopes.

The lunar *wrinkle ridges* are again quite different – rather like the rilles in reverse – but differing from them in several important respects. They represent, in fact, very low ridges running like shallow submerged veins – sometimes simple, sometimes very ramified – underneath the surface of the vast plains of the lunar maria (and preferentially near their shores), for several hundred kilometres in the direction parallel broadly with the meridians. This latter fact may, however, be due largely to observational selection; for these ridges prove to be so low and gently-sloping that they can be discerned only when the Sun stands very low above the lunar horizon. Ridges running in the direction of the meridian are detectable more easily than those running parallel with the latitude circles because (unlike in this latter case) they cast shadows on the surrounding plains rather than on themselves.

Representative examples of such wrinkle ridges in the Mare Tranquilitatis (near

Fig. 16-26. Wrinkle ridges in Mare Tranquilitatis, with the "ghost" crater Lamont near the center. The impact place of Ranger 8 on February 20, 1965 is marked with a white cross. Note also the rilles north of craters Sabine and Ritter, a part of which can be seen with much greater resolution on the Ranger 8 photograph reproduced on Figure 16-18. Observatoire du Pic-du-Midi photograph, taken with the 24-inch refractor (Manchester Lunar Programme).

the crater Lamont) can be seen on Figure 16-26; while the well-known Serpentine ridge in Mare Serenitatis is shown on Figure 16-27. Whether or not the crater situated on the crest of the northern end of this latter ridge (near Posidonius) – well seen on Figure 16-28 – is due to a chance impact or is generically related with the ridge is as yet impossible to say.

Detailed topographic properties of such ridges were studied recently by DALE (1962) by the shadow and photometric methods outlined in Section 14. His work revealed that the actual slopes of the ridges are of the order of one degree, and scarcely ever exceed 3°; their transverse dimensions of 10–20 km correspond, therefore, to elevations of no more than 200–300 metres above the surrounding landscape. Such very flat formations are much more easily seen from above, at the time of sunrise or sunset, rather than from the ground itself; and astronauts standing on the Moon might find it hard to recognise such a ridge when they come to face it.

Lastly, several craters on the Moon – though not a very great number – are sur-

Fig. 16-27.  Sunrise over the Serpentine wrinkle ridge in Mare Serenitatis, with the crater Posidonius (below). Photograph taken with the 24-inch refractor of the Observatoire du Pic-du-Midi (Manchester Lunar Programme).

rounded by systems of *bright rays*, resembling the ejecta patterns around terrestrial explosion craters. Bright rays associated with the craters Tycho, Copernicus and Kepler, well shown on Figures 16-29 and 16-30, represent perhaps the most conspicuous examples of such systems; but several others can be discerned on Figure 16-1 as well. Some (like those diverging from Proclus on the western shores of Mare Crisium, shown on Figure 16-31) spread fanwise over only a little more than half a circle; while other craters (like Messier, shown likewise on the same figure) possess only two parallel rays reminiscent of a cometary tail. Large ray systems are seldom characterized by complete radial symmetry, and may include rays which do not intersect the common focus. The most extensive system of rays is that diverging from the crater Tycho; and someof its rays – like the one traversing the Mare Serenitatis (see again Figure 16-1) exceed 2000 km in length. The ray system associated with the crater Copernicus extends over 500 km from its divergent point; and all other patterns of this type are smaller.

The rays accompanying lunar craters consist essentially of loop-shaped streaks of brighter material than that of the maria over which they have been splashed. By their

Fig. 16-28. The northern end of the Serpentine wrinkle ridge near Posidonius in Mare Serenitatis. Note the position of a crater on the brow of the ridge near its northern bifurcation. Photographed with the 24-inch refractor of the Observatoire du Pic-du-Midi (Manchester Lunar Programme).

reflectivity, the rays appear to be an extension of the crater rims and cannot be sharply distinguished from them. Moreover, their major arcs and loops can be often locally resolved into a system of feather-shaped elements, ranging between 15–20 km in length, radially spreading from the crater.

The rays do not exhibit any measure vertical relief (i.e., cast no shadows); and the variation of light reflected from their surface in the course of the lunar day suggests that even their micro-relief is essentially the same as that of the surrounding darker landscape – from which they differ mainly by their relatively high albedo. They represent probably nothing else than thin layers of ejecta from the crater around which they are distributed; and distinct ray elements can be interpreted as splashes of crushed rock derived from the impact of individual large fragments. We shall return to this interpretation in Chapter 17, when examining the evidence recently furnished by Ranger 7 on the fine structure of some of the Tycho rays (cf. Figure 17-15).

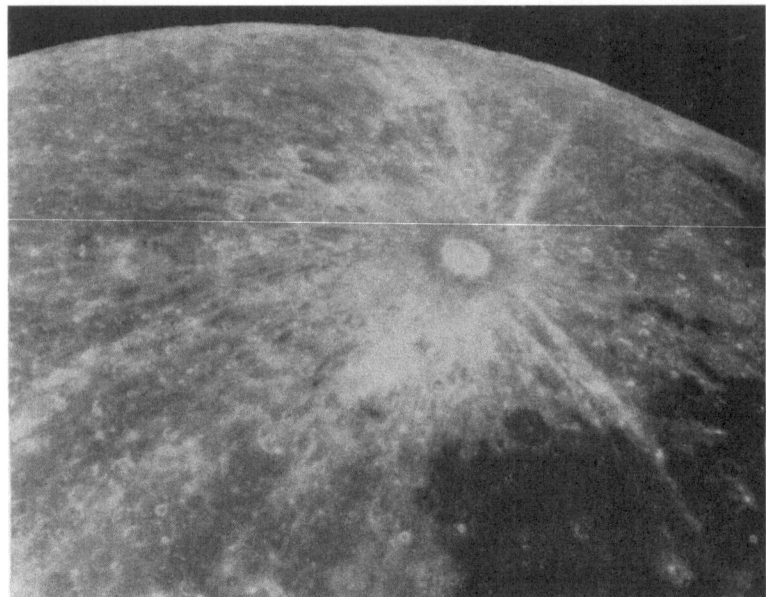

Fig. 16-29.   The bright ray system surrounding the crater Tycho near the Moon's south pole. Note the dark annulus surrounding the ramparts of the crater, and also rays which are tangential to the walls. Photograph taken with the 24-inch refractor of the Observatoire du Pic-du-Midi (Manchester Lunar Programme).

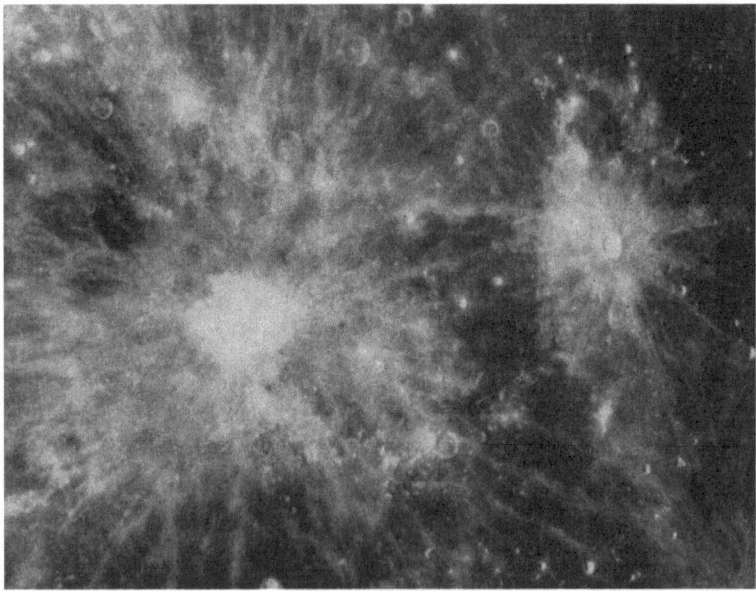

Fig. 16-30.   Bright ray systems surrounding the craters Copernicus (left) and Kepler (right), as photographed by the 24-inch refractor of the Observatoire du Pic-du-Midi (Manchester Lunar Programme).

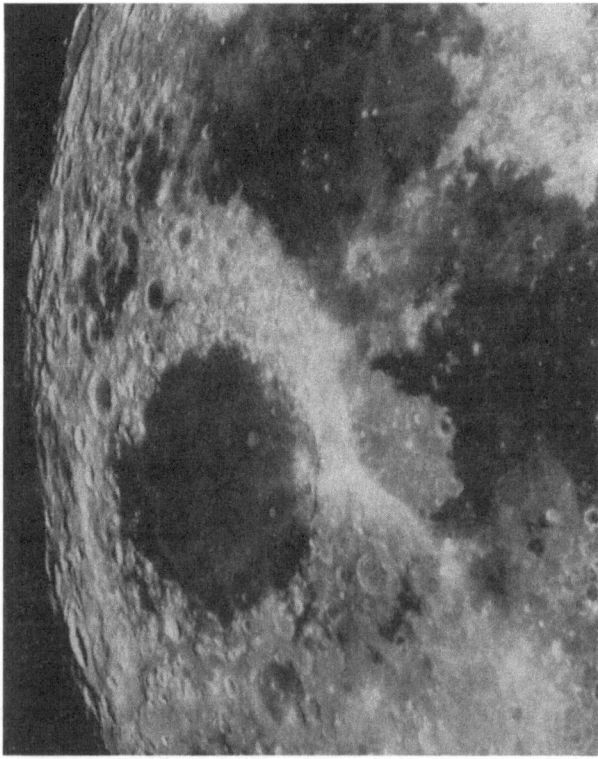

Fig. 16-31.   Bright rays from Messier in Mare Foecunditatis (above) and Proclus on the Western shores of Mare Crisium (Observatoire du Pic-du-Midi photograph; 24-inch refractor).

## CHAPTER 17

# ORIGIN OF THE LUNAR FORMATIONS

The preceding section of this survey contained a general description of the more common types of formations encountered on the lunar surface. What can we say at present about the *origin* of such a relief – and, in particular, of its characteristically dominant feature, the so-called *craters*? A glance at the almost bewildering array of such formations of all sizes – as shown on many photographs accompanying this chapter – makes it perhaps unlikely that all of them originated in the same way, or at the same time; and a more detailed analysis of their features suggests that a suspicion of different origin is probably well founded. In fact, the most reasonable approach to this problem can be made if we ask ourselves the following question: what are all the processes which could have conceivably cooperated in shaping up the surface of our satellite? And once we thus formulate our problem, we find ourselves facing two principal contending theories of crater origin: namely, the external theory – invoking the effects produced by' *impacts* of other celestial bodies (asteroids, meteorites, or comets) on the lunar surface – and the alternative theory relying on the *internal processes* connected with convection, gradual defluidization and degassing of the lunar globe consequent upon its build-up of internal heat due to spontaneous decay of radioactive elements, or any other activity which could be loosely termed as "volcanic". In point of fact, *the entire surface of the Moon must be regarded as the outer "boundary condition" of all internal processes which may have been going on in lunar interior since the origin of our satellite, as well as an "impact counter" of external events which may have visited it from outside*. In no other sense can an interpretation of the lunar surface possess any physical meaning.

That the surface of the Moon is "full of inequalities" was recorded already by Galileo GALILEI, the first telescopic observer of the Moon, in his *Sidereal Messenger* (1610); and this fact caused at first no small commotion among peripathetic philosophers of that time. Fifty-seven years later, Robert HOOKE – who, like Newton, was also interested in the Moon – dropped bullets into a pipe clay and water mixture and, behold, saw formations arise which one could call "impact craters". But being an inquisitive soul, Hooke did not stop here; but as he tells us in his *Micrographia* (1667) he also boiled a mixture of powdered alabaster with water, and observed that this too produced transient crater-like structures on the surface of the liquid. Thus he started a very interesting controversy about the origin of lunar craters which has not been definitely settled ever since.

Hooke himself rejected the impact analogy because, "it would be difficult to imagine whence those bodies should come". This opinion was formed, to be sure, more than a century before the extra-terrestrial origin of any meteor was recognized; and by that time a volcanic origin of lunar craters appears to have been almost universally accepted by the leading scientists of the time. Thus a volcanic origin had been championed by KANT (1785); and William HERSCHEL even reported in 1787 what he believed to be observations of volcanic eruptions on the Moon. The impact hypothesis for the origin of the lunar craters was revived, to be sure, by GRUITHUISEN (1829), who also appears to have anticipated the planetesimal accretion hypothesis for the origin of the Moon. Nevertheless, many of Gruithuisen's views were extreme (such as his speculations about the inhabitability of our satellite); and most of the principal selenologists continued to accept some form of volcanism as the crater-forming process on the Moon.

The pendulum of the scientific thought on the origin of lunar surface features did not begin to swing away from volcanism and back to impacts until the work of the American geologist G. K. GILBERT (1893), who reviewed the characteristics of lunar craters together with those of the various types of terrestrial volcanoes; and concluded that the differences in form (and, to a lesser extent, in size) between the respective lunar and terrestrial objects were so great that a volcanic origin for the lunar craters seemed improbable. From this Gilbert went on to develop a consistent impact hypothesis for the origin of the lunar craters, which was based on some acute telescopic observations of the Moon as well as upon laboratory experiments.

Gilbert's views appeared, however, to be too advanced for his time; and by the end of the 19th century the consensus among a large majority of astronomers and geologists was still firmly in favor of volcanic origin of lunar craters. It was not till throughout the first half of the 20th century that the opinion began to shift significantly towards external impacts – mainly among the geologists – a fact which led, in 1926, the American geologist W. M. Davis to record (in his biography of Gilbert) candidly that "...It has been remarked that the majority of astronomers explain the craters on the Moon by volcanic eruptions – that is, by an essentially geological process – while a considerable number of geologists are inclined to explain them by the impact of bodies falling upon the Moon – that is, by an essentially astronomical process. This suggests that each group of scientists finds the craters so difficult to explain by processes with which they are professionally familiar that they prefer recourse to a process belonging to another field than their own, with which they are probably imperfectly acquainted and with which they therefore feel freer to take liberties."

Since the time when these words were written, the astronomers too began to gravitate towards the impact hypothesis in increasing numbers – until, by 1956, the distinguished dean of lunar investigators, Harold C. UREY, spoke probably for the majority when saying that, "...It is a characteristic of science that different objective observers studying the same evidence come to the same conclusions, and that the over-whelming majority of such observers agree substantially. When this occurs, we regard the conclusions of such scientists as true. For this purpose... I am concluding that the volcanic hypothesis (of the origin of lunar surface features) is false and that the collision one is true..." (UREY 1956b, p. 1674). And, somewhat later, Urey remarked that "Astro-

nomers have required almost a century of discussions to recognise that the structure
of the lunar surface is produced chiefly by collisions" – presumably referring to
the time that elapsed since Alexander von Humboldt expressed himself with equal
firmness in favour of the volcanic hypothesis.

Urey's words, written in 1956, reflected no doubt the view of most investigators
of the subject at that time. Since then, however, the accumulating evidence confronted
us with many new facts bearing on the effects of gradual desiccation and degassing
of the lunar globe (cf., e.g., KOZYREV, 1959) which will be reviewed in this chapter.
Partly as a result of this evidence, and partly as a result of our increasing understanding of the physical processes operative in the lunar interior, the two principal
protagonist theories of the origin of the lunar surface features continue to face each
other in contest for due recognition. In order to assess their relative merits and
drawbacks as objectively as can be done at the present time, let us, in what follows,
outline their principal arguments and supporting evidence.

To begin with the theory of *external impacts*, we must remember that the interplanetary space through which the Earth with the Moon continue to circle around
the Sun is not entirely empty. Far from it; for it contains a wide variety of ingredients
of all weights and sizes: from the elementary particles (essentially hydrogen plasma)
of solar wind, through microscopic specks of dust and larger meteorite debris (representing probably the left-overs from the time of formation of the whole solar system),
to major meteorites, asteroids, or comets whose orbits through space may intersect
the path of the Moon and occasionally collide with it. The frequency with which the
surface of the Moon – like that of the Earth – undergoes direct hits by asteroids or
comets is as yet difficult to assess with any accuracy (cf., e.g., SHOEMAKER and HACK-
MAN, 1962) – let alone what it could have been in more remote times. As, however,
in the course of the long lunar past such hits must undoubtedly have been scored,
it is important to realize the consequences which such events would entail; and these
are indeed bound to be spectacular.

In order to visualize them at least to some extent, consider a moderately large
meteorite – of the size and mass of a rock weighing one million tons ($m = 10^{12}$g) and
impinging on the lunar surface with a velocity $v$ of (say) 30 km/sec – equal to that of
the Earth in its relative orbit around the Sun. The kinetic energy

$$E = \tfrac{1}{2} mv^2 \qquad (17\text{-}1)$$

of such a missile would be a quantity of the order of $10^{25}$ erg; and would enable it
to penetrate into the lunar crust like a bullet to get buried well underneath the surface
before coming to a complete stop. The kinetic energy which the meteorite possessed
before impact must, moreover, be conserved and its entire amount reappear in other
guises to which it was converted in accordance with the prevalent laws of physics –
mainly as mechanical energy of shock and fracture, thermal energy, and seismic energy of elastic waves. If, for the sake of argument, the entire kinetic energy of the
meteorite were converted in heat absorbed by it, its temperature $T$ would be given

by

$$T = \frac{v^2}{2C_v},\tag{17-2}$$

where $C_v$ denotes the specific heat of the meteoritic material. As, for stony meteorites, $C_v \sim 10^7$ erg/g·deg, this mechanism should generate a temperature $T$ of the order of several million degrees – i.e., sufficient to volatilize completely the whole impinging mass and convert it into an extremely hot bubble of gas imprisoned at a depth of several diameters of the original body beneath the lunar surface. In reality, the actual temperature of the impinging body would be considerably lower, because a large part of the original energy would be spent on other processes than heating; but still it is difficult to escape conclusion that the impact of such a body would create local temperatures of the order of a few – possibly several – hundred thousands degrees for a very short time.*

Needless to say, such hot gas bubbles could not be contained by the weight of the overlying debris for time-intervals longer than microseconds. They would immediately expand with great violence; and the effects of this expansion should severely affect regions that are very large in comparison with the size of the original missile. The main effect of the explosion should, therefore, be essentially that of a point-charge; and the initial direction of the impinging body could not have had much influence on the symmetry of the resultant surface markings. The probable result is schematically shown on the accompanying Figure 17-1, which represents the expected cross-section of an impact crater. The amount of the actual solid material left around by the intruder should be negligible; most of it should have evaporated and escaped back into space, or become dispersed over a large part of the adjacent lunar surface.

Let us attempt now to describe such a situation in more quantitative terms and confine, to begin with, our attention to the *interior ballistics* of such a situation. When a meteorite strikes the ground with a velocity of the order of a few km/sec or more, the event will give rise to two *shock waves* propagating in opposite directions from the impact interface: one advances into the lunar ground, and the other recedes back

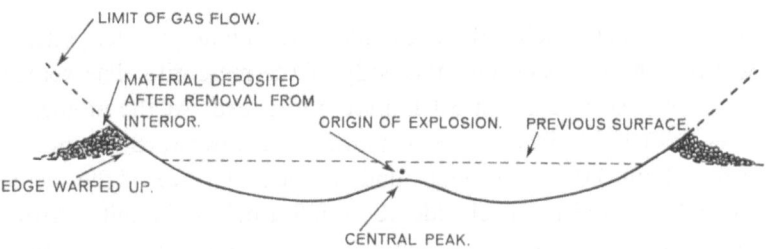

Fig. 17-1. A schematic profile of a lunar impact crater (After Gold, 1955).

* In order to make us appreciate this fact, suffice it to recall that a body moving at a speed of mere 3 km/sec possesses a kinetic energy approximately equal to that released by explosion of an equal weight of TNT.

through the body of the impinging meteorite.* The material between these shock
fronts is compressed, and its velocity altered by their fronts is compressed, and its
velocity altered by their passage: the ground its accelerated downward, and the
meteorite itself decelerated. A part of the kinetic energy of the meteorite to heat which
will increase the temperature of the mass of the meteorite as well as of the surrounding
rocks, and a part emerges as the kinetic energy of the shocked rocks ahead of the
meteorite.

A precise calculation of the partition function of the total energy would depend
on the specific shape and kind of material of the meteorite as well as of its target area,
and is beyond the means of mathematical analysis for all but the simplest cases. A
solution in a closed form may, however, be obtained for the one-dimensional case
of a semi-infinite meteorite, impinging normally on a semi-infinite meteorite target:
in this case, the two shock fronts are planes, and the material between them is at
uniform pressure and velocity, expressible as function of the initial velocity of impact,
and the density of both the meteorite and of the target.

Consider, with SHOEMAKER (1962), a plane-parallel meteorite of thickness $d$ impinging on a semi-infinite ground with properties of silicate rocks with a velocity of
15 km/sec. Such an impact will give rise to a pressure between the two shock fronts
of the order of $4\frac{1}{2}$ megabars; at such a pressure, the rocks will be compressed by 58%
and the meteorite, to 43%. At the moment when the rear shock propagating in the
reverse direction reaches the back side of the plate, the front shock is moving into
the ground at a speed of 17.2 km/sec and will have advanced a distance of 1.48 $d$ from
the initial ground level; the leading face of the meteorite will have penetrated a distance of 0.87 $d$ below the original ground; and its back side, a distance of 0.30 $d$; the
center of gravity of the compressed system will be 0.78 $d$ beneath the surface, and the
whole meteorite will be moving with a velocity of 10.0 km/sec into the ground. At
this moment, the meteorite will suffer its maximum compression. The kinetic energy
of the compressed system will be two-thirds, and the internal energy one-third, of the
original energy of the impinging meteorite. About two-thirds of the internal energy
will be in the compressed rock, and one-third in the compressed meteorite; about
53% of the total energy will have been transferred from the meteorite to the compressed rock.

When the retro shock reaches the back side of the plane-parallel plate, a rarefaction wave will be reflected back into the body of the meteorite. The velocity of the
rarefaction wave will still be about 5.1 km/sec downward. At the moment the rarefaction reaches the leading face of the meteorite, the downward-moving shock will
have penetrated about 3.0 $d$ into the ground; the leading face of the meteorite will
have penetrated 1.8 $d$, and the back side about 0.8 $d$ below the initial ground level;
the center of gravity of the moving system will be 1.8 $d$ underground. About 88% of
the original kinetic energy of the meteorite will have been transferred to the compressed

---

* In order to produce such shocks, it is *not* necessary that the velocity of the impinging body exceeds
the acoustic velocity $[(\lambda + 2\mu)/\rho]^{\frac{1}{2}}$ of compression waves in the target or meteorite material of density $\rho$ and Lamé parameters $\lambda$ and $\mu$.

rock ahead of the meteorite, where the energy will be equally divided between its internal and kinetic part. The compressed rock will still keep moving downward at 10.0 km/sec.

An actual meteorite may differ from a plane-parallel plate approximating it in the above example in many essential respects, but the foregoing results demonstrate several major qualitative facts about hypervelocity impacts. First, they show that an iron meteorite moving at average cosmic velocity would by compression alone penetrate and bury itself beneath the surface of a target consisting of ordinary silicate rocks. Secondly, they show that even after reflection of a rarefaction wave from its rear side, the meteorite will not necessarily fly apart, but may continue moving bodily into the ground. Third, it may be seen that the major part of the original energy of the meteorite is transferred to the rocks in front of the meteorite at an early stage of penetration. And finally, in the example considered above, the internal energy of the meteorite at no time exceeds one-seventh of its original kinetic energy, and only a fraction of this internal energy will be converted into heat. The major part of it is used up by expansion of the meteorite behind the rarefaction wave, and contributes to further propagation of the shock front into the rock. In the case of a real meteorite, rarefaction waves are also reflected from its sides, which permits lateral expansion of the body as well as longitudinal compression. A rarefaction wave is also reflected from free ground level, which permits lateral displacement of the rocks belived the shock.

The foregoing remarks based on quantitative discussion of a special case of plane-parallel impact document – and to some extent qualify – the earlier verbal description of the consequences of a hypervelocity impact of a meteorite on the lunar surface. In particular, they demonstrate the limits of the extent to which we may be entitled to regard the creation of an impact crater as an "explosion". In particular, the volatilization of the impinging body at some depth below the surface does not use up all its initial kinetic energy – the latter is bound to be partitioned among several tasks – and although very high energies are produced by hypervelocity impacts, these are the consequences rather than the cause of the shocks which produce the craters. In point of fact – for a given total energy – the higher the initial specific energy, the smaller the resultant crater; for vaporization of the meteorite (or of the rocks of the target) will not facilitate the extension of an impact crater (except, possibly, in rocks which are especially rich in volatile constituents). We may properly speak of meteorite impact craters as "explosion" craters in the sense that a large amount of material is flung out of them at relatively high speed. The same is true of nuclear explosion craters which simulate the appearances of lunar craters to a large extent (cf., e.g., SHOEMAKER, 1960; ROBERTS and FULLMER, 1963; ROBERTS, 1964); but, in this sense, the pits formed by raindrops on soft mud are also "explosion" craters; and similarity in form is likewise considerable (cf., e.g., SABANEYEV, 1962; and others).

In order to investigate quantitatively the changes in the pressure, density, and temperature across a shock front created by a hypervelocity impact, let $P_0$, $P_1$, $\rho_0$, $\rho_1$ and $e_0$, $e_1$ denote the pressure $P$, density $\rho$ and specific internal energy $e$ in front, and

immediately behind, a shock wave advancing at a velocity $U$; while the particle velocity, zero in front of the shock, be $u$ immediately behind it. If so, the well-known
Rankine-Hugoniot conditions (cf. e.g., COURANT and FRIEDRICHS, 1948, pp. 121–126),
expressing the conservation of mass, momentum, and energy, assert that

$$U\rho_0 = (U - u)\rho_1, \tag{17-3}$$

$$P_1 - P_0 = \rho_0 u U, \tag{17-4}$$

$$e_1 - e_0 = \frac{P_0 + P_1}{2}\left\{\frac{1}{\rho_0} - \frac{1}{\rho_1}\right\}; \tag{17-5}$$

the last condition implying that the increase in internal energy across a shock front
is due to the work done by the mean pressure in performing the compression.

For hypervelocity impacts of meteorites moving relative to the lunar surface with
their cosmic speeds of the order of 10 km/sec, the pressure $P_0$ in front of the shock
wave becomes negligible in comparison with $P_1$; and if so, an insertion of (17-3) and
(17-4) in (17-5) reveals at once that

$$e_1 - e_0 = \tfrac{1}{2}u^2. \tag{17-6}$$

In order to determine $u$, let us consider the *exterior ballistics* of particles ejected from
a surface point $P$ due to the impulse received when the shock-wave due to a "deto-
nation" of the meteorite at a point 0, located at a depth $d$ below the surrounding
undisturbed surface which we shall provisionally approximate by a plane. If so, the
trajectory, in the $xy$-plane, of a particle ejected from $P$ at an angle $\varphi$ to the horizontal
will be governed by a pair of well-known differential equations

$$\frac{d^2x}{dt^2} = 0 \quad \text{and} \quad \frac{d^2y}{dt^2} = -g, \tag{17-7}$$

of free flight in non-resisting medium, where $g$ denotes the gravitational acceleration
on the lunar surface (equal to 167 cm/sec$^2$); and the dependent variables $x(t)$, $y(t)$
are subject to the initial conditions

$$\left.\begin{array}{ll} x(0) = 0 & y(0) = 0, \\ x(0) = u\cos\varphi, & y(0) = u\sin\varphi \end{array}\right\} \tag{17-8}$$

The nature of the trajectories defined by the foregoing Equations (17-7)–(17-8)
depends on the magnitude of the initial velocity $u$ (i.e., the velocity of a particle immediately behind the shock). If $u$ is greater than the parabolic velocity of escape $V_{esc}$
of (approximately) 2.38 km/sec, the corresponding trajectories will become hyper-
bolic; and particles travelling along them will leave the lunar surface for ever. If
$V_{esc} > u > V_{esc}/\sqrt{2}$, the corresponding particles will describe elliptic orbits around the
Moon; and only if $u < V_{esc}/\sqrt{2} \sim 1.69$ km/sec will such particles fall back on its surface.

Suppose that this is the case and that, moreover, $u \ll V_{esc}/\sqrt{2}$ represents a strong inequality justifying the replacement of the actual lunar surface by a plane tangent to it at the point of ejection. The integral of the Equation (17-7) subject to the initial conditions (17-8) clearly obtains in the form

$$x = ut \cos \varphi, \tag{17-9}$$

$$y = ut \sin \varphi - \tfrac{1}{2} gt^2. \tag{17-10}$$

The $y$-coordinate of the ejected particle is zero at $t=0$ (i.e., the time of ejection) as well as at

$$T = (2u/g) \sin \varphi \tag{17-11}$$

following from (17-10) for the time of impact, for which the horizontal range

$$X = uT \cos \varphi = (u^2/g) \sin 2 \varphi \tag{17-12}$$

by (17-9); and the greatest range

$$R = \max (X) = u^2/g \tag{17-13}$$

for $\varphi = 45°$.

What should be the distribution of ejecta within this range? An element of mass $dm$ of the total mass $m$ ejected within an interval $d\varphi$ for a given value of $\varphi$ should be given by the expression

$$dm = m \cos \varphi \, d \varphi, \tag{17-14}$$

and fall at a distance $x$ on an elemental area

$$dS = 2\pi x \, dx, \tag{17-15}$$

covering it with a layer of thickness

$$h = \frac{dm}{\rho dS} = \frac{m \cos \varphi \, d \varphi}{2\pi \rho x \, dx}, \tag{17-16}$$

where $\rho$ stands for the mean density of the ejected debris. Since, however, by (17-12)

$$dx = (2n^2/g) \cos 2 \varphi \, d \varphi, \tag{17-17}$$

it follows that

$$h = \frac{m g \cos \varphi}{4\pi \rho u^2 X \cos 2 \varphi} = \frac{m}{4\pi \rho R^2} \left\{ \frac{1 \pm \sqrt{1 - \lambda^2}}{2\lambda^2 (1 - \lambda^2)} \right\}^{\frac{1}{4}} \tag{17-18}$$

where

$$\lambda = \sin 2 \varphi = \frac{X}{R} \tag{17-19}$$

and the $\pm$ signs refer to the intervals $0° < \varphi < 45°$ and $45° < \varphi < 90°$, respectively. As values of $\varphi$ in both these ranges contribute to the total infall at any particular value of $X < R$, it follows that the total accumulation debris should be given by

$$h = \frac{m}{4\pi \rho R^2}\left\{\frac{\sqrt{1 + \sqrt{1 - \lambda^2}} + \sqrt{1 - \sqrt{1 - \lambda^2}}}{\lambda\sqrt{2(1 - \lambda^2)}}\right\}. \qquad (17\text{-}20)$$

This formula can be regarded as only crudely approximate, for in deriving it we have tacitly assumed that all mass particles are ejected with the same velocity. This is very unlikely to be true; for large boulders will no doubt be ejected with lower velocities than small debris or dust. In order to refine the above result, a closer knowledge of the energy partition of the ejecta will be prerequisite, which we do not yet possess. Nevertheless, an inspection of (17-20) as it stands reveals that the height $h$ should exhibit two maxima: namely, for $\lambda = 0$ and 1. The former corresponds to the accumulation of debris ejected vertically and falling back on the same spot; the latter, to debris accumulating at the maximum range. The former may well correspond to the formation of "central mountains" in impact craters; but any identification of the latter with their "ramparts" would probably be wrong, as these are mostly too small in comparison with $R$, and due probably to more direct effects of central explosion in uplifting the lips of the pit.

The actual values of the distance $R$ up to which the debris flung out of $P$ has been deposited on the lunar surface can better be estimated from (say) the extent of the bright ray systems surrounding impact craters, which may represent the limits of the original "splash effects". To give a concrete example – the greatest distance to which bright rays diverging from the crater Copernicus (cf. Figures 16-1 or 16-30) can be traced on the lunar surface is a little more than 500 km. If so, it follows at once from (17-12) that the particles which traveled so far must have received an impulse at $P$ corresponding to the velocity

$$u = (gR)^{\frac{1}{2}} = 0.91 \text{ km/sec} \qquad (17\text{-}21)$$

immediately behind the shock.

If so, however, it follows also from (17-6) that the change in internal energy across the shock should correspond to a temperature difference which can be calculated from the equation

$$C_v(T_1 - T_0) = \tfrac{1}{2}u^2 = 4.18 \times 10^9 \text{ cm}^2/\text{sec}^2, \qquad (17\text{-}22)$$

where $C_v$ denotes the specific heat of the rocks. If we adopt $T_0 \sim 240\,°\text{K}$ and $C_v = 7 \times 10^6$ erg/g·deg as a good average for silicate material, the foregoing equation (17-22) leads to $T_1 \sim 850\,°\text{K}$ – certainly elevated, but insufficient to bring about any melting of the silicate material which is being splashed out on the surrounding landscape.

Does this result imply that the terrestrial tektites – which have frequently been mentioned as possible artifacts of lunar origin – could not be melted debris ejected from the lunar craters? Hardly so; for our above result pertains to particles which fell back on the Moon. If, instead, we were to identify $u$ in Equation (17-22) with the escape velocity of 2.38 km/sec, the resulting value of $T_1$ would be in the neighbourhood of 4000 °K – certainly sufficient for melting. It is, in fact, probable that *particles ejected from the gravitational field of the Moon as a result of accelerations imparted by the shock waves would be thrown out in molten stage*; but this is not the place to follow the implications in more detail.

In order to proceed further, let us return to Equations (17-3) and (17-4), which for $P_0 = 0$ and together with (17-6) reduce to the following two relations

$$\rho_1 = \frac{\rho_0 U}{U - u},\tag{17-23}$$

$$P_1 = \rho_0 u U,\tag{17-24}$$

for three unknowns $\rho_1$, $P_1$ and $U$. In order to make their solution determinate, another relation between these unknowns must be sought; and this may be provided by the Birch-Murnaghan equation of state of the form

$$P_1 = \frac{3}{2\beta_0} \left\{ \left(\frac{\rho_1}{\rho_0}\right)^{\frac{7}{3}} - \left(\frac{\rho_1}{\rho_0}\right)^{\frac{4}{3}} \right\},\tag{17-25}$$

where $\beta_0$ is a constant characteristic of different materials (cf., e.g., BRIDGMAN, 1948; HUGHES and MCQUEEN, 1958; or others), for which a round value of $10^{-12}$ cm²/dyne may be tentatively adopted. If so, and if $\rho_0 = 3.28$ g/cm³, it follows from the preceding equation that, for $u = 0.91$ km/sec,

$$\left.\begin{array}{l} P_1 = 2.5 \times 10^5 \text{ atm} \\ \rho_1 = 3.68 \text{ g/cm}^3 \\ U = 8.3 \text{ km/sec.} \end{array}\right\}\tag{17-26}$$

These are, to be sure, conditions obtaining at the point $P$ at which the actual ejection has taken place by the passage of the shock. The conditions obtaining at the epicenter are likely to be much more extreme, as we already mentioned, and should depend essentially on the kinetic energy of the impinging meteorite. The same is true of the sub-surface depth $d$ at which the epicenter is buried; semi-qualitative theories of this phenomenon as have been developed (cf., e.g., SHOEMAKER, 1962; STANYUKOVICH and BRONSHTEN, 1962) so far are still in a rather rudimentary stage.

A few further remarks may, however, be added concerning the exterior ballistics of the particles ejected during the crater formation. Equations (17-7) control their flight only in the field of constant gravity, over a plane surface at rest. The Moon is, however, a sphere of mean radius $a = 1738$ km, and rotating with an angular velocity $\omega_{\mathbb{C}} = 2.6617 \times 10^{-6}$ sec⁻¹ in a period of 27322 days. The equations of motion of a mass-point ejected from the surface of a rotating sphere have been set up exactly by MOULTON (1926; Section 4), but are rather complex. If, however, all their terms involving $\omega_{\mathbb{C}}^2$, the deviation $z$ of the trajectory from the $xy$-plane as well as the time-derivative $\dot{z}$ are ignored except when multiplied by $a$, the exact equations simplify to

$$
\left.
\begin{aligned}
\frac{d^2x}{dt^2} &= \quad\quad -g\frac{x}{a} + 2\,\omega_{\mathrm{C}}\,\sin\lambda\,\cos\beta\,\frac{dy}{dt}, \\[4pt]
\frac{d^2y}{dt^2} &= -g + 2g\frac{y}{a} - 2\,\omega_{\mathrm{C}}\,\sin\lambda\,\cos\beta\,\frac{dx}{dt}, \\[4pt]
\frac{d^2z}{dt^2} &= \quad\quad +2\,\omega_{\mathrm{C}}\,\cos\lambda\,\cos\beta\,\frac{dy}{dt},
\end{aligned}
\right\}
\tag{17-27}
$$

where $\beta$, $\lambda$ denote – as before – the selenographic latitude and longitude; and $g$, the lunar gravitational acceleration at the point $P$ of ejection. The terms multiplied by $\omega_{\mathrm{C}}$ represent all the sensible effects of the Moon's rotation on the flight trajectories; and since (17-27) represents a linear system of differential equations with constant coefficients, their solution can be expressed in a closed form in terms of elementary functions; but its actual construction is rather lengthy and can be left as an exercise for the interested reader. It may be noted that (within the scheme of our approximation) the first two of the equations (17-27) are independent of the third; but the latter depends on the solution of the first two.

In the case of long trajectories tantamount to sub-orbital flights of mass particles in the gravitational field of the Moon, advantage can be taken of the existence of certain known integrals of the problem of two bodies relevant to the situation. Thus for any mass-particle ejected from $P$ with a velocity $u$ in a direction inclined by an angle $\varphi$ to the horizontal plane $x = 0$ tangent to the lunar sphere at $P$ (cf. Figure 17-2)

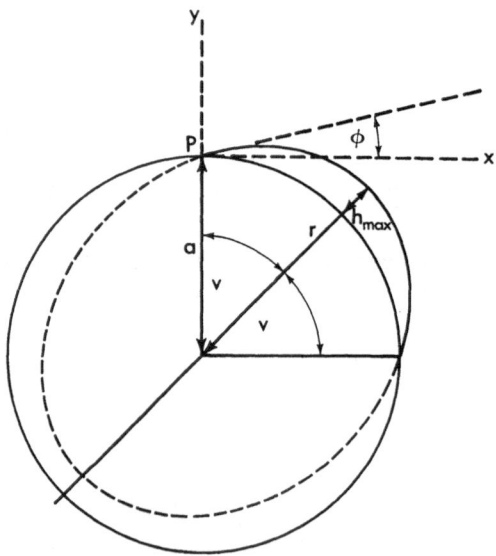

Fig. 17-2. Exterior ballistics of the ejecta over curved lunar surface.

the distance $r$ to the Moon's center will be given by the radius-vector

$$r = \frac{p}{1 - e \cos v}$$  (17-28)

of a Keplerian ellipse, with the semi-latus rectum

$$p = (u^2/g) \cos^2 \varphi;$$  (17-29)

and if so, the "true anomaly" $v$ between the point of launch and the aposelenium of the trajectory (cf. again Figure 17-2) will follow from

$$\cos v = -\frac{p - r}{re}.$$  (17-30)

On the other hand, the initial radial velocity

$$y(0) = u \sin \varphi = (a e u/p) \cos \varphi \sin v$$  (17-31)

by Kepler's second law particularized for $r = a$. Dividing (17-28) by (17-31) we arrive then at the relation

$$p \cot v = (a - p) \cot \varphi$$  (17-32)

or, more explicitly,

$$\tan v = \frac{u^2 \sin \varphi \cos \varphi}{ga - u^2 \cos^2 \varphi},$$  (17-33)

where, it may be remembered, the product $ga$ represents the square of the velocity of a particle in a circular orbit around the Moon and is, therefore, greater than $u^2$.

The foregoing Equation (17-33) defines the half-angle $v$ of a sub-orbital trajectory whose actual length $R$ along the lunar surface is equal to

$$R = 2 av = 60.67 v^{(0)} \text{ km}$$  (17-34)

if $v^{(0)}$ is expressed in degrees; attaining the maximum altitude

$$h_{\max} = a \left\{ \frac{\bar{u}^2 + \sqrt{1 - \bar{u}^2(2 - \bar{u}^2) \cos^2\varphi} - 1}{2 - \bar{u}^2} \right\}$$  (17-35)

above the lunar surface (where $\bar{u}^2 = u^2/ga$); and the corresponding time of flight is given by

$$T = 2\left(\frac{a}{g}\right)^{\frac{1}{2}} \left\{ \frac{1 - e \cos v}{1 - e^2} \right\}^{\frac{3}{2}} \left\{ \sqrt{1 - e^2} \tan \varphi + 2 \tan^{-1} \sqrt{\frac{1 + e}{1 - e}} \tan \frac{v}{2} \right\}$$  (17-36)

where the eccentricity $e$ of the sub-orbital trajectory is found to be given by

$$e = \frac{\sin \varphi}{\sin (\varphi + v)}.$$  (17-37)

From the foregoing results it transpires that the velocities of ejection $u$ capable of producing the spectacular ray system associated with the crater Tycho (the longest rays of which are more than 2000 km in length) are between 1.4–1.5 km/sec – i.e., quite moderate judged by terrestrial standards – and the corresponding times of flight, about $3 \times 10^3$ seconds or a little less than one hour. Velocities of this order are attained by average guns or rifles used in terrestrial warfare against men or animals. Armchair strategists should note the tremendous distances at which one could trade blows on the surface of the Moon; but also bear in mind that the times of flight of such shots are so long as to enable moving targets to displace themselves from the line of fire with relative ease before the actual arrival of the projectile.

Tycho's rays do not seem to diverge from a single point; and some give an impression as though they were "wound up" on the lunar sphere after more than revolution. UREY (1952) pointed out that the relative displacements of such rays deposited after more than one complete circumnavigation could have been caused by centrifugal force. Deposits descending on the lunar surface after one or more orbits are indeed dynamically possible as a result of perturbations arising from departure of the Moon's main mass from a sphere (though not otherwise). Should this actually be the case, the observed ray displacements in the $z$ direction due to centrifugal force could enable us to determine the actual velocity of the Moon's axial rotation at the time when the respective crater was made.

Let us, however, return again to the more local features of impact phenomena mentioned earlier in this section. Diverse terrestrial experiments with impact of metallic or stony particles in brittle media – or with explosive charges ranging from microscopic dimensions to underground nuclear detonations in megaton range of explosive power – are indeed found to produce local effects simulating the ramparts of certain

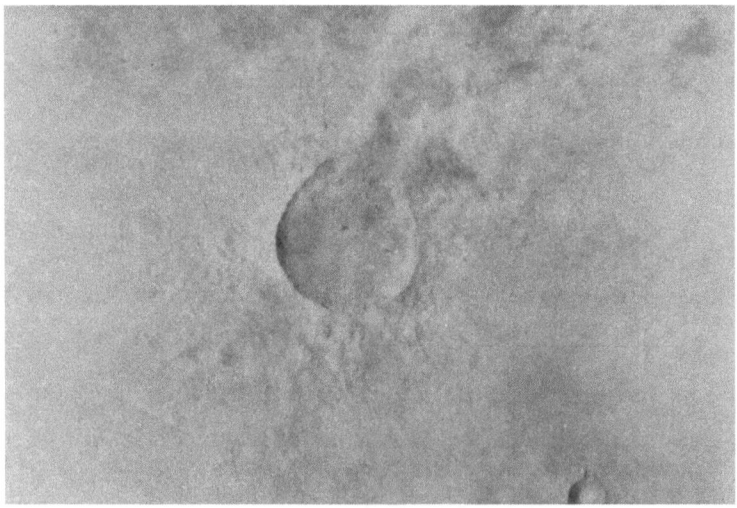

Fig. 17-3.   Aerial view of the Sedan nuclear crater in Nevada, U.S.A. produced by an explosion of a large equivalent to 0.1 megaton of TNT. (By courtesy of the U.S. Atomic Energy Commission.)

types of lunar craters to at least a superficially astonishing degree (cf. Figure 17-3). Moreover, since the total energy requisite for producing terrestrial craters of given size in laboratory experiments, aerial bombardment of ground, or nuclear explosions, are known, an extrapolation of this trend should permit us at least to estimate the energies likely to be involved in the formation of much larger craters on the Moon by impacts of meteorites or other cosmic bodies.

Estimates of the *minimum* energy requisite for crater formation can evidently be obtained if we evaluate the amount of work necessary for raising the observed walls of a crater from the depressed level of its floor. Such minimum estimates, based on the observed topographic features, for some craters, were recently carried out by SUDBURY (1965), who found that for Theophilus (Figures 16-8 or 16-11, of the mean diameter of 104 km), the requisite energy turns out to be close to $7 \times 10^{26}$ ergs; for Copernicus (Figures 16-7 or 16-10; 90 km across), it is $2 \times 10^{26}$ ergs; for Tycho (Figure 16-9 or 16-12; 86 km across), $1 \times 10^{26}$ ergs; while for the smaller Eratosthenes (60 km diameter) it is less than $6 \times 10^{25}$ ergs and for Kepler (34 km diameter) only $1 - 2 \times 10^{25}$ ergs. In actual fact, the energies requisite for crater formation are bound to be substantially larger; as a part of the kinetic energy of the impinging body will be spent by other processes than rampart formation – such as the heating of the ground, production of elastic waves; or ejection of debris far beyond the confines of the crater walls (some of it outside the gravitational confines of the Moon altogether).

Extensive terrestrial experiments ranging from laboratory models to nuclear explosions of yield in the range of megatons of TNT (Scooter) have revealed (cf., e.g., NORDYKE, 1962) that *the diameter D of an impact crater formed by a shallow sub-surface explosion should be proportional to a little less than the cube-root of the total energy E expended*; or, in more specific terms,

$$\log_{10} D = 0.294 \log_{10} E - 6.92 \qquad (17\text{-}38)$$

if $D$ is expressed in kilometers and $E$ in ergs. This equation, based on terrestrial material, holds good statistically for explosion craters formed in the terrestrial gravity field. Accordingly, the energy required to form the well-known impact crater near Cañon Diablo in Arizona (see Figure 17-4), for which $D$ is slightly more than 1 km, turns out to be a quantity of the order of $3 \times 10^{23}$ ergs, which for a typical impact velocity of 15 km/sec would correspond, by (17-1), to a mass $m \sim 3 \times 10^5$ tons.

On the Moon, where the gravitational acceleration $g$ is about one-sixth of its terrestrial value, Equation (17-34) for the range of the ejecta might lead us to believe that, for a given explosive energy, the dimensions of impact craters should be six times as large as on the Earth. This simple extrapolation would, however, ignore the contribution of atmospheric phenomena (gas venting) to the formation of terrestrial craters – where a good part of the momentum is imparted to the ejecta through air motion. Taking the latter phenomena duly into account, HESS and NORDYKE (1961) concluded recently that the lack of gas around the Moon should just about offset the effects of lower gravity – so that the resultant dimensions $D$ of lunar craters caused

Fig. 17-4.    Meteor crater near Cañon Diablo, Arizona, U.S.A., due to an impact of a large meteorite in prehistoric times (10–50 thousand years ago).

by a sudden release of energy $E$ should still be approximately the same as on Earth, and continue to conform statistically to the foregoing equation (17-38).

Its application to the problems of lunar macroscopic topography requires, to be sure, an extrapolation of its validity by 2–3 orders of magnitude in $D$, and up to 8 orders in $E$, beyond the limits supported by the terrestrial evidence. Nevertheless, if we do so and put our faith in the results, it should follow that in order to produce a lunar crater 20 km in diameter by impact, a kinetic energy of the order of $10^{28}$ ergs would have to be expended in the effort; and to double or quadruple its size, energies 10 or 100 times as large would be a prerequisite. The mass of such a body impinging with a velocity (say) of 20 km/sec would be $10^{16}$ grams (ten milliard tons) for a total energy of $10^{28}$ crgs, and proportionally larger for energies 10 or 100 times the amount considered. The diameters of solid spheres of such a mass (varying, in general, as the cube-root of the energy of impact) would, moreover, be approximately 1200, 2500 and 5400 meters if they were of stony material of average density 3 g/cm³, and 20% smaller if their principal constituent were nickel-iron (of density close to 7 g/cm³). Thus impact production of craters as large as Clavius or Deslandres would call for energies of the order of $10^{30}$ ergs (i.e., ten million megatons of TNT) associated with high-velocity impacts of small asteroids – of the size of Adonis, Hermes, or Eros (to name a few which paid rather close calls on the Moon in recent decades) – whose dimensions are estimated to 10 km, and masses to $10^{13}$ tons.

The sudden expenditures of such prodigious amounts of energy would, however, trigger off a chain of events which would not only be devastating beyond imagination on (and around) the actual point of impact; but whose consequences should be felt, to a different degree, all over the Moon through the medium of *seismic waves*. In order to assess the intensity of disturbances which can be caused by such waves, let us return to our impinging meteorite as it comes to a complete stop at some depth beneath the lunar surface – an event necessitating the conversion of its entire kinetic energy into other forms.

How will this large energy store be apportioned? While its actual partition would depend rather critically on the actual depth of penetration, such knowledge as we possess indicates that only about one-half of the kinetic energy of the incident body may be converted into heat and bring about its volatilization. Most of the other half should be absorbed in a ground shock and largely used up for irreversible heating (though not volatilization) of underlying rocks at a very short distance. Only when the shock pressures have degraded to the order of the elastic strength of these rocks (which should be about 1 kilobar) will the remaining energy be available for propagation as seismic waves; and its amount is estimated to a few per cent of the total kinetic energy of the impinging body.

Thus we are led to conclude that a seismic energy of the order of $10^{26}$ ergs would emanate from an impact capable of producing a lunar crater 20 km in diameter, with proportionate modification for craters of other sizes; and the seismic waves excited by it should carry in their train the message of the event throughout the lunar globe to all parts of its surface. In other words, any impact of a body capable of producing a lunar crater would also be bound to set off a "moonquake", characterized by a very shallow epicenter; and their effects on the Moon as a whole require some attention.

In order to appreciate more fully the seriousness of such "moonquakes", let us recall that the most destructive earthquakes experienced on our planet within the memory of mankind entail energy expenditures of the order of $10^{26}$–$10^{27}$ ergs only – i.e., thousand to ten thousand times *less* than that of a hypothetical moonquake accompanying the origin of a crater 100–200 km in diameter. When one considers the fact that there are almost fifty craters of this size or larger on the visible hemisphere of the Moon alone, and that the total number of those exceeding 1 km in size is several hundred thousand, one is bound to inquire with some doubt: if all, or even a majority, of these formations have been produced by impact, how could any old mountain or wall anywhere on the Moon have survived such a long series of sudden and devastating seismic disturbances which would have been caused by each new arrival of an impacting body from space?

Of the different types of destructive seismic waves which are bound to get excited by impacts of external bodies on the solid globe of the Moon, the surface (Rayleigh or Love) waves – representing the high-frequency end of the spheroidal or toroidal oscillations of the lunar globe discussed in Chapter 9 – should be of particular interest because of their relatively low damping. On the Earth, their amplitudes are found to be reduced approximately to one-third at a distance of 5000 km; and in the feeble gravitational field of the Moon they may be damped still less. As, moreover, the circumference of the lunar globe is only 10 921 km in length, the Rayleigh waves produced by meteoritic impacts should converge on their antipodal points from all directions carrying a total energy of the order of one-thousandth of that of the original impact. This fraction would probably be sufficient to damage (by fragmentation and other symptoms of terrestrial earthquakes) the antipodal region to some extent. While such antipodal points of all visible craters are, of course, bound to be located on the far side of the Moon, one wonders what kind of phenomena may have been produced on the visible lunar face by impacts on its far side.* The individual and cumulative effects of all such impacts have not so far been taken into account at all by those that defend the purely meteoritic origin of the lunar craters. What the eventual verdict of an appropriate investigation of their significance will turn out to be, no one as yet can say. But until the seismic effects of meteoritic impacts on the Moon have been duly considered – in addition to all more local events we cannot be sure that such impacts represent the only, or even the principal, clue for deciphering the enigmatic hieroglyphs of the lunar face.

This should be all the more true of much larger (and proportionally more devastating) impacts with planetesimals which many investigators have invoked to account, not only for the craters themselves, but also for the existence of extended lunar plains of roughly circular form – such as Mare Crisium, Mare Serenitatis, or even the largest of them – Mare Imbrium – which are indeed regarded by many investigators as huge craters, produced by nearly grazing collisions of the lunar surface with low-velocity planetesimals in the early days of the existence of our solar system.

This view, proposed first by Gilbert in 1893, has more recently been ably defended by Urey. According to Urey, the origin of the Mare Imbrium can best be explained by an assumption that, in the early days of the lunar history, a solid object about 200 km in size (and weighing some $1.5 \times 10^{16}$ tons) made a nearly grazing plunge into the present Mare Imbrium, from the direction of its Sinus Iridum (well shown on Figure 4-5), with a relative velocity of 2.4 km/sec (i.e., just above that of escape from the gravitational field of the Moon). The kinetic energy of such a planetesimal would have been equal to about $4 \times 10^{32}$ ergs – equivalent to ten milliard megatons of TNT or a hundred milliard atomic bombs of the Hiroshima calibre (i.e., one for each 1100 square meters of the entire surface of the Earth), or again about a million times that of the largest and most destructive earthquake ever experienced on our own planet. This formidable display of figures should make it easy to agree that a collision of this order of magnitude would constitute a catastrophe beyond any imagination – not only if it occurred on Earth, but all the more if it happened to the much smaller lunar globe. The aftermath of such a catastrophe – not only the scar on the surface, but also its effect on the axial rotation of the Moon and its orbit around the Earth – would be profound. As, however, it is difficult to visualize the results – and, in particular, the dynamical effects of such planetesimal bombardment of the Moon have never been quantitatively assessed – it would be perhaps premature to attempt pronouncing, at this stage of lunar exploration, any final verdict.

Moreover, throughout all our discussion we have so far been concerned with the effects of collisions of the lunar surface with *solid* bodies – like meteorites, asteroids, or planetesimals. Any effort to explain the origin of the principal features of the lunar

---

* It may be of interest to note, in this connection, that the Mare Muscoviae on the Moon's far side appears to be antipodal to the crater Tycho.

Fig. 17-5. Sunrise over Sinus Iridum in Mare Imbrium; 24-inch refractor of the Observatoire du Pic-du-Midi (Manchester Lunar Programme).

surface by impacts of solid bodies would be seriously incomplete without a parallel consideration of effects which could be wrought on the lunar face by collisions with other known types of denizens of the interplanetary space – namely, the *comets*. According to the statistics available to astronomers at the present time, comets appear to be at least as frequent at our present distance from the Sun as are meteorites of comparable masses (i.e., $10^{15}$–$10^{18}$ g); and a wide distribution of the elements of cometary orbits is bound to render high-velocity collisions (in the range of 30–70 km/sec) with the Moon much more frequent than would be the case with the asteroids.

On the Earth, examples of both types of collisions and of their after-effects appear to be preserved; for, while the well-known Barringer crater at Cañon Diablo in Arizona (see Figure 17-4) was without doubt produced by the impact of a major meteorite in prehistoric times (and a few scores of other similar formations are now on record); the Siberian Tunguzka crater of 1908 was apparently produced (judging from a well-nigh complete absence of metallic component in the debris and other structural characteristics) by a collision of the Earth with a comet. For it is known that cometary heads – the only part of their anatomy which matters in the case of collision – represent but loose conglomerations of frozen hydro-carbons, with appreciable admixture of unstable chemical compounds (such as solid hydrogen peroxide, or azides); and these on impact would behave like high explosives – thus releasing chemical energy in addition to the kinetic energy of the head as a whole. Now, unlike a solid meteorite, cometary heads possess no tensile strength, and their impact on the Moon would scarcely indent the lunar surface to any appreciable extent. Instead, they would be completely volatilized at once and envelop for a short time the surrounding region in a stream of hot gas rapidly dispersing into vacuum.

Quite apart from the chemical binding energies, the kinetic energy of a major comet – such as Halley's, for instance – represents alone a quantity of the order of $10^{31}$ ergs; and the latter, if it could be converted totally into heat, would be equivalent to $2 \times 10^{23}$ calories. If we assume, reasonably enough, that 2000 calories are necessary to melt one gram of lunar surface matter into fluid lava, a single cometary impact of this calibre could provide, for example, some $10^{20}$ grams of lava capable of covering the 400000 sq km of Mare Imbrium to a uniform depth of some 100 meters. This may possibly furnish another explanation of the origin of lunar maria, compatible with the fact that no maria show any deformation in their central regions (where impact should have taken place) and that their ramparts (as represented by the Alps and Apennine chains in the case of Mare Imbrium) are well below the horizon at the center of these great plains. No destructive seismic effects of the impacts of planetesimals of masses postulated by Gilbert or Urey need to be feared in this hypothesis.

Comets with kinetic energies of the order of $10^{31}$ ergs are, to be sure, relatively rare. On the other hand, the number of plains on the Moon of the size of Mare Imbrium is also limited; and the probability that, in the past 4500 million years, the Moon may have suffered a sufficient number of collisions with comets of requisite masses may even be considerable. Collisions with smaller comets may, in turn, have produced craters of the type of Archimedes or Plato (see Figure 18-6) whose floors, surrounded by very low ramparts, bear a striking similarity to the surrounding maria. So far, however, any such suggestion can be put forward as a tentative possibility; and further investigations will be necessary to see it can be placed on a more solid basis.

Having taken stock thus of the principal *external agents* – namely, impacts of various celestial bodies which can mutilate the lunar face over long intervals of time – let us turn next to examine the *internal processes* whose action can affect the surface of our satellite in a similar manner. These latter processes are, in general, connected with the gradual build-up of internal heat by radioactive decay (which we discussed already in the chapter 8), producing a super-adiabatic gradient of temperature in the interior and thus possibly giving rise to a slow convection in the lunar interior. As a by-product of the secular heating and aided by the convection currents, the gases and other volatile elements initially present in the interior of the lunar globe can make their way to cooler layers underneath its surface and accumulate there or escape into space. This defluidization and degassing, which must be continuously operative in the Moon as it has been in the Earth, can – in its last stage – produce (by upwelling and withdrawal of a molten rock column) local areas of surface depression or subsidence, which the geologists on our Earth refer to as the "calderas".

A schematic view of some such well-known terrestrial calderas is shown on the accompanying Figure 17-6. Their essential features bear indeed a rather striking qualitative similarity to those of many large lunar craters, whose most important characteristic – and this cannot be emphasised too strongly – are not their (often inconspicuous) ramparts, but rather the *general depression of their floors* below the level of the surrounding landscape. Thermal expansion and contraction phenomena accompanying defluidization could have brought about the occurrence of fracture patterns in the relatively solid lunar crust, which may have provided ducts of escape for the molten magma beneath.

Many typical mountain-walled plains on the Moon appear to be distinctly polygonal – in fact, hexagonal. Note, in particular, the group of large craters in the central parts of the Moon's apparent disk consisting of Ptolemy and Alphonsus, Arzachel,

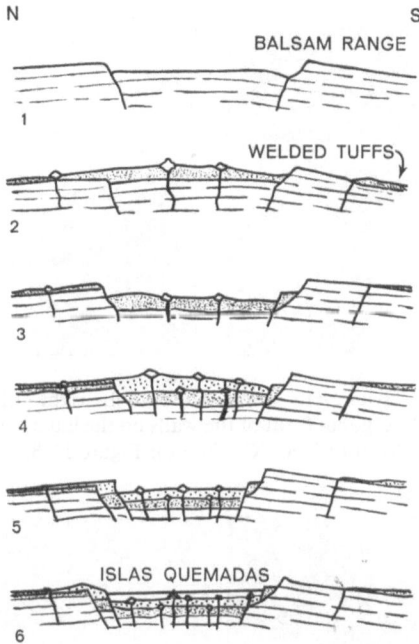

Fig. 17-6.   A schematic example of a terrestrial "caldera of subsidence" in Central America.

Flammarion; Hipparchus and Albategnius; or Purbach, Regiomontanus and Walter to the south; and many others. Not only these, but even the largest formations of this type – such as Clavius near the south pole, or Mare Crisium on the Moon's eastern limb, show distinct hexagonal outlines when their foreshortening near the limb is rectified by the projection on a sphere (cf. Figures 17-7 and 17-8).

The hexagonal pattern by itself does not rule out impact origin – witness the polygonal shape of the terrestrial meteor crater in Arizona (Figure 17-4), or of the lunar crater Copernicus (Figure 16-7) which is no doubt the result of an impact. However – and this would be much more difficult to explain on impact hypothesis – many of the hexagonal craters (in particular, the one near the centre of the Moon's apparent disk) are similarly orientated and their sides more or less parallel – whether they are adjacent or separated by some distance. On the other hand, it is well known that hexagonal shape is characteristic of the convection pattern which develops in a parallel layer of viscous liquid heated from below (cf. BÉNARD, 1900, 1901; or, more recently, CHANDRASEKHAR, 1961). Possible relevance of such a convection to the origin of lunar craters has been discussed by PUISEUX (1907); or WASIUTYNSKI (1946); and although the argument cannot as yet be definitely settled (on account of our incomplete knowledge of the distribution of heat sources in lunar interior), it is open

Fig. 17-7.   The lunar crater Clavius: left – direct photograph; right – photograph rectified by a projection on a sphere. Note the hexagonal form of the walls on the latter, and compare with the rectified form of Mare Crisium on Figure 17-8.

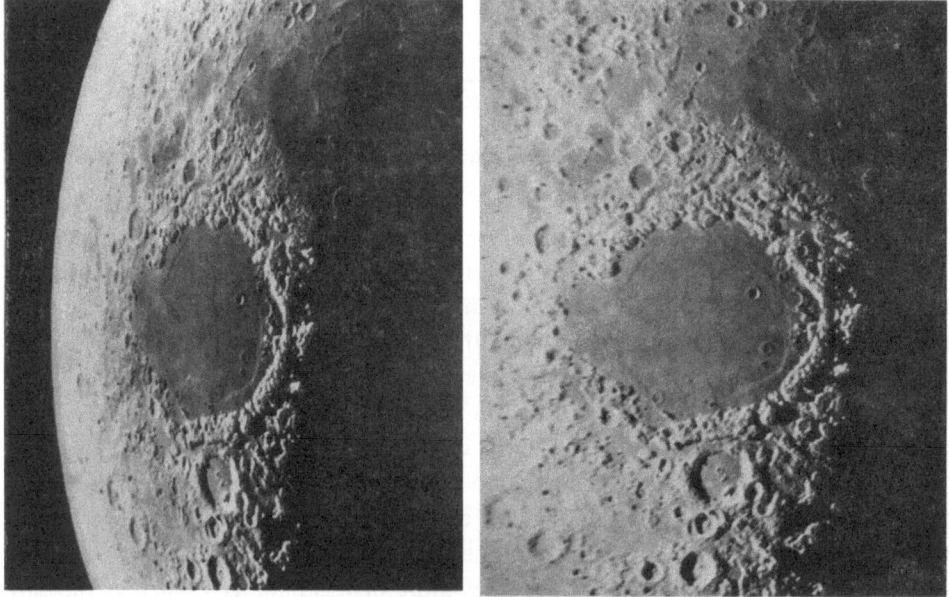

Fig. 17-8.   Mare Crisium: left – direct photograph; right – photograph projected on the sphere. Note the distinctly polygonal outline of the rectified image.

to surmise that some (possibly the largest) craters on the Moon may have originated as a result of internal processes – side by side with many others which are due to impacts; and while the number of the latter keeps accruing in the course of the time; those which may have been produced by convection originated probably during one distinct epoch of the Moon's history which need not have been far removed from the time of the origin of our satellite.

The escape of volatile by degassing of the interior constitutes an inevitable long-drawn cosmic process, which is bound to occur in any solid body radioactively heated within; and should not be identified too closely with the *volcanic processes* as we know them on the Earth. The latter represent essentially superficial phenomena, of much smaller order of magnitude and lateral extent. Yet volcanic activity on Earth follows, in general, as an aftermath of defluidization; and should the same be true on the Moon as well, it may have left its imprint inside the depressions of lunar calderas as it did in their presumed terrestrial homologues. These are likely to be much less conspicuous than the calderas themselves by their size, but still easily demonstrable by their particular characteristics.

In an attempt to identify, if possible, such formations on the lunar surface, let us inspect a photograph reproduced on Figure 16-6. Within the walls of a crater – possibly a caldera – called Regiomontanus we find a small "hill-top"crater desig-nated as Regiomontanus A which, (like others of this type,) probably did not originate by impact; for the likelihood that an impinging meteorite would happen to strike just the top of so small a hill appears to be negligible. According to the measurements by TURNER (1959), the whole hill rises barely 650 meters above its surroundings and the crater on the top is but 5½ km across (see Figure 17-9). Its base of some 100 km in circumference renders the whole formation not unlike (and, in fact, a little smaller than) the well-known terrestrial volcano Krakatoa off the coast of Java. This is indeed what Krakatoa would look like to us if we could observe it at the distance of the Moon!

The relatively large size of such presumably volcanic lunar craters transpiring from this comparison can, at least to some extent, be rationalized by reference to a much lower gravity prevailing on the Moon, which should enable a given force to disgorge volcanic ejecta more than six times farther than on the Earth. In addition, there is no atmosphere on the Moon to weigh down on the boiling magma from above, or to decelerate the ejecta in flight. Hypothetical volcanic eruptions on the Moon would, in fact, represent discharges of gas and magma into vacuum; and this could

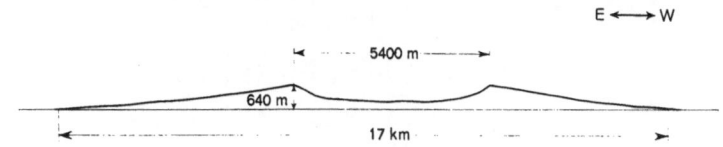

## REGIOMONTANUS A

Fig. 17-9.   East-West profile of the "hill-top crater" Regiomontanus A. Its external characteristics recall strikingly those of the terrestrial volcano Krakatoa.

possibly enable such phenomena to attain considerably larger relative proportions than they do on the Earth. For other examples of lunar formations where volcanic activity is at least indicated, cf. also Figure 24-1 in the last chapter of this book.

In the face of these and other facts it is very difficult to escape the conclusion that *both* types of formative processes – external as well as internal – had a hand in shaping up the face of the Moon as we see it to-day. Detailed analysis of all aspects of circumstantial evidence accessible so far from the Earth suggests that craters like Copernicus, Theophilus, or Tycho are almost certainly due to impacts of solid bodies, on the surface which must likewise have been solid down to a considerable depth. Such craters are relatively deep for their size, and are distinguished by their hummocky rims. The rims of Copernicus (Figure 16-7) are closely simulated in structure by those of other craters equal or similar in size; and, in general, the ratio of the width of the hummocky terrain to the diameter of the crater diminishes with decreasing dimensions of the formation. Around some craters virtually all the rim terrain is made up of a nearly random arrangement of hummocks typical of the crest of the ramparts of Copernicus; around others the rim exhibits a pronounced radial pattern of low

Fig. 17-10.   Impact craters Eudoxus and Aristoteles in Mare Frigoris, photographed with the 120-inch reflector of the Lick Observatory.

Fig. 17-11. Sunset over the craters Atlas and Hercules on the shores of Mare Frigoris, as photographed with the 43-inch reflector of the Observatoire du Pic-du-Midi (Manchester Lunar Programme) on November 23, 1964.

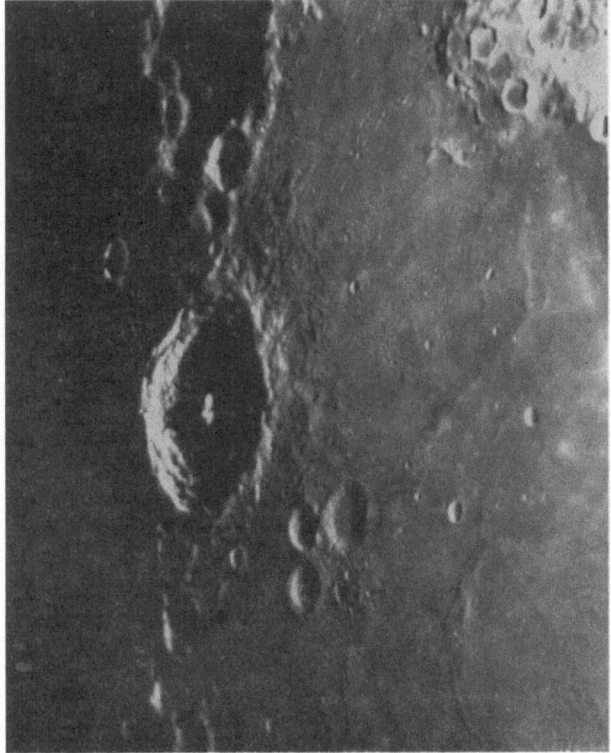

Fig. 17-12. Sunset over the crater Langrenus on the eastern limb of the Moon, photographed with the 120-inch reflector of the Lick Observatory. Note the large number of secondary craters in the plains of Mare Foecunditatis, surrounding Langrenus.

ridges typical of the periphery of Copernicus. The interior walls of these craters are almost invariably terraced; the floors are irregular; and nearly all possess a single or multiple *central peak* (cf. Figures 17-10 to 17-12).

However, by far the most revealing identification of primary impact features are the *secondary craters* and (for younger formations) the *bright ray systems* radially diverging from it. Earlier in this chapter we outlined the process by which a lunar crater can originate by an impact of a cosmic intruder; and the essential feature of this process was, the ejection of a large amount of solid debris by the shock released on impact; and the mass-distribution of this debris may range from fine dust to rocks weighing millions of tons (of size limited only by internal cohesion), depending on the nature of the ground. Some of this matter may be ejected with velocities sufficient for escape from the gravitational field of the Moon; but most of it undoubtedly falls back on the surface of the Moon, at a distance depending on the initial velocity. It is plausible that the light dust travel farthest – and its spread on the Moon may well give rise to the radial ray systems surrounding the crater, and remaining visible by contrast as bright streaks on a darker background, until they too are darkened in time by radiation damage to the reflectivity of the underlying strata. On the other hand, heavy rocks may not travel that far, but fall back on the surface in the neighbourhood of the parent formation.

Fig. 17-13.   Secondary impact craters of the Copernican system (after SHOEMAKER, 1962).

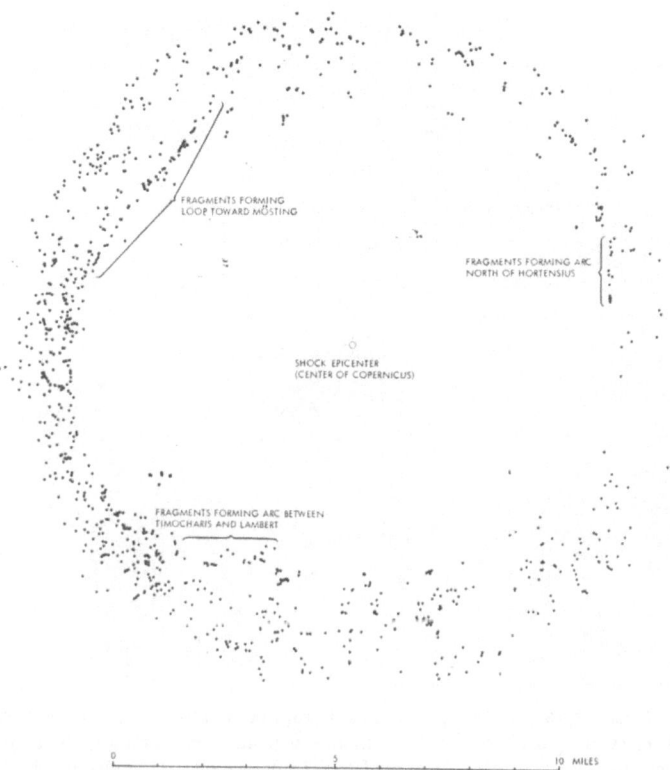

Fig. 17-14. Detailed distribution of the secondary impact craters of Copernicus in the immediate neighbourhood of its ramparts (after SHOEMAKER, 1962).

The clustering of, not hundreds, but thousands of such "secondary craters" discernible by existing telescopes around the craters Copernicus or Langrenus has been demonstrated by SHOEMAKER (1962); for their distribution, cf. the accompanying Figures 17-13 and 17-14; those attending Theophilus can be partly seen on Figure 17-15 on the next page; and the remarkable photographs secured by the American Ranger 7 spacecraft on July 31, 1964, in close proximity to the lunar surface (cf. Figures 17-16 or 23-2) reveal thousands of others, too small to be visible from the Earth, overlying rays of the craters Copernicus or Tycho, 610 or 1000 km distant from the point of impact. A glance at these photographs reveals that, on a close view, the landscape overlaid by a bright ray must once have received a very heavy bombardment; and the surmise is without doubt well founded.

The actual process of formation of secondary craters of this type should closely paralled that of its primary impact feature, as discussed earlier in this chapter, but with one important systematic difference: namely, the velocity of secondary impacts must have been very much less than that of the primary intruder. We have mentioned already that the impact velocity of the latter may have been anywhere between 2.4 and 72 km/sec – perhaps 20–30 km on the average – while the velocity of secondary

Fig. 17-15.   Mare Nectaris, east of Theophilus, photographed on November 23, 1964, with the 43-inch reflector of the Observatoire du Pic-du-Midi (Manchester Lunar Programme). Note the large number of secondary craters (and also a faint system of rays) surrounding Theophilus; the "ghost" crater Daguerre and others can be seen on the left.

impacts must clearly be less than that of a circular orbit around the Moon – i.e., 1.68 km/sec (otherwise they would not have fallen down) – and perhaps around 1 km/sec on the average. If so, however, *the impinging secondary particles on the Moon will possess, on the average, about 500–1000 times less kinetic energy per unit mass than the primary intruder*; and this fact should leave a distinct mark on the feature of the resulting surface formation: secondary craters should be not only much smaller, but also *shallower* than the primary craters in which the impinging particle has buried itself deeper below the surface; and their walls should be lower, giving them a characteristic washed-out appearance. These distinctions were difficult to spot from the distance of the Earth; but the Ranger photographs such as those reproduced on Figures 17-16 or 23-2 leave no room for doubt that this is actually the case.

Apart from the craters like Copernicus, Tycho and others which are without doubt of impact origin, there are other and even larger craters on the Moon – such as Clavius, Ptolemy or Alphonsus, where such an explanation of origin is doubtful. In spite of their vague similarity to impact formations, they lack most of their distinct characteristics (hummocky walls or floors, central peaks); and, most revealing of all, the attendant hosts of secondary craters – let alone bright rays – are completely missing from their surroundings – facts which could scarcely be all accounted for by ageing.

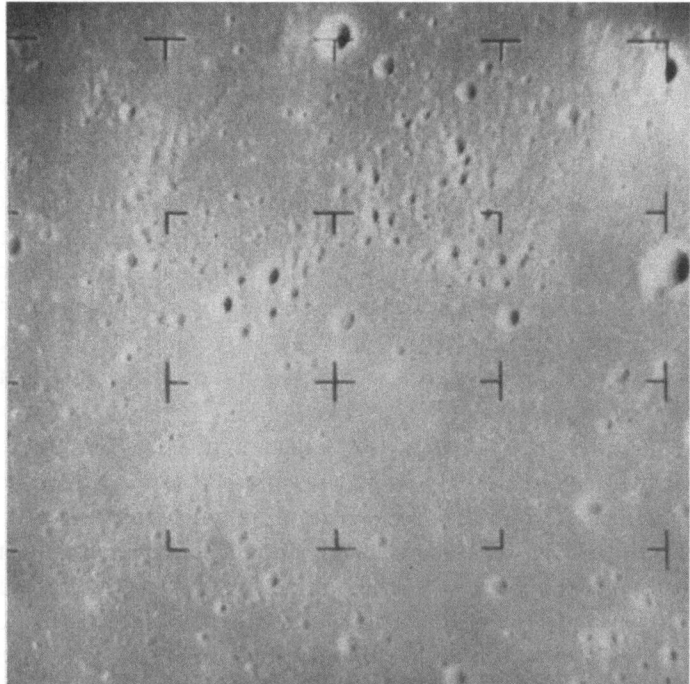

Fig. 17-16. A close-up of the lunar surface overlaid by one of Tycho's rays, as photographed by Ranger 7 on July 31, 1964, from an altitude of 55 km. The field is approximately 26 km across; and the smallest details discernible on the lunar ground are 50 metres in size.

Moreover, the outlines of their ramparts show distinct polygonal (hexagonal) symmetry and common orientation – which led, many years ago, Puiseux to conjecture that... "La croûte solide de la Lune, à l'époque plus ancienne où nous puissions remonter, a été constitué, dans tous ses parties, par un assemblage de cases polygonales juxtaposées et imparfaitment soudées." (PUISEUX, 1906). Puiseux, and following him WASIUTYNSKI (1946), had in mind the effects of crustal convection – effects whose mechanism is still far from being adequately investigated.

It should be reiterated that polygonal structure of lunar craters cannot, by itself, be invoked to rule out their impact origin; for the ramparts of Copernicus and other craters of undoubted impact origin are mildly polygonal. What are, however, the chances that two neighbouring craters of distinctly hexagonal outline like Ptolemy and Alphonsus (Fig. 15-24) would fit in neatly together as is observed, *sharing one confluent side,* if they were produced by impacts? Their formation must obviously have been simultaneous; but even if – a most improbably event – two large meteorites struck the Moon at exactly the same time and so close to each other to give rise to Ptolemy and Alphonsus, what are the chances that the ramparts raised in this way would share one side, undamaged by destructive interference (which we clearly see at work in the impact pair of Theophilus and Cyrillus on Figure 16-8 for instance)?

No; Ptolemy and Alphonsus, together with other lunar craters with confluent sides, were scarcely produced by external impacts – and neither was the enigmatic crater Wargentin near the western limb of the Moon (cf. Figure 17-17), a structure more than 86 km across, whose floor is not depressed below the level of the surrounding landscape but, on the other hand, raised to the level of its rims – thus giving an impression of a crater whose interior has been filled to the brim. Wargentin represents, to be sure, a unique formation in this respect on the entire visible face of the Moon – but, nevertheless, one which clearly could not have been produced by impact, but rather by the action of internal processes associated with convection and the accompanying uplifts and subsidence. To close our eyes to this fact in the face of photographic evidence as shown on Figure 17-17 would be as unreasonable as an attempt to deny the impact origin of the Arizona crater on the basis of the existence of Krakatoa or Katmai.

Unfortunately, the internal processes which may have produced Ptolemy or Alphonsus, or have arrested Wargentin in its present development, are physically much more complicated than the mechanism of external impacts; and their details cannot as yet be described mathematically with the same accuracy. Indeed, most geological – and, probably, most selenological – processes are too complex to be safely deduced from first principles of the underlying physical sciences. On the other hand, the lunar environment must be subject to the same laws of physics and chem-

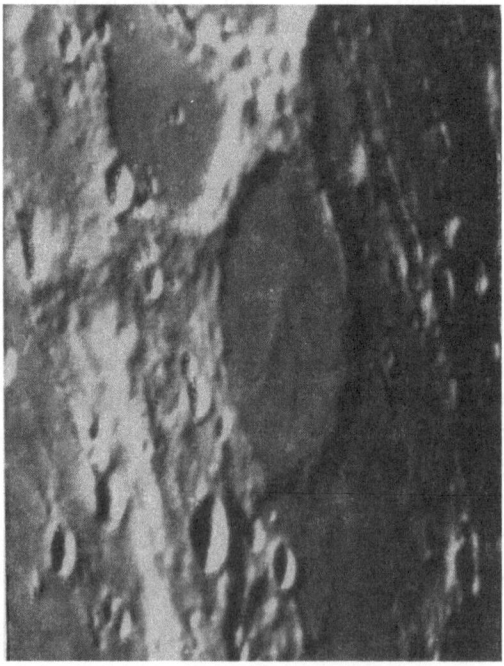

Fig. 17-17.   The lunar crater Wargentin. Lick Observatory photograph with the 120-inch telescope.

istry as we know them on Earth; and the rocks on the Moon must, therefore, share at least some of the range of physical and chemical properties as are known among the terrestrial rocks. Undoubtedly there are differences – possibly critical differences – in the physical, chemical and historical setting of the surfaces of the Earth and the Moon; as there are, indeed, striking differences in their topography. But if the problem of the origin of the lunar surface features is approached by analogy – a method followed in lunar studies since the time of Galileo Galilei – and attention focussed on those features which are similar in the two different environments, a critical comparison (with due regard to scale) of individual lunar features with their possible terrestrial homologues may prove fruitful in the future – until the last remaining arguments are settled by the geologist's hammer on the spot.

As far as the origin of the lunar maria are concerned, a similar uncertainty still persists. Many earlier students of the subject regarded them as solidified lava flows. In spite of the superficial resemblance of the surface features in lunar maria with what a lava flow on the Moon may look like, it is increasingly difficult to explain where the heat necessary for the melting of so much lava came from *at the time when they were formed*. The stratigraphy of the lunar surface giving us a clue to its time-scale will be considered in more detail in the next chapter. For the present, let us point out the obvious fact that, inasmuch as the surface of the maria contains records of fewer external impacts per unit area than the continental land masses, it must be younger – on a cosmic scale – than the continents. Yet it is extremely doubtful if the Moon could ever have been molten (let alone near the surface) except at the very beginning (say $10^7$ years) of its existence, provided that it initially possessed an adequate concentration of radio-active elements like $Al^{26}$ or other radioactive elements with half-lives of the order of $10^6$–$10^7$ years; and if the Moon ever melted in the interior as a result of radiogenic heat liberated by spontaneous disintegration of $K^{40}$ or other elements with half-lives of the order of $10^9$–$10^{10}$ years, it would be molten still now and would remain so several thousand million years hence. In order to escape this difficulty, UREY (1952, 1956, 1962) sought the origin of the lava in large-scale surface melting caused by low-velocity impacts of solid planetesimals; and the present writer (1959a) pointed out possible relevance of cometary impacts in this connection. On the other hand, GOLD's views (1955) seeking to explain lunar maria as large reservoirs of fine dust which has drifted to the lowlands from continental plateaus were much weakened by the evidence of Ranger photographs showing the existence of vertical differentiation on a meter scale on a typical mare ground (cf. Figure 23-2); while another view expressed recently by WILSON (1962) that mare ground possesses its particular characteristics as a result of extrusion (or admixture) of hydrocarbons ("asphalt lakes") exuded by gradual defluidization remains likewise still hypothetical.

Of such defluidization processes, by far the most important in this connection should be the gradual extrusion of *water* from the lunar interior. The reasons which lead us to expect that such extrusion may indeed take place we discussed already in the preceding section, when we pointed out that secular warming up of the Moon's interior by spontaneous disintegration of long-lived radioactive elements like $K^{40}$,

$Th^{232}$ or the two well-known isotopes of uranium is bound to expel water from solid hydrates and drive it – in the form of superheated steam – along any crack or fissure opening outwars.

Unless this argument contains some flaw (which is very difficult to imagine), there seems but little room for doubt that most part of the deep interior of the Moon (where the prevailing temperatures are in excess of 1000°) must now be rather completely desiccated, and its outer crust greatly enriched with water. But how far can this water penetrate to the surface? For the sub-surface layers of the Moon are again quite cold (how we know this will be explained in more detail in subsequent chapter 20) – some 35° C below the freezing point of water. Therefore, we should expect that, before reaching the surface, the hot steam seeping outwards will gradually condense into liquid and eventually form ice. Along crustal fissures, which can serve as ducts for more rapid escape, the formation of ice may in fact take place very close to the surface and give rise there to formations which can possibly reveal the presence of sub-surface lunar glaciers to the external observer.

It was recently suggested by SALISBURY (1961) and others that such sub-surface lunar glaciers (covered by dust and other debris) are observed by us as the lunar "domes" we described already earlier in the preceding chapter – formations which bear indeed more than a superficial resemblance to the "pingoes" found in Alaska and northern Canada, and which may indeed represent their terrestrial homologues; but a closer examination will be necessary before such a suggestion can be placed on a more secure basis. Another lunar formation which might be indicative of the presence of sub-surface moisture is the "wrinkle ridges". SALISBURY (1961) suggested recently that these ridges may have been produced by hydration of the beds of anhydrid minerals (such as olivines) – a process which is generally accompanied by an increase of volume just about sufficient to produce the observed bulging. But such a view represents again a hypothesis which is highly tentative so far; and further work (mainly on thermal radiation from such ridges) remains to be done before we can be more positive about its merits.

But coming back to water finding its way to the lunar surface – suppose that, in an extreme case, juvenile water escaping from the hot interior may reach the surface in occasional spurts as a geyser. No such phenomenon has, to be sure, so far been actually observed on the Moon (though it would be very difficult to detect, except under special circumstances). As we mentioned, however, before, certain of the largest lunar "domes" show some evidence of central depressions, (see Figure 16-17) which may indeed represent possible springs of such geysers. If hot water occasionally spurts through them into space, what should happen to it afterwards?

The exact answer depends, of course, on the actual temperature of the geyser and the altitude which it may attain. But there is little room for doubt that, should such an outburst occur during the lunar night (or in places shielded from direct sunshine) a good part of the squirting water would not evaporate, but rather would condense into ice. The mean lifetime of ice thus formed and exposed directly to the conditions of interplanetary space would depend on the prevailing temperature, which varies widely between day and night. While a piece of ice on the sunlit surface would not, in general, outlast a (lunar) day, the mean lifetime of ice in areas shielded from sunlight could, however, be surprisingly long.

In such areas, the prevailing temperature is around 120 °K; and at so low a temperature the rate of sublimation of ice into vacuum is known to be only about $3.1 \times 10^{-14}$ g/cm$^2$ sec, giving rise to a vapor pressure of $1.4 \times 10^{-12}$ mm of Hg. At this rate, only $4.6 \times 10^3$ g/cm$^2$ – or about 46 meters of ice – would have evaporated during the entire age of the Moon! WATSON, MURRAY and HARRISON BROWN (1961) estimate that a few per cent of the mountainous areas near the poles of the Moon remain permanently in shade; and over a somewhat larger area the Sun never rises totally above the horizon. Such areas are exposed to lunar night temperatures most of the time; and it is there that one can expect to encounter ice on the surface. Below the surface, at a depth of not more than a few feet, the prevailing temperature is much higher (close to $-35°$ C) and remains constant day and night. At such a depth the ice, shielded by the surface dust and intermingled with it, may persist for long periods of time as permafrost, several hundred feet deep, representing a fossil reservoir from which water may one day be tapped by human hand.

Moreover, water is not the only volatile compound on which attention should be focussed in this connection. Another group of compounds which should have been present in the primordial mass of the Moon in a finite concentration are the *hydrocarbons*. Moreover, heavy molecules of this type should have been thermally "cracked" and driven outside in the course of time. WILSON (1962) proposed a suggestion – supported by specific references to observed lunar features – that dark lunar flatlands are covered by residues of such hydrocarbon materials from which lighter molecules escaped into space (i.e., not unlike the vacuum-reduced crude oil or asphalt). May, perhaps, the plains of the lunar maria (or the floors of such craters as Archimedes or Plato) represent formations of this type, to which the well-known "asphalt lake" in Trinidad constitutes a terrestrial homologue? Even very small cosmic abundances of hydrocarbons – Wilson argues – could modify the lunar surface considerably (as the relatively minute amount of water modified the landscape of our Earth).

A temporary escape of carbon gas from the central peak of the crater Alphonsus, observed spectroscopically by Kozyrev and Ezerski in 1958 (cf. KOZYREV, 1959; 1962; KALINYAK and KAMIONKO, 1962) offers some circumstantial evidence in support of the view that hydrocarbons may constitute a significant component in at least certain parts of the lunar crust; for the $C_2$ molecules, whose luminescent spectrum Kozyrev observed in the form of the emission Swan bands originated almost certainly by photo-dissociation of heavier hydro-carbons under the influence of the UV-radiation from the Sun. The nature of these parent molcules can so far only be guessed at (cf. UREY, 1961). Yet the presence of at least traces of certain types of hydrocarbons in the Moon's crust seems thereby attested by more than theoretical speculation (cf. also KOZYREV, 1963); and, in the present state of research, its implications should continue to be kept in mind.

## CHAPTER 18

# LUNAR SURFACE AS AN IMPACT COUNTER, AND ITS STRATIGRAPHY

In the preceding two chapters we gave a descriptive survey of the principal types of formations encountered on the lunar surface, as well as a brief account of the processes by which such individual formations could have originated. The aim of the present chapter will be to consider now the *collective* aspects of the observed surface features, their distribution and density over different types of lunar ground; and to draw some conclusions from this evidence which may throw light on the "partition function" of different processes considered in the preceding chapter, or on the conditions prevailing in interplanetary space.

Earlier in this work we already stressed the fact that the origin of all features visible on the surface of our satellite must go back to either the internal processes, for which this surface represents a boundary condition; or to external impacts on the Moon of particles of all sizes – from micrometeorites to comets or asteroids – bombarding it from outside; for *tertium non datur*. In the present chapter, we wish to elaborate further this thesis and to follow up in more detail the consequences of an *assumption* that *the dominant process responsible for the shaping up of the lunar relief is the cratering by external impacts*. As the Moon is completely devoid of any atmosphere, and its surface directly exposed to impacts of all particles which happen to be obstructed by it in their heliocentric orbits, the bombardment by meteorites and other particles found in interplanetary space *must* demonstrably occur on the Moon. If so, however, the cumulative effect of such encounters – the fossil record of which has been preserved on the lunar surface with a permanence greatly transcending that of any geological markings on the Earth – should enable us to translate the observed size-frequency distribution of lunar craters into a space density (present as well as past) of bodies whose impacts would be capable of giving rise to the observed formations. In other words, what we propose to do hereafter is to regard the solid crust of our satellite as an astronomical impact counter, whose cumulative record is open to telescopic inspection at a distance, and to investigate the consequences of its interpretation on this basis.

An adoption of this point of view should not imply, to be sure, that the probability of parallel operation of other processes of internal origin in making their contribution to the formation of the lunar landscape is not recognised. In particular, we wish to record our present conviction that impacts may not have anything to do with some of the largest craters found in some of the oldest parts of the lunar surface – in particular, with pairs of quasi-hexagonal formations with confluent sides (like Ptolemy-Alphonsus on Figure 15–24). Impacts had also nothing to do with the origin of Wargentin (Figure 17-17); and most probably not with such hill-top craters as Regiomontanus A (Figure 16-6 or 17-9), or the central depressions of some of the larger domes (such as those seen on Figure 16-17 in the neighbourhood of the Cauchy rilles); for the probability of so central a hit of such a small formation is vanishingly small. Lastly, external processes are probably not basically responsible for such maar-like formations as disclosed inside the crater Alphonsus on photographs secured recently by Ranger 9 (see Figure 24-1). However, all these exceptions (and many others) constitute a very small minority of formations which can be distinguished on the lunar surface from the Earth, or even from closer proximity; so that, for statistical purposes, the ensemble of lunar craters can be treated with impunity as if they were all of impact origin – regardless of how wrong this may be in individual cases so long as these are relatively few.

Within the scope of these reservations, let us proceed now to inspect the statistics on the size-frequency distribution of lunar craters, and compare them with numbers to be expected from such independent knowledge as we possess on the density of solid particles of all sizes in interplanetary space at a distance of one astronomical unit of the Sun.

In doing so, several considerations of practical nature must be borne in mind. First, not all parts of the lunar surface lend themselves for such statistics with equal ease: in particular, the continents are too broken and irregular to permit crater counts down to a sufficiently small size. The mare ground is much more suitable for this purpose; but even there we are still very largely dependent on telescopic observations (visual or photographic) from the distance of the Earth. Such data permit complete crater counts to the size of 1 km on the lunar surface with ease, and down to 400–500 m in size with the aid of large telescopes under really good seeing conditions; the limit of ground-based detection for optimum conditions being close to 300 m. Below this limit, we possess so far only three glimpses – albeit of greatest value – of crater distribution down to 1 m size in two relatively minute portions of Mare Cognitum and Tranquilitatis obtained recently by Rangers 7 and 8; and down to a substantially smaller size (about 30 cm on the Moon) inside the crater Alphonsus as recorded by Ranger 9. Such data enable us to extend the telescopic crater statistics by at least three orders of magnitude in diminishing size; but so far only over a minute fraction of the lunar surface.

Secondly, in considering the use of the mare ground as an impact counter, we should keep in mind the possibility that not all maria are of equal age. Attempts at differential dating of the individual maria by their size-frequency crater counts (cf., e.g., KREITER, 1960; ÖPIK, 1960; DODD, SALISBURY and SMALLEY, 1964; HARTMANN, 1965) have, however, led so far to inclusive results; suggesting similarity between different maria which makes it possible to combine the crater counts over different maria into averages representative of the typical mare ground on the Moon.

That this averaging should be also dynamically legitimate as far as external impacts are concerned has recently been shown by TURSKI (1963), who proved that (at least for the present Earth-Moon configuration) the gravitational shielding of the

Moon by the Earth, which could produce unequal distribution of impacts over the
lunar surface, is too weak to be of any significance.

After these preliminaries, let us proceed now with the presentation of the actual
data as summarized in the following Table 18-1. The first column of this table lists
the (decimal) logarithm of the diameter $D$ of the respective formation in kilometers;
while the second and third columns contain the cumulative numbers $N$ of craters of

TABLE 18-1

Cumulative size-frequency distribution of post-mare lunar craters

| $\log D$ (in km) | $\log N$ (per $10^4$ km²) (after SHOEMAKER) | $\log N$ (per $10^4$ km²) (after HARTMANN) |
|---|---|---|
| 2.00 | − 2.5 | − 2.4 |
| 1.78 | − 2.0 | − 2.0 |
| 1.48 | − 1.4 | − 1.6 |
| 1.30 | − 1.0 | − 1.2 |
| 1.00 | − 0.4 | − 0.6 |
| 0.70 | + 0.2 | + 0.1 |
| 0.30 | 1.0 | 1.0 |
| 0.00 | 1.6 | 1.8 |
| − 0.30 | 2.2 | 2.5 |
| − 0.70 | 3.2 | 3.5 |
| − 1.00 | 4.1 | 4.3 |
| − 1.30 | 5.0 | 5.0 |
| − 1.70 | 6.4 | 6.1 |
| − 2.00 | 7.4 | 6.9 |
| − 2.30 | 8.4 | 7.7 |
| − 3.00 | 10.7 | 9.0 |

the corresponding size per unit area of $10^4$ km² of typical mare grounds, according
to the recent counts of SHOEMAKER (1965) and HARTMANN (1965). The results of these
two independent investigators are in a sufficiently close agreement to command confidence in their significance; and down to a size of approximately 1 km they can be
well represented by an empirical relation of the form

$$\log N = 1.6 - 2 \log D, \qquad \log D > 0; \tag{18-1}$$

while below the size of 1 km, the above relation should be replaced by

$$\log N = 0.3 - 3.6 \log D, \qquad \log D < 0. \tag{18-2}$$

The change in slope of this size-frequency distribution around $D \sim 1$ km is probably
due to the incipient preponderance of secondary craters: for sizes in excess of 1 km
secondary craters are too few to influence significantly the statistics; but below this
size they rapidly become so numerous as to dominate the counts based on the recent
Ranger data. An extension of the size-frequency distribution to the range of
$-1 > \log D > -2$, according to Hartmann, further unevenesses (cf. the accompanying

Figure 18-1) which may be due to the perishability (a lifetime shorter than that of the Moon) of very small craters. Neglecting these, however, Hartmann showed that all his data can be also reasonably well satisfied by a simple straight line of the form

$$\log N = 2.0 - 2.4 \log D \tag{18-3}$$

for the entire range $2 > \log D > -2$.

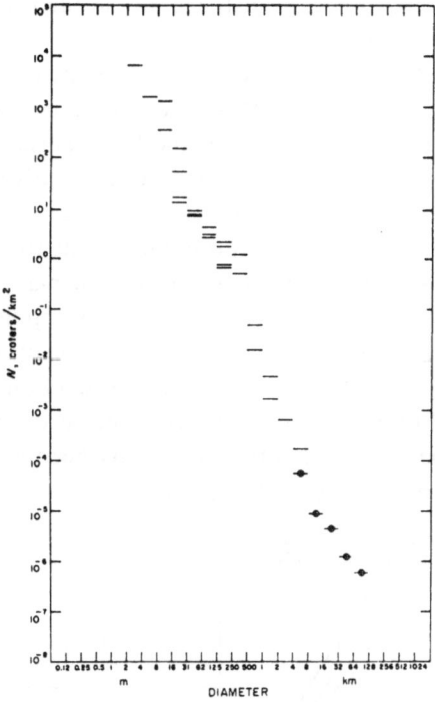

Fig. 18-1.   Cumulative size-frequency distribution of lunar craters (after HARTMANN, 1965).

The data summarized in Table 18-1 lend themselves to several interesting deductions. First, the frequency of lunar craters increases so rapidly with diminishing size that, below a certain limit, each element of the surface should have been disturbed by impacts. The limits for which this should be true will obviously be given by the equation

$$D^2 N = 10^{10} \text{ m}^2, \tag{18-4}$$

which for the data tabulated in Table 18-1 leads to a solution close to $D \sim 3$ metres. In other words, for constant flux of meteoritic bombardment, a contiguous cover of craters roughly 3 m in size should have been formed everywhere on the Moon during its long astronomical past; and still more numerous smaller craters should have been formed more than once in the same place, thus obliterating successively each other.

If further extrapolation of our data is of any significance, micro-craters of 1 cm average size should have been formed, destroyed, and created anew some 2000 times on each $cm^2$ of lunar ground – each having an average lifetime of 2 million years; while craters 1 mm in size (produced by impact of cometary micrometeorites weighing, on the average, $10^{-9}$ grams) the mean lifetime should be close to 100000 years.

Incidentally, the observed statistics reveal that if all visible craters of diameters in excess of 1 km originated by a uniform process lasting $4\frac{1}{2} \times 10^9$ years, new observable craters should accrue to the lunar surface at a rate of about one per 50 thousand years. No wonder, then, that no such instance has so far been established since the advent of telescopic astronomy!

Having surveyed the essential features of our present knowledge of the size-frequency distribution of craters on the Moon, let us set out to enquire whether the observed density of craters per unit area of the lunar surface is compatible – as it should on impact hypothesis – with known size-frequency distribution of solid bodies in interplanetary space. The ensemble of our knowledge of the present particulate contents of space at a distance of one astronomical unit from the Sun has recently been summarized by HAWKINS (1964) in the form of a mass-frequency distribution diagram shown on the accompanying Figure 18-2. If we multiply now the abscissae of this diagram by $4 \times 10^9 \times 10^4$, we are led to the data summarized in the first two columns of Table 18-2, listing the log of mass $m$ as a function of the logarithm of the cumulative number $N$ of impacts to be expected to occur over an area of $10^4$ $km^2$ in $4 \times 10^9$ years. It should be stressed that these data have been obtained by astronomical methods which have nothing to do with the Moon.

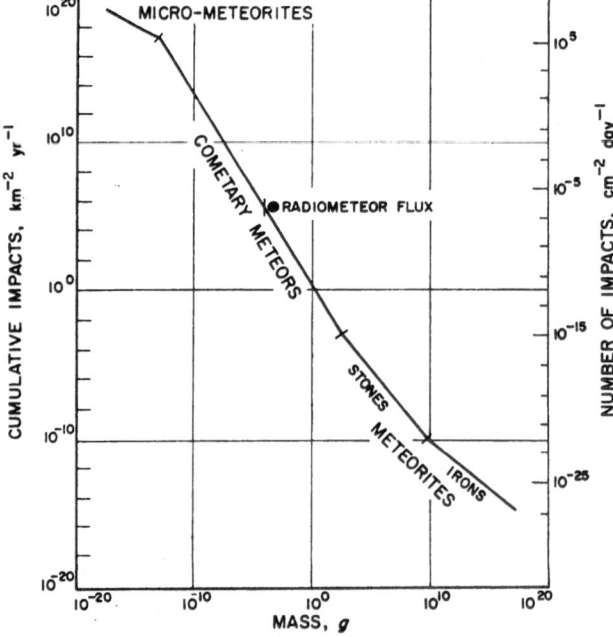

Fig. 18-2.   Frequency of meteoritic objects of different mass in interplanetary space (after HAWKINS, 1964).

TABLE 18-2

Frequency of meteoritic objects of different mass in space

| log $m$ (in grams) | log $N$ (per $10^4$ km$^2$ per $4 \times 10^9$ years) | log $D$ (in km) | | |
|---|---|---|---|---|
| | | $v = 2.3$ km/sec | 10 km/sec | 30 km/sec |
| 18 | $-$ 1.7 | 1.43 | 1.81 | 2.09 |
| 16 | $-$ 0.4 | 0.84 | 1.22 | 1.50 |
| 14 | $+$ 0.9 | 0.25 | 0.63 | 0.91 |
| 12 | 2.2 | $-$ 0.34 | 0.04 | 0.32 |
| 10 | 3.6 | $-$ 0.93 | $-$ 0.55 | $-$ 0.27 |
| 6 | 7.1 | $-$ 2.11 | $-$ 1.73 | $-$ 1.45 |
| 2 | 11.1 | $-$ 3.29 | $-$ 2.91 | $-$ 2.63 |
| $-$ 2 | 16.1 | $-$ 4.47 | $-$ 4.09 | $-$ 3.81 |
| $-$ 6 | 21.4 | $-$ 5.65 | $-$ 5.27 | $-$ 4.99 |
| $-$ 10 | 28.0 | $-$ 6.83 | $-$ 6.45 | $-$ 6.17 |

A possession of these data invites their comparison with the results recorded on the lunar "impact counter" of known size during the same interval of time; and the clue to it is our equation (17-38) of Hess and Nordyke, relating the kinetic energy of the impinging particle with the size of the resulting crater. This kinetic energy is, in turn, proportional to the mass $m$ as well as to the square of the velocity $v$ of impact; being expressible as

$$\log E = \log m + 2 \log v + 9.70 \qquad (18\text{-}5)$$

if $E$ is expressed in ergs; $m$, in grams; and $v$, in km/sec. A combination of Equations (17-38) and (18-5) then yields an equation of the form then yields an equation of the form

$$\log D = 0.294 \log m + 0.588 \log v - 4.07, \qquad (18\text{-}6)$$

relating statistically the diameter of an impact crater with the mass and velocity of the impinging body.

For primary impacts, this latter velocity may lie anywhere between 2.4 km/sec and 72 km/sec (i.e., the parabolic velocity of escape from the Moon and from the solar system, respectively); though non-cometary particles in retrograde orbits moving at a speed close to 72 km/sec with respect to the Moon are apt to be excessively rare. For the vast majority of impacts their relative velocity should be less than 30 km/sec. If we adopt, for the sake of illustrative example, the values of $v = 2.38$, 10, and 30 km/sec, the mass $m$ of column (1) in Table 18-2 can be converted by use of (18-6) into the corresponding values of log $D$ as listed in columns 3–5 of the same table; and a plot of log $D$ against log $N$ of column (2) is then diagramatically shown on the accompanying Figure 18-3.

On the other hand, the empirical data of Shoemaker and Hartmann from Table 18-1 on the relationship between $D$ and $N$, based on actual counts of the lunar craters, are plotted on the same figure as circles and open crosses, respectively. A glance at Figure 18-3 reveals that these purely lunar data fit in satisfactorily with theoretical expectations, based on the known density of particulate contents of interplanetary space, in the entire size or mass range covered by available observations. If the craters which have been so counted (or, at any rate, a large majority of them) were not of impact origin, the fact that all observed points follow so closely the theoretical curves would have to be regarded as fortuitous.

Moreover, all observed points appear to lie between theoretical curves corresponding to $v = 2.4$ and 10 km reveals that *a large majority of impacts of solid particles*

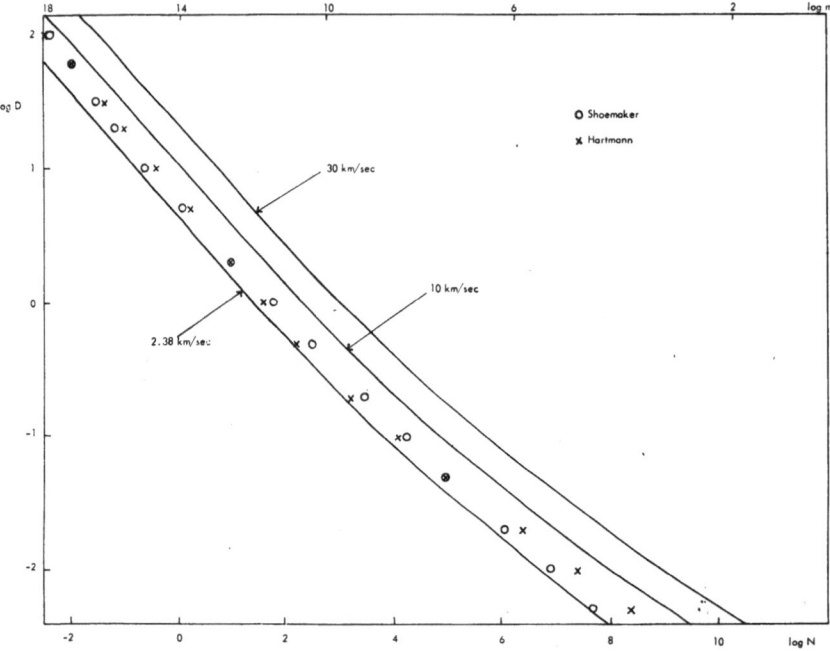

Fig. 18-3.  Predicted size-frequency distribution of lunar craters from the known density of the meteoritic objects in space, and its comparison with the observations.

*of all sizes on the lunar surface must have occurred with relatively low velocities* – between 2.5 and 5 km. This view has long been advocated by UREY (1952 and later); and the outcome of the present analysis bears out this contention in a very convincing manner. However, another feature which likewise transpires from our diagram is less satisfactory: namely, the fact that the size-frequency distribution of lunar craters, as counted by Shoemaker or Hartmann on the Ranger material down to 10 metre-size can apparently be accounted for by *primary* impacts of the particles of requisite size, without leaving much room for secondary craters.

However, down to a size of approximately 1 km, *the mean areal size-frequency distribution of lunar craters in the maria is quite consistent with the number of impacts to be expected for the present density of solid particles in space*; and that in order to explain the origin of a large majority of lunar craters on impact hypothesis, *the space density of solid debris need not have been any higher than now since the formation of the maria.* If this density has been appreciably higher in the past, the number of craters on the Moon should be greater than is actually the case. This is, in particular, true of small craters; for if Shoemaker and others are right in their belief that the large majority of craters less than one kilometre in size are secondaries, it would appear that Hawkins's abundance of particles capable of producing on impact primary craters of the same size may have been overestimated.

This is, however, not the place to pursue this problem in further detail; but we wish rather to return to the general problem of the history of lunar surface features. Can we, in particular, fit the origin of such features into some kind of *stratigraphic time-sequence*? There are, indeed, three different and independent (though indirect) indications which can be invoked to this end: namely, (a) the principle of overlap; (b) the degree of ruggedness; (c) the ground reflectivity; and in what follows we shall discuss them in turn.

Of the three, the "principle of overlap" is perhaps the most obvious and dependable. If there are two craters that overlap each other – and the photographs reproduced in this article show many examples of such a situation – then the one with unbroken rim must be more recent than the one whose rim was damaged or entirely removed. In the case of two overlapping craters of comparable dimensions – such as the pair of Theophilus and Cyrillus on Figure 16-8, for instance – an application of this principle requires the formation of Theophilus to have been posterior to that of Cyrillus. But the principle can be applied also to a situation in which a large crater contains smaller ones within its enclosure (for an illustrative example, cf. a photograph of the crater Clavius reproduced on Figure 16-5). No kind of process that raised the ramparts must represent creations subsequent to that of the large configuration. On this reasoning Clavius must, therefore, be considerably older than all other craters which can at present be seen on its floor. The greater the number of small craters inside a large one, the greater should be the disparity between their ages.

Crater overlaps of great multiplicity can be found in certain part of the lunar surface; and in some places it is possible to arrange thus five or six craters in a time sequence. Another relative age criterion (supplementing overlap) is afforded by the existence of the streaks of relatively bright material (the "rays") which are seen diverging in all directions from certain impact craters. As these rays must have originated by ejection at the same time as the parent crater, the latter must obviously be younger than any features overlaid by its rays. Such rays represent, indeed, a system of tentacles spreading widely over certain parts of the lunar surface and enabling us to extend our system of relative dating as far as they reach.

The main importance of the preceding age criteria lies, however, in their application to the dating of the maria. If we accept the foregoing premises, there seems no escape from the conclusion that *the oldest parts of the visible lunar surface are those which are most rugged*, and contain the greatest number of craters or other types of mountains per unit area. For irrespective of whether the rate of operation of both the external or internal processes of crater formation on the Moon has been uniform or diminishing with time, the oldest parts of its surface should obviously have accumulated the greatest number of scars. If this is indeed so, then the oldest parts of the visible surface of the Moon are no doubt the regions surrounding its south pole, which may indeed contain a continuous record of events that have affected it since the time of the formation of our satellite (or, at any rate, since the solidification of its crust). But as the mean density of craters in most of the great dark plains of the lunar surface is much less than that encountered near the south pole, it follows that the maria should be younger than the mountainous continental regions; and some of the great craters – like Copernicus, Kepler, or Aristarchus – which spread the tentacles of their bright rays over large parts of the surrounding maria, must be younger still.

In point of fact, the brighter any element of the lunar surface, the more recent its present relief is likely to be; for the gradual infall of interplanetary dust alone is bound to darken the ground and lessen the differences of its reflectivity in the course of the time. This process has, in turn, an interesting application to the chronology of impact craters. As we mentioned already before, many craters of this group are the foci of prominent ray patterns, but others are entirely unaccompanied by rays. Thus Eratosthenes (Figure 18-4) is a good example of a crater that exhibits all the principal topographic features of Copernicus and is surrounded by a well-developed pattern of gouges, but completely lacks rays. Moreover, where not overlapped by the Copernican ejecta, the walls as well as the floor of Eratosthenes exhibit relatively low reflectivity.

All gradations may be observed in the apparent brightness of the rims and asso-

Fig. 18-4.   The craters Eratosthenes (left) and Copernicus (right), photographed with the 200-inch Hale telescope of the Palomar Observatory. The "ghost" crater Stadius is in the middle.

ciated rays among craters of presumably impact type. Copernicus, Aristillus and Theophilus represent a sequence of craters accompanied by rays of diminishing reflectivity. The rays of Aristillus (Figure 18-5) are plainly visible, but not as bright as those of Copernicus; the rays of Theophilus are already very faint, though its second-

Fig. 18-5.   Eastern part of Mare Imbrium, containing the craters Archimedes, Aristillus and Autolycus. Photographed with the 120-inch reflector of the Lick Observatory. White cross marks the position of the impact of Lunik 2 on September 13, 1959.

ary impact craters (cf. Figure 17-15) are as widely distributed and as numerous as those of Copernicus. The reflectivity of the walls of Theophilus approaches that of Eratosthenes.

It is highly probable that this sequence reflects the increasing age of the respective formations. Wherever a Copernican-type crater without rays (or a crater with very faint rays) occurs in an area traversed by rays from some other crater, the bright rays are in all cases superimposed on the darker crater or the fainter ray pattern; no single instance of a converse case if known. Some process, or combination of processes, must then evidently be at work on the lunar surface that causes the gradual fading of the rays and other surface elements of higher reflectivity. The infall of dark grey cosmic dust is indubitably one – and probably the most important one – in the course of time; but darkening of the material itself by radiation damage and mixing of the thin layer of ray material with underlying darker base by micrometeoritic bombardment may well contribute to the general process of fading.*

Extensive reconnaisance studies of the *stratigraphy of the lunar surface* undertaken in recent years by Shoemaker and his colleagues in the United States (cf. SHOEMAKER et al., 1961, 1962) on the basis of different superposition principles just described, led to a gradual realization that – like on the Earth – the surface of the Moon is built up of an overlapping series of deposits, constituting a definite succession from which the relative sequence of events in the history of the lunar surface can be determined. In essence, these deposits can be grouped in the following five main stratigraphic systems: (1) Pre-Imbrian, (2) Imbrian, (3) Procellarian, (4) Eratosthenian, (5) Copernican, named so after their prototypes, or the principal regions where such strata can be seen widely exposed; and corresponding to five consecutive intervals of time.

The *Pre-Imbrian* (or Archaic) sculpture on the Moon can best be seen in the large mountainous regions surrounding the lunar south pole, and reaching in places quite far towards the equator. Thus the impressive group of large craters near the center of the apparent lunar disc – Ptolemy, Alphonsus, Arzachel and others – are clearly located on pre-Imbrian ground; and so are all the other numerous and large craters (Clavius, Deslandres) to the south of them. It is this pre-Imbrian period which we should identify with Puiseux's "l'époque plus ancienne où nous puissions remonter...". Over the northern hemisphere of the Moon (and in the west) the pre-Imbrian strata have been largely overlaid by layers deposited subsequently in four more recent periods, from the Imbrian to the Copernican; but imprints of the archaic layers protrude through many of the younger deposits (like, for instance, the Carpathian mountains in the neighbourhood of the crater Copernicus), or are only partly submerged by them.

The beginning of the *Imbrian* period can be defined as the time when the lowest strata of the Mare Imbrium region were deposited. In contrast with the rugged archaic structure, the Imbrian system is characterized by gently rolling topography, deposited on a surface of considerable relief including ridges, valleys and craters in strata which may be from several hundred up to a few thousand meters thick where they fill pre-existing depressions, but are evidently quite shallow where they cover underlying rough ground – as witnessed by the numerous "ghost craters" (an example of which can be seen on the northern slopes of Aristillus on Figure 18-5, and also on Figure 18-8) and isolated half-submerged peaks abundant in this region. The Imbrian strata are largely found in the north-west quadrant of the apparent lunar disk, and appear to thin out rapidly south of the Copernicus region to expose the pre-Imbrian strata predominant over much of the entire southern hemisphere.

Incidentally, these "ghost craters" – which are very numerous in the region of Mare Imbrium and represent nothing else than rims of the ramparts of ancient craters

---

* It may be noted, in this connection, that the albedo of interplanetary dust constituting the zodiacal cloud, which is constantly exposed to solar radiation damage at a much closer range, is only about 0.02 – i.e., more than three times less than the mean albedo of the Moon, and twice less than that of the darkest spots on the surface of our satellite.

protruding from subsequent deposits – provide us with the means for gauging the actual depth of the overlying strata from the measured dimensions of such ghosts. In the preceding part of this chapter we mentioned that the diameters of typical craters are statistically correlated with the heights of their ramparts. If we now assume that the submerged craters conformed to the same relation at the time of their form-

Fig. 18-6.  Sunset over the crater Plato and the Teneriffe mountains in the northern part of Mare Imbrium. Photograph taken with the 24-inch refractor of the Observatoire du Pic-du-Midi (Manchester Lunar Programme).

ation, a difference between the present height of their ramparts and that corresponding to their visible dimensions should represent the depth of the overlying strata. This depth ranges, in general, from a few hundred meters up to one or two kilometers in the Imbrian system – i.e., is very shallow in comparison with the lateral extent of these strata.

In addition, the Imbrian system contains other craters of considerable interest in this connection – such as Archimedes (Figure 18-5) or Plato (Figure 18-6) – which can be legitimately call "half-submerged"; for while their visible ramparts possess the characteristic hummocky structure of impact formations, their floors are smooth, dark, and resembling in most respects the surrounding mare ground. The possibility

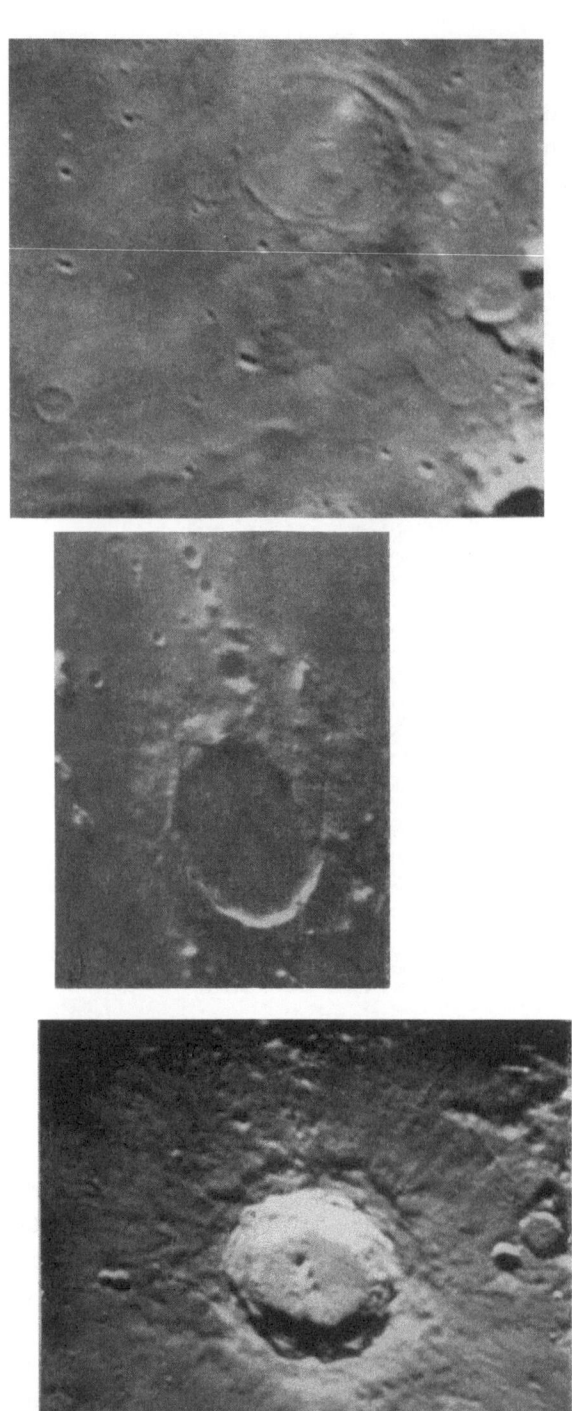

Fig. 18-7. Time sequence of the evolution of lunar craters. Left: Copernicus – an impact crater of relatively recent origin; central peak and hummocky structure of the walls well preserved. Center: Plato – crater with flooded floor, resembling the surrounding maria; central peak absent. Right: ghost craters Daguerre and others in Mare Nectaris, of pre-marial origin, for which not only the floors, but also ramparts are partly submerged.

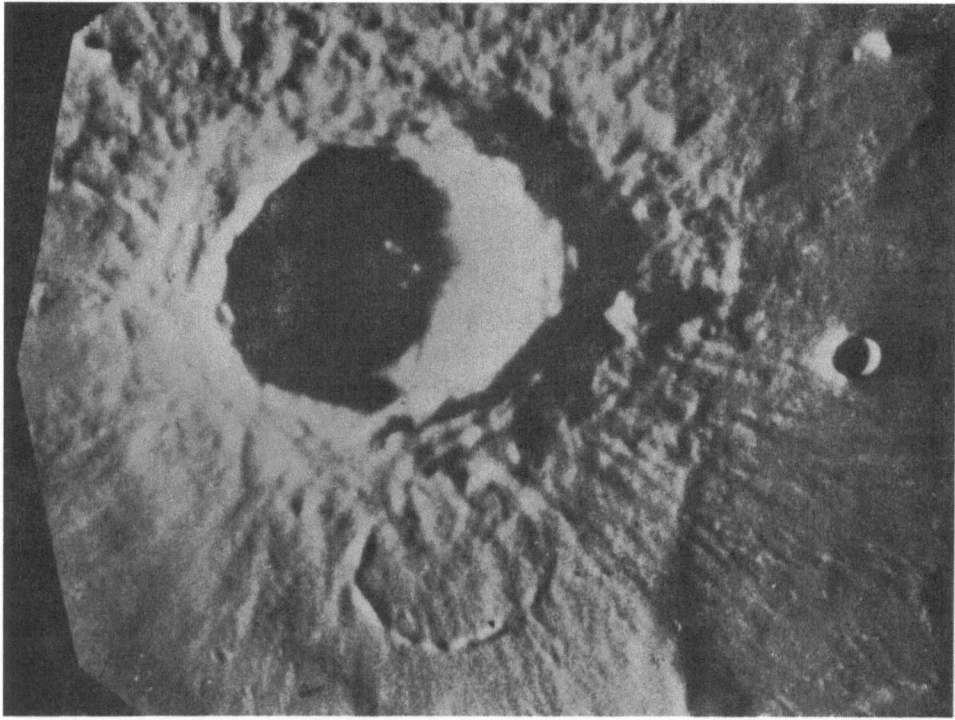

Fig. 18-8.  A closeup view of the crater Aristillus with a ghost ring on its northern slopes, photographed by Bernard Lyot on 21 March 1945 with the 24-inch refractor of the Observatoire du Pic-du-Midi.

that their interiors were flooded by mare material obtrudes persistently on one's mind; but it seems equally clear that the external material did not get to the interior across the walls of the crater which remain unbroken. It seems, therefore, probable that craters like Archimedes or Plato are (like the typical "ghost craters") of pre-mare origin; while the neighbouring Aristillus (Figure 18-5) represents a post-mare formation. A remarkable photograph of an ancient "ghost crater" preserved on the northern slopes of Aristillus is reproduced above on Figure 18-8.

The *Procellarian system* rests stratigraphically on the Imbrian, and forms the relatively smooth dark floors of the Oceanus Procellarum, Mare Imbrium, or Sinus Aestuum. Its topographic relief is smoother than that characteristic of any other period of evolution of the lunar surface; its typical landmarks being the "wrinkle ridges" described already in Chapter 16; and the overall depth of the Procellarian strata (judging from the extent and distribution of partly buried pre-Procellarian features) is probably nowhere greater than a few thousand meters.

The *Eratosthenian system* is represented by the deposits surrounding the crater Eratosthenes (Figure 18-4), Reinhold, Landsberg, and a number of others, together with the material which covers their floors; but the most extensive sheet of such deposits appears to be associated with Eratosthenes – hence their name. Eratosthenian deposits rest on the Imbrian and Procellarian systems, and locally even on pre-Imbrian strata; a pattern of numerous gouges on the Procellarian in both Mare Imbrium and Sinus Aestuum on either side of Eratosthenes shows conclusively that ejecta from this crater are superimposed on the Procellarian. The reflectivity (albedo) of the Eratosthenian ground is somewhat higher than that of the Procellarian (which represents the darkest of all the deposits), but not as high as that of the subsequent Copernican strata; and its depth does not seem to exceed anywhere a few hundred meters.

The Eratosthenian period came to its end at the time of deposition of the faintest "rays" that can be distinguished by their higher reflectivity on the underlying darker background, and which have not yet faded out completely since. Implicit in this definition is an assumption that all impact craters formed during the preceding periods (through the Eratosthenian) were once foci of bright-ray systems, which gradually weakened and disappeared in the course of the time. The emergence of such bright rays heralds the commencement of a new age – the *Copernican period* – which extends up to the present. Its principal characteristics are the large craters of the class of Copernicus and Tycho, Aristarchus and Kepler, together with a host of smaller formations of the similar type, whose ejecta – both bright and dark – cover large areas in the far neighbourhood of such craters, superposed upon layers of all preceding periods.

An approximate idea of the *absolute time-scale* of this tentative system of lunar stratigraphy (and its comparison with the corresponding geological time-scale on the Earth) may be obtained by a comparison of the areal density of impact structures on the lunar surface with the expected rates of celestial bombardment. HACKMAN (cf. SHOEMAKER, 1962) examined the frequency of craters thought to be of impact origin, distinguishable on the vast planes of the Procellarian deposits, and found that those exceeding 1 km in size occur in a density ranging from about 0.24 per 1000 square kilometers in Mare Crisium to 0.53 per $10^3$ km$^2$ in Mare Nubium, and average about 0.45 per $10^3$ km$^2$ for all the readily visible mare surfaces. If we suppose that the Procellarian period ended early in the history of the Earth-Moon system, the mean rate of impact capable of producing observable craters should, accordingly, be about one per 1000 km$^2$ per $10^9$ years.

This rate may be compared with the one calculated from the areal density of known probable impact structures in the terrestrial strata, the absolute age of which can be determined by radioactive dating. In the region of the central United States and Canada (an area geologically well explored and favourable for the recognition of impact structures), it is found to amount to about 0.01 per 1000 square kilometers in the past 300 million years. This rate represents probably an underestimate, because our knowledge of the geology of this region is still far from complete, and also because of the loss of information due to erosion. Account must also be taken of the fact that impacts on the Earth should be more numerous than those on the Moon, because of the stronger gravitational attraction of our planet. Nevertheless, when all these factors

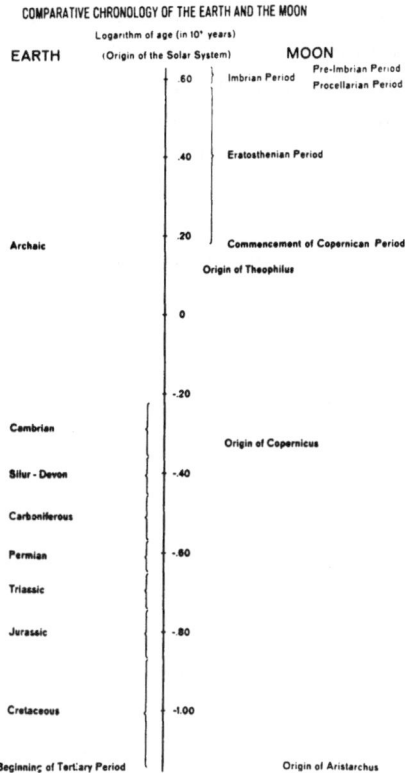

Fig. 18-9.   Comparative Chronology of the Earth and the Moon.

and uncertainties are considered, the data on hand appear consistent with a hypothesis that *the end of the Procellarian period is not far removed from the beginning of the geological time on the Earth*, and that the rate of impacts has been fairly constant since that time. The Procellarum period itself may have been relatively short; but considerable evidence indicates that the preceding Imbrian period occupied a significant interval of time in the entire history of our satellite.

If the rate of impacts has indeed remained steady since the end of the Procellarian period, we may attempt to estimate on this basis the absolute age of certain craters of the Eratosthenian or Copernican periods which are sufficiently large to have become targets for subsequent impacts. For example, within the ramparts of Copernicus, in an area of some 50000 km², there are at least two recognizable craters of probable impact origin; and on a rate of one crater per 10000 km² per thousand million years, the age of Copernicus should be of the order of 500 million years. In terms of the terrestrial geologic time-scale, this would place the origin of Copernicus in the early Paleozoic (cf. Figure 18-9). Thus it would have been the dim and uncomprehending eye of the early trilobites and their contemporaries, rather than of the great reptiles of the Mesozoic age, that could have beheld this stupendous event; and of other larger craters, only Aristarchus may have been formed within the Tertiary times. But, need-less to say, a determination of more exact dates will have to await manned exploration on the spot in the (hopefully) near future. The actual beginning of the Copernican time (i.e., the age of its faintest rays) may well antedate the terrestrial pre-Cambrian; for the fading of the rays is a very slow process. Nevertheless, even so it is hard to escape the conclusion, by dead reckoning from both ends of our time-scale, *the interval which elapsed between the end of the Procellarian period and the commencement of the Copernican era – i.e., the Eratosthenian time – occupied somewhat more than a half of the entire age of our satellite.*

# BIBLIOGRAPHICAL NOTES

## Chapter 13

A presentation of the stereoscopic method for the determination of the form of the lunar surface, as given in this chapter, follows largely GOUDAS (1965a); cf. also GOUDAS (1966a). For a similar harmonic analysis of the shape of the Earth, cf., e.g., PREY (1922), VENING MEINESZ (1959), HOF-SOMMER et al. (1959). Of other earlier references, cf., SAUNDER (1900, 1901, 1905).

Concerning the methods and results of a determination of the lunar limb profiles, further information can be found in CHEVALIER (1917), JOKSCH (1957), GAVRILOV (1959, 1961), POTTER (1960), BROCKHAUS and JOKSCH (1960), BROCKAMP (1960), POTTER and BYSTROV (1962), GORYNIA and DROFA (1962) or GOUDAS (1965b, 1966a).

For a determination of the shape of the Moon from measurements of the terminators, cf., MAINKA (1901), RITTER (1934), YAKOVKIN and BELKOVICH (1935) or HOPMANN (1964).

## Chapter 14

The definition of the lunar coordinates as introduced in this Chapter goes back to Tobias MAYER (1750); who together with Schröter and Lohrmann can be regarded as fathers of scientific selenodesy. The choice of the crater Mösting A as the fundamental zero point of lunar coordinates is due to BESSEL (1839).

The detection of vertical irregularities on the surface of our satellite belongs among the first telescopic discoveries of Galileo GALILEI (1610), who was also the first to attempt estimates of the heights of lunar mountains from the distance at which they become sunlit beyond the terminator. Needless to stress, Galileo was in no position to perform actual measurements with his rudimentary perspicill; and the altitudes assigned by him to some (unidentified) peaks – rendering them rivals in height of our Mount Everest – represented gross overestimates of the actual situation, as was pointed out only a little later by HEVELIUS (1647).

The first investigator actually to measure the extent of the visibility of individual lunar peaks beyond the terminator was William HERSCHEL (1780), using a micrometer at his 6-foot telescope magnifying 222 times. Although Herschel customarily exaggerated the precision of his micrometric measurements (listing them to $0''.001$, while their actual errors must have been several hundred times as large), he was correct in a realization that the lunar peaks are, in general, much lower than was thought by Galileo or even Hevelius; the majority of them being "between $\frac{1}{2}$ and $1\frac{1}{2}$ miles in height".

Herschel's work was soon followed by SCHRÖTER (1791, 1802) and, in the 19th century, by BEER and MÄDLER (1837) and SCHMIDT (1878), who abandoned the Galilei-Herschel method of watching for a beyond-the-terminator appearance, and set out to determine the relative altitudes of the lunar mountains from the observed lengths of their shadows cast by the individual peaks on the surrounding landscape at the time of lunar sunrise or sunset. The geometrical basis of this method had been credited to OLBERS (see, e.g., GRAFF, 1901); while Beer and Mädler together with Schmidt, have provided (from observations with their modest-size telescopes) the bulk of the data on lunar altitudes available until the inception of the USAF-Manchester lunar mapping work in 1959.

In more recent times, the Olbers technique, developed further by GRAFF (1901), MACDONALD (1929, 1931, 1932, 1940) and CROSS (1954, 1955) of FIELDER (1958) was adapted to photographic cine-technique by MCMATH, PETRIE and SAWYER (1937). This method has since been exhaustively elaborated by Kopal and his collaborators of the Manchester Lunar Programme (cf., e.g., KOPAL et al., 1961;

KOPAL, 1959c; 1960a, 1961b,c, 1962d, 1963e; KOPAL and RACKHAM, 1962; RACKHAM, 1962; SUDBURY, 1965; JONES, 1965) who have jointly brought its "state of art" to the level at which it is presented in this volume. One particular contribution of Manchester astronomers to this field has been their recognition and appropriate treatment of the penumbral phenomena on the Moon. The same is true of the method of reduction of the shadow observations made from abroad a spacecraft approaching the lunar surface, as presented in this section; of the use of "over-exposed" terminator photography, etc.

Of subsequent work in this field, c.f., e.g., POHN, MURRAY and BROWN (1962), POHN (1963), ARTHUR (1963), HOPMANN (1963) etc. A programme for an automatic transformation of the celestial and lunar coordinates by means of an electronic computer has been published by WILDEY (1964).

## Chapter 15

Several recent historians of science (e.g., HOUZEAU in his *Vade-Mecum de l'astronome*; or WOLF in *Handbuch der Astronomie, ihrer Geschichte und Literatur*) referred to the existence of drawings of the lunar surface in the volume *De Phenomenis in Orbe Lunae, etc.* by Iulio Caesare LA GALLA (Venice 1612) which would be second in age only to those of Galileo. In actual fact, however, none of the copies of La Galla's book preserved in the Bibliothèque Nationale de Paris contains any lunar maps; nor do the copies extant in the National Library of Florence. However, according to MAFFEI (1962), the copy in possession of the National Library of Rome contains drawings of the Moon attributed to La Galla – but these prove to be identical with those published previously by Galileo in his *Nuntius Sidereus*! The presumed drawings by La Galla are, therefore, in reality those of Galileo.

The same conclusion has been arrived at also by EMANUELLI (private information) who pointed out that the publisher of La Galla's work was the same Tommaso Baglioni, in Venice, who published Galileo's *Nuncius* two years before. It is, therefore, probable that the lunar drawings (obviously Galileo's) were inserted in at least a part of the edition of La Galla's book (otherwise lacking any illustrations altogether) by the publisher himself, perhaps in order to increase its attractiveness for the reader.

Of other contemporary drawings of the Moon, those of MALAPERT – based on the observations made on November 29, 1619, and brought to light in 1910 by P. Bosmans – are in no way superior to the previous work of Galileo or Scheiner.

The lunar maps reproduced in RICCIOLI's *Almagestum Novum* were not the actual work of the author of that book, but rather of his Jesuit confrère P. Francesco GRIMALDI. Note the delightful inscription on the top of his "Figure pro nomenclatura et libratione lunare", reproduced on our Figure 15–11: "Nec homines Lunam incolunt, nec Anime in Lunam migrant" – an observation in which Riccioli was well ahead of many of his successors.

For earlier histories of lunar mapping cf., e.g., KOPAL (1962) and MAFFEI (1962).

## Chapter 16

The literature devoted to a description of the prolific range of formations and markings observable on the lunar surface is truly enormous (in many languages); and any attempt at a compilation of even partial bibliography of it would fill many pages of this book. As, however, a large part of such a literature of older vintage is completely out of date, and of interest to the historian rather than to a more critical reader of the subject, anyone desirous to peruse it must do so at his own risk.

Most ground-based lunar photographs illustrating different types of lunar surface formations in this section have been secured with the 43-inch reflector and the 24-inch refractor of the Observatoire du Pic-du-Midi, in the course of a collaborative programme in lunar photography between the University of Manchester and the United States Air Force.

Of investigations of the far side of the Moon, cf. LIPSKI (1960, 1962, 1963, 1965), BARABASHEV, MIKHAILOV and LIPSKI (1960), BREIDO and SHCHEGOLOV (1962), MARKOV (1962), MARKOV and SHCHEGOLOV (1963), with comments by KATZ (1960) or WHITAKER (1963).

## Chapter 17

A more comprehensive presentation of the crater-forming processes by external impacts than that given in this section has been made by SHOEMAKER (1962) in a contribution which has since become standard, and which contains an almost exhaustive list of references to previous literature. A similar (though in places less critical) presentation of the case for the internal origin of lunar formations has recently been summarized by sixty-one different authors in the 'Geological Problems in Lunar Research', published as Vol. **123** (pp. 367–1257) of the *Annals of the New York Academy of Sciences* in 1965.

Of individual investigations of more recent date concerned with the mechanism of the origin of the lunar craters by impacts of external bodies cf., e.g., DALY (1946), DIETZ (1946), BALDWIN (1949, 1963), KUIPER (1954, 1955, 1959), UREY (1955, 1956b,c,d), GILVARRY and HILL (1965a,b) ALTER (1956, 1957, 1958), GILVARRY (1960), LEVIN (1964), JEAN-PIERRE (1964), KVÍZ (1964), CROSS (1965), SALISBURY, SMALLEY and RONCA (1965), SIDA (1965), BRINKMANN (1966), MILLER (1966), and many others.

For more specific aspects of the mechanism of crater formation cf., e.g., STANYUKOVICH and FEDYNSKY (1947), PATRIDGE and VAN FLEET (1958), *et al.* Of investigations concerned with empirical size-energy relation for impact craters cf. BALDWIN (1949, Appendix D; or 1963, Chapter 8); HILL and GILVARRY (1956); HESS and NORDYKE (1961); NORDYKE (1962); HAWKINS (1963); etc. For the topographic profile of the impact formations cf., e.g., GOLD (1955); or, more recently, FULMER and ROBERTS (1963), ROBERTS (1964), CARLSON and JONES (1965), etc. Investigations of the displacement and loss of mass accompanying the formation of primary impact craters have been undertaken by WARNER (1961) or GILVARRY (1964).

Collisions of the Moon with the comets were first considered by KOPAL (1959a) as a possible explanation of the origin of certain types of flat-floored craters (like Plato) or maria; and, more recently, KUIPER (1965) as well as UREY (1965) invoked the same mechanism to account for the origin of ray-craters. But all these tentative identifications of differences between meteoritic and cometary impacts must still be viewed with considerable reserve.

The problem of the origin of the lunar bright rays, and the ballistics of their ejection, have in more recent years been considered by HACKER and STEWART (1935), LENHAM (1955), ALTER (1955), O'KEEFE (1957), GIAMBONI (1959), FIELDER (1961, 1962), DEVADAS (1962) and, most recently, by KOPAL (1966) in connection with the Tychonic ejecta photographed by Ranger 7 (cf., Figure 17-16). This latter work has shown that such ray material could have been deposited only by ejection in nearly circular orbits grazing the lunar surface.

Of recent literature concerned with descriptive properties and possible origin of lunar rilles, domes or wrinkle ridges cf., e.g., FIELDER (1960, 1962a), SALISBURY (1961), RAE (1963), CAMERON (1964), PITHER (1964), BRUNGART (1965) or QUAIDE (1965).

For diverse views of the nature of lunar maria expressed in recent years cf., e.g., GOLD (1955) with comments by UREY (1956d), KUIPER (1954, 1959), UREY (1956b,c), GILVARRY (1957, 1958), KOPAL (1959a), WARNER (1961a,b), WILSON (1962), NASH (1963) or FIELDER (1963). This latter paper errs, however, in its basic assumptions; for the Moon could not have been heated to its melting point by the action of long-lived radioactive elements in the past – if it vere had been so molten, it would be so at the present, and would continue to grow hotter still for at least $3 \times 10^9$ years in the future (cf., KOPAL, 1962a).

## Chapter 18

For recent investigations of the particular contents of space, and the mass-distribution of solid particles at a distance of one astronomical unit from the Sun, cf., PIOTROWSKI (1953), ÖPIK (1956, 1958), UREY (1960), BROWN (1960), HAWKINS (1960, 1963, 1964) or HARTMANN (1965).

For statistical studies of crater distribution in different parts of the lunar surface cf., ARTHUR (1954), ÖPIK (1960), KREITER (1960), McGILLEM and MILLER (1962), DODD, SALISBURY and SMALLEY (1963), PALM and STROM (1963), BALDWIN (1964, 1965), MARCUS (1964, 1965), RONCA (1965) *et al.* The latest contribution to such studies by FIELDER (1965) contains conclusions biased in favour of lunar volcanism which, as was shown by MARCUS (1966), are unwarranted by the data on which they are based.

Fundamental papers on absolute dating of lunar stratigraphy, based on a combination of the relevant lunar and interplanetary data, are those by SHOEMAKER, HACKMAN and EGGLETON (1961), SHOEMAKER (1962, pp. 347–348), SHOEMAKER and HACKMAN (1962) or EGGLETON and MARSHALL (1963).

PART FOUR


RADIATION OF THE MOON

# INTRODUCTION

The fourth and concluding part of the present volume will be devoted to a topic which we have largely avoided so far in previous parts of this book: namely, to a discussion of the *radiation* received from the lunar surface in diverse parts of the spectrum, of its various properties and of their bearing on the specification of the physical structure of the lunar surface from its top-most layer down to a depth of many metres.

Whenever astronomers wish to study the properties of any remote celestial body inaccessible to direct approach, all information we wish to obtain must reach us across the intervening gap of space through two (and only two) different channels: the effects of gravitational attraction of their mass, and their radiation. The gravitational attraction governs the motion of the celestial bodies in accordance with certain laws of far-reaching exactitude which are well known and understood; and the consequences of their application to the Earth-Moon system have already been discussed in the first part of this book. On the other hand, the laws governing the light emission, or the interaction with matter (absorption, scattering, etc.) are much more complex than those controlling the motion of celestial bodies; and, in addition, the measurements of the intensity or spectral distribution of the radiant energy are very much less accurate than those of the time or position underlying dynamical astronomy.

As a result, any information which we may hope to extract from an analysis of moonlight will be bound to be more limited in accuracy than the properties of the lunar globe which we deduced from the motion of the Moon. As, however, this new avenue of approach is bound to disclose a wide range of entirely new facts we wish, in what follows, to outline in some detail the ways in which such information can be obtained. Thus Chapter 19 which follows these introductory remarks will be concerned with a brief survey of the photometric information on the Moon which reaches us through the conventional optical window (between the wavelengths of 2900 Å and approximately $1\mu$) of atmospheric transparency. As we shall see, this component of the total moonlight is dominated by sunlight incident on the lunar surface and *scattered* from it by a process which proves to be very largely independent of the frequency. The Moon is, however, on the whole a pretty poor reflector; and less than 10 per cent of incident sunlight gets scattered from it in this way. The rest (i.e., more than 90 per cent of it) gets absorbed, and re-emitted, as *thermal radiation*, observable to us on Earth through the infra-red atmospheric window between $\lambda = 8$–$12\mu$, and again in the microwave domain for $\lambda > 1$ mm; the properties of this thermal radiation of the Moon and deductions which can be drawn from it will be discussed in Chapter 20. Chapter

21 will then be concerned with a survey of the *electromagnetic properties* of the lunar surface, as deduced from its microwave radiation as well as its ability to reflect radar waves. The next section will then summarize our present knowledge of another – and not insignificant – source of transient non-thermal radiation of the Moon but recently discovered: namely, the *luminescence* of its surface. In Chapter 23 an attempt will be made (which, we feel, is no longer premature) to integrate our present knowledge obtained through all these different spectral channels into a consistent model of the lunar surface layers; and the concluding Chapter 24 will then contain a summary of our present knowledge of the Moon – of its interior as well as surface – which is available to us on the eve of the manned exploration of our satellite.

## CHAPTER 19

# PHOTOMETRY OF SCATTERED MOONLIGHT

The aim of the present chapter will be to unfold for the reader some of the facts and arguments by which one can attempt to deduce, from photometric measures in different parts of the spectrum, the probable structure of the Moon's surface, obscured otherwise to direct observation by an impenetrable haze of diffraction blurring, coupled with photographic plate grain and less than perfect atmospheric "seeing". In embarking on this quest, let us classify in appropriate physical terms of what the "light" of the Moon actually consists.

As has been known (or at least conjectured) since the days of Anaxagoras, most part of the moonlight visible to the human eye is really *sunlight*, incident directly on the Moon and *scattered* from its surface in all directions – including that of the Earth. This is, to be sure, true of the sunlit lunar hemisphere – when the Sun is visible above its horizon, exhibiting the phases described already in Chapter 5; for, in addition, the Moon is also illuminated by the Sun indirectly – via the Earth, which constitutes the second most important source of light (the only one of importance which can illuminate the lunar night hemisphere).

In order to appreciate the relative importance of these two light sources, let us recall that the apparent photovisual magnitude of the Sun, as seen from the surface of the Earth, is equal to $-26.8$ magn on the international scale (cf., e.g., MARTYNOV, 1959; FESSENKOV, 1962); and on the Moon – in the absence of any atmosphere – it would be increased to $-27.2$ magn. The apparent photovisual magnitude of "full Earth" as seen from the Moon at its mean distance is known (from the observed intensity of the "ashen light") to be $-17.2$ magn (DANJON, 1933). A difference of 10 magn between the Sun and the Earth as seen from the Moon corresponds to a ratio of 10000 in the intensity of illumination by the two bodies; and while the apparent visual stellar magnitude of the sunlit Moon at a phase close to "full" is known to be $-12.7$ (ROUGIER, 1933; GALLOUET, 1963), that of the earthlit full-Moon would be $-2.7$ magn. – i.e., more than a magnitude brighter than the brightest of the fixed stars (Sirius). This is why, near the new Moon, we can see its earthlit portion distinctly as the "old Moon in the arms of the new" without any difficulty or optical aids.

The "light cruve" of the Moon – i.e., the time variation of its total visible light with the *phase* $\alpha$ – defined as the selenocentric angle between the directions to the Sun and the Earth* – is of high interest to the students of the surface of our satellite. Its

---

* This is the same angle as the one denoted in Chapter 5 by $m$ (cf. Figure 5-1) and defined by Equations (5-8) or (5-20). Its present designation by $\alpha$ is motivated by a desire to conform to a custom prevalent among the students of lunar photometry.

most characteristic feature is its steepness – in particular, a rapid rise of light towards full Moon, and an equally rapid diminution of it after the full phase has been passed (cf. Figure 19-1). The visible illuminated surface of a full Moon is only twice as large as that of the first quarter; yet at a phase angle $\alpha$ only a few degrees from zero the Moon is more than eleven times as bright than at the quadrature ($\alpha = \pm 90°$). Moreover, as has recently been shown by Gehrels and his collaborators (GEHRELS, COFFEEN, and OWINGS, 1964), in the close vicinity of theoretical full-Moon (for the phase-angles

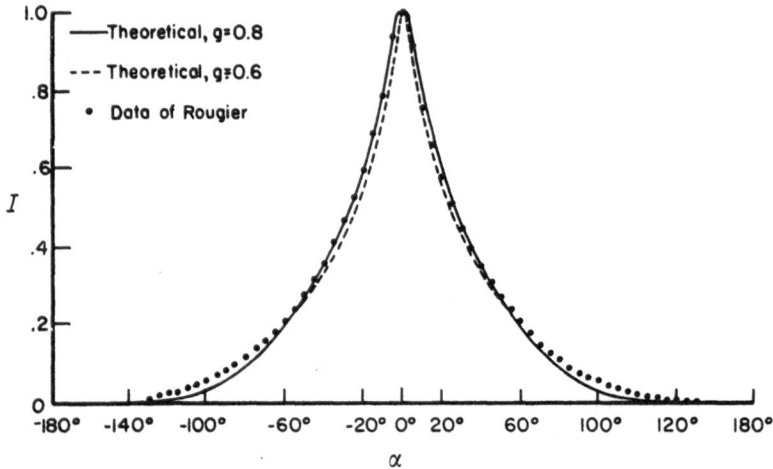

Fig. 19-1. A curve showing the variation of the global light intensity of the Moon with the phase. The theoretical curves have been computed on the assumption of two different values of the "compaction parameter" $g$. (After HAPKE, 1963.)

$0 < |\alpha| < 5°$) the lunar light curve steepens conspicuously – see Figure 19-2 so that the extrapolated apparent brightness at $\alpha = 0°$ (which cannot, of course, be measured directly, as the Moon enters then the shadow cone of the Earth and becomes eclipsed; see Chapter 5) would seem to be close to $-13.55 \pm 0.06$ vis. magn. – i.e., 19 times as great as at the time of the first or last quarter (the "opposition effect"). A further photometric peculiarity at the other extreme end of the phase-range may be noted: namely (as was pointed out by DANJON, 1933) the visible crescent of the Moon seems to disappear completely when its angular distance from the Sun becomes, not zero, but as large as $\pm 7°$. Therefore, the photometric behaviour of the Moon around *both* extremes of the phase range appears to be anomalous, and in need of explanation.

Let us attempt to express now the variation of the brightness of the Moon with the phase in a more mathematical form. In order to do so, let the luminosity of the illuminating source (i.e., the Sun or the Earth) be $L_\odot$ or $L_\oplus$, respectively, and their respective distances from the Moon be $R$ and $r$. Let, moreover, $a$ denote as before the mean radius of our satellite. If so, any element $ds$ of the lunar surface scattering in

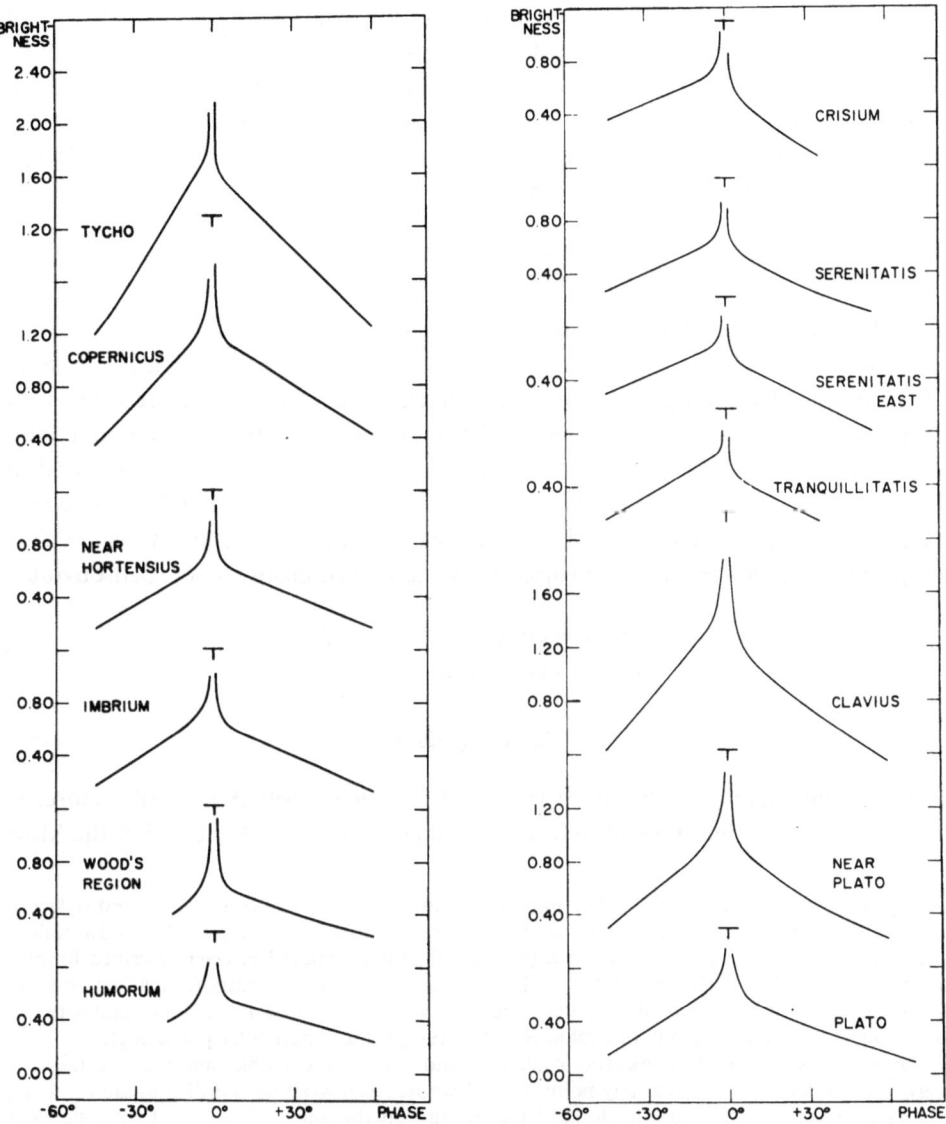

Fig. 19-2. The variation in brightness of different regions of the lunar surface, observed photo-electrically between 1956–1959 by GEHRELS *et al.* (1964), as a function of the phase, demonstrating the extent of the photometric "opposition effect". The *T* symbols indicate the extrapolated brightness at zero phase. (After GEHRELS, 1964.)

accordance with Lambert's law should reflect an amount of sunlight $dl$ equal to

$$dl = (A/\pi) \cos i \cos \phi \, ds, \tag{19-1}$$

where $i$ denotes the angle of *incidence* of sunlight (i.e., the zenith distance of the Sun) at the respective place; $\phi$, the angle of *reflection* (i.e., one between the position of the observer and the surface normal to $ds$); and $A$, the spherical (Bond's) *albedo* – defined as the fraction of the light of a parallel beam scattered by the surface in all directions. If, on the other hand, the scattering takes place in accordance with Lommel-Seeliger law, Equation (19-1) should be replaced by

$$dl = \frac{B}{\pi} \left( \frac{\cos i \cos \phi}{\cos i + \cos \phi} \right) ds, \tag{19-2}$$

where $B$ stands for another normalizing constant.

The total light of sunlit Moon as seen from the Earth at any particular phase will then be obtained by integrating the light element $dl$ over the entire crescent visible at any particular phase. In order to do so, let us adopt a rectangular system of coordinates, two axis of which lie in the "equator of illumination" (cf. Figure 5-2), and the third coincides with the line of the cusps. Let, moreover, $\psi$, $\omega$ denote the latitude and longitude of any point referred to the equator of illumination; $\omega$ being measured from the "principal meridian" whose projection is the line of cusps. If so, then obviously

$$\left. \begin{aligned} \cos i &= \cos \psi \cos (\omega - \alpha), \\ \cos \phi &= \cos \psi \cos \omega; \end{aligned} \right\} \tag{19-3}$$

and the surface element

$$ds = a^2 \cos \psi \, d\psi \, d\omega, \tag{19-4}$$

where $a$ stands again for the mean radius of the lunar globe.* If, furthermore, we assume the illuminating sunlight to form a parallel beam**, the light $l_t(\alpha)$ of the Moon

---

* The geometrical formulae for the determination of the angles $i$ and $\phi$ of incidence and reflection, needed in the reductions of any photometric observations of the lunar surface, are obtained as follows. If the position of a particular point (on a photograph, for instance) has been described by plane coordinates referred to the lunar equator 'of illumination' and the line of cusps (as marked on Figure 5-2), the polar coordinates $\psi$, $\omega$ of such a point are given by $x = a \sin \omega$, $y = a \sin \psi$; and with their aid the angles $i$ and $\varepsilon$ can readily be evaluated from (19-3) for any particular phase angle $\alpha$.

Suppose, however, that the measurements of $x$ and $y$ are not available; and that our task is to ascertain the values of $i$ and $\phi$ at any point $S(\beta, \lambda)$ whose selenographic coordinates are $\beta_S$, $\lambda_S$ and at any particular time. In order to do so, let us recall that the angle $i$ is identical with the zenith distance of the Sun at $S$; and as such it should be identical with the angle $90° - v$ on Figure 14-1, with $v$ as defined by Equation (14-35). Next, the angle $\phi$ is evidently one between the topocentric vector $rs'$ on Figure 14-2 and the normal to the lunar surface at $S$. The selenocentric direction cosines of the former follow, moreover, from Equations (14-40)–(14-42) in which the position of the spacecraft $\Delta$ is identified (if this be the case) with the position O of the observer on Earth; while the direction cosines of the surface normal at $S$ follow directly from (14-46). A cross-multiplication of the respective direction cosines then furnishes the desired instantaneous value of $\cos \phi$.

** Effects arising from the convergence of this beam are utterly too small in planetary or satellite cases; but the reader desirous to assess their magnitude is referred, e.g., to KOPAL (1959), Section IV.6.

at any particular phase $\alpha$ follows, on the assumption of Lambert's law (19-1), as

$$\left.\begin{aligned}
l_{\mathbb{C}}(\alpha) &= \frac{AL_\odot}{\pi}\left(\frac{a}{R}\right)^2 \int\limits_{-\pi/2}^{\pi/2} \cos^3\psi\, d\psi \int\limits_{\alpha-\pi/2}^{\pi/2} \cos(\omega-\alpha)\cos\omega\, d\omega \\
&= \tfrac{2}{3}AL_\odot\left(\frac{a}{R}\right)^2 \frac{\sin\alpha+(\pi-\alpha)\cos\alpha}{\pi};
\end{aligned}\right\} \tag{19-5}$$

while for the Lommel-Seeliger law of the form (19-2) we obtain

$$\left.\begin{aligned}
l_{\mathbb{C}}(\alpha) &= \frac{BL_\odot}{\pi}\left(\frac{a}{R}\right)^2 \int\limits_{-\pi/2}^{\pi/2} \cos^2\psi\, d\psi \int\limits_{\alpha-\pi/2}^{\pi/2} \frac{\cos\omega\cos(\omega-\alpha)}{\cos(\omega-\alpha)+\cos\omega}\, d\omega \\
&= \tfrac{2}{3}\left(\frac{a}{R}\right)^2 BL_\odot\left\{1-\sin\frac{\alpha}{2}\tan\frac{\alpha}{2}\log\cot\frac{\alpha}{4}\right\}.
\end{aligned}\right\} \tag{19-6}$$

The foregoing equations govern the variation of global brightness of the Moon with the phase for two different assumed laws of light scattering on their surface, and their comparison with the observations affords a check of the validity of the underlying Equations (19-1) or (19-2). A comparison with the actual observations shown on Figure 19-1 reveals that, in both cases, the agreement is exceedingly poor – showing that the actual photometric functions which characterise light scattering on the lunar surface must be quite different from those represented by Equations (19-1) or (19-2). It may, however, be noted that (by a somewhat heuristic procedure) HAPKE (1963) found the lunar light changes with the phase to be reproduced much better by a *product* of the Lambert and Lommel-Seeliger phase laws (19-5) and (19-6) – at least outside a narrow region of a few degrees in $\alpha$ around full Moon, where Gehrels's "opposition effect" causes additional complications of the analysis.

Another complication inherent in the use of the Lommel-Seeliger law is the fact that its normalizing constant $B$ is not identical with the conventional definition of the spherical albedo. If the latter is defined, as usual, as the ratio of the total amount of sunlight reflected from the Moon in all directions to the amount of light incident upon it, its proper value $A$ cannot obviously he determined from the observations at any one phase, but becomes rather a measure of an integral of the variation of the intensity of moonlight with the phase.

In more specific terms, let

$$A = p \times q, \tag{19-7}$$

where

$$p = \frac{L_{\mathbb{C}}}{L_\odot}\left(\frac{r}{a}\right)^2 \tag{19-8}$$

denotes the fraction of light reflected at the center of full Moon in the direction of the

Earth under normal illumination, and

$$q = 2 \int_0^\pi f(\alpha) \sin \alpha \, d\alpha, \qquad (19\text{-}9)$$

denotes the amount of light reflected normally and expressed in terms of the total light reflected in all directions, if the phase-variation $f(\alpha)$ of moonlight has been normalized so that $f(0) = 1$.

In order to evaluate $p$, let us recall that, according to the best contemporary measurements, the apparent photovisual magnitudes of the Sun and the Moon (before the onset of the opposition effect) at its mean distance from us are $m_\odot = -26.78 \pm 0.04$ and $m_{\mathbb{C}} = -12.75 \pm 0.02$, respectively (FESSENKOV, 1962); so that, to a sufficient precision, the difference $m_\odot - m_{\mathbb{C}} = -14.0 \pm 0.03$. If so, however,

$$\frac{L_{\mathbb{C}}}{L_\odot} = (2.512)^{-14.0} = 2.51 \times 10^{-6}; \qquad (19\text{-}10)$$

and since, moreover,

$$\frac{r}{a} = \frac{384.4 \times 10^3 \text{ km}}{1.738 \times 10^3 \text{ km}} \qquad (19\text{-}11)$$

it follows that

$$p = 2.51 \times 10^{-6} \times (221.2)^2 = 0.123. \qquad (19\text{-}12)$$

On the other hand, numerical integrations of ROUGIER's (1933) light curve $f(\alpha)$ led FESSENKOV (1962) to adopt the value of

$$q = 2 \int_0^\pi f(\alpha) \sin \alpha \, d\alpha = 0.585, \qquad (19\text{-}13)$$

subject to an uncertainty of not more than a few units of the last place. Accordingly, the mean albedo of the lunar globe proves out to be

$$A = 0.072 \qquad (19\text{-}14)$$

at photovisual wavelengths ($\lambda_{\text{eff}} \sim 5400$ Å).

The lunar surface turns out, therefore, to be a pretty poor reflector of visible light – much less effective than most known terrestrial substances – for it scatters only about 7.2% of the incident beam, as compared with 56% for solid limestone, 24% for granite, 14% for basalt, or 6% for volcanic lava. On the other hand, it should be remembered that the above value of 0.072 for the mean lunar albedo represents a *global average* of reflectivity over the entire visible hemisphere; and not all parts of it reflect light so poorly. Indeed, a glance at the disc of a full Moon through a telescope reveals that its individual regions differ in reflectivity, which ranges from about 5–6 per cent for dark maria or crater floors to almost 18 per cent in the crater Aristarchus. The darkest region of the Moon of appreciable extent (excepting certain small "dark spots", such as found on the floor of the crater Alphonsus, whose reflectivity seems as low as 0.03) is Sinus Medii, characterized by a photovisual albedo of 0.054, followed by Mare Nubium with 0.062; and the floors of certain large craters (Grimaldi, Riccioli) appear to be almost equally dark. On the other hand, floors of other well-known craters – such as Archimedes or Ptolemy – possess albedoes between 0.095–0.102; while in lunar mountainous regions, or on crater walls, $A$ is generally larger than 0.1. The walls or rays of Copernicus are characterized by $A = 0.12$. For the crater Tycho, $A = 0.137$, and for some of its rays it attains 0.163. The brightest large crater on the visible face of the Moon is undoubtedly Aristarchus, for the floor of which $A = 0.163$; and even this relatively high value of $A$ is exceeded by the albedo of its central peak attaining 0.183. Still higher values of local reflectivity will undoubtedly be found on the Moon, but over small areas; and their detection from the Earth is hampered by limited resolving power of most telescopes used so far for photometric studies.

Extensive measurements of the reflectivity of a great many regions of the lunar surface reveal that the ratio of intensity of the brightest and darkest details on the Moon of measurable size (exceeding a few kilometres in size) does not, therefore, seem to exceed much a factor 3; and the mountainous areas are, on the average, not more than 1.8 times brighter than the maria. If these ratios appear to be small in comparison with the range encountered among terrestrial rocks examined in the laboratory, the reader should keep in mind, first, that even the local lunar albedos are characteristic of the *average* reflectivity of not less than several square kilometers of the lunar terrain (in which different reflectivities of several distinct constituents are indistinguishably merged – a fact which by itself is bound to reduce the extremes); and, secondly, that on the Moon we are without doubt dealing with the reflectivity of different materials in their *pulverized* state. We mentioned before that, in the terrestrial laboratory, the reflectivity of a limestone is four times as large as that of an average piece of granite, and more than nine times as large as that of volcanic lava. This, however, is true only of solid specimena of these minerals. If we pulverize them, however, to a diminishing grain size of the material, the difference in reflectivity between so resulting powders will rapidly tend to diminish.*

So far we have been mainly concerned with the brightness of any particular element of the lunar surface at the time of full Moon. Before coming to consider the variation of such brightness with the wavelength, let us stress another fact of cardinal importance for the interpretation of the structure of the lunar surface: namely, the *photometric homogeneity* of the entire face of the Moon. By this we mean the fact – ennunciated first by BARABASHEV (1923) and MARKOV (1924) – that not only does the light curve of the Moon as a whole exhibits a steep maximum at zero phase, but *the apparent brightness of every detail of its face attains maximum at full Moon*, regardless of its relative position or angular distance from the center of the lunar disk – be it a part of the maria, continents, or of the bright rays.

---

* The author is indebted to Dr. J. Conel of the Jet Propulsion Laboratory, California Institute of Technology, for this remark.

The apparent disk of full Moon exhibits, therefore, *no limb-darkening*; and its average brightness is the same near the center as it is near its edge. This remarkable photometric property of the Moon – verifiable readily with the naked eye as well as by a glance at any photograph of full Moon – was noticed already by Galileo GALILEI, and emphasized by him in his *Dialogues on the Great World Systems* (Florence, 1632). It is truly remarkable, because a sphere exhibiting diffuse reflection of the light of a distant source and observed near zero phase should be brighter near the center than towards the limb. The planet Mars, for instance, whose surface is also covered by dust (stirred occasionally by winds to give rise to dust storms) exhibits such a limb-darkening to marked degree; but the Moon at optical frequencies shows no trace of it – a fact which already BOUGUER (in his *Traité d'optique*, Paris, 1729) attempted to explain by a considerable degree of surface roughness. A difference in photometric behaviour of the Moon and of Mars in this respect is probably a consequence of their different surface micro-structure (going back to the fact that the surface of Mars is shielded from direct impact of small micrometeorites by its atmosphere, while on the Moon any such protection is completely lacking).

Before, however, we follow (in Chapter 23) the implications of this suggestion in more detail, let us stress that the observed absence of lunar limb-darkening alone rules out the possibility that either Lambert's or Lommel-Seeliger's law of scattering as embodied in Equations (19-1) or (19-2) – which both imply strong limb-darkening – could represent the light scattering on the lunar surface to any approximation; and little wonder then that the phase-laws (19-5) or (19-6) based upon them failed to represent the lunar monthly light changes with any adequacy. The same has, moreover, been true of most other laws of scattering proposed to account for the photometric peculiarities of the lunar surface (cf., e.g., ÖPIK, 1924; FESSENKOV, 1928; SCHOENBERG, 1929; BENNETT, 1938, *et al.*) – save for very artificial models which are most unlikely to be encountered on the Moon (such as the lichens of VAN DIGGELEN, 1959; or the "fairy castles" of HAPKE and VAN HORN, 1963). The repeated failures of all these efforts to approximate the reality to any tolerable degree have gradually led us to a realization that *the photometric properties of the Moon are determined primarily by the shadow phenomena exhibited by untold millions of surface irregularities; and that the precise form of the diffuse reflection law is relatively unimportant in comparison with the effects of surface micro-relief* (MINNAERT, 1961; p. 229).

Let us, therefore, abandon any further effort in this direction at this stage and turn our attention to the *spectral composition* of the light of the Moon. If the latter consisted of sunlight scattered on solid particles which are large (or again small) in comparison with the wavelength – i.e., outside the region in which the Mie law of scattering becomes frequency-dependent – the spectrum of moonlight should be an exact replica of that of the Sun – with profiles of all the absorption lines identical in contour, and equal in intensity. The extent to which this is actually the case for spectral lines will be discussed later in Chapter 22; but even for the distribution of light in the continuous spectrum this is not true; for the light of the Moon as a whole is distinctly *redder* than the illuminating sunlight.

In more precise terms, the B-V colour index of the Sun is known to be $+0.62$ magn; while that of moonlight is equal to $+0.94$ magn (GALLOUET, 1963); the difference of $+0.32$ magn between them renders the "photographic" albedo of the Moon to be equal to $A=0.055$. We may add that, in the extreme UV accessible so far only from rockets or satellites, the mean albedo of the Moon becomes less than 0.03; but increases again well above 0.10 in the near IR opened up recently by stratospheric balloons (cf. WATTSON and DANIELSON, 1965).

According to extensive ground-based measurements by Gehrels and his associates (GEHRELS *et al.*, 1964), the colour index of the integrated moonlight varies, moreover, somewhat with the phase, in such a way that

$$B\text{-}V = 0.838 + 0.0017\,|\alpha|\ \text{magn}\,. \tag{19-15}$$
$$\pm\ \cdot\ 3\pm\ \cdot\ 2$$

and

$$U\text{-}B = 0.397 + 0.0015\,|\alpha|\ \text{magn}\,. \tag{19-16}$$
$$\pm\ \cdot\ 8\pm\ \cdot\ 3$$

for $-50° < \alpha < +60°$, being bluest at the time of the full-moon and growing slightly redder towards both quarters.

This general reddening of moonlight in comparison with that of the Sun should, in accordance with Mie's theory, be indicative of the fact that the average size of the light-scattering particles on the lunar surface should be close to $0.8 \pm 0.1$ microns; and their (complex) refractive index $\hat{n} = 1.34 - (0.02 \pm 0.02)i$. The latter is somewhat smaller than the value which we shall later (Chapter 21) deduce from observations in the microwave domain of the spectrum (cf. Equation 21-16); and the average grain size seems too small to account satisfactorily for observed polarization properties of the lunar surface, of which more will be said presently.

The differences in colour index between different parts of the lunar surface range over only 0.08 of a magnitude in B-V (COYNE, 1965) or 0.09 (GEHRELS *et al.*, 1964). Therefore, while the integrated moonlight is distinctly redder than the illuminating sunlight, its colour is everywhere rather monotonously the same. Differences of less than one-tenth of a magnitude in colour index manifest themselves on smooth ground (in the maria, for instance) across sometimes quite sharply defined boundaries – see Figure 19-3 – and doubtless signify an actual change in composition or structure of the outermost lunar surface layer. For example, the plains of the Sinus Iridum are distinctly redder than the surface of Mare Imbrium with which it merges; or the shores of Mare Serenitatis are bluer than the interior of the mare. The colour index of any surface element seems to be statistically correlated with its albedo – in the sense that the brighter the feature, the redder it appears to be. The maria as a whole are certainly bluer than the continents; the only conspicuous exception to this statistical rule is the crater Aristarchus – the brightest large spot on the Moon – which appears also to be distinctly bluer than it should.

The scattering of sunlight from the lunar surface entails one more important

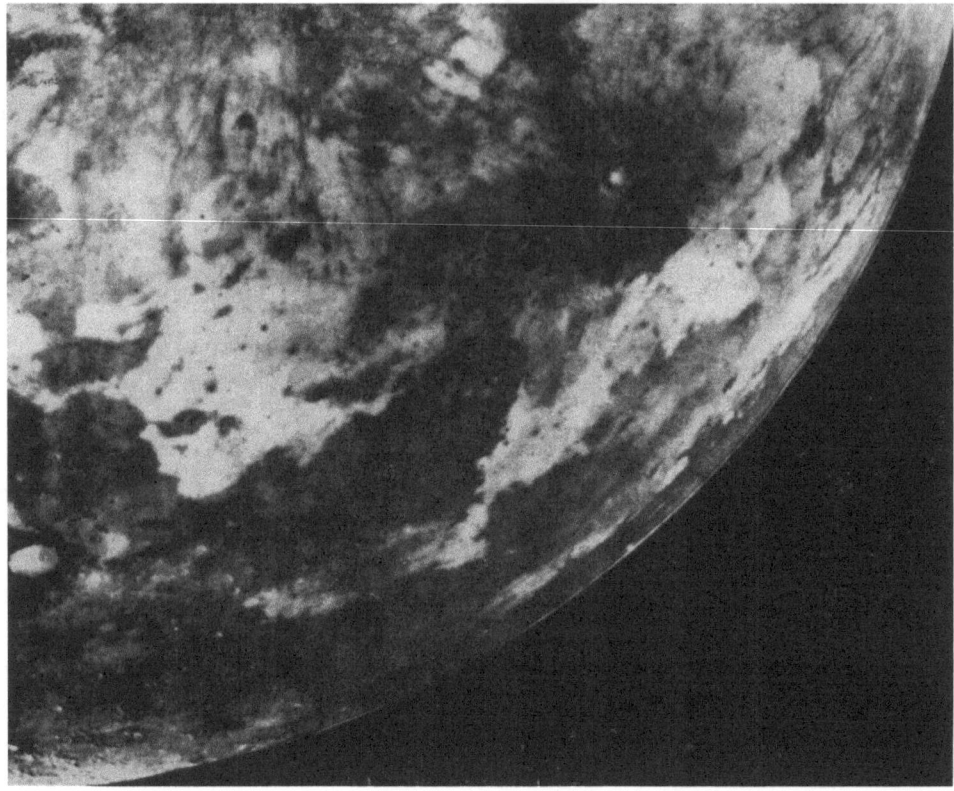

Fig. 19-3.   Northwest quadrant of the Moon in the region of Mare Imbrium, exhibiting enhanced colour contrasts (darker is redder) by a superposition of a positive taken at an effective wavelength of 7800 Å on to a negative at 3500 Å, due to Whitaker. (After GEHRELS *et al.*, 1964.)

consequence: namely, that the light so scattered should become also *polarised* in a manner (and to a degree) depending on the nature of the ground. That white moon-light is indeed partly polarised was noted as early as 1811 by ARAGO; but it was SECCHI (1859) who proved that this polarisation was the lunar (rather than the terrestrial atmospheric) phenomenon; and Lord ROSSE who established that the amount of polarisation varies with the phase (cf. Figure 19-4).

However, the first precise and detailed analysis of the polarisation of moonlight we owe to LYOT (1929). It was he who found that, not only the degree, but also the plane of polarisation is predominantly either parallel with, or perpendicular to, the "equator of illumination" defined by the direction of the incident and reflected rays. The phase at which the plane of polarisation rotates so by 90° coincides, moreover, for most lunar objects with the quadratures.

Subsequent contributors to such studies included DOLLFUS (1957), KOHAN (1962, 1964, 1965), CLARKE (1962, 1963, 1965) and – most extensively – GEHRELS and his

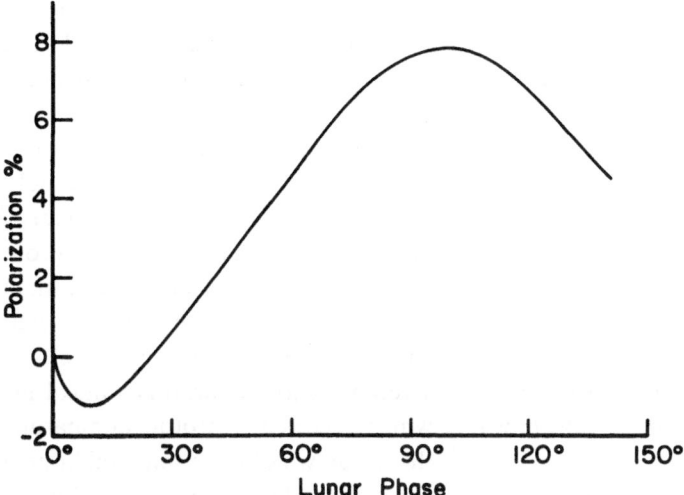

Fig. 19-4. Mean lunar polarization curve. (After Lyot.)

associates (1964). Gradual rotation in position angle of the plane of polarization – reported by Kohan or Clarke – was not confirmed (within the limits of accuracy of $\pm 3°$) by Gehrels who, in agreement with Lyot, found the maximum of the electric vector of moonlight to be always either parallel with, or perpendicular to, the equator of illumination (a fact which, according to Mie's theory, would imply an effective absence of multiple scattering).

The actual amount of polarisation of light scattered by different formations on the lunar surface varies from spot to spot; and is, on the whole, the greater, the lower the reflectivity (albedo) of the lunar ground – dark maria polarise light more strongly than the continental regions. The polarisation of light scattered from the dark plains of Mare Imbrium (in particular, Sinus Iridum), Mare Crisium or Mare Foecunditatis attains 15–16%, while the brighter types of ground polarise light noticeably less. Exception to this statistical rule is again the crater Aristarchus, which polarises light to 12% in spite of its high albedo; and the amount of this polarization may have undergone a significant change in the past decade (DZHAPIASHVILI and KSANFOMALITI, 1962). The mean polarization of moonlight attains a maximum of approximately 6 per cent in the visible part of the spectrum, and increases slowly towards the violet.

Extensive comparisons of the scattering and polarizing properties of lunar ground with terrestrial samples of silicate materials have been carried out by many investigators (BARABASHEV, FESSENKOV, LYOT, and more recently by DOLLFUS, HAPKE, and others). Until quite recently, the results of such comparison have failed to match the observed lunar properties with those of any specific terrestrial rocks closely enough. In particular, none of the terrestrial samples was able to reproduce the negative polarization of moonlight at small phase angles. However, by elimination of alternative possibilities – already most of the earlier investigators of the subject arrived at the conclusion that *the polarizing properties of lunar ground are those of a finely pulverized dust* of opaque grains, the average size of which was estimated by Lyot to be of the order of a few microns (i.e. *larger* than those required for the reddening of scattered moonlight).

The conclusions of Lyot and his contemporaries were based on a comparison of lunar observations with optical measurements of scattering and polarization by freshly ground mineral powders in the laboratory. In recent years, however, a realization has gained ground that material on the lunar ground – unprotected by any atmosphere – must have suffered an appreciable *radiation damage* due (mainly) to solar corpuscular radiation bombarding the lunar surface for an astronomically long time, and causing a gradual darkening of the top-most surface layer as well as a change in its polarization properties. When, in recent years, laboratory optical measurements were repeated on silicate powders which received a proton bombardment causing damage equivalent to the effects of the 'solar wind' over some millions of years, the results of DOLLFUS and GEAKE (1965) as well as HAPKE (1965) leave but little room for doubt that radiation-damaged silicate dust matches the observed optical properties of the lunar ground almost to perfection – and that the relatively low albedo of the lunar surface, as well as the negative polarization of moonlight in the range of 0° and 30° can be understood only in terms of the radiation damage produced on the Moon by solar wind over long intervals of time.

The full significance of this result will be further elaborated in Chapter 23 of this book. For the present we wish to stress that all facts reported so far in this chapter concern the scattering and polarization, on the lunar surface, of sunlight which represents the principal illuminating source. The second most important source illuminating the Moon is, of course, our Earth; and although this latter source, as seen from the Moon, is approximately ten thousand times fainter than sunlight, 'full Earth' would appear to be almost a hundred times brighter from the Moon than full Moon appears to us, because of the larger angular size and higher albedo of the apparent disk of our planet. In other words, earthlit landscape of the 'new' Moon would be about a hundred times as bright as the terrestrial landscape at full Moon; and this represents a not inconsiderable source of illumination. Only when both the Sun and the Earth are below the horizon on the Moon can lunar night be regarded as really dark; but no portion of such ground is ever visible from the surface of the Earth.

The variation of brightness of the 'ashen light' of the Moon with the phase can be observed only within a very limited range of the phase angle; for (although the brightness of the sunlit Earth as seen from the Moon exhibits, of course, the same cycle of changes as the Moon does to us, only shifted in phase by 180°), a geocentric terrestrial observer would see the earthlit Moon always at zero phase angle; while for an observer situated on the surface of the Earth, this angle may vary only within the limits of the equatorial lunar parallax (i.e. less than ±1°). Such a range may be just sufficient to detect an indication of the 'opposition effect' in the light curve of earthlit Moon, but little else.

And, moreover, there is one additional difference in the nature of the illuminating source: namely, while the Sun emits light which is (almost completely) unpolarised, sunlight scattered from the Earth in the direction of the Moon is fairly strongly polarized because of its prior multiple scattering in the terrestrial atmosphere. The lunar surface re-scatters some of this light back to the Earth after having partially depolarised it. The polarization of the 'ashen light' of the Moon reproduces, therefore, that of earth light as seen from space, only to a diminished degree; and this reduction in polarization allows us again to investigate the nature of the lunar ground from the manner in which earthlight is depolarized by it.

Light changes at optical frequencies $(\lambda < 1\mu)$ exhibited by the Moon during *lunar eclipses* do not, by themselves, constitute a proper subject for this chapter, as their details bear largely on the optical properties of the terrestrial atmosphere rather than on any physical properties of our satellite (cf., e.g., LINK, 1956, 1962, 1963). However, in one respect the photometric observations of lunar eclipses revealed a result of considerable interest: namely, a careful rediscussion of the residual brightness of totally-eclipsed Moon as observed between 1583 and 1920 led DANJON (1920) to a discovery of the fact that such brightness appears to be strongly correlated with the phase of the 10.8-year cycle of solar activity – in the sense that the eclipses preceding its minimum are brightest; and those following it, faintest (cf. Figure 19-5). Since 1920, Danjon's evidence was supplemented by DE VAUCOULEURS (1944) on the basis of new results obtained from 47 eclipses observed between 1893–1943; and further confirmed by LINK (1956, 1963), or BELL and WOLBACH (1965).

In contrast with Danjon's result based on the observations of lunar eclipses, Gehrels and his associates (cf. GEHRELS *et al.*, 1964) found that the global brightness of the *uneclipsed* Moon also showed a certain dependence on the solar cycle – in the

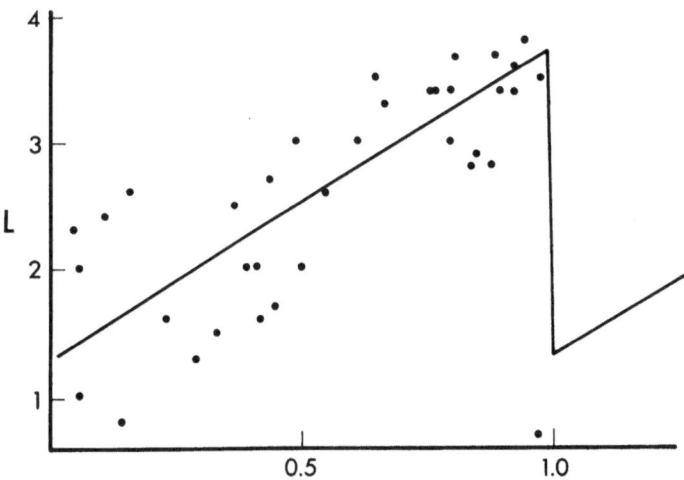

Fig. 19-5.   The variation of residual brightness of the eclipsed Moon with the solar cycle, according to DANJON (1920). *Abscissae*: fractional solar-cycle phase; *ordinates*: residual brightness on Danjon's luminosity scale.

sense that in 1956–1959 (i.e. near the recent maximum of solar activity) the lunar surface brightness appeared to be 10–20% higher than near its minimum between 1963 November – 1964 January. Moreover – and most revealingly – the increase in brightness near the maximum was accompanied by a diminution in the total degree of polarization of visible moonlight – as though the additional source of light making the Moon brigther were non-polarized. The brightness of the visible sunlight between maximum and minimum of solar activity is known to remain constant within less than one per cent of its mean value (cf. e.g., STERNE et al., 1940); so that the observed fluctuations in the intensity of moonlight appear to be many times as large as possible fluctuations of the illuminating sunlight. The origin of increased lunar brightness must then obviously be sought elsewhere; but a further discussion of this problem is being postponed to Chapter 22.

# THERMAL EMISSION OF THE LUNAR SURFACE

In the preceding chapter we have been concerned with the total amount of sunlight reflected from the Moon in the direction of the Earth, and its variation with the phase. We noted that, on the whole, the Moon is a pretty poor reflector of visible light; for only about 7.2% of incident sunlight is scattered from it when the average is taken over the entire disk. What happens, however, to the rest of the incident light? The balance must obviously be *absorbed* by the lunar surface and converted into heat. But any physical body possessing a finite absolute temperature must also *emit* radiation of its own; and if this body behaves as "grey" (as is very much the case of the lunar surface) its emission $B_\lambda(T)$ at a wavelength $\lambda$ is bound to be governed by Planck's formula

$$B_\lambda(T) = \frac{8\pi hc^2}{\lambda^5}\left\{e^{\frac{hc}{k\lambda T}} - 1\right\}^{-1},$$ (20-1)

where $h$ denotes the Planck constant; $k$, the Boltzmann constant; and $c$, the velocity of light. Moreover, the total radiation (integrated over all wavelengths) of such a body will be clearly given by

$$B(T) = \int_0^\infty B_\lambda(T)\,d\lambda = \frac{2\pi^4(kT)^4}{15c^2h^3} = \frac{\sigma}{\pi}T^4,$$ (20-2)

where

$$\sigma = \frac{2\pi^5 k^4}{15c^2 h^3}$$ (20-3)

denotes the Stefan-Boltzmann constant.

For the Sun, whose absolute temperature can be sufficiently approximated by $T = 5700\,°K$, most part of the radiation emitted in accordance with (20−1) will lie between the violet and the red end of the visible spectrum (with a maximum in the yellow, around $\lambda \sim 0.5\,\mu$, in accordance with Wien's displacement law $\lambda_{max}T = 0.29$ cm·deg), the ensemble of which gives the impression of "white light". This is also (essentially) the colour of reflected moonlight; but the lunar radiation proper is of a very different kind. The sole – and sufficient – reason is the fact that the temperature of the lunar surface is so much lower that than of the Sun. The Moon – like the Earth – receives all its heat by radiation from the Sun*; but as their average distance from the

---

* Exception to this statement is represented by the minute trickle of radiogenic heat of the Moon which makes its way to the surface and is radiated away from there. The amount of heat due to this source was already investigated in Chapter 8, and found to give rise to a flux of the order of a few ergs/cm² sec, corresponding (by 20-2) to a temperature of the order of a few degrees of Kelvin only.

Sun amounts to 214 solar radii, each unit area of lunar surface receives only about one $214^2$-th or a 46000th part of the heat flux passing through each square centimeter of the surface of the Sun. As, for a black body, this flux is proportional (cf. Equation 20-2) to the fourth power of the absolute temperature, it follows that the mean temperature of the Moon should be $\sqrt{214}$ or 14.6 times lower than that of the Sun – or approximately 390 °K (i.e., 117 °C) – at the subsolar point of the lunar surface.

However, with increasing zenith distance $z$, the oblique incidence should attenuate the flux in such a way as to make the ground temperature towards the limb to vary as $\cos \frac{1}{4}z$ – the thermal radiation of the Moon should (unlike scattered light) be expected to exhibit very pronounced "limb-darkening" (cf. Figure 20-1). During the night, when the source of heat has set below the horizon, the lunar surface can radiate only by cooling (i.e., drawing on the heat supply stored during the day). However, as heat is

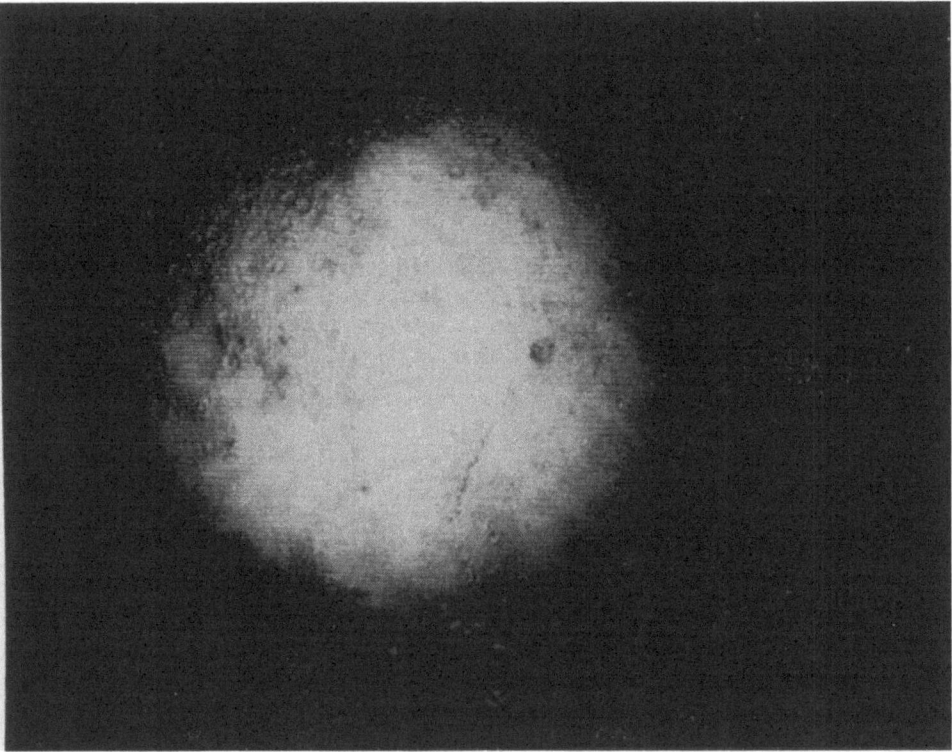

Fig. 20-1.   A synthesis of infrared scans of the Moon (through a passband between 10 and 12 $\mu$ in wavelength), showing a view of the lunar face in the light of its thermal radiation (after Shorthill and Saari). Note a conspicuous darkening to the limb of the lunar disk (completely absent on photographs taken through the optical atmospheric window in the scattered light of the Moon); and also the presence of several dark areas and spots (bordering, for instance, the Apennine mountain chain; or the positions of the craters Copernicus, Tycho, Manilius, and many others). These are not due to any shadow phenomena – there are no shadows on the Moon in the light of its thermal radiation – but indications or local differences in surface temperature.

emitted at night at a much lower temperature than that at which it was absorbed during day-time, the rate of its night-time loss will be very much reduced – and, as a result, the ground temperature should remain finite throughout the lunar night till the time of sunrise; its actual value depending on the thermal capacity and heat conducting properties of the lunar crust.

These elementary considerations lead us to expect that even in the lunar tropics the noon temperature of the Moon will not exceed 390 °K (and could, in fact, be somewhat less; as only about nine-tenths of incident sunlight becomes converted into heat; the balance being back-scattered). If so, however, Planck's law (20-1) discloses that its radiation should be very different from "white light" – most part of it being emitted in infra-red (with a maximum around $\lambda = 10\ \mu$ or 0.01 mm). Light of this colour is, of course, quite invisible to the human eye, and incapable of impressing the photographic plate. It will, in addition experience considerable difficulty in penetrating through our terrestrial atmosphere in between the interlocking absorption bands of water vapour and carbon dioxide. However, that part of it which gets through the atmospheric window between 8–13 $\mu$ can be detected – and, in fact, measured quite accurately – by its thermoelectric effect.

The first investigator actually to measure the amount of heat received from the Moon was Lord Rosse (1869, 1870, 1872), followed by Langley (1884, 1887), Very (1898, 1906); and, in more recent times, by Pettit and Nicholson (1930), Pettit (1935, 1940), Sinton (1959, 1960) and, in particular, Shorthill and Saari (1963) whose accurate and extensive recent work covered the field in a very exhaustive manner. From the days of Lord Rosse up to about 1950, the standard technique of such measurements was to employ sensitive thermocouples; these have only quite recently been superseded by semi-conductor thermistors (such as the mercury-doped germanium crystals) cooled to the temperature of liquid hydrogen.

The general trend of diurnal temperature variation on the Moon, resulting from such measurements, is diagramatically illustrated on the accompanying Figure 20-2, based on the work of Low and Davidson (1965). A glance at it reveals that, at the subsolar point, temperatures as high as 390 °K are attained each day, in accordance with our expectation. During the afternoon – as the zenith distance of the Sun is increasing the temperature begins to decline slowly at first; but very rapidly near the sunset – until a temperature of approximately 120 °K has been reached. Thereafter the temperature continues to decline steadily below 100 °K to less than 90 °K in the second half of the night; but its recovery after sunrise is even more dramatic. The extremes of its total range of almost 300 °K correspond to the terrestrial temperatures of boiling water and of liquid air, respectively.* As the period of this change is equal to 29.53 days or 709 hours of our time, the corresponding mean gradient of temperature is less than one degree per hour; though the actual gradient at the time of sunrise or sunset can attain 7°/h. Even this may not seem, perhaps, too much out of the ordi-

---

* This is, of course, the maximum range encountered in the lunar equatorial belt. Near the poles, where the Sun never rises very high above the horizon, nor sets completely for very long, the temperature variations should become correspondingly smaller.

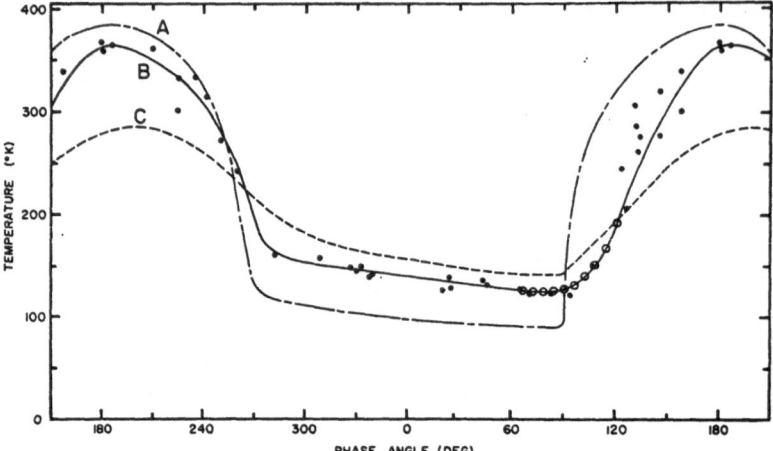

Fig. 20-2. Graphs exhibiting the variation of the central temperature of the Moon versus the phase. Curve *A* represents a variation of the surface temperature based on infrared measurements (between 8 and 12 $\mu$) of the flux of thermal radiation; curve *B*, the sub-surface temperature measured at $\lambda = 1$ mm in the microwave domain of the radio spectrum; curve *C*, temperature measured at $\lambda = 3$ mm. (After Low and Davidson, 1965.)

nary judged by the terrestrial standards; for similar variations can be experienced in desert climate. It is the persistence of such gradients over so many hours which make the lunar extremes in temperature so wide.

What do these facts reveal about the structure of the lunar surface layer? In order to investigate their consequences, let us return to the equations of heat conduction which served us in good stead in Chapter 8 for investigation of temperatures in the lunar interior. When we emerge from it to the surface, all effects of curvature can be safely ignored; and Equation (8-4) of conductive energy transfer reduced to

$$\frac{\partial T}{\partial t} = K \frac{\partial^2 T}{\partial r^2},$$    (20-4)

where (as in Chapter 8) $K$ denotes the coefficient of thermal diffusivity. However – unlike in Chapter 8 – the variable $r$ will denote now the sub-surface depth, starting from zero on the surface and increasing inwards.

The outer boundary condition associated with Equation (20-4) should be of the form

$$\kappa \frac{\partial T}{\partial r} = (1 - A)\pi S \cos z - \sigma T^4$$    (20-5)

at $r = 0$, where $\kappa$ denotes, as before, the coefficient of heat conduction; $\pi S$ denotes the flux of the Sun's heat (i.e., total radiation in all wavelengths, equal to the "solar constant") incident normally on the lunar surface; being zero during the lunar night, and changing continuously between 0 and $\pi S$ during sunrise or sunset (when only a part of the solar disk is visible over the lunar horizon) in accordance with Equation (14-65). Moreover, if $A$ denotes the local albedo of the lunar surface as defined in Chapter 19, $(1-A)\pi S$ stands for that fraction of normally incident light which is absorbed by the surface; and on oblique incidence is further alternated by the factor $\cos z$, where $z$ denotes the lunar zenith distance of the illuminating Sun (as given by Equation 3-22). If the source of the energy is $\pi S$, its flux gets attenuated downwards until, for $r = \infty$,

$$T(\infty, t) = 0, \qquad (20\text{-}6)$$

The boundary-value problem represented by the foregoing Equations (20-4)–(20-6) is again nonlinear in $T$, and as such can be solved only by numerical means (cf. WESSE-LINK, 1948; JAEGER, 1953; KROTIKOV and SHCHUKO, 1963, and others). In doing so, it is of advantage to change over from $r$ and $t$ to new independent variables $x$ and $\tau$, normalized in accordance with the equations

$$x = \left(\frac{\rho C_v \omega}{2\kappa}\right)^{\frac{1}{2}} r \quad \text{and} \quad \tau = \omega t, \qquad (20\text{-}7)$$

where

$$\omega = \frac{2\pi}{P}; \qquad (20\text{-}8)$$

$P$ representing the period of a sideric month. Equations (20-4) and (20-5) then reduce to

$$\frac{\partial T}{\partial \tau} = \frac{1}{2}\frac{\partial^2 T}{\partial x^2}, \qquad (20\text{-}9)$$

subject to the boundary condition

$$\sqrt{\kappa \rho C_v}\left(\frac{\partial T}{\partial x}\right)_0 = \sqrt{\frac{2}{\omega}}\{(1 - A)\pi S \cos \tau - \sigma T^4\} \qquad (20\text{-}10)$$

on the equator. Moreover, for an average point of the lunar surface, $A = 0.072$ (cf. Eq. 19-14); while $\pi S = 2.00$ cal/cm$^2$ min $= 1.39 \times 10^6$ erg/cm$^2$sec, $\omega = 2.6617 \times 10^{-6}$ sec$^{-1}$ and $\sigma = 5.669 \times 10^{-5}$ erg/cm$^2$sec deg$^4$. Under these conditions, the results of numerical integrations of the equations (20-9)–(20-10) are shown for different values of the parameter $(\kappa \rho C_v)^{\frac{1}{2}}$; and a comparison of these results with the observations reveals that the best average fit requires the numerical value of this parameter to be close to $10^{-3}$ cal/cm$^2$ deg sec$^{\frac{1}{2}}$. For a solid surface consisting of silicate rocks such as we meet on the Earth we should expect (cf. BIRCH, 1952) that $\kappa \sim 2 \times 10^5$ erg/cm sec deg and $C_v \sim 7 \times 10^6$ erg/g deg, while (cf. Eq. 7-23) $\rho \sim 3.2$ g/cm$^3$ – leading to a value of $(\kappa \rho C_v)^{\frac{1}{2}} = 2 \times 10^6$ erg/cm$^2$ deg sec$^{\frac{1}{2}}$ or 0.05 cal/cm$^2$ deg. sec$^{\frac{1}{2}}$, which is about 50 times as large as that indicated by the observations!

This glaring discrepancy was interpreted by the early investigators of the subject as a sufficient indication of the fact that *the surface of the Moon* to which the temperature measures refer *cannot consist of exposed solid rocks* (for no rocks are known on

Earth which could match the thermal properties of the lunar surface by a wide margin), *but is probably covered by loosely packed dust* conducting heat only through the corners at which the individual dust grains are in actual contact. In this way the inward flow of heat can be almost arbitrarily impeded by reducing the effective area of contact (i.e., by increasing the looseness of dust packing); and all subsequent work which will be referred to in this chapter has confirmed that this constitutes the only possible approach to the problem. That there must be dust on the Moon – partly native, partly of cosmic origin – we have stressed in several places in this book (cf. Chapters 17 and 18); and the present result is but another argument pointing to the same conclusion.

If this is indeed the case, how rapidly does the diurnal heat wave on the Moon propagate inward through the surface dust layer? The answer can again be obtained by numerical integration of Equation (20-9) in the direction of increasing $x$. A more analytic insight can, however, be obtained if advantage is taken of the fact that the surface temperature $T(0, \tau)$ can be actually measured, and is known to behave in a manner indicated by the curve $A$ on Figure 20-2. Suppose that its variation is represented analytically by an *empirical* Fourier expansion of the form

$$T(0, \tau) = \sum_{n=0}^{\infty} T_n \cos(n\tau - \varepsilon_n), \qquad (20\text{-}11)$$

replacing the physically correct condition (20-5), where the $T_n$'s and $\varepsilon_n$'s are the empirical amplitudes and phase coefficients. If so, the particular solution of (20-9) satisfying the *linear* boundary conditions (20-6) and (20-11) can be found by standard methods to assume the form

$$T(r, t) = \sum_{n=0}^{\infty} T_n e^{-x\sqrt{n}} \cos(n\tau - x\sqrt{n} - \varepsilon_n) \qquad (20\text{-}12)$$

where the variables $x$ and $\tau$ continue to be defined by (20-7). The exponential terms on the right-hand side of (20-12) then represent the diminution of the temperature with the depth; while the term $-x\sqrt{n}$ in the argument of the periodic terms represents the *phase-lag*, due to the fact that the diurnal heat wave on the surface takes finite time to penetrate inwards.

According to recent calculations by KROTIKOV and SHCHUKO (1963), the explicit forms of the empirical equation (20-11) in equatorial regions of the lunar surface for different values of the parameter $(\kappa \rho C_v)^{-\frac{1}{2}}$ are as given in Table 20-1*: the last column contains the corresponding temperatures $T$, at the subsolar point (i.e., for $\tau = 0$), which are determined by the amount of incident solar energy and virtually independent of the value of $\kappa \rho C_v$. When the coefficients $T_n$ as well as $\varepsilon_n$ listed in the foregoing tabu-

---

* In order to obtain the best possible fit to the observations, Krotikov and Shchuko were led to adopt the value of $1.92 \times 10^6$ erg/cm² sec for the solar constant, in place of $(2.00 \pm 0.02) \times 10^6$ as measured from the Earth. The difference between the two may represent the fraction of incident sunlight which is neither back-scattered nor re-emitted as thermal radiation, but appears as luminescent light – about which more will be said in Chapter 22.

## TABLE 20-1

### Coefficients of the Expansion

$$T(0, \tau) = T_0 + \sum_{n=1}^{3} T_n \cos(n\tau - \varepsilon_n)$$

of lunar surface temperature for different values of thermal conductivity (after KROTIKOV and SHCHUKO, 1963)

| $(\kappa\rho C_v)^{-\frac{1}{2}}$ in cm² deg sec$^{\frac{1}{2}}$/cal | $T_0$ | $T_1$ | $\varepsilon_1$ | $T_2$ | $\varepsilon_2$ | $T_3$ | $\varepsilon_3$ | $T(0,\tau)$ |
|---|---|---|---|---|---|---|---|---|
| 20 | 276 °K | 83 °K | 12° | 26 °K | − 3° | − 6 °K | 33° | 378 °K |
| 125 | 247 | 132 | 5 | 34 | − 6 | − 19 | 11 | 394 |
| 250 | 237 | 146 | 4 | 35 | − 7 | − 23 | 6 | 395 |
| 400 | 230 | 156 | 3 | 36 | − 7 | − 26 | 6 | 396 |
| 500 | 227 | 159 | 3 | 36 | − 7 | − 28 | 5 | 395 |
| 700 | 223 | 165 | 2 | 36 | − 7 | − 30 | 4 | 394 |
| 1000 | 219 | 170 | 2 | 36 | − 6 | − 31 | 3 | 394 |
| 1200 | 217 | 173 | 2 | 36 | − 6 | − 32 | 3 | 394 |

lation are inserted in (20-12), this latter equation should permit us to compute the corresponding temperature variation at any depth $x$ *below* the surface (normalized in accordance with 20-7); but the explicit evaluation of such changes can, at this stage, be left as an exercise for the interested reader.

It should also be stressed that the foregoing result holds good only if the transfer of energy is purely conductive. In the case of *radiative transfer* (discussed already in Chapter 8) such as would occur in a loosely packed dust across the empty, space separating individual dust grains, if can be shown that Equation (20-9) continues to hold good to a satisfactory approximation (cf. KOPAL, 1964a) for the total emissivity $B(T)$ in place of $T$ as the dependent variable, provided that the expressions (8-3) and (8-51) are used for the coefficients of heat conduction and specific heat of photon gas. Inserting these in (20-5) we find that, under these circumstances,

$$\frac{1}{c}\frac{\partial B}{\partial t} = \frac{1}{3k\rho}\frac{\partial^2 B}{\partial r^2} + \tfrac{1}{4}(1 - A)k\rho S e^{-k\rho r \sec z} \qquad (20\text{-}13)$$

where $k$ denotes the mean coefficient of opacity of lunar surface layers, defined by the equation

$$k = \frac{\partial B}{\partial r} \div \int_{\infty}^{0} \frac{1}{k_\lambda}\frac{\partial B_\lambda}{\partial r}\frac{\sigma\lambda}{\lambda}; \qquad (20\text{-}14)$$

the last term on the right-hand side of (20-13) representing the downward exponential attenuation of the incident flux (for its derivation cf., e.g., KOPAL 1959, p. 165). The boundary condition (19-10) on the surface rewritten in terms of the emissivity $B$ assumes then the linear form

$$\frac{\partial B}{\partial r} = \tfrac{3}{4} k\rho \{(1 - A) S \cos z - B\} \tag{20-15}$$

at $r=0$; while Equation (20-6) together with (20-2) implies that

$$\lim_{r \to \infty} B = 0. \tag{20-16}$$

In order to normalize the variables, let us set

$$B + \{(1 - A) k\rho c\, \omega S\}'\, \Theta \tag{20-17}$$

and

$$r = \left(\frac{2c}{3k\rho\,\omega}\right)^{\frac{1}{2}} x \tag{20-18}$$

in accordance with (20-7) if (8-3) and (8-51) are inserted for $\kappa$ and $C_v$; while the normalization of the time $t$ likewise follows (20-7). Moreover, let the product $k\rho$ be likewise hereafter regarded as constant. If so, Equations (20-9)–(20-10) can be rewritten as

$$\left. \begin{aligned} \frac{\partial \Theta}{\partial t} &= \frac{1}{2}\frac{\partial^2 \Theta}{\partial x^2} + e^{-h'x}, \\ \left(\frac{\partial \Theta}{\partial x}\right) &= h\{\Theta(0, t) - \phi(t)\} \end{aligned} \right\} \tag{20-19}$$

for $0 < x < \infty$, $\tau > 0$, where we have abbreviated

$$\phi(t) = \frac{\cos z(t)}{k\rho c\,\omega} \tag{20-20}$$

and

$$h = \tfrac{3}{4} k\rho a, \qquad h' = \tfrac{4}{3} h \sec z. \tag{20-21}$$

In addition, we shall require (in accordance with 20-6) that $\Theta(\infty, t) = 0$.

Unlike (20-9)–(20-10), Equations (20-19) are *linear* in the dependent variable, and their solution can be expressed (cf. LOWAN, 1935; KOPAL, 1964a) in the form

$$\Theta(x, t) = \sum_{j=1}^{3} \Theta_j(x, t), \tag{20-22}$$

where

$$\Theta_1(x, t) = \frac{h}{\sqrt{2\pi}} \int_0^\infty (x + \rho) e^{-h\rho} \int_0^t \frac{\phi(\tau)}{\sqrt{(t - \tau)^3}} \exp\left(-\frac{(x + \rho)^2}{2(t - \tau)}\right) d\rho\, d\tau, \tag{20-23}$$

$$\Theta_2(x, t) = \frac{h}{\sqrt{2\pi t}} \int_0^\infty \left\{f(\xi) - \frac{1}{h}\frac{\partial f}{\partial \xi}\right\} d\xi$$

$$\times \int_0^\infty \left\{\exp\left(-\frac{(x + \rho - \xi)^2}{2t}\right) - \exp\left(-\frac{(x + \rho + \xi)^2}{2t}\right)\right\} e^{-h\rho}\, d\rho, \tag{20-24}$$

and

$$\Theta_3(x,t) = \frac{h}{\sqrt{2\pi t}} \int\limits_0^\infty d\xi \int\limits_0^\infty e^{-h\rho} d\rho \int\limits_0^t \left\{ 1 - \frac{1}{h} \frac{\partial}{\partial \xi} \right\} e^{-h'\xi}$$

$$\times \left\{ \exp\left( -\frac{(x+\rho-\xi)^2}{2(t-\tau)} \right) - \exp\left( -\frac{(x+\rho+\xi)^2}{2(t-\tau)} \right) \right\} \frac{d\tau}{\sqrt{t-\tau}} . \qquad (20\text{-}25)$$

The expression for $\Theta_1(x,t)$ represents the effects of insolation on the surface; $\Theta_3(x,t)$, their attenuation with the depth; while $\Theta_2(x,t)$ stands for the attenuation of the initial distribution of the emissivity $\Theta(x,0) \equiv f(x)$.

The explicit evaluation of the first of these three integrals reveals that, for slowly-varying $\phi(t)$,

$$\Theta_1(x,t) = \phi(t) \{ \text{erfc } \alpha - \exp(\beta^2 - \alpha^2) \text{erfc } \beta \}, \qquad (20\text{-}26)$$

where we have abbreviated

$$\alpha = \frac{x}{\sqrt{2t}} \quad \text{and} \quad \beta = \frac{x = ht}{\sqrt{2t}} . \qquad (20\text{-}27)$$

At the surface $(x=0)$ Equation (20-26) reduces to

$$\Theta_1(0,t) = \phi(t) \{ 1 - e^{\frac{1}{2}h^2 t} \text{erfc}(\tfrac{1}{2}h^2 t)^{\frac{1}{2}} \}, \qquad (20\text{-}28)$$

which for $t=0$ becomes equal to zero, while for $t=\infty$ it reduces to $\phi(t)$. Therefore, if an initially cold surface is irradiated at $t=0$ by a source of flux $\pi S$, its emissivity $\Theta_1$ will gradually rise from zero to its equilibrium value $\phi(t)$ in a time necessary for the factor $\exp(\tfrac{1}{2}h^2 t) \text{erfc}(\tfrac{1}{2}h^2 t)^{\frac{1}{2}}$ to diminish from 1 to 0.

If we turn now to the equation (20-24) for $\Theta_2$, its partial integration leads to

$$\Theta_2(x,t) = \frac{1}{\sqrt{2\pi t}} \int\limits_0^\infty f(\xi) \left\{ \exp\left( -\frac{(x+\xi)^2}{2t} \right) + \exp\left( -\frac{(x-\xi)^2}{2t} \right) \right\} d\xi \qquad (20\text{-}29)$$

$$- h e^{hx + \frac{1}{2}h^2 t} \int\limits_0^\infty f(\xi) e^{h\xi} \text{erfc} \frac{x+ht+\xi}{\sqrt{2t}} d\xi .$$

If, moreover, the initial distribution $f(x)$ of emissivity at $t=0$ could be regarded as a constant (say, $f$) – i.e., if the layer in question were initially isothermal – then, for $t>0$, the foregoing equation simplifies to

$$\Theta_2(x,t) = f \{ \text{erf } \alpha + \exp(\beta^2 - \alpha^2) \text{erfc } \beta \}, \qquad (20\text{-}30)$$

where $\alpha$ and $\beta$ continue to be given by (20-27). When $t \to \infty$, the asymptotic behaviour of erf $\alpha$ or erfc $\beta$ as $\alpha \to \beta \to \infty$ reveals that

$$\Theta_2(\infty, t) = \Theta_2(x,0) = f . \qquad (20\text{-}31$$

On the other hand, at the surface $(x=0)$ we have

$$\Theta_2(0, t) = f\, e^{h^2 t/2}\, \mathrm{erfc}\,(h^2 t/2)^{\frac{1}{2}};\tag{20-32}$$

and, lastly,

$$\Theta_2(x, \infty) = 0.\tag{20-33}$$

The evaluation of the expression (20-25) for $\Theta_3(x, t)$ proves to be the most troublesome of the three (for fuller details, cf. KOPAL, 1964a); but after a certain amount of rather troublesome algebra we eventually establish that

$$
\begin{aligned}
\Theta_3(x, t) =\ & \frac{1}{h'^2}\left\{\frac{h'+h}{h'-h}\right\} e^{(\alpha'+\beta)^2 - \beta^2}\, \mathrm{erfc}\,(\alpha'+\beta) \\
& - \frac{2 e^{(\alpha+\beta)^2 - \beta^2}}{h(h'-h)}\, \mathrm{erfc}\,(\alpha+\beta) + \frac{2(1+hx)}{hh'}\, \mathrm{erfc}\,\beta \\
& + \frac{e^{(\alpha'-\beta)^2 - \beta^2}}{h'^2}\{(2-4\alpha'^2)\,\mathrm{erfc}\,\alpha' - (1-4\alpha'^2)\,\mathrm{erfc}\,(\alpha'-\beta)\} \\
& + \frac{4\alpha'}{h'^2\sqrt{\pi}}\{e^{-2\alpha'\beta} - e^{-\beta^2}\},
\end{aligned}\tag{20-34}
$$

where we have abbreviated

$$\alpha' = h'\,(t/2)^{\frac{1}{2}}.\tag{20-35}$$

For $x=0$, the foregoing equation reduces to

$$\Theta_3(0, t) = \frac{2}{hh'}\left\{1 + \frac{h e^{\alpha'^2}\,\mathrm{erfc}\,\alpha' - h'\, e^{\alpha^2}\,\mathrm{erfc}\,\alpha}{h'-h}\right\},\tag{20-36}$$

while, as $t\to\infty$,

$$\Theta_3(x, \infty) = \frac{2(1+hx)}{hh'}\quad\text{for}\quad x<\infty,\tag{20-37}$$

so that, in steady state on the surface,

$$\Theta_3(0, \infty) = \frac{2}{hh'} = \frac{3}{2h^2}\cos z;\tag{20-38}$$

this corresponds to the "law of limb-darkening" in our particular case.

As $x\to 0$, Equations (20-26) as well as (20-34) reduce to

$$\Theta_1(\infty, t) = \Theta_3(\infty, t) = 0\tag{20-39}$$

for $t<\infty$, as they should in accordance with (20-6); while $\Theta_2(\infty, t)$ has already been given by (20-31).

Numerical calculations of the relative efficiency of radiative transfer of heat in lunar surface layers, based on the foregoing theory, have recently been undertaken by

KOPAL (1964a). Their outcome revealed that, even for a high degree of porosity of the material, the heat transport by radiation will be quite negligible in comparison with conduction under lunar night-time conditions, and is likely to remain at most times – except, possibly, at noon temperatures in the lunar tropics. But even then the contributions of conduction and radiative transfer could be distinguished by one particular feature: namely, unlike in the case of conduction, *radiative transport of heat should exhibit no appreciable time-lag*; and sub-surface temperatures controlled by it should be essentially in phase. The fact that (as we shall explain in more detail when we come to discuss the thermal emission of the Moon in the microwave domain) such a phase-lag is definitely indicated by the observations demonstrates alone that radiative transfer of heat in lunar surface layers is likely to be unimportant. At substantially higher temperatures (such as attained, for instance, on the surface of the planet Mercury), however, the opposite may be the case.

Throughout all our foregoing discussion we have been concerned with the *diurnal* variation of temperature on the Moon, caused by the fluctuating height of the Sun above the lunar horizon in the course of each month. However, there are times when these temperature variations become very much more rapid: and that is during the relatively brief intervals of *total eclipses* of the Moon when, for a few hours, our satellite passes through the shadow-cone cast into space by the Earth (cf. Chapter 5). The relative dimensions of the Sun, Earth, Moon, and the present characteristics of their orbits are such that this event is likely to occur about once each terrestrial year (i.e., twelve days on the Moon); and although the duration of the eclipses is relatively short (around four hours), the eclipsed lunar surface in the shadow cone of the Earth experiences almost as large a variation in temperature as between day and night. In particular, the fall in temperature as the Sun gradually disappears behind the apparent disk of the Earth is so precipitous that a drop from 390° to 180 °K may occur within 40–50 minutes of our time; and the emergence of Sun at the opposite limb by the end of totality brings about an even more frantic a "heat wave" than the cold wave which preceded it by about an hour. The actual temperature gradient during such times may attain 5 deg/min – a value 40 to 50 times as large as that characterising the maximum diurnal temperature changes on the Moon; but even so steep a gradient does not seem capable of doing any appreciable damage to rocks which are exposed to it for no more than two hours per year (cf., e.g., RYAN, 1962).

Early observers of the infrared thermal emission of the Moon PETTIT and NICHOLSON, 1930; PETTIT, 1935, 1940, 1945) lacked sufficient resolving power to localise precisely the type of the ground from which radiation was received; and, accordingly, their results referred to the average thermal properties of the lunar surface rather than to those of any particular type of ground. The same was still true of the lunar isotherms published by GEOFFRION, KORNER and SINTON (1960). However, with increasing sensitivity of the IR-detectors (which permitted the isolation of smaller areas of the lunar surface by the use of narrower diaphragms) new measurement – both during a lunation and within eclipses – revealed that local temperatures on the Moon are by no means functions of the instantaneous zenith distance alone.

During the total lunar eclipse of March 13, 1960, SHORTHILL, BOROUGH and CON-
LEY (1960) discovered that several ray craters on the Moon (Aristarchus, Copernicus
and Tycho) cooled off less rapidly than their surroundings and that, in particular, the
crater Tycho remained by about 40 °K warmer than its environs an hour after it
entered the umbra. This phenomenon was promptly verified during the next eclipse
of the Moon on September 8 of the same year by SINTON (1960); and further extensive
measurements of such thermal anomalies – during the same eclipse as well as within
a lunation – were undertaken by SAARI and SHORTHILL (1963). In the course of this
work these investigators discovered a further important fact: namely, that many spots
on the lunar surface – in particular, the *ray craters* Aristarchus, Copernicus, Tycho
(but also smaller ones, such as Dawes, Proclus or Manilius), *which remain warmer at
night (or during eclipse) than the surrounding landscape, are cooler again in daytime*;
and vice versa. Moreover, by use of more refined techniques during the lunar eclipse
of December 19, 1964, SAARI and SHORTHILL (1965) discovered the existence of virtual-

Fig. 20-3.  A composite image of the full Moon, as reconstructed from photoelectric scans of the
lunar surface in the visible part of the spectrum, with a spatial resolution of approximately ten
seconds of arc. (By courtesy of Drs. J. Saari and R. W. Shorthill,
Boeing Scientific Research Laboratories.)

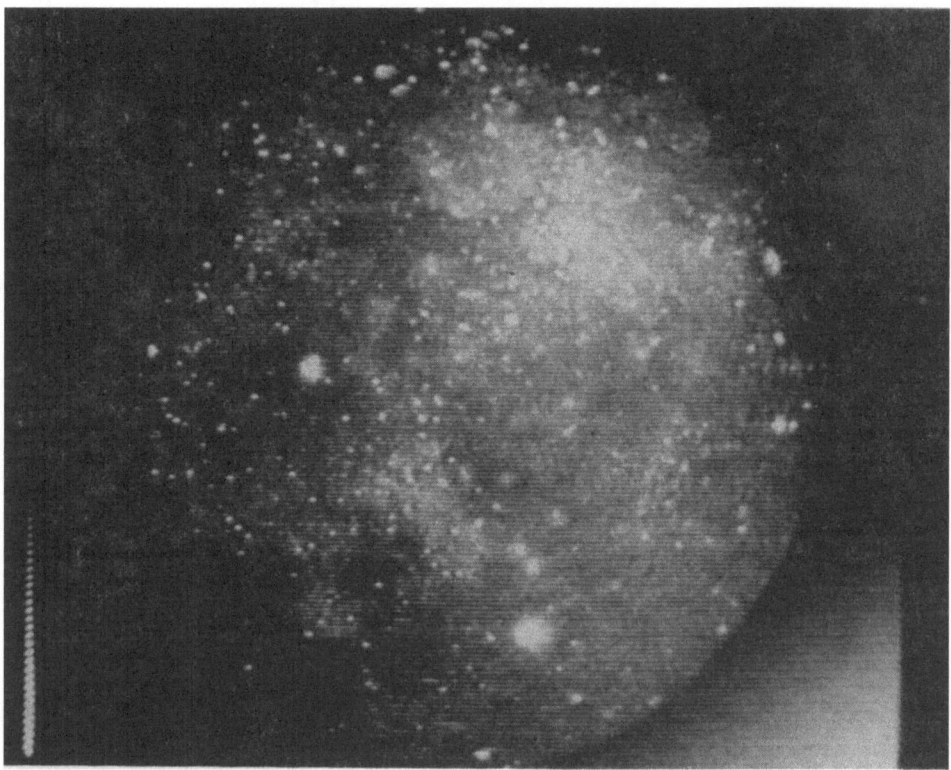

Fig. 20-4a.   A composite image of totally eclipsed Moon, based on infrared scans (through 10–12 $\mu$ window) of the lunar surface, made during the eclipse of December 19, 1964, by Drs. J. Saari and R. W. Shorthill of the Boeing Scientific Research Laboratories (cf. SAARI and SHORTHILL, 1965). North is on the top, and West to the left.

Each white spot of the image corresponds to a region of enhanced thermal emission (indicative of less rapid rate of cooling); the three most conspicuous ones being identical in position with the craters Tycho (bottom), Copernicus (left) and Langrenus (right limb). A fuller key for identification is given on the accompanying Figure 20-4b.

The Earth's shadow progressed from the West to the East on the Moon. Therefore, the enhanced emission from the East quadrant as compared with the West is due to the fact that, at the time when the underlying scans were made, the eclipse was nearly over and the western quadrant had a longer time to cool off. For the same reason the eastern (right) limb still remains visible in the infrared, while the western (the first one to be eclipsed) radiated already too little to register.

Note also the relative thermal enhancement of certain maria (e.g., Humorum) in comparison with the adjacent landscape.

ly thousand of localised thermal anomalies (most of which – though not all – were found to coincide in position with known ray-craters or other crater formations of relatively high albedo) which remained noticeably warmer than their surroundings during the brief interval of totality. All these results have demonstrated conclusively that *the photometric homogeneity of the lunar face*, stressed in the preceding chapter in connection with the photometry of the scattered light of our satellite, *no longer obtains in the infrared*; and conspicuous departures from it are found in great numbers.

Fig. 20-4b.   An identification key to infrared image of the eclipsed Moon as shown on Figure 20-4a. (After SAARI and SHORTHILL, 1965.)

In order to demonstrate the extent to which this is true, let us compose a composite image of the full Moon in visible (scattered) light, as shown on the accompanying Figure 20-3 and obtained by Shorthill and Saari with the same technique as their infrared scans, with Figure 20-4a showing an infrared view of the eclipsed Moon on 1964 December 19. This latter figure shows what an eclipsed Moon would look like to us if our eyes were sensitive to infrared, rather than visible, light! The bright (i.e., 'hot') spots reminiscent of a starry sky cover the lunar face in great numbers; and the reader should have no difficulty to spot among them the ray craters Tycho, Copernicus, Kepler or Langrenus (i.e., precisely those which appeared 'dark' on Figure 20-1 in daytime); and a fuller identification key is provided on Figure 20-4b; but their total number is almost beyond the means of counting; and many are apparently smaller than the resolving power of the 74-inch telescope of the Helwan Observatory at Kottamia, Egypt, with which the work of Saari and Shorthill was carried out. The actual range of thermal anomalies revealed by anomalous infrared emission from the craters Tycho (or Heinsius A) at the time of the eclipse is well shown on a single scan passing through these formations, and reproduced on Figure 20-5.

In order to appreciate the full extent of thermal diversity of the lunar surface as revealed by the recent work of Saari and Shorthill, compare the isophotes of visible ight in the central part of Mare Tranquilitatis (containing the impact point of the

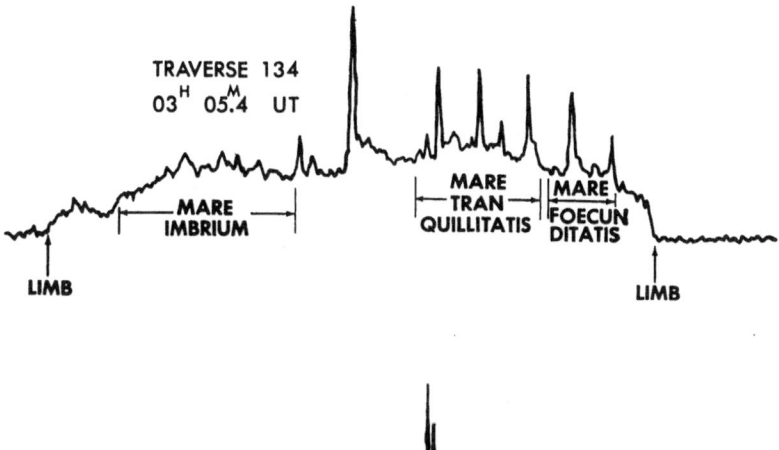

Fig. 20-5.   Tracings of infrared signals from the eclipsed Moon in the directions indicated on Figure 20-4b. These scans were made near the end of totality, and show that e.g. Mare Humorum remained approximately 10 °K warmer than its surroundings. Note also the structure in the thermal profile of the crater Tycho. (After SAARI and SHORTHILL, 1965.)

American Ranger 8), just before full Moon (age 14.93 days) one day before the 1964 December 19 eclipse, as shown on the accompanying Figure 20-6 with the isothermal contours of the same region, measured at the same time (Figure 20-7); and with thermal contours of the region during the total eclipse the next day (Figure 20-8). The diversity of the records could not be greater; and reveals that the thermal properties of the Moon's crust are anything but uniform over even relatively small sections of the lunar surface.

How to account for these phenomena? First, let us stress a fact which should already be obvious from the preceding records: namely, that the observed thermal anomalies are much too large to be accounted for by any variations in albedo at optical frequencies: the lunar regions which remain cooler in daytime and warmer at night (or during eclipse) reflect, indeed, more light (and, therefore, absorb less) than the average; but quantively this difference alone would be insufficient to account for the

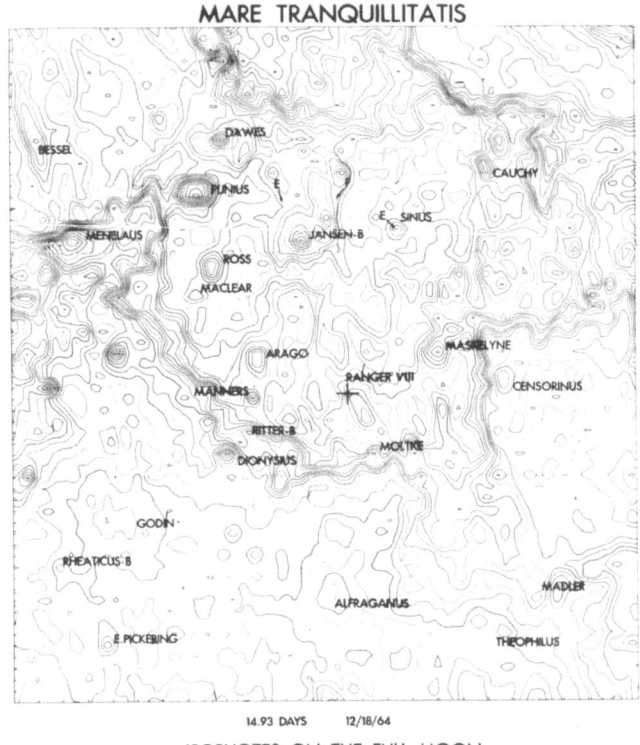

MARE TRANQUILLITATIS

14.93 DAYS  12/18/64

ISOPHOTES ON THE FULL MOON

Fig. 20-6. Isophotes on the full Moon in Mare Tranquillitatis. (By courtesy of Drs. J. Saari and R. W. Shorthill, Boeing Scientific Research Laboratories.)

measured thermal anomalies. In order to do so, we must also recognise that, in the regions concerned, either the coefficient $C_v$ of specific heat, or $\kappa$ of heat conduction – possibly both – possess different values than in the adjacent ground. Whether this difference arise from a chemical (or mineralogical) difference in composition of the material of the "hot spots" – i.e., real change in the values of $\kappa$ and (or) $C_v$ per unit mass – or whether it is caused by a different degree of compactness of essentially identical substances – i.e., a change in the effective values of $\kappa$ and $C_v$ per unit volume, is as yet impossible to say; both factors may play their role and influence the observed outcome. But whatever may be the case, the results by Shorthill and Saari leave no room for doubt that the actual local temperature prevailing at any spot of lunar surface is by no means uniquely determined by the zenith distance of the Sun or the time of the day or night; and anomalies due to local conditions may exceed 60 °K.

Thus far in our survey of the thermal properties of the lunar surface we have limited ourselves to a discussion of the information which can be extracted from the measurements of thermal radiation of the Moon in the near infrared (mainly at wavelengths between 8–13 $\mu$'s). Observations in infrared are greatly hampered by absorption of

MARE TRANQUILLITATIS

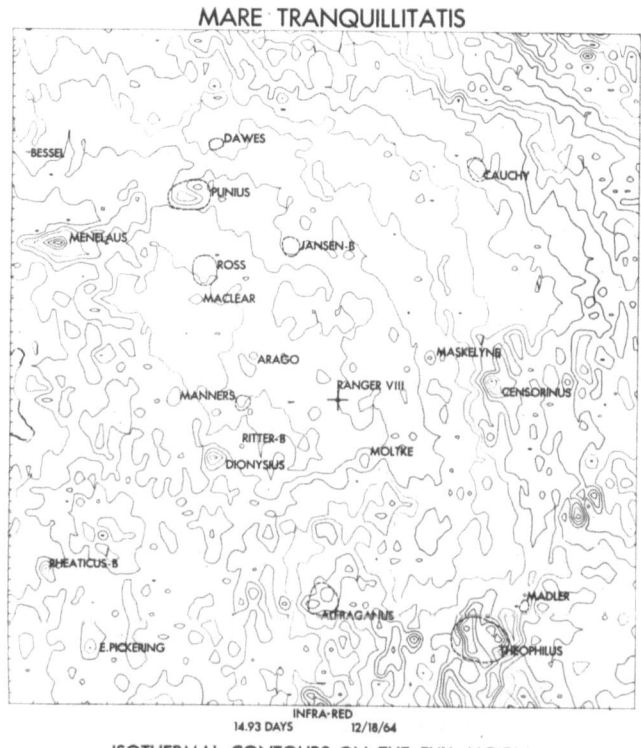

INFRA-RED
14.93 DAYS    12/18/64
ISOTHERMAL CONTOURS ON THE FULL MOON

Fig. 20-7.   Isothermal contours on the full Moon in Mare Tranquillitatis. (By courtesy of Drs.
J. Saari and R. W. Shorthill, Boeing Scientific Research Laboratories.)

light in our atmosphere, which effectively closes down this infrared window for $\lambda > 13.8$
$\mu$; and although a glimpse of the celestial bodies can be obtained around $\lambda \sim 20\mu$ (cf.
Low, 1965) further penetration in the infrared is virtually prohibited by the interlocking absorption bands of water vapour and carbon dioxide, which become so opaque
in this part of the spectrum that even the Sun becomes completely invisible through
them – until wavelengths of the order of 1 mm (i.e., $1000\mu$) are reached in the domain
of *radio-frequencies*. The wavelength range between 1 mm and about 10 meters constitutes the second "atmospheric window" through which celestial bodies can be observed (albeit by very different techniques); its upper limit being imposed, not by absorption of any gas, but by the reflecting properties of the terrestrial ionosphere.

Observations of the Moon in the domain of radio frequencies are of relatively
recent date. Thermal radiation of the Moon reaching us through this window was
first detected by DICKE and BERINGER (1946), but systematic measurements were initiated by PIDDINGTON and MINNETT (1949) working at the wavelength of 12.5 mm
(i.e., frequency of 24000 Mc/sec) and using an antenna of 0°.075 beamwidth. In the
years which elapsed since that time this part of the lunar spectrum has rapidly been

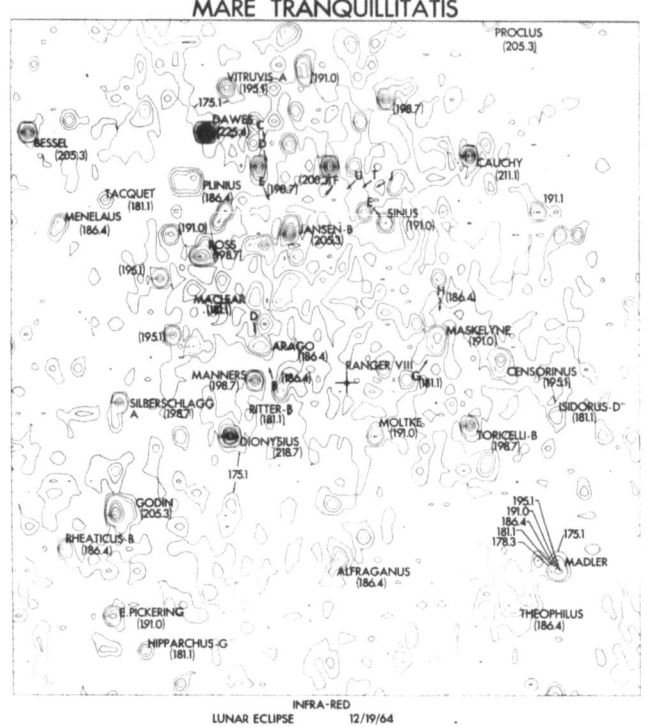

Fig. 20-8.   Thermal Contours during totality in Mare Tranquillitatis. (By courtesy of Drs. J. Saari
          and R. W. Shorthill, Boeing Scientific Research Laboratories.)

explored by a number of investigators, working at wavelengths ranging from 1 mm
(Low and DAVIDSON, 1965) to 168 cm (BALDWIN, 1961). A list of such investigations
of the monthly variation of the thermal emission of the Moon in the microwave domain and their principal results are listed in the accompanying Table 20-2. A glance
at them discloses that the range of the corresponding temperatures is systematically
*smaller* than that measured in the infrared – becoming the more reduced, the longer
the wavelength. Secondly, the maxima and minima of microwave emission do not
follow in phase the altitude of the Sun above the horizon, but *lag behind* the infrared
thermal emission by amounts increasing again with the wavelength.

     Before we proceed to discuss these results in more detail and correlate them with
the outcome of infrared measurements, a few words may be in place to explain the
method underlying their interpretation. The first point to be realized in this connection
is the fact that (due to the nature of the radio receivers) *the observations in the microwave domain* – unlike those in the infrared – *are highly monochromatic.* Therefore, in
place of the *total emissivity B(T)* obtainable (after a number of corrections have been
applied for atmospheric as well as instrumental absorption) from infrared measure-

ments, they yield the *intensity* $I_\lambda$ of thermal radiation *in a very narrow bandwidth*. Moreover, as the lunar surface layer become increasingly transparent to microwaves (i.e., such radiation leaving the Moon originates at an increasing depth below the visible surface), we no longer have a sharp boundary condition equivalent to (20-5). Instead, we have to depart from the time-dependent monochromatic equation of *radiative equilibrium* in a plane-parallel layer, of the form

$$\left\{\frac{1}{c}\frac{\partial}{\partial t} - \cos\theta\frac{\partial}{\partial r}\right\}I_\lambda = k\rho(B_\lambda - I_\lambda), \tag{20-40}$$

where $\theta$ denotes the angle between the incident and emitted ray; and the depth $r$ is (as in 20-4) measured positively downwards.

As is well known (cf., e.g. FORSYTH, 1929) this partial differential equation can be formally solved to yield for the intensity $I_\lambda(0, \theta; t)$ of radiation emerging from the surface an expression of the form

$$I_\lambda(0,\theta;t) = \int_0^\infty B_\lambda(r,t)e^{-k\rho r\sec\theta}k\rho\sec\theta\,dr, \tag{20-41}$$

where, for a time-dependent equation of transfer,

$$t = \frac{r-a}{c\cos\theta} \tag{20-42}$$

represents its subsidiary integral (in which $a$ stands for an arbitrary constant).

Let us assume next that the layers in question are in local thermodynamic equilibrium. If so, the emissivity $B_\lambda$ should continue to be given by the Planck law (20-1) which, in the long-wave domain, can be satisfactorily approximated by the Rayleigh-Jeans formula

$$B_\lambda(r,t) = \frac{8\pi c\bar{k}}{\lambda^4}T(r,t) \tag{20-43}$$

which, unlike (20-1), is linear in the temperature $T$. Now let us approximate the latter by means of our previous Equation (20-12) in terms of the coefficients $T_n$ and $\varepsilon_n$ related to the variation of temperature at $r=0$*. If so, then by a combination of (20-41) and (20-43) with (20-12) it follows immediately that

$$I_\lambda(0,\theta,t) = \frac{8\pi c\bar{k}m}{\lambda^4}\sum_{n=0}^\infty T_n\int_0^\infty e^{-(m+\sqrt{n})x}\cos\{n\tau - x\sqrt{n}(1-\mu\sqrt{n}) + \varepsilon_n\}\,dx \tag{20-44}$$

* In doing so we assume, in effect, that the solar heat propagates downwards by conduction only (i.e., ignore the heating effect of the microwave tail of the solar spectrum).

TABLE 20-2

Coefficients of the Equation $T = T_0' + T_1' \cos(\tau - \psi)$

| Authority | $\lambda$ | $T_0'$ | $T_1'$ | $T_1'/T_0'$ | $\psi$ |
|---|---|---|---|---|---|
| 1. Low and Davidson (1965) | 0.10 | $248 \pm 25$ | $118 \pm 13$ | $0.48 \pm 0.03$ | 12 |
| 2. Fedoseev (1963) | 0.13 | $219 \pm 35$ | $120 \pm 18$ | $0.55 \pm 0.04$ | 16 |
| 3. Naumov (1963) | 0.18 | $240 \pm 48$ | $115 \pm 23$ | $0.48 \pm 0.05$ | 14 |
| 4. Gary, Stacey and Brake (1965) | 0.33 | $108 \pm 28$ | $74 \pm 7$ | $0.36 \pm 0.04$ | 24 |
| 5. Kislyakov (1961) | 0.40 | $230 \pm 23$ | $73 \pm 7$ | $0.32 \pm 0.03$ | 24 |
| 6. Kislyakov and Salomonovich (1963) | 0.40 | $228 \pm 34$ | $85 \pm 13$ | $0.34 \pm 0.05$ | 27 |
| 7. Kislyakov and Plechkov (1964) | 0.40 | $204 \pm 9$ | $56 \pm 3$ | $0.27 \pm 0.03$ | 23 |
| 8. Coates (1959) | 0.43 | $230 \pm 50$ | 60 | 0.26 | |
| 9. Staelin, Barrett and Kusse (1964) | | | | | |
| 10. Mitchell and Whitehurst (1958) | 1.75 | 150 | 25 | 0.17 | |
| 11. Salomonovich (1958) | 0.8 | $197 \pm 20$ | $32 \pm 5$ | $0.16 \pm 0.03$ | 40 |
| 12. Amenitssky, Noskova and Salomonovich (1960) | 0.8 | $210 \pm 32$ | $42 \pm 6$ | $0.20 \pm 0.02$ | $30 \pm 5$ |
| 13. Salomonovich and Lozovsky (1962) | 0.8 | $211 \pm 31$ | $40 \pm 6$ | $0.19 \pm 0.02$ | 30 |
| 14. Gibson (1958) | 0.86 | $180 \pm 18$ | 30 | 0.33 | 39 |
| 15. Piddington and Minnett (1949) | 1.25 | $249 \pm 13$ | 52 | 0.21 | 45 |
| 16. Zelinskaya, Troitsky and Fedoseev (1959) | 1.63 | $224 \pm 28$ | $36 \pm 5$ | $0.16 \pm 0.02$ | $37 \pm 3$ |
| 17. Kamenskaya, Semenov, Troitski and Plechkov (1962) | 1.63 | $208 \pm 7$ | $37 \pm 2$ | $0.18 \pm 0.01$ | 30 |
| 18. Dimitrienko and Kamenskaya (1963) | 1.63 | $207 \pm 7$ | $32 \pm 2$ | $0.15 \pm 0.01$ | |
| 19. Salomonovich and Koschenko (1961) | 2.0 | $190 \pm 29$ | $19 \pm 3$ | $0.10 \pm 0.01$ | $45 \pm 5$ |
| 20. Grebenkamper (1958) | 2.0 | $200 \pm 10$ | | | |
| 21. Kaidanovsky, Ihsanova, Apushinski and Shivris (1961) | 2.3 | | $14 \pm 4$ | | 35 |
| 22. Mayer, McCullough and Sloanaker (1961) | 3.15 | $195 \pm 25$ | $12 \pm 5$ | $0.062 \pm 0.025$ | $44 \pm 15$ |
| 23. Troitsky and Zelinskaya (1955) | 3.2 | $170 \pm 34$ | $< 12$ | $< 0.07$ | |
| 24. Koschenko, Losovsky and Salomonovich (1961) | 3.2 | $223 \pm 33$ | $17 \pm 3$ | $0.076 \pm 0.011$ | $45 \pm 5$ |

TABLE 20-2 (Continued)

| Authority | $\lambda$ | $T_0'$ | $T_1'$ | $T_1'/T_0'$ | $\psi$ |
|---|---|---|---|---|---|
| 25. Strezhneva and Troitsky (1961) | 3.2 | 255 ± 38 | 16 ± 3 | 0.065 ± 0.012 | 50 |
| 26. Krotikov, Porfirev and Troitsky (1961) | 3.2 | 210 ± 5 | 14 ± 2 | 0.067 ± 0.010 | 55 |
| 27. Bondar, Zelinskaya, Porfirev and Strezhneva (1962) | 3.2 | 215 ± 2 | 15 ± 2 | 0.070 ± 0.009 | |
| 28. Medd and Brotten (1961) | 9.4 | 220 ± 11 | < 11 | < 0.05 | |
| 29. Koschenko, Kuzmin and Salomonovich (1961) | 9.6 | 230 ± 35 | < 5 | < 0.02 | |
| 30. Krotikov (1962) | 9.6 | 218 ± 5 | 7 | 0.032 | 40 |
| 31. Akabane (1955) | 10 | 315 ± 50 | 36 | 0.11 | 45 |
| 32. Castelli, Ferioli and Aarons (1960) | 10 | 256 ± 38 | | | |
| 33. Mezger and Strassl (1959) | 20.5 | 250 ± 30 | < 5 | < 0.02 | |
| 34. Waak (1961) | 20.8 | 205 | 5 | | |
| 35. Davies and Jennison (1960) | 22 | 270 ± 54 | | | |
| 36. Castelli, Ferioli and Aarons (1960) | 23 | 254 ± 38 | < 6 | | |
| 37. Ko (1961) | 32 | 246 ± 12 | | | |
| 38. Razin and Fedorov (1963) | 32.3 | 233 ± 5 | | | |
| 39. Denisse and LeRoux (1957) | 33 | 220 ± 6 | | | |
| 40. Krotikov and Porfirev (1963) | 35–36 | 236 ± 6 | | | |
| 41. Krotikov (1963) | 50 | 241 ± 12 | | | |
| 42. Seeger, Westerhout and Conway (1957) | 75 | 185 ± 20 | | | |
| 43. Baldwin (1961) | 168 | 236 ± 8 | | | |

## REFERENCES

1. Low, F. J. and Davidson, A. W.: 1965, *Astrophys. J.* **142**, 1278.
2. Fedoseev, L. N.: 1963, *Izv. Vysših Učebn. Zavedenii Radiofiz.* **6**, 6.
3. Naumov, A. I.: 1963, *Izv. Vysših Učebn. Zavededenii Radiofiz.* **6**, 848.
4. Gary, B. L., Stacey, J. and Drake, F. D.: 1965, *Astrophys. J. Suppl.* No. 108.
5. Kislyakov, A. G.: 1961, *Izv. Vysših Učebn. Zavedenii Radiofiz.* **4**, 433.
6. Kislyakov, A. G., and Salomonovich, A. E.: 1963, *Izv. Vysših Učebn. Zavedenii* **6**, 431.
7. Kislyakov, A. G., and Plechkov, V. M.: *Izv. Vysših Učebn. Zavedenii Radiofizika* **7**, No. 1.
8. Coates, R. J. 1959, *Astrophys. J.* **64**, 326.
9. Staelin, D. H., Barrett, A. H. and Kusse, B. R.: 1964, *Astrophys. J.* **69**, 69.
10. Mitchell, F. F. and Whitehurst, R. N.: 1958, *Univ. of Alabama Radio Astr. Lab. Rept.*, No. 1, pp. 5–10.
11. Salomonovich, A. E.: 1958, *Astron.* **35**, 129.
12. Amenitsky, N. A., Noskova, R. I., and Salomonovich, A. E.: 1960, *Astron. Ž* **37**, 185.
13. Salomonovich, A. E. and Losovskii, B. Ya.: 1962, *Astron. Z.* **39**, 1074.
14. Gibson, J. E.: 1958, *Proc. Inst. Radio Engrs.* **46**, 280.
15. Piddington, J. H. and Minnett, H. C.: 1949, *Austr. J. Sci. Res.* (A) **2**, 63.
16. Zelinskaya, M. R., Troitskii, V. S., and Fedoseev, L. N.: 1959, *Astron. Ž* **36**, 643.
17. Kamenskaya, S. A., Semenov, B. I., Troitskii, V. S., and Plechkov, V. M.: 1962, *Izv. Vysših Učebn. Zavedenii Radiofiz.* **5**, 882.
18. Dimitrienko, D. A. and Kamenskaya, S. A.: 1953 *Izv. Vysših Učebn. Zavedenii Radiofiz.* **6**, 655.
19. Salomonovich, A. E., and Koschenko, V. N.: 1961, *Izv. Vysših Učebn. Zavedenii Radiofiz.* **4**, 425
20. Grebenkamper, C. J.: 1958, *U.S. Naval Res. Lab. Rept.*, No. 5151.
21. Kaidanovsky, N. L., Ihsanova, V. N., Apushinski, G. P., and Shivris, O. N.: 1961, *Izv. Vysših Učebn. Zavedenii Radiofiz.* **4**, 428.
22. Mayer, C. H., McCullough, T. P., and Sloanaker, R. M.: 1961 in *Planets and Satellites* (Univ. of Chicago Press), pp. 448.
23. Troitskii, V. S., and Zelinskaya, M. R.: 1955, *Astron. Ž.* **32**, 550.
24. Koschenko, V. N., Lozovsky, B. Ya., and Salomonovich, A. E.: 1961, *Izv. Vysših Učebn. Zavedenii Radiofiz.* **4**, 596.
25. Strezhneva, K. M. and Troitskii, V. S. (1961), *Izv. Vysših Učebn. Zavedenii Radiofiz.* **4**, 600.
26. Krotikov, V. D., Porfirev, V. A., and Troitskii, V. S.: *Izv. Vysših Učebn. Zavedenii Radiofiz.* **4**, 1004.
27. Bondar, L. I., Zelinskaya, V. A., Porfirev, V. A., and Strezhneva, K. M.: (1962), *Izv. Vysših Učebn. Zavedenii Radiofiz.* **8**, 802.
28. Medd, W. J. and Broten, N. W.: 1961, *Planet. Space Sci.* **5**, 307.
29. Koschenko, V. N., Kuzmin, A. D., and Salomonovich, A. E.: 1961 *Izv. Vysših Učebn. Zavedenii Radiofiz.* **4**, 425.
30. Krotikov, V. D.: 1962, *Izv. Vysših Učebn. Zavedenii Radiofiz.* **5**, 604.
31. Akabane, K.: 1955, *Proc. Imper. Acad. Japan* **31**. 161.
32. Castelli, J. P., Ferioli, C. P., and Aarons, J.: 1960, *Astrophys. J.* **65**, 485.
33. Mezger, P. G. and Strassl, H.: 1959, *Planet. Space Sci.* **1**, 213.
34. Waak, J. A.: 1960, *Astrophys. J.* **65**, 565.
35. Davies, R. D. and Jennison, R. C.: 1960 *Observatory* **80**, 74.
36. Castelli, J. P., Ferioli, C. P., and Aarons, J.: 1960, *Astrophys. J.* **65**, 485.
37. Ko (1961) as quoted i n *Planets and Satellites* (Univ. of Chicago Press), p. 448.
38. Razin, V. A. and Fedorov, V. T.: 1963, *Izv. Vysših Učebn. Zavedenii Radiofiz.* **6**, No. 5.
39. Denisse, J. F. and LeRoux, E.: 1957 as quoted in *Astrophys. J.* **126**, 585.
40. Krotikov, V. D. and Porfirev, V. S.: 1963, *Izv. Vysših Učebn. Zavedenii Radiofiz.* **6**. 245.
41. Krotikov, V. D.: 1963, *Izv. Vysših Učebn. Zavedenii Radiofiz.* **6**, No. 6.
42. Seeger, C. L., Westerhout, G., and Conway, R. G.: 1957, *Astrophys. J.* **126**, 585.
43. Baldwin, J. E.: 1961, *Monthly Notices Roy. Astron. Soc.* **122**, 513.

where

$$m = \left(\frac{2\kappa\rho}{C_v\omega}\right)^{\frac{1}{2}} k \sec\theta \tag{20-45}$$

and

$$\mu = \left(\frac{2\kappa\rho}{\rho C_v}\right)^{\frac{1}{2}} \frac{\sec\theta}{c} \tag{20-46}$$

are nondimensional constants. The extra term $\mu n x$ in the argument of the periodic function on the right-hand side of (20-44) stems from the fact that, in view of (20-42), the time $t$ involved in $B_\lambda(r, t)$ cannot be treated as constant when integrating with respect to $r$ (or $x$); and the same effect will also slightly change the phases $\varepsilon_n$.

A term-by-term integration on the right-hand side of (20-44) reveals that, more explicitly,

$$I_\lambda(0, \theta, t) = \frac{8\pi c k}{\lambda^4} \sum_{n=0}^{\infty} \frac{T_n \cos(n\tau - \phi_n + \varepsilon_n)}{\sqrt{(1 + \delta_n)^2 + (\delta_n - d_n)^2}}, \tag{20-47}$$

where we have abbreviated

$$\delta_n = \frac{\sqrt{n}}{m} = \left(\frac{nC_v\omega}{2\kappa\rho}\right)^{\frac{1}{2}} \frac{\cos\theta}{k}, \tag{20-48}$$

$$d_n = \frac{n\mu}{m} = \frac{n\omega}{k\rho c} \tag{20-49}$$

and

$$\phi_n = \tan^{-1} \frac{\delta_n - d_n}{\delta_n + 1}. \tag{20-50}$$

In view of the fact that the right-hand side of (20-49) contains the velocity of light $c$ as a divisor, the numerical magnitude of the factor $\mu$ is likely to be quite small; and ignoring it we can set $d_n = 0$.* To this degree of approximation the foregoing results have first been established by PIDDINGTON and MINNETT (1949); but their generalized form as given above was subsequently obtained by KOPAL (1964a).

In order to demonstrate the essential features of our present problem, let us assume that the $d_n$'s are indeed ignorable and that, moreover, the expansion on the right-hand side of (20-47) can be effectively truncated at $n = 1$. If so, then (by a recourse to the Rayleigh-Jeans formula 20-43) the "radio" temperature $T_e$ of the lunar disk can be approximated by an expression of the form

$$\begin{aligned} T_e &= (1 - R)\left\{T_0 + \frac{T_1 \cos(\tau - \phi_1 - \varepsilon_1)}{\sqrt{1 + 2\delta_1 + 2\delta_1^2}} + \cdots\right\} \\ &= T_0' + T_1' \cos(\tau - \psi), \end{aligned} \tag{20-51}$$

---

* This is tantamount to the neglect of the derivative $\partial I/\partial t$ on the left-hand side of the equation (20-40) of radiative transfer.

where $R$ stands for the power reflection coefficient (Fresnel loss) of the lunar surface (of which more will be said later on), and the values of $T_{0,1}$ as well as $\varepsilon_1$ can be regarded as known from (20-11). If so, however, then (in view of 20-50) the variable part of Equation (20-51) contains only one unknown – namely,

$$\delta_1 = \left(\frac{C_v \omega}{2\kappa\rho}\right)^{\frac{1}{2}} \frac{\cos\theta}{k};$$
(20-52)

and can be solved for it from the equations

$$\frac{T_1/T_0}{\sqrt{1 + 2\delta_1 + 2\delta^2}} = \frac{T_1'}{T_0'}$$
(20-53)

and

$$\tan^{-1}\frac{\delta_1}{1 + \delta_1} = \psi - \varepsilon_1,$$
(20-54)

following from (20-51), where the observed ratios $T_1'/T_0'$ as well as the values of the phase-lag $\psi$ can be found tabulated in the last two colums of Table 20-2.

The foregoing equations (20-53) and (20-54) constitute two independent relations for determining $\delta_1$; and if our theory is applicable to the observational data listed in Table 20-2 (and the latter are free from any important systematic errors), they both must lead to the same value of $\delta_1$. That this is indeed so is evident from the accompanying Figures 20-9 and 20-10, where individual dots represent the values of $T_1'/T_0'$ or

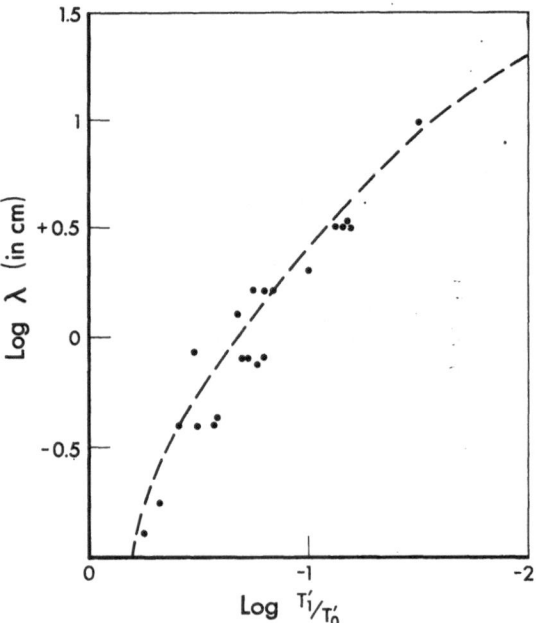

Fig. 20-9.   Observed decrease in diurnal temperature changes on the lunar surface with increasing wavelength in the microwave domain of its spectrum.

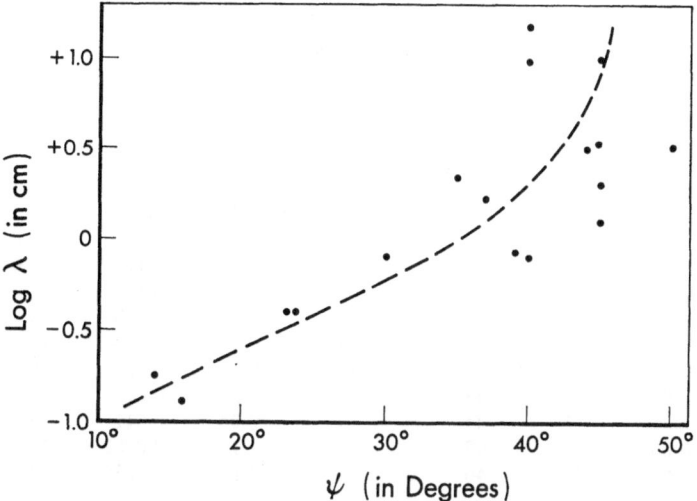

Fig. 20-10.   The phase-lag of diurnal temperature changes on the lunar surface, observed at different wavelengths in the microwave domain of its spectrum.

$\psi$ actually observed, and the dotted lines represent the variation of $\delta_1$ with $\lambda$ for the adopted ratio of

$$\frac{T_1}{T_0} = \frac{170°}{219°} = 0.776 \tag{20-55}$$

and $\varepsilon_1 = 2°$, corresponding (from the computations by Krotikov and Shchuko) to the value of $(\kappa\rho C_v)^{-\frac{1}{2}} = 1000$. The numerical values of $\delta_1$ for different wavelengths ranging from 0.4 cm to 15.85 cm, as represented by the dotted curves, are then listed in column 3 of the accompanying Table 20-3. They are seen to be closely proportional to $\lambda$ throughout most part of this range, and to satisfy the relation

$$\frac{\delta_1}{\lambda} = 1.98 \pm 0.05 \text{ cm}^{-1}. \tag{20-56}$$

Incidentally, the data listed in Table 20-2 reveal that, at wavelengths longer than about 10 cm, all monthly variations in temperature die down completely, and the constant component of the radio-emission corresponds to a temperature $T_0' = 240° \pm 6$ °K (i.e., $-33° \pm 6$ °C) prevalent at the corresponding depth irrespective of the phase. But, according to the computations by Krotikov and Shchuko, for the adopted value of $(\kappa\rho C_v)^{\frac{1}{2}} = 0.0010 \pm 0.0001$ the constant component of the temperature should be equal to $T_0 = 219° \pm 3$ °K. If we insert this in the relation

$$T_e' = (1 - R) T_0 \tag{20-57}$$

following likewise from (20-51), we find that

$$R = 0.087 \pm 0.024, \tag{20-58}$$

TABLE 20-3

Thermal Profile of the Mean Lunar Crust

| log $\lambda$ | $\lambda$ (cm) | $\delta_1$ | $\psi$ (deg) | $T_1'$ (deg K) | $k\rho r$ | $k\rho$ (cm$^{-1}$) | $r$ (cm) | $k$ (cm$^2$/g) |
|---|---|---|---|---|---|---|---|---|
| $-0.8$ | 0.16 | 0.32 | 15 | 118 | 1.00 | 0.44 | 2.3 | 0.73 |
| $-0.6$ | 0.25 | 0.50 | 20 | 93 | 0.99 | 0.28 | 3.6 | 0.46 |
| $-0.4$ | 0.40 | 0.80 | 25.5 | 70 | 0.94 | 0.17 | 5.4 | 0.29 |
| $-0.2$ | 0.63 | 1.25 | 31 | 52 | 0.85 | 0.11 | 7.7 | 0.18 |
| 0 | 1.00 | 2.0 | 35 | 37 | 0.71 | 0.068 | 10.4 | 0.11 |
| 0.2 | 1.58 | 3.1 | 38 | 24 | 0.58 | 0.044 | 13 | 0.073 |
| 0.4 | 2.51 | 4.7 | 41 | 16 | 0.46 | 0.029 | 16 | 0.048 |
| 0.6 | 3.98 | 7.8 | 43 | 11 | 0.34 | 0.018 | 19 | 0.029 |
| 0.8 | 6.31 | 13 | 44 | 7 | 0.24 | 0.011 | 23 | 0.018 |
| 1.0 | 10.00 | 22 | 45 | 4 | 0.17 | 0.0062 | 27 | 0.0103 |
| 1.2 | 15.85 | 39 | 46 | 2 | 0.11 | 0.0035 | 32 | 0.0058 |

revealing that even in the domain of radio-frequencies the Moon continues to behave pretty much like a black body.

At what depth does the sub-surface temperature of the Moon become constant and equal to 240 °K in the course of a month? In order to answer this question, we must ascertain the mean depth from which the 10-cm radiation actually emerges – i.e., the transparency of the lunar crust to the microwaves – and this can be accomplished in the following manner. It is intuitively obvious that, with decreasing transparency, Equation (20-51) based on radiative transfer should tend to become identical with (20-12) of conductive energy transfer. In order to prove the circumstances under which this will be so, a comparison of the first of Equations (20-7) with (20-52) reveals that

$$x = \delta_1(k\rho r)\sec\theta \simeq \delta_1(k\rho r), \qquad (20\text{-}59)$$

where $\sec\theta = 1$ at the center of the disk and only about 1.05 for observations extending over the entire disk.* Moreover, for highly opaque media the mass-absorption coefficient $k$ will be very large, rendering $\delta_1$ small; and as, for small values of $\delta$,

$$e^{-\delta} = 1 - \delta + \tfrac{1}{2}\delta^2 - \tfrac{1}{6}\delta^3 + \cdots \qquad (20\text{-}60)$$

* It should be kept in mind that the regions of microwave emission cannot be located with any precision on the apparent lunar disk, due to the low resolving power of existing radio-telescopes. Thus, in accordance with Rayleigh's well-known diffraction limit of 1.22 $(\lambda/D)$, an aperture of $D = 24$ inches (such as employed by Sinton, 1955) will at $\lambda = 1.5$ mm resolve less than 10 minutes of arc (i.e., less than a half of the apparent lunar radius); while at $\lambda = 10$ cm, an aperture of 15 meters would be required to resolve half a degree. For this reason, all radio-temperatures $T_e$ measured so far for $\lambda > 3$ cm are averages taken over the entire disk; these being related with the central temperature $T_c$ by the approximate formula, $T_e = 0.96\, T_c$, where 0.96 stands for the mean value of $\cos^{\frac{1}{4}}\theta$ averaged over the entire disk.

while

$$\frac{1}{\sqrt{1 + 2\delta + 2\delta^2}} = 1 - \delta + \tfrac{1}{2}\delta^2 + \tfrac{1}{2}\delta^3 + \cdots \qquad (20\text{-}61)$$

and also

$$\phi = \tan^{-1} \frac{\delta}{1 + \delta} = \delta - \delta^2 + \cdots. \qquad (20\text{-}62)$$

the reader may easily verify that, correctly to the squares of $\delta$, Equations (20-12) and (20-51) are indeed identical provided that $k\rho r = 1$ – i.e., that *the mean depth of penetration is inversely proportional to $k\rho$.** 

How far does this proportionality extend? In order to ascertain this, let us inquire at which depth $r$ does a temperature variation of the form (20-12) becomes equal to that inferred from the intensity of microwave emission? Such a requirement provides us with an additional relation

$$T_1' = (1 - R) T_1 e^{-x} = 0.913 \times 170° e^{-\delta_1(k\rho r)}; \qquad (20\text{-}63)$$

where $\delta_1$ is already a known quantity; and if the values of $T_1'$ are taken from the observations, this equation can be used to compute the corresponding values of the product $k\rho r$. A plot of the individual values of $T_1'$ observed at the wavelength $\lambda$ (cf. column 4 of Table 20-3) is shown on the accompanying Figure 20-11; and their smoothed means (as represented on the figure by a dotted line) are listed in column (5) of Table 20-3. Their insertion (together with the $\delta_1$'s from column 3 of the same table) in Equation (20-63) leads to the values of $k\rho r$ listed in column (6) of the same table. On the other hand, the value of the product $k\rho$ alone can be solved for from (20-52) in the form

$$k\rho = \left(\frac{\omega}{2}\right)^{\frac{1}{4}} \frac{\rho C_v}{\delta_1 (\kappa \rho C_v)^{\frac{1}{4}}}, \qquad (20\text{-}64)$$

where $(\omega/2)^{\frac{1}{4}} = 1.153 \times 10^{-3} \sec^{-\frac{1}{4}}$ and $(\kappa \rho C_v)^{\frac{1}{4}} = 10^{-3}$ cal/cm$^2$ deg sec$^{\frac{1}{4}}$. Moreover, we mentioned before that $C_v \sim 0.2$ cal/g deg while, for a pumice-like surface, we venture to set $\rho = 0.6$ g/cm$^3$. If so, the above equation (20-64) reduces to $k\rho = 0.138/\delta_1$ and leads to the numerical values listed in column (7) of Table 20-3. A division of columns (6)÷(7) then yields the corresponding mean depth $r$ of emergent radiation (col. 8); while the ultimate column (10) lists the corresponding mass-absorption coefficients $k$ (in cm$^2$/g), evaluated from the data of column (7) for the adopted value of $\rho = 0.6$ g/cm$^3$.

An inspection of the data collected in Table 20-3 reveals several noteworthy facts. First, the mass-absorption coefficient $k_\lambda$ appears to be inversely proportional to $\lambda$, so that the nondimensional product

$$\lambda k\rho = 0.069 \pm 0.002 \qquad (20\text{-}65)$$

---

* A reciprocal of $k\rho$ is identical with the depth at which the energy density of the incident wave has been reduced to $1/e$.

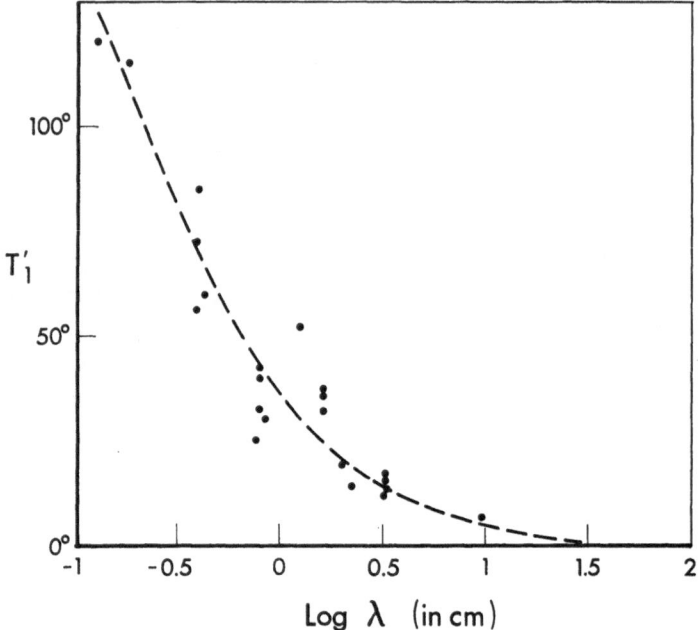

Fig. 20-11. Dependence of the observed amplitude of microwave temperature variation on the wavelength.

throughout the range $0.4 < \lambda < 15$ cm; though the behaviour of $k$ for $\lambda \gg 15$ cm can no longer be traced from microwave observations (as the Moon then continues to radiate at a constant rate). A proportionality of this type was indeed to be expected from known properties of silicate materials in the laboratory (cf., e.g., VON HIPPEL, 1954); but the fact that this is so convincingly borne out by purely astronomical evidence strengthens our belief in the correctness of our interpretation of the underlying data.

Secondly, while the mean depth $(k\rho)^{-1}$ from which thermal radiation at different wavelengths emerges from the lunar crust attenuated to $e^{-1}$ is very considerable (1.6 m for $\lambda = 10$ cm, for instance), *the mean sub-surface depth r below which a constant temperature of $240° \pm 6$ °K prevails throughout the whole lunation appears to be of the order of one foot only*. Microwave and infrared observations combined reveal, therefore, that the diurnal heat wave penetrates only centimeters rather than decimeters (let alone meters) into the lunar crust; and even this with a time-lag of many days. Thermal radiation at wavelengths of the order of one meter undoubtedly emerges from layers which may be several meters in depth; but as it exhibits no monthly variation (and the data on $k\rho r$ from column 6 are not easily extrapolable), it is impossible to assign to this depth a more specific value. Besides, it is also irrelevant; for only global averages could be obtained at these wavelengths; as even a 250-foot radiotelescope working at meter waves possesses only the same resolving power as a pinhole of 0.04 mm free aperture in visible light!

Lastly, the coefficient $\kappa$ of heat conduction consistent with the values of $(\kappa\rho C_v)^{\frac{1}{2}}$

TABLE 20-4

Microwave Observations of Lunar Eclipses

| Oppolzer Number | Date of Eclipse | Wavelength | Reference |
|---|---|---|---|
| 4890 | 1953 Jan. 29 | 0.86 cm 3.2 | GIBSON, J. E.: 1958, *Proc. I.R.E.* **46**, 280 KAIDANOVSKY, N. L., TURUSBEKOV, M. T., and CHAIKIN, S. E.: 1956, *Publ. 5th Confer. on Cosmogony*, Moscow, p. 347 |
| 4892 | 1954 Jan. 19 | 0.15 0.86 | SINTON, W. M.: 1956, *Astrophys. J.* **123**, 325 GIBSON, J. E.: 1958, *Proc. I.R.E.* **46**, 280. |
| 4896 | 1956 Nov. 18 | 0.75 | MITCHELL, F. H. and WHITEHURST, R. N.: 1958, *Univ. of Alabama Phys. Dept.*, OOR Rept. |
| 4897 | 1957 May 13 | 21 | MEZGER, P. G. and STRASSL, H.: 1959, *Planet. Space Sci.* **1**, 213 |
| 4901 | 1960 March 13 | 0.86 10 23 | GIBSON, J. E.: 1961, *Astrophys. J.* **133**, 1072 CASTELLI, J. P., FERIOLI, C. P. and AARONS, J.: 1960, *Astron. J.* **65**, 485 |
| 4902 | 1960 Sept. 5 | 0.86 | TYLER, W. C. and COPELAND, J.: 1961, *Astron. J.* **66**, 56 |
| 4906 | 1963 Dec. 30 | 0.32 | EPSTEIN, E. E. and STACEY, J. M.: 1964, *Astron. J.* **69**, 541. |
| 4908 | 1964 Dec. 19 | 0.10 | BALDOCK, R. V., BASTIN, J. A., CLEGG, R. E. EMERY, R., GAITSKELL, J. R. and GEAR, A. E.: 1965, *Astrophys. J.* **141**, 1289 |
| | | 0.10 | LOW, F. J. and DAVIDSON, A. W.: 1965, *Astrophys. J.* **142**, 1278 |

$= 0.001$ cal/cm$^2$ deg sec$^{\frac{1}{2}}$, $C_v = 0.2$ cal/g deg, and $\rho = 0.6$ g/cm$^3$ turns out to be equal to

$$\kappa = 8 \times 10^{-6} \text{ cal/cm sec deg}, \tag{20-66}$$

which is several hundred times smaller than that of solid silicate rocks, but comparable with the coefficient of heat conduction of powders in vacuum (WESSELINK, 1949; GARSTANG, 1958).

Microwave observations carried out during *total eclipses* of the Moon confirm, in general, the extremely poor thermal conductivity of lunar surface layers, as inferred from monthly fluctuation of thermal flux in the infrared as well as microwave domain of the spectrum. We mentioned already that, as the Moon gradually enters the shadow of the Earth (or emerges again from it), variations of surface temperature are indeed precipitous (three-quarters as large as those between day and night, but telescoped into a time span of a few hours). However, at the wavelength of 1.25 cm PIDDINGTON and MINNETT (1949) found no variation whatever in the intensity of thermal emission of totally-eclipsed lunar globe; and neither did GIBSON (1958) record any at $\lambda = 0.86$

cm. It was not till at $\lambda = 0.32$ cm that EPSTEIN *et al.* (1964) found some response of the thermal radiation to the advancing shadow – as SINTON did, in 1956, at $\lambda = 0.15$ cm; but again reduced in amplitude (265–290 °K at 0.32 cm, or 160°–300 °K at 0.15 cm) and delayed in phase by approximately one hour (i.e., one-quarter of the eclipse cycle).

A more detailed interpretation of such observations as are summarized in the accompanying Table 20-4 has so far bear impossible by the lack of the solutions of the underlying thermal-conduction problem. The reason of this lack in our knowledge is the fact that, during an eclipse, the amount of sunlight incident on any part of the lunar surface varies rapidly with the phase and, accordingly, the boundary condition (20-5) should be replaced by

$$\kappa \frac{\partial T}{\partial r} = (1 - A)\pi S \{1 - \alpha(t)\} \cos z - \sigma T^4, \qquad (20\text{-}67)$$

where $\alpha(t)$ represents the fractional loss of sunlight at the particular place and time (being zero at the moment of the first contact of the eclipse, and one during totality). For the mathematical formulation of this function taking account of solar limb-darkening cf., e.g., KOPAL (1959; sec. IV.4); but save for a limited attempt by JAEGER (1953) no numerical integrations of Equation (20-4) satisfying the above boundary condition have so far been carried out; and any quantitative interpretation of the eclipse results in the infrared or microwave domain of the spectrum will have to await their outcome.

## CHAPTER 21

# ELECTROMAGNETIC PROPERTIES OF THE LUNAR SURFACE

The results of the studies of the radiation of our satellite in the microwave domain of its spectrum, as summarized in the preceding chapter, open the way for investigating not only the thermal properties of the outermost crust of the Moon down to some depth below the visible surface, but also certain of its electromagnetic characteristics, related to the former by Maxwell's equations governing the propagation of electromagnetic waves in dielectric media.

In order to introduce this subject, let

$$\hat{n} = n(1 + i\kappa) \tag{21-1}$$

represent the (complex) coefficient of refraction at the interface of the lunar surface with free space, such that

$$\hat{n}^2 = \mu\hat{\varepsilon} = \mu\left(\varepsilon + \frac{2i\sigma}{v}\right), \tag{21-2}$$

where $\mu$ denotes the magnetic permeability; and $\hat{\varepsilon}$, the complex dielectric constant whose real part $\varepsilon$ represents the permittivity, and the imaginary part consists of the coefficient of conductivity $\sigma$ and $v$ the frequency of the respective electromagnetic radiation of wavelength $\lambda$ – such that

$$\lambda v = c \tag{21-3}$$

stands for the velocity of propagation of electromagnetic waves in free space.

If so, a comparison of Equations (21-1) and (21-2) reveals that

$$\left.\begin{array}{c} n^2(1 - \kappa^2) = \mu\varepsilon \\ vn^2\kappa = \mu\sigma \end{array}\right\} \tag{21-4}$$

whence

$$n^2 = \frac{\mu\varepsilon}{2}\left\{\sqrt{1 + \frac{2\sigma^2}{\varepsilon v}} + 1\right\} \tag{21-5}$$

and

$$n^2\kappa^2 = \frac{\mu\varepsilon}{2}\left\{\sqrt{1 + \frac{2\sigma^2}{\varepsilon v}} - 1\right\}. \tag{21-6}$$

The *power reflection coefficient R* of a medium of permittivity $\varepsilon$, permeability $\mu$, and conductivity $\sigma$ for radiation of frequency $v$ incident normally on the surface is then defined (cf., e.g., BORN and WOLF, 1959, p. 617) as

$$R = \left| \frac{\hat{n} - 1}{\hat{n} + 1} \right| = \frac{n^2(1 + \kappa^2) + 1 - 2n}{n^2(1 + \kappa^2) + 1 + 2n}, \tag{21-7}$$

where $\kappa$ and $n$ are given in terms of the electromagnetic parameters $\varepsilon$, $\mu$, and $v$ by Equations (21-5)–(21-6). Moreover, the *absorption coefficient* $\chi$ of the medium is likewise defined (cf. Born and Wolf, *op. cit.*, p. 611) by

$$\chi = \frac{4\pi v n \kappa}{c} = \frac{4\pi n \kappa}{\lambda}. \tag{21-8}$$

If the medium were non conducting (i.e., $\sigma = 0$) it follows at once from (21-4) that $\kappa = 0$ and $n^2 = \mu\varepsilon$, in which case $\chi = 0$ and (21-7) would reduce to

$$R = \left( \frac{n - 1}{n + 1} \right)^2. \tag{21-9}$$

If the conductivity $\sigma$ is nonvanishing but small enough for the ratio $\sigma/\varepsilon v$ to be regarded as a small parameter (as should be true of a strong dielectric, like the lunar surface), an expansion of the right-hand sides of (21-7) and (21-8) leads to

$$R = \left( \frac{n - 1}{n + 1} \right)^2 + \frac{n(3n^2 - 1)}{2(\chi + 1)^4} \left( \frac{2\sigma}{\varepsilon v} \right)^2 + \cdots \tag{21-10}$$

and

$$\chi = \frac{4\pi\sigma}{c} \sqrt{\frac{\mu}{\varepsilon}} + \cdots \tag{21-11}$$

correctly to the squares of $\sigma$.

The question of the units in which the individual electromagnetic parameters involved in the foregoing expressions are to be expressed has so far been left open. If, however, we are concerned with the absorption or reflection from the lunar surface bordering on free space whose permittivity $\varepsilon_0$ as well as permeability $\mu_0$ are given by

$$\varepsilon_0 = 1 \text{ e.s.u.} = 8.354 \times 10^{-12} \text{ farad/meter} \tag{21-12}$$

and

$$\mu_0 = c^{-2} \text{ e.s.u.} = 4\pi \times 10^{-7} \text{ henry/meter}, \tag{21-13}$$

the constants $\varepsilon$ and $\sigma$ of our theory characterizing the lunar surface must be expressed in terms of $\varepsilon_0$ taken as the unit; and $\mu$, in terms of $\mu_0$. Hence, the real part $n$ of the complex refraction index occurring on the right-hand side of (21-10) is to be understood as

$$n = \sqrt{\frac{\mu\varepsilon}{\mu_0\varepsilon_0}} \tag{21-14}$$

a nondimensional quantity; and, similarly, Equation (21-11) should be rewritten as

$$\chi = \frac{4\pi\sigma}{c\varepsilon_0}\sqrt{\frac{\mu}{\mu_0}\frac{\varepsilon_0}{\varepsilon}} \tag{21-15}$$

which (since the dimension of $\sigma$ is $\varepsilon\ \sec^{-1}$) is of the dimension of $cm^{-1}$.

Let us proceed now to apply the above theory to further interpretation of some results obtained in the preceding chapter. Thus, by use of Equation (20-58), we found the power reflection coefficient $R$ of the lunar surface in the microwave domain ($\lambda \sim 0.4$–10 cm) to be equal to $0.087 \pm 0.024$; and when we identify it with the expression $R$ as given by the simplified equation (21-9), we find it to correspond to a value of

$$n = 1.84 \pm 0.13 \tag{21-16}$$

leading to

$$\begin{aligned}\mu\varepsilon &= (3.4 \pm 0.6)\mu_0\varepsilon_0 \\ &= (7.6 \pm 1.3) \times 10^9\ \text{ohm}^2\,.\end{aligned} \Biggr\} \tag{21-17}$$

If, moreover, we identify $\mu$ with $\mu_0$, the corresponding value of the dielectric constant $\varepsilon$ characteristic of the lunar surface comes out to be equal to

$$\varepsilon = 3.4\varepsilon_0 = 3.0 \times 10^{-11}\ \text{farads/m}. \tag{21-18}$$

This is considerably *less* than dielectric constants of most solid rocks or materials; but because we consider this results still as preliminary we shall postpone a discussion of its meaning to a later part of this chapter; and, in the meantime, let us turn to a determination of the electrical conductivity $\sigma$ of the lunar surface.

In order to do so, let us recall that the absorption coefficient $\chi$ as defined by Equation (21-8) is identical with the product $k\rho$ of the absorption coefficient $k$ per unit mass times the density $\rho$; and in the preceding chapter (Equation 20-65) we found that, in the first foot or so of the lunar surface, the product $\lambda k\rho$ appears to be sensibly constant and equal to $0.069 \pm 0.002$. By virtue of the identity $k\rho \equiv \chi$ Equations (20-65) and (21-15) now lead to a relation of the form

$$\lambda k\rho = \frac{4\pi}{\sqrt{3.4}}\left(\frac{\lambda\sigma}{c\varepsilon_0}\right) = 0.069 \pm 0.002, \tag{21-19}$$

yielding

$$\sigma = 0.0101\nu\ \text{e.s.u.} \tag{21-20}$$

Consistent with (20-65), *the electrical conductivity of the material of the lunar crust varies proportionally to the frequency* $\nu$; and for $\lambda = 1$ cm ($\nu = 30\,000$ Mc/sec), $\sigma = 3 \times 10^8$ e.s.u. $= 0.033$ ohm/m; for $\lambda = 10$ cm ($\nu = 3000$ Mc/sec), $\sigma = 3 \times 10^{-3}$ ohm/m; and for $\lambda = 100$ cm ($\nu = 300$ Mc/sec), $\sigma = 3 \times 10^{-4}$ ohm/m. The general smallness of the conductivity $\sigma$ resulting from (21-19) or (21-20) renders the quadratic term of the expansion (21-10) for the power reflection coefficient virtually negligible, thus proving the worth of the approximate expression (21-9).

One final word may be added concerning another significant electromagnetic property of thermal emission of the Moon in the microwave domain: namely, its *polarization* arising from its emergence from a medium characterized by the dielectric constant $\varepsilon$ into free space, where $\varepsilon$ drops to $\varepsilon_0$. In order to establish the extent of such a polarization, consider a smooth homogeneous dielectric sphere of uniform surface temperature, negligible conductivity ($\sigma = 0$) and magnetic permeability $\mu = \mu_0$, surrounded by free space whose permittivity $\varepsilon_0$ can again be taken our unit of $\varepsilon$. If so, the power reflection coefficient $R$ for a plane electromagnetic wave incident on a surface element of our dielectric sphere at an angle $\theta$ between the direction of propagation and the surface normal will be given (cf. again BORN and WOLF, 1959; pp. 612–617) by the modulus

$$R_{11} = \left| \frac{\varepsilon \cos\theta - \sqrt{\varepsilon - \sin^2\theta}}{\varepsilon \cos\theta + \sqrt{\varepsilon - \sin^2\theta}} \right|^2 \tag{21-21}$$

for linear polarization with the electric vector parallel with the plane of incidence, and

$$R_{12} = \left| \frac{\cos\theta - \sqrt{\varepsilon - \sin^2\theta}}{\cos\theta + \sqrt{\varepsilon - \sin^2\theta}} \right|^2 \tag{21-22}$$

for orthogonal linear polarisation. The emissivity $B(\theta)$ for thermal radiation leaving the surface at an angle $\theta$ is, by Kirchhoff's law, then proportional to

$$\left.\begin{aligned} B_{11}(\theta) &= 1 - R_{11}(\theta), \\ B_{12}(\theta) &= 1 - R_{12}(\theta); \end{aligned}\right\} \tag{21-23}$$

and a radio-telescope accepting only one linear polarisation will thus reveal a temperature distribution which is dependent on both the position of the emitting surface element as well as on the direction of polarisation.

This polarisation of the thermal radiation of the Moon in the microwave domain was first observed by SOBOLEVA (1962) at $\lambda = 3.2$ cm, and confirmed by HEILES and DRAKE (1963) at $\lambda = 21$ cm (1413 Mc/s). The ratios of the dielectric constants $\varepsilon$ of the lunar surface to that of free space, established in this manner, proved to be ($\varepsilon/\varepsilon_0 = 2.1 \pm 0.3$ (Heiles and Drake), and 1.7 (Soboleva) – somewhat lower than those resulting from (21-17), but the agreement is fair.

This, as well as all other foregoing conclusions on probable electromagnetic characteristics of the lunar surface have been based on the observed properties of *thermal* emission of the Moon in the microwave domain of its spectrum. However, in the past twenty years continued advances in the field of radio transmission have provided the astronomers with a new and powerful technique for actively probing the electromagnetic and other properties of the lunar surface at a distance: namely, by the method of *radar echoes*, consisting of the transmission of high-frequency radar pulses in the direction of the Moon and observing the characteristics of their echoes returning to the Earth after a time-lapse of the order of 1.28 seconds.

The first group of investigators to detect radio echoes from the Moon were members of the U.S. Army Signal Corps Laboratory at Belmar, N.J., under the direction of Lieut. Colonel J. H. DeWitt. On January 10th, 1946 – another memorable landmark in the study of our subject – and using a mere 3 kW transmitter to generate a series of pulses of 0.2–0.5 sec duration spaced by 4 seconds at the frequency of $115\frac{1}{2}$ Mc/sec they succeeded in detecting their lunar echoes after 2.56 seconds, during which the signals have completed their round-trip of more than three-quarters of a million kilometres (cf. WEBB, 1946; MOFENSON, 1946); but a full account of this work was not published until three years later (DEWITT and STODOLA, 1949). These observations only just preceded those of BAY (1946) in Ujpest, Hungary, who detected his first lunar echoes on February 6th, 1946*, using much the same kind of equipment as the American investigators. In 1949, workers at the Australian Commonwealth Scientific and Industrial Research Organization (KERR, SHAIN and HIGGINS, 1949) commenced a series of observations at frequencies of 17.94 Mc/s and 21.54 Mc/s using much larger power and shorter pulses; and subsequently the subject was further advanced by contributions from Jodrell Bank Experimental Station of the University of Manchester (cf. MURRAY and HARGREAVES, 1954; BROWNE et al., 1956; EVANS, 1956, 1957a, b) Naval Research Laboratory in Washington (YAPLEE et al., 1958; TREXLER, 1958); Royal Radar Establishment at Malvern (HEY and HUGHES, 1958) Lincoln Laboratories of the Massachusetts Institute of Technology (PETTENGILL, 1960; EVANS and PETTENGILL, 1963), and many others.

A study of the radio-echoes reflected from the Moon has materially advanced the solution of several problems – both astronomical and geophysical. First, the time-lag between the transmitted pulse and its returning echo (which can nowadays be timed by standard electronic techniques with a precision of the order of one part in a million) combined with the velocity $c$ of the propagation of radio-waves in vacuum (known to the same order of accuracy) can be used to compute the absolute value of the instantaneous radius-vector between the reflecting element of the lunar surface and the transmitter-receiver on Earth within an error of less than half a kilometer (cf. YAPLEE, BRUTON, CRAIG, and ROMAN, 1958); and thus contribute to our knowledge of the motion of the Moon in space (see Chapters 1 and 2).

A second interesting fact which transpired from the studies of radar Moon echoes has been their variable intensity: the returning pulses were found to oscillate in strength in the period of a few seconds, and also between 20–30 minutes. The former (more rapid) fading was noted already by DEWITT and STODOLA (1949) and later by KERR and SHAIN (1951) who attributed it to the Moon's libration (cf. Chapter 4) – an explanation confirmed by EVANS and THOMSON (1959). On the other hand, the second type of fading (in period of 20–30 minutes) is of terrestrial origin, and due to the Faraday rotation of the plane of polarization of the radio waves in the Earth's ionosphere in the presence of the terrestrial magnetic field (cf. MURRAY and HARGREAVES, 1954).

* Bay commenced work on this project during the summer of 1944, but because of the events of the second world war his efforts were not crowned with success until 1946, by which time he had been forced to rebuild his equipment three times.

Third, a study of the total power return of lunar echoes and their time-profiles – i.e., the widening of the original sharp pulse by reflection from different parts of the lunar surface at different distance from the observer* – can disclose much on the physical and even topographic structure of the surface of our satellite; and these interest us in the first place. An example of a typical oscillographic record of echoes of this kind is shown on the accompanying Figures 21-1 and 21-2; and the total power returned should be proportional to the area subtended by it above the noise level. Its profile shows invariably a steep (virtually instantaneous) rise to maximum peak, followed by a more gradual decline in power as echoes from more distant (i.e., limb) zones of the lunar surface will reach the receiver; the latter being responsible for the tails of the observed profiles.

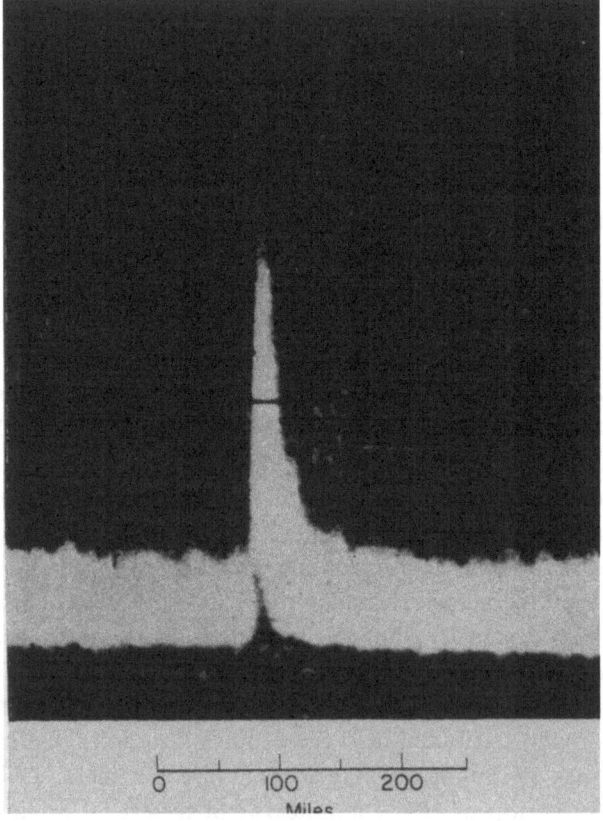

Fig. 21-1.   The amplitude versus time and range of a typical lunar radar echo  is shown, as obtained by TREXLER (1958) using transmitter pulses of 12 microsecond duration.

* The time-lag of echoes reaching us from the central and limb portions of the apparent lunar disk amounts to approximately 11.6 milliseconds.

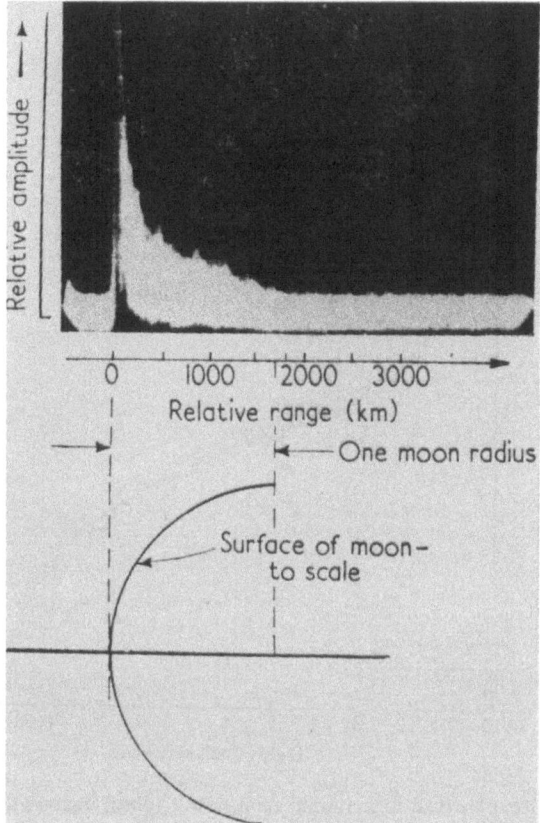

Fig. 21-2. The amplitude-range display of lunar echoes, as observed by LEADABRAND *et al.* (1960). A large echo due to specular reflection is observed at the leading edge of the Moon; while a weaker echo decaying towards the limb of the Moon is due to diffuse reflection.

An analysis of the shapes of typical echoes – such as shown on the accompanying Figure 21-3 – shows a large measure of agreement and discloses several noteworthy facts. First, it transpires *that more than a half of the total echo power is accounted for by the part of the profile immediately adjacent to its steep peak,* reaching us in the first 50 microseconds after the onset of the leading edge of the echo; and since a delay of 50 $\mu$sec corresponds to a difference on only 8 km along the radius of the Moon, this part of the echo must be reflected from a meniscus of not more than 340 km in diameter (i.e., about one-tenth of the diameter of our satellite).

Secondly, the steepness in the rise of the leading edge of the echo and its almost as rapid initial decline reveal that they are due to a *quasi-specular reflection* from a surface which must be smooth or gently undulating, and only sparsely covered with objects of the order of a wavelength in size. EVANS and PETTENGILL (1963) conclude that, at $\lambda = 3.6$ cm, some 14% of the surface appears to be covered by irregularities of

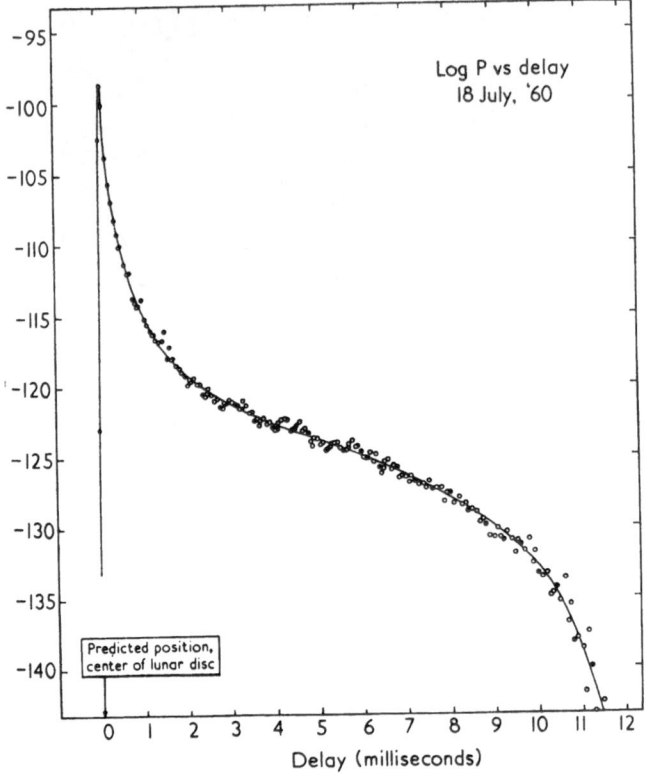

Fig. 21-3.   A plot of the observed logarithmic echo power (dbm) against the time delay (in milliseconds) for signals received from the Moon on 1960 July 18 at the Millstone Radar Station of M.I.T. Lincoln Laboratory, using 0.1 msec transmitted pulses and receiver bandwidth of 35 kc/sec. (After PETTENGILL and HENRY, 1962a.)

the order of the wavelength in size; whereas at $\lambda = 68$ cm, only 8% of the surface is that rough; the mean gradient of the ground thus seems about 1 in 7 for points spaced 3.6 cm, and 1 in 11 for points spaced 68 cm* – compare this with the actual low gradients inferred from the shadow studies for the majority of macroscopic features of the lunar surface in Chapter 16 (p. 251)!

Lastly, beyond a range bounded by a circle of about three-quarters of the Moon's apparent radius, weaker tails of the echoes have been observed (LEADABRAND *et al.*, 1960; PETTENGILL, 1960) falling off as $\cos \theta_n$, where $1.5 > n > 1$ (cf., Figure 21-4); in such regions the reflection becomes clearly *diffuse* and can eventually be described by Lambert's law. By comparing the intensity of these tails with that of the specular echoes in the neighbourhood of the leading edge, we conclude that only about one-tenth of the surface of the apparent lunar disk gives rise to diffuse reflection (EVANS,

* A deduction fully confirmed by the first close-up photographs of the lunar surface, secured by the Russian soft-lander Luna 9 (see Fig. 23-2).

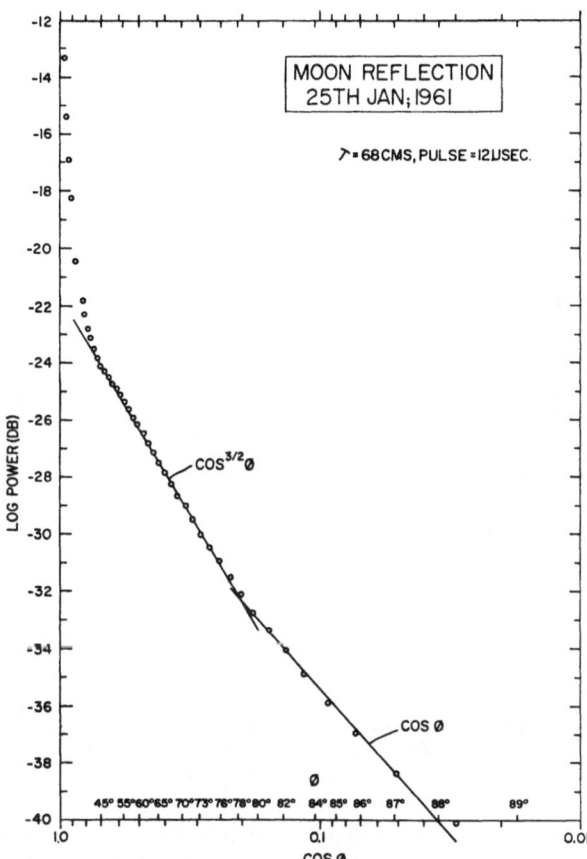

**Fig. 21-4.** The anatomy of the tails of lunar radar echoes, demonstrating a gradual transition from specular to diffuse reflection between the center and the limb of the apparent disk of the Moon (after Evans, 1966).

1962). The presence of the echo tails caused by diffuse reflection (indicative of the 'limb-darkening' of the Moon at radar frequencies) was noted first by Trexler in 1951, but his results remained unpublished until an independent re-discovery of the same effect by Evans (1957a).

Subsequent work on wavelengths between 0.86 cm (Lynn *et al.*, 1964) to 6 m and beyond (Evans and Pettengill, 1963a; Klemperer, 1965) has, moreover, revealed that *the rate at which the echo intensity falls off to the limb depends strongly on the frequency;* being the more rapid, the longer the wavelength. At metre or decametre waves, the 'limb-darkening' of the Moon illuminated by radar flashes of microsecond durations is extreme – see Figure 21-5 – but diminishes rapidly with decreasing wavelength; and the limits of no darkening at optical frequencies would be represented by a horizontal abscissa at the top of the diagram.

The characteristics of limb-darkening at radar frequencies is of high interest for the interpretation of the micro-structure of the lunar surface on the cm-scale (see the

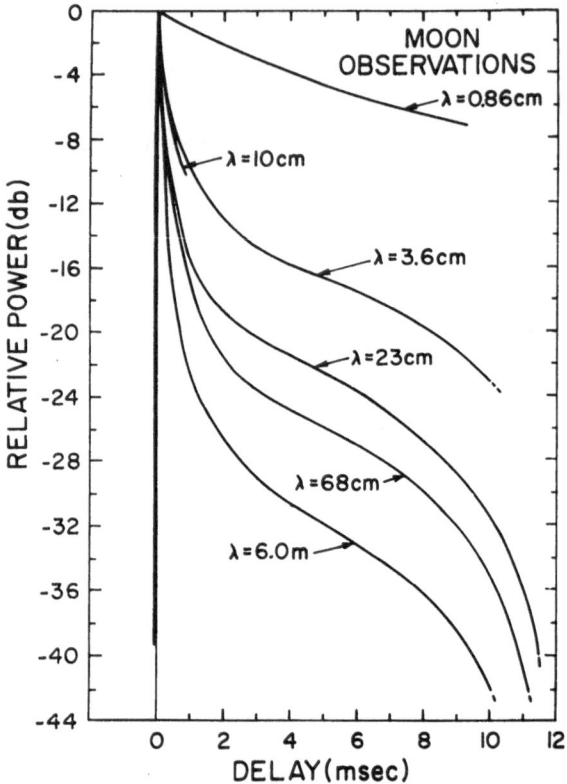

Fig. 21-5. A schematic diagram showing gradual diminution of the "limb-darkening" of the Moon illuminated by instantaneous radar pulses of dininishing wavelength. (After EVANS, 1966.)

forthcoming Chapter 23); but for our present purposes we are primarily concerned with the total intensity of the returning echo, which should clearly depend on the electromagnetic properties of the lunar surface. Let $s$ denote the scattering cross-section of a sphere of radius $r$; if so, it is evident that

$$s = \pi a^2 \rho = \pi a^2 g R, \tag{21-24}$$

where $\rho$ denotes the fractional cross-section obtainable from the intensity of the radar echoes, and $g$ stands for the "gain factor" of the scattering surface, defined by the ratio

$$g = \frac{4\pi \int_0^\pi P(i, i, 0) \sin i \, di}{\int_0^{\pi/2} di \int_0^{\pi/2} d\phi \int_0^{2\pi} P(i, \phi, \alpha) \sin i \sin \phi \, d\alpha}, \tag{21-25}$$

where $P(i, \phi, \alpha)$ stands for the scattering cross-section, per unit area and solid angle of scattering, for the angle of incidence $i$, reflection $\phi$, and phase $\alpha$.

At optical frequencies, this function is proportional to the derivative $dl/ds$ as introduced by Equations (19-1) or (19-2) before for Lambert or Lommel-Seeliger law, respectively. In the domain of radio frequencies it cannot, however, be determined from the observations alone because, for ground-based radar pulses, $i = \phi$ and $\alpha = 0$. Thus a rigorous evaluation of the gain factor (and, therefore, of $R$ from the observed values of $\rho$) becomes possible only for the assumed forms of $P$. For a perfect sphere scattering isotropically, $g = 1$. For a sphere scattering is accordance with Lambert's law (19-1), $g = 8/3$ (cf., GRIEG et al., 1948); while for the Lommel-Seeliger law (19-2) EVANS and PETTENGILL (1963a) obtained $g = 2.68$ (indistinguishable from the Lambert value). Since, however, diffuse reflection accounts for only about 10% of the returning echo strength and 90% of it is specular (for which $g = 1$), it follows that the effective value of the gain factor should be closely approximated by

$$g = 0.9 \times 1 + 0.1 \times 2.67 = 1.17 \qquad (21\text{-}26)$$

which we shall hereafter adopt.

The values of the fractional cross-sections $\rho$ as deduced from the measurements of many investigators in a wavelength range between 0.86 cm (LYNN et al., 1965) and 18 metres (DAVIS and ROHLFS, 1964) are plotted on the accompanying Figure 21-6.

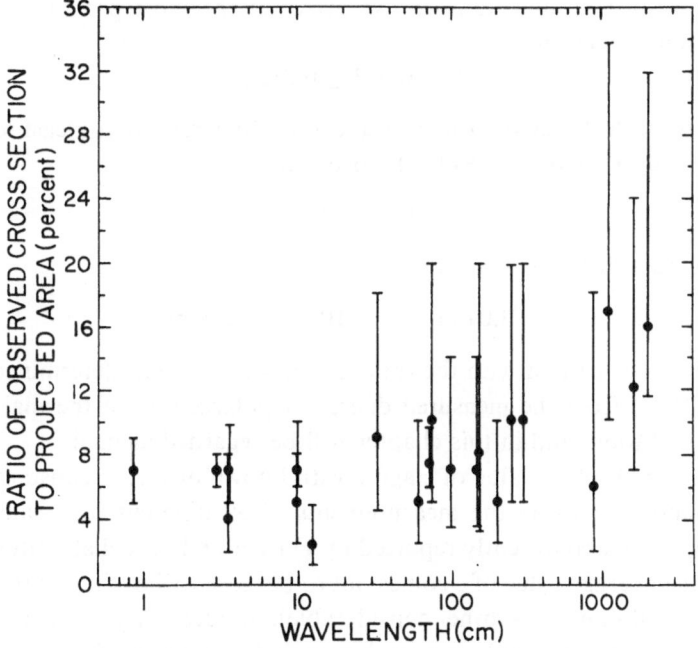

Fig. 21-6. A plot of the observed values of the radar-scattering fractional cross-sections $\rho$ of the lunar surface, as defined by Equation (21-24) and multiplied by a factor $10^4$, versus the wavelength. (After EVANS, 1966).

Up to wavelengths close to 10 metres these data furnish no evidence for any variation of $\rho$ with $\lambda$, which seems to remain constant within the limits of observational errors. Beyond $\lambda = 10$ m, the recent work by Davis and Rohlfs detected an upward trend in $\rho$ which may, in principle, be interpreted either as an indication of the effects of electrical conductivity (which should, in accordance with Equation 21-10, be essentially quadratic in $\lambda$), or the fact that, at the depth from which long wavelength radar pulses are returned, the effective dielectric constant $\varepsilon$ becomes larger than it is closer to the surface (indicating a greater degree of volumetric compression of the material).

At present it is still impossible to separate clearly these two alternatives; but the small electrical conductivity of the lunar surface which has already transpired from Equation 21-20 makes the former alternative unlikely and favours the latter. Whatever the case may be, however, for $\lambda > 10$ m the fractional cross-section $\rho$ appears to be constant and equal to

$$\rho = 0.074 \pm 0.021, \tag{21-27}$$

which combined with (21-24) and (21-26) yields

$$R = 0.063 \pm 0.018 \tag{21-28}$$

a value less than $0.087 \pm 0.024$ inferred for the same power reflection coefficient from thermal emission of the Moon in the microwave domain in Chapter 20 (Equation 20-58), but consistent with it well within the limits of observational errors of both independent determination of this quantity. Taking their mean (weighted in accordance of their respective uncertainties) we arrive at the most probable value of the power reflection coefficient

$$R = 0.073 \pm 0.015, \tag{21-29}$$

consistent with both the thermal microwave and radar reflection measurements, and corresponding (by Equation 21-9) to the product

$$\varepsilon\mu = (3.0 \pm 0.4)\,\varepsilon_0\mu_0 \tag{21-30}$$

or, for $\mu = \mu_0$, to a dielectric constant

$$\varepsilon = 3.0\,\varepsilon_0 = 2.7 \times 10^{-11} \text{ farads/m}. \tag{21-31}$$

This value is also consistent with the ratio of $\varepsilon/\varepsilon_0 = 2.1 \pm 0.3$ as determined by HEILES and DRAKE (1963) from the measured degree of polarization of thermal microwave emission of the Moon, and in this chapter will be regarded as final.

How does the absolute value of $\varepsilon$ agree with known dielectric constants of terrestrial rocky materials? From the measurements of 39 different rock samples whose dielectric properties were recently reported by Fensler and his collaborators (FENSLER et al., 1962), the average value of $\varepsilon$ comes out to be close to 9.6 (though the dispersion between individual samples is rather considerable), decidedly higher than the observed lunar value of 3.0. The two values may, however, be reconciled if we stop to realize that the terrestrial measurements refer to *solid* rocks, while lunar material is *porous*. Different formulae have been proposed by several authors (ODELEVSKY, 1951; BOETT-

CHER, 1952; LEVIN, 1954; TROITSKY, 1962) to relate the effective dielectric constants of porous materials with those of the same solid substances for different degrees of porosity; and if, in our case, we accept the observed value $\varepsilon = 3.0\ \varepsilon_0$ as characteristic of porous material whose dielectric constant in solid (compressed) state would be $\varepsilon_1 = 9.6\ \varepsilon_0$ the fraction of unit volume actually occupied by matter according to all above-quoted authors turns out to be very much the same and equal to $0.54 \pm 0.15$: Moreover, in accordance with our premises this should be regarded as the mean value of the porosity not only on the surface, but down to a mean level from which radar pulses of wavelengths up to 10 meters are actually reflected; and this could be from the depth of several dozen meters. *The relatively low mean value of the dielectric constant of the lunar surface makes it, therefore, necessary for the pulverized* (porous) *surface structure to extend,* on the whole, *to a depth of many meters.* Unless there is something seriously wrong with our measurements or interpretation of the lunar radar echoes, *solid rocky surface on the Moon must be located beyond the depth of penetration of radar pulses emitted at wavelengths up to 10 meters.* Neither, we may add, is the observed ratio $\varepsilon/\varepsilon_0$ low enough to be consistent with the "dendrite growth" on the lunar surface, postulated by HAPKE and van HORN (1963) to explain the peculiarities of its photometric function, extending to any appreciable depth.

The average fractional porosity of $0.54 \pm 0.15$ by volume, necessary to reconcile the "astronomical" and laboratory values of the dielectric constant, entails one other consequence. In Chapter 7 we established (cf. Equations 7-23) that the density of the solid sub-surface layers of the Moon should be close to 3.28 g/cm$^3$. If so, the density of the topmost porous layer should be equal to $3.28 \times 0.54 = 1.8$ g/cm$^3$ – i.e., about three times as large as the pumice-like value of 0.6 g/cm$^2$ adopted in the latter part of the preceding chapter in our analysis of the thermal profile of the lunar surface. As the product $\rho C_v$ is independent of the volume, the values of the depth of penetration $(k\rho)^{-1}$ or $r$ would remain unaffected by this change (and so would, of course, be the product $\lambda k\rho$ as given by 20-65). However, the value of the coefficient $\kappa$ of heat conduction as given by Equation (20-66) of the preceding chapter should then be divided by three and diminished to approximately $3 \times 10^{-6}$ cal/cm$^2$ sec deg – rendering fine pulverization all the more necessary.

It should also be stressed that, until quite recently, it has not proved possible to associate radar echoes from the Moon with any particular features of its surface. It is true (cf. Figure 21-7) that the time resolution of the observed pulse echoes permit us to distinguish reflection from successive annuli inclined to the line of sight by an angle

$$\phi = \cos^{-1}\left(1 - \frac{ct}{2a}\right), \tag{21-32}$$

where $c$ denotes the velocity of light; $a$, the lunar radius of 1738 km; and $t$, the time-lag measured from the leading edge of the echo. However, the Moon also rotates about a fixed axis; and this rotation will give rise to relative *Doppler shifts* of the echoes returned from the approaching and receding hemispheres; and such shifts will cause the returning echoes to drift not only in time, but also in frequency. Suppose that the

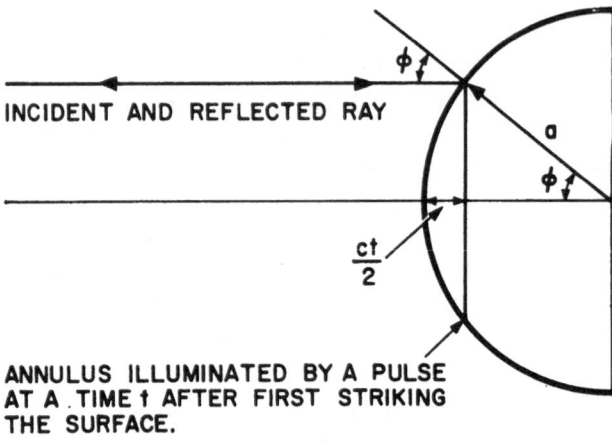

INCIDENT AND REFLECTED RAY

$$\frac{ct}{2}$$

ANNULUS ILLUMINATED BY A PULSE
AT A TIME t AFTER FIRST STRIKING
THE SURFACE.

a = RADIUS OF THE MOON
c = VELOCITY OF LIGHT

RANGE DELAY t AND ANGLE OF INCIDENCE φ
OF THE RADIO WAVES

Fig. 21-7.

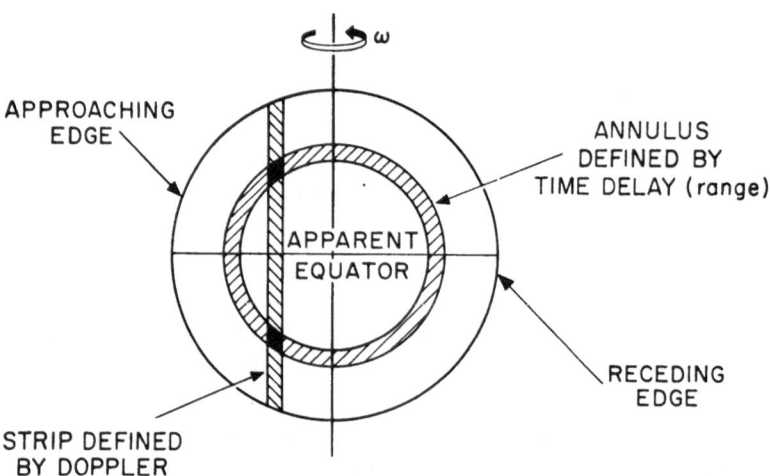

Fig. 21-8.  Range-frequency technique for radar scanning of the lunar surface.

characteristics of the returning echoes are analyzed in both range *and* frequency; as
the loci of constant range are concentric circles, while the loci of constant frequency
are straight lines parallel with the projected axis of rotation, their intersections should
define *two points* on the lunar surface (cf. Figure 21-8), the positions of which can be
located the more precisely, the greater the accuracy of observations; and, moreover,
the phenomena of lunar librations (chapter 4) should permit us to discriminate between the two in the course of one month.

The observational precision attainable in this way is indeed considerable; for the two-way depth delay between the centre and limb reflections amounts to 11.6 milli-seconds (i.e., about a hundred times as long as the pulse width), while the limb-to-limb Doppler spread corresponds to 25c/sec for axial rotation in the period of one month, and can likewise be measured to a small fraction of its absolute amount. Simultaneous range-frequency observations of the Moon were initiated by PETTENGILL (1960), and led (cf., PETTENGILL and HENRY, 1962) to the discovery of a most interesting fact: namely, that anomalously strong echoes (by a factor of about 10) are obtained from the crater Tycho or its immediate surroundings. As, moreover, the same difference was observed in both polarised and depolarised component of the signals, it cannot be explained by reflection from large flat facets favourably orientated with respect to the direction of the incoming signals (since these would not depolarise). As the measurements give only the magnitude of the product $\rho = gR$, it is impossible to distinguish which part of the anomalous magnitude of $\rho$ is due to an anomalously large value of $g$ (i.e., to an increased surface roughness), or to that of $R$ (i.e., higher dielectric constant). It is, however, possible that in the vicinity of Tycho the surface material is both denser and rougher (on the scale of the wavelength) than it is elsewhere on the Moon.

This work has since been continued at the Arecibo Ionospheric Observatory by employing 0.1 millisecond pulses and 0.1 cps Doppler resolution, which permit us to resolve by the range-Doppler technique an area of $20 \times 40$ km on the lunar surface. As a result of this work, THOMPSON (1965) has found that virtually all ray craters (Copernicus, Eratosthenes, Kepler, Langrenus, Theophilus and other craters of higher than average albedo at optical frequencies) are also anomalously strong radio reflectors; though none are so pronounced as Tycho. It is noteworthy that this same crater (and others like it) were also found to exhibit conspicuous thermal anomalies during eclipse (SHORTHILL, BOROUGH and CONLEY, 1960; SINTON, 1960; SAARI and SHORT-HILL, 1965) as well as during the lunation (SHORTHILL and SAARI, 1963). The two anomalies are probably connected; and as there are virtually thousands of spots known now on the lunar surface to be thermally anomalous (cf., SAARI and SHORTHILL, 1965), it is probable that a similar number of them will exhibit radar reflection anomalies as well; but the march of progress has not caught up with all of them yet.

Incidentally, the near-specular nature of reflections from the lunar surface entails one more practical consequence: and that is for the Moon to play the role of a convenient screen provided by Nature in outer space for bouncing radio messages from one distant part of the Earth to another, on short enough wavelengths to penetrate the Earth's ionosphere and thus be immune to interference from sunspots. This was originally thought to be impossible (cf., e.g., GRIEG et al., 1948), because reflections from large rough areas could have introduced enough distortion to make the signals unintelligible. However, the quasi-specular nature of the lunar echoes has opened y dramatic possibility of modulating the outgoing waves, not only by pulses, but ba human speech and receiving intelligent echoes from the Moon. Thus already in November 1951 the U.S. National Bureau of Standards, in collaboration with the Collins Radio Company, succeeded in transmitting intelligible signals from a 30 kW radio station broadcasting at the frequency of 417 Mc/sec (72 cm wave length) between Cedar Rapids, Iowa and Stirling, Virginia – separated by a distance of 1200 km – using the Moon as a reflector and thus inaugurated a new epoch in long-range radio transmission in UHF. The first sentence sent out by the American radio-engineers via the Moon paraphrased deliberately the first message sent on 24 May, 1844 by Samuel Morse over his new telegraph line between Washington and Baltimore: "What hath God Wrought" – a message which may become the forerunner of future relays of transatlantic television programmes of probably a more frivolous nature.

Speaking of echoes, one more recent and spectacular achievement of human science should be reported in this connection: and this is the reflection of *laser* beams from the lunar surface. The term itself (acronym for light amplifier with stimulated emission of radiation) calls for a few words of explanation; and perhaps the best way to do so will be by comparing certain properties of laser beams with those of ordinary light. As is well known, ordinary white light is produced by electromagnetic vibrations in arbitrary plane, with arbitrary phase; and arbitrary mixture of wavelengths. It is, in general, possible to lessen the mixture of wavelengths by appropriate filters or other devices; and atomic processes are known to emit light which is pretty nearly monochromatic (i.e., in which all vibrations possess the same wavelength). It is, furthermore, possible, by means of suitable devices to segregate waves vibrating in the same plane (and thus obtain light which we call polarised); but even in such a beam the phases of individual vibrations may still be entirely uncorrelated (and thus constitute light which lacks coherence).

Sunlight, or artificial light of electric bulbs, fluorescent tubes and many other well-known sources, is not coherent because the atoms which emit such a light do so independently of each other. On the other hand, at lower frequencies, electromagnetic waves used to transmit sound radio or television, are coherent because the electrons whose oscillation is separately responsible for producing the radio energy are made to move in concert with each other within the appropriate electrical circuits. However, coherent light in optical frequencies – the laser – was not excited by human hand until in the last few years.

The essential idea of any laser is to ensure that excited atoms of suitable paramagnetic substance (such as the crystals of synthetic ruby, gallium arsenide, or even certain inert gases) in a magnetic field will emit their light quanta, not haphazardly, but in unison. This can indeed be achieved by bathing them in a radiation capable of inducing a chain reaction: a single pulse of light, emitted by a single atom, will emerge from the laser material as a shower. In the ruby and gas lasers this shower of pulses can be amplified into a cloudburst by making the light bounce several times between parallel mirrors, so as to traverse the breadth of the material more than a hundred times, and be intensified each time.

As a result, the intensity of the light flashes emitted by lasers can be truly enormous. One flash of this kind given out by a ruby crystal can produce one calory (i.e., about $10^7$ ergs) of energy in $10^{-8}$ second. Unfortunately, this flow cannot as yet be sustained for much longer intervals of time; but while it lasts, one small ruby crystal can emit energy at a rate comparable with that of the largest existing nuclear power stations. Indeed, laser constitutes the most efficient way for squeezing out energy from atoms without disturbing their nuclei.

Although the coherence as well as strictly monochromatic nature of laser beams can make them eminently useful in many branches of science, so far they have been used mainly for the intensity of light which they can produce; and the first spectacular proof of its power has been the successful attempt of the physicists from the Lincoln Laboratories of the Massachusetts Institute of Technology in the United States of America in May 1962 to observe the reflection of its light waves from the Moon. In this pioneer experiment, a ruby laser was made to send out pulses of half a millisecond duration at the wave length of 6943 Å (i.e., in the red part of the spectrum). The light of a flash repeated each minute, and collimated by a 12-inch mirror, formed so parallel a beam that the area of the Moon's surface illuminated by it was not more than 2–3 km across. The light reflected back was naturally too faint to be visible by the naked eye; and in order to detect it, a red photomultiplier had to be placed in the focus of a telescope of 50 inches free aperture. Out of some $2 \times 10^{23}$ photons sent out to the Moon, the telescope recorded the return of about 12 only, after a round-trip of some 770000 km accomplished in 2.56 seconds. The fact that so few returning photons could be identified we owe entirely to their coherence – just as the same property of radio waves enables us to identify the signals sent out from deep-space probes with emergies of a few watts at distances of tens or even hundreds of millions of kilometers.

The Lincoln scientists repeated this experiment on three successive days (May 9–11, 1962). The first night they sent out 13 flashes to the region of the crater Albategnius; the second night 22 flashes were beamed on Copernicus; and the third night, 48 were directed towards Tycho. Accurate timing of the echoes reflected from so small a segment of the lunar surface opens up an entirely new avenue of exploration of the Moon at a distance. Unlike radar, laser beam should optically resolve on the lunar surface the same elements as ordinary light of the same wavelength with a given telescope. However, because of its incomparably higher frequency (some $5 \times 10^8$ Mc/sec for the ruby laser), the searching powers of its beam constitute potentially a vastly greater source of information than radar concerning the topography and motion of the Moon (both in space and around its own center of gravity), and will no doubt be fully utilized as soon as the techniques of producing sufficiently intense laser beams have been mastered.

In conclusion of the present chapter dealing with the electromagnetic properties of the Moon we wish to discuss one more subject closely connected with our main topic: namely, possible *electrostatic charge* of the lunar surface. In order to introduce this problem, let us recall some of the pertinent basic facts mentioned earlier in this volume. Thus in Chapter 12 concerned with the vestigial lunar atmosphere we mentioned that the complete absence of any optical twilight phenomena on the Moon relegated the upper limit of a possible gas density above the lunar surface to $10^{-12}$ g/cm$^3$. In the terrestrial atmosphere, an air density equal to this limit is attained at about 180 km above sea level – in the midst of the F1-layer – and the solar radiation at that height is known to be intense enough to bring about ionization producing some $10^5$ free electrons per cm$^3$. We do not know, of course, just how much less gas there is around the Moon below the optical upper limit – i.e., how much of it could survive the "knocking-off" power of the solar wind blowing constantly past the sunlit lunar face. However, we mentioned also that this wind may provide for the Moon an external source of gas; for most of its protons after striking the surface must rebound as neutral hydrogen atoms. By charge exchange with the incident protons, such atoms are then converted to slow protons bringing about an increase in density, in the immediate neighborhood of the lunar surface, to a level well above that of the incident solar wind.

Accreted, accommodated, and re-emitted solar wind should account for an average permanent density of about 80 hydrogen atoms per cm$^3$ (ŐPIK, 1962) – rising 10 to 100 times after solar eruptions. However, even so the solid surface of the Moon continues to be exposed wellnigh completely to ambient space conditions and illumination by undiluted sunlight. The consequences of such an exposure can be very important in many respects: not only (as we shall see in the next chapter) will the lunar "auroral zone" be relegated on to the solid surface, but the "lunar ionosphere" should be immediately above it as well. For, on the surface, the solar UV- and X-ray radiation is bound to produce ionization not only from atomic ground states (which, by recombination, may give rise to a visible luminescence), but from all atomic levels. This ionization should, in fact, be relatively much more complete than that encountered in the E- and F-layers of our own atmosphere; for there sunlight meets only gases characterized by relatively high ionization potentials; whereas, on the lunar surface, elements possessing weakly bound electrons are very much more abundant. In the terrestrial ionosphere, at an altitude of 300 km, the electron density is close to $10^6$ particles per cm$^3$; but the density of neutral atoms is still 10000 times greater. Around the airless Moon, however, the situation may be reversed: ŐPIK and SINGER (1960) estimate that the electron density may be close to $10^4$ particles per cm$^3$ (i.e., 100 times greater than that of the neutral hydrogen atoms).

In addition to photo-ionization, we must keep in mind the ionizing effects of the solar corpuscular radiation, as well as of the continuous flux of primary cosmic rays striking the lunar surface at an average rate of one per cm$^2$ every ten seconds (cf. BUETTNER, 1952). On the whole, however, all three processes – namely: (1) photoionization, due to solar UV radiation and X-rays; (2) solar corpuscular radiation (essentially protons); (3) primary cosmic rays (again mainly protons); are bound to keep charging the surface of the Moon *positively* in the course of time – the first source acting only during daytime; the second (in the absence of magnetic field) also in daytime; and the last, continuously.

To what level can this charge actually accumulate? It is obvious that a growing positive surface charge is bound again to retard (or prevent) the escape of electrons below a certain energy threshold by electrostatic attraction* – thus gradually slowing down the growth of the charge. Solar UV protons impinging on the lunar surface should liberate photoelectrons at a rate close to $10^{11}$ cm$^{-2}$sec$^{-1}$. Those of highest energies will escape into interplanetary space, leaving behind them a surface charged positively to a potential of some 20–30 volts. But the bulk of the photoelectrons will not have enough energy to escape. They may spend a little time above the lunar surface forming a transient "electron cushion", but will eventually return to it under the influence of the field of electrostatic space charge.

Besides, this surface charge must also be lessened by a capture of free electrons from the interplanetary space. The density of interplanetary electron gas at the Moon's average distance from the Sun is not yet definitely established, and neither is it constant. It seems to fluctuate with the solar activity at least between 10–100 particles per cm$^3$; and perhaps more. The actual density as well as extent of a lunar "electron cushion" above the surface must be the result of a contest between the growth of a positive surface charge by the above processes (1)–(3) – which, however, is by itself bound eventually to inhibit further electron escape – and a discharge into space through interplanetary electron gas. The resultant effect should, however, be to envelop the sunlit face of theMoon with a space charge or transient atmosphere of free electrons, whose density and scale height should fluctuate in the course of each lunation. Its quantitative analysis is, unfortunately, made difficult by several uncertain factors (such as the electron density prevailing in free space, or the "quantum efficiency" of the lunar surface to UV light), for which even their order of magnitude is as yet difficult to narrow down within reasonable limits; but that the phenomenon itself exists on the Moon can scarcely be in doubt.

Could the existence of such an "electron cushion" over the sunlit hemisphere of the Moon interfere with the observed microwave emission or radar reflections from the surface of our satellite as discussed earlier in this chapter? Is it, in particular, possible that the reflections of the radar pulses actually occur on the top of this electron cushion rather than on the solid surface beneath? Several reasons can be advanced to show that this is most unlikely to be the case. First, such reflections could take place only if the lunar ionosphere were critically dense at the respective frequency $v$ – i.e., that its electron density $N$ be given by

$$N = 1.24 \times 10^4 v^2 \qquad (21\text{-}33)$$

if $v$ is expressed in Mc/s. As, on the Moon, $N \approx 10^4$, it follows that the critical frequency for the Moon is probably of the order of 1 Mc/s, while radar pulses of not less than 10 Mc/s must be used to penetrate the terrestrial ionosphere and reach the Moon. At 3000 Mc/s (i.e., 10 cm wavelengths) the critical density becomes of the order of $10^{11}$

---

* Needless to say, the Moon's gravitational attraction is quite powerless to alter this situation. The lunar gravitational potential is equivalent to only about 0.02 eV for a hydrogen atom, so that hydrogen ions will be repelled by the Moon if it has a positive space potential of only 0.02 volts; and even singly-charged xenon ions will be repelled if the lunar potential rises to mere 2.5 volts.

electrons/cm$^3$ – much too high to be produced by the illumination of the lunar surface by the Sun.

But even if (for the sake of argument) the lunar ionosphere could become so dense at noontime, its power reflection coefficient $R$ would not only have to be close to one (i.e., more than 10 times larger than actually observed), but would also have to fluctuate in the course of each lunation and virtually disappear at night. The radar observations do not, however, show any marked variation in intensity of returning echoes in the course of a month. Besides, in the presence of a critically dense ionosphere the microwave temperatures as deduced from the intensity of thermal emission of the Moon would have to be very much higher (referring as they would to the sky brightness reflected in the ionosphere, and not to the lunar surface), and also depend strongly on the frequency. From the absence of all these phenomena we are virtually certain that the electron density of the lunar ionosphere is at no time close enough to its critical value as given by Equation (21-33) to enable it to reflect radar echoes; and that these (as well as the microwave thermal emission) do come to us from the lunar solid surface.

Let us, however, pursue this line of inquiry one step further and ask ourselves the following question: what temperature should a (black-body) illuminating source possess to endow the Moon with a critically dense ionosphere. In order to obtain the answer, let us depart from the well-known equation

$$c^2 \frac{d^2\mathbf{E}}{dr^2} + (\omega^2 - \Omega^2)\mathbf{E} = 0 \qquad (21\text{-}34)$$

for the transmission of radiation through an electron gas, where $\mathbf{E}$ represents the electric vector and $\omega = 2\pi\nu$ while

$$\Omega^2 = 4\pi \, Ne^2/m \qquad (21\text{-}35)$$

where $N$ denotes the electron density and $e$, $m$, the electronic charge and mass. When $\omega < \Omega$, the wave is attenuated in the direction of increasing $r$, and $\mathbf{E}$ will vary as $\exp(-\kappa r)$, where $c^2\kappa^2 = \Omega^2 - \omega^2$. As $\kappa < \Omega/c$, the characteristic attenuation distance $\kappa^{-1}$ is always greater than $c/\Omega$.

Now let $D$ be the characteristic thickness of the hypothetical electron cushion near the lunar surface. If so, the corresponding charge density would be $NeD$; the electrostatic field, $4\pi NeD$; and the potential, $4\pi NeD^2$. The average kinetic energy $\chi$ of the electrons should, accordingly, be given by

$$\chi = 4\pi \, Ne^2 \, D^2. \qquad (21\text{-}36)$$

For appreciable attenuation of radio waves we require that

$$D > \frac{c}{\Omega} \qquad (21\text{-}37)$$

and this implies that

$$\chi > 4\pi \, Ne^2 c^2 \Omega^{-2} = mc^2; \qquad (21\text{-}38)$$

the average kinetic energy of the electrons should, therefore, be at least as large as their rest mass.* But if the electrons were to be ejected from the solid surface by the absorption of photons, then the energy of these photons must also have been of the order of $mc^2$ or larger; and this would, in turn, necessitate for them to originate in a source of temperature

$$T > \frac{mc^2}{k} = 6 \times 10^9 \text{ deg Kelvin}, \qquad (21\text{-}39)$$

where $k$ denotes the Boltzmann constant. The flux of photons (solar of galactic) impinging on the lunar surface with the requisite energies is decidedly too small to be of any consequence; and hence, such ionosphere as may exist around the Moon cannot possibly interfere with the transmission or reflection of radio waves from its surface to any appreciable extent.

There exists, however, another way in which such an ionosphere could produce observable effects; and that is by refraction of the waves passing tangentially to the surface – the effects of which may be detected from the lunar occultation of cosmic radio sources. As is well-known, the refractive index $n$ of an electron gas of density $N$ is given by the equation

$$1 - n^2 = 8.06 \times 10^{-5} v^{-2} N; \qquad (21\text{-}40)$$

and for frequencies $v$ which are not too high the departure of $n$ from unity will *prolong* (cf., e.g., LINK, 1956) the observed occultation of a radio source by the Moon, by amounts which may become measurable for values of $N$ that are far below the critical limit expressed by (21-33).

Perhaps the best-known instance of an observation of this kind was afforded by the occultation of the Crab nebula on January 24, 1956 which was observed by Costain, Elsmore and Whitfield in Cambridge (COSTAIN *et al.*, 1956). These investigators noted that the radio waves emanating from this source did not vanish and reappear at exactly the same moments as the visible light of the Crab nebula became hidden by the lunar limb or emerged from behind it: the observed occultation of the radio source of $59.6 \pm 0.26$ min was longer by 0.4 min than the predicted occultation of the nebula. This difference would correspond to a refraction by $13''.4 \pm 8''.7$ on the Moon's limb; and with a scale-height of about 50 km it would require that $N \sim 10^3$ electrons/cm$^3$. The significance of this result is marginal; but values of $N$ substantially larger than $10^3$ seem ruled out. Future observations of the occultations of other known radio sources, sufficiently near the ecliptic to be in the Moon's path, will doubtless lessen its uncertainty and place our empirical knowledge of the electron density above the lunar surface on a more secure basis.

---

* Strictly speaking, in the face of this result we should re-calculate $\chi$ with the aid of the relativistic form of (21-34). However, this could lead only to a further increase of $\chi$ (because relativistic electrons possess greater inertia than nonrelativistic ones); and the above inequality (21-38) would be only strengthened.

## CHAPTER 22

# LUMINESCENCE OF THE LUNAR SURFACE

After the digression of the preceding chapter concerned with the electromagnetic properties of the lunar surface, let us return to the basic question raised at the commencement of this part (Chapter 19): namely, what are all the constituents of moonlight?

As has been known (or at least suspected) since the days of Anaxagoras, most of the moonlight is really sunlight, incident directly on the lunar surface (or indirectly, via the Earth) and absorbed or scattered by it in all directions. The only component of moonlight which is not of solar origin (and whose sources may antedate, in fact, the birth of the Sun or the Moon) is a minute flux of thermal radiation due to the gradual escape of radiogenic heat built up in the interior of our satellite (cf. Chapter 8); but as we have seen there, this source is utterly insignificant (scarcely capable of warming up the lunar surface to a few degrees Kelvin); and Anaxagoras knew, of course, nothing about it.

The *visible* light of our satellite is, therefore, essentially sunlight *scattered* from the lunar surface by a process which leaves its spectral composition largely unchanged (no doubt – as we shall discuss in more detail in the next chapter – because the scattering particles of surface dust are mostly large in comparison with the wavelength); and in the preceding Chapter 19 we have printed out that only about 7 per cent of incident light is scattered in this way. Moreover, in Chapter 20 we learned that most part of incident sunlight which is not scattered must be *absorbed* by the lunar surface, converted into heat, and *re-emitted* in the infrared to which our atmosphere is only partially transparent. This kind of light is emitted at wavelengths generally too long to be visible to the human eye to contaminate the white colour of the "silvery Moon" of our songs and romances, due to the scattered component, even though it represents some nine-tenths of transformed sunlight.

The question can, however, be asked: does scattered sunlight, plus that absorbed and re-emitted at generally lower frequencies, represent the sum total of the observable moonlight? The answer to this question is now known to be in the *negative*; and the aim of the present chapter will be to elaborate it in more detail. It is, of course, well known that – apart from visible sunlight – the lunar surface (unprotected by any atmosphere to speak of) is continuously exposed to high-energy UV – and X-ray quanta, as well as to all corpuscular radiation from the Sun. This surface is, in turn, then bound to give rise to X-ray emission ("bremsstrahlung") at wavelengths mostly between 10–100 Å* – radiation which cannot, of course, be observed from ground, but has

---

* In other words, the Sun can be likened to a "hot cathode" and the surface of the Moon, to the anti-cathode, of an "ion tube" of cosmic dimensions, whose glass walls (i.e., our atmosphere) enclose the observer rather than the apparatus.

already been measured (albeit with low resolution) from high-altitude rockets (JUDAY, 1965).

Suppose, however, that the recombination of atoms ionized by energetic particles or quanta does not occur (like in the bremsstrahlung) by a single transition, but by a cascade process giving rise to *luminescent emission* at longer wavelengths, which can penetrate our atmosphere and becomes observable on ground. Indications that such an emission may indeed make an appreciable contribution to the total light of the Moon have been forthcoming from different directions for many years. To begin with the most elementary manifestations – it has been noted by experienced observers in the past (cf., e.g., ROUGIER, 1933) that the apparent brightness of the Moon at a given phase (when due regard is paid in reductions to its instantaneous distance from the Earth, libration, etc) is not quite the same from month to month or a year to year; and these fluctuations appeared, moreover, to be correlated with the cycle of solar activity (though their amplitude seemed larger than that manifested by the solar constant). According to the most recent and reliable photometric studies by Gehrels and his associates (cf. GEHRELS *et al.*, 1964), between 1956–59 (i.e., near the maximum of the last cycle of solar activity) the lunar surface was between 10–20 per cent brighter in visible light than between 1963 November–1964 January, when solar activity was near its minimum.

Another indication of solar influence on global brightness of lunar eclipses, brought to light first by DANJON (1920), was mentioned already at the end of Chapter 19. By a discussion of the data extending over $3\frac{1}{2}$ centuries, Danjon proved that the brightness of the eclipsed Moon was strongly correlated with the cycle of solar activity: the residual brightness at the time of totality was found to increase with advancing cycle and to drop abruptly at the time of the minimum of solar activity in such a way that eclipses just preceding a minimum are brighter than those following it (cf. Figure 19-5). One other feature comes to mind which also changes abruptly at the minimum of solar activity: namely, the location of sunspots and associated disturbed areas on the apparent solar disk, which are known to expire near the equator at the end of a cycle to re-appear in high latitudes at the commencement of the next one.

Inasmuch as the corpuscular streams emanating from the disturbed regions of the Sun do not follow the same optical path as visible light, they could reach the Moon during eclipse and invoke luminescence; and as the spots move to lower latitudes (coinciding more nearly with the ecliptic) their luminescent effect may increase. However, when the sunspot latitudes change suddenly (as they do around the time of the minimum of the cycle) the excitation caused by them may be expected to drop in intensity, and the residual brightness of a lunar eclipse around that time should be markedly smaller. This is just what Danjon found; and the case for the corpuscular excitation of the luminescence of the lunar surface receives thereby indirect support.

Evidence for it was strengthened further by the results of the photometric studies by LINK (1946) and others (CIMINO and FORTINI, 1953; CIMINO and FRESA, 1958) of partial phases of lunar eclipses, during which the loss of light was found to be less than that expected from the geometry of the Sun-Earth disks – indicating that signifi-

cant effects are being produced by illumination of the Moon by uneclipsed parts of
the solar corona. The latter emits but negligible amount of visible light, but (on account of its high temperature of 1–2 million degrees) a much larger proportion of
X-rays; and Link conjectured that this hard radiation produces luminescence on the
lunar surface which accounts for the optical anomaly.

Photometric studies of the global light of the Moon clearly foreshadowed the existence of a relationship between these optical phenomena and solar activity, but revealed little or nothing about possible *spectral composition* of the hypothetical luminescence. In order to learn more about it, we must obviously resort to some spectro-
metric method; and this can be done in the following manner. Consider the profile of
an absorption line of the solar spectrum – preferably a deep one (i.e., of small residual
intensity) – as shown schematically on Figure 22-1. The scattering of sunlight on the
lunar surface depends but very little on the frequency. Therefore, if moonlight consisted only of scattered sunlight, the line-profiles in its spectrum should be exact replicas
of those in the parent sunlight. If, on the other hand, the surface of the Moon (intercepted by the slit of the spectrograph) luminesces, the lunar line-profile should become
correspondingly *shallower* than those in the solar spectrum. In more specific terms, if
$D_{M,S}$ denote the ratio of the intensity of any particular point of a line profile to that
of the adjacent continuum in the spectrum of the Moon and the Sun, respectively,
then the ratio

$$d = \frac{D_M - D_S}{1 - D_M} \tag{22-1}$$

denotes the fractional intensity of the luminescent radiation superimposed on scattered
moonlight, and expressed again in terms of the intensity of the adjacent continuum
taken as the unit.

There exists, furthermore, another way in which the luminescence of the lunar

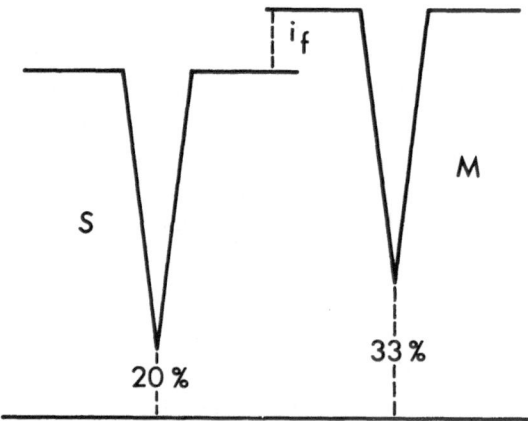

Fig. 22-1.   Central intensity of an absorption line in the solar (S) and lunar (M) spectrum in the
presence of luminescence of intensity $i_f$.

surface sould be established from the observations: namely, by measures of the *polarization* of moonlight, and its dependence on the wavelength. The basic idea goes back to the fact that, whereas the fraction of the moonlight which represents scattered sunlight becomes (as we already discussed in Chapter 19) distinctly polarized by this process – and the degree as well as the position of the plane of polarization varies with the phase – the luminescent emission is, of course, non-polarized. Therefore, any temporary anomalies on the phase-polarization curves which exceed the limits of observational errors – manifested so clearly on Lyot's excellent measurements (LYOT, 1929) – may (as was recently pointed out by Gehrels) have been caused by variable admixture of non-polarized light of luminescent origin. Moreover, the polarization of scattered moonlight is known to vary but slowly with the wavelength. Therefore, any irregularities on the curve of polarisation versus wavelength should indicate – no less distinctly than the filling-up of the absorption lines of the lunar spectrum – the presence of luminescent bands at the respective frequency (TEYFEL, 1960).

The "method of line-depths" for the determination of the fractional luminescence $d$ of the lunar surface from the measured ratios $D_{M,S}$ by means of Equation (22-1) was proposed by LINK (1951), and employed in practice first by KOZYREV (1956) and DUBOIS (1957, 1959). Both found indications of luminescence to be present in different parts of the lunar surface, and in light of different frequencies. According to Kozyrev, the maximum value of $d$ observed for the crater Aristarchus in the profile of the H-line of ionised calcium ($\lambda$ 3970 Å) amounted to 0.13 on Oct. 4, 1955; but less than a month later (1955 Oct. 28 and Nov. 4) it diminished to 0.03. Since, however, both Kozyrev and Dubois employed photographic techniques which were of marginal accuracy for the purpose, their results were regarded by contemporary astronomers as suggestive rather than conclusive; and it became obvious that in order to obtain results of indubitable significance, the methods of photographic spectroscopy would have to give way to photoelectric spectrometry.

This was first done at Manchester between 1960–62 (as a part of systematic program of lunar luminescence studies sponsored by the U.S. Air Force), when Grainger and Ring built a scanning-type photoelectric spectrograph which was used in connection with the 50-inch reflector of Padua University Observatory at Asiago. The actual observations of the Moon commenced in 1961, and were largely confined to the scans of the profile of Ca II H-line ($\lambda$ 3970 Å), for reasons which previously led Kozyrev to the same choice: namely, the large half-width of this line (9 Å) and its low residual intensity – both factors facilitating the detection of luminescence. Spectroscopic work carried out between 1961–62 confirmed, on the whole, the previous results by Kozyrev or Dubois, and established the reality of lunar luminescence beyond reasonable doubt (cf. GRAINGER and RING, 1962).

The strongest luminescence – amounting to $10\pm1$ per cent of intensity of the adjacent continuum – was at that time found to be emitted by the bright ray traversing Mare Serenitatis through the crater Bessel (the mare itself appeared to luminesce no more than to $2\pm0.7$ per cent); and other regions (in particular, a region in the neighbourhood of Plato at $\lambda = 3\,°W$, $\beta = 56\,°N$) proved to luminesce to almost the same extent. However, the actual value of $d$ changed abruptly with any shift in position of the slit (indicative of pronounced localization of the luminophor on the lunar surface), and also with the time; but the data were too few to indicate any real correspondence between the lunar and solar phenomena. Moreover, Spinrad working at the Dominion Astrophysical Observatory at Victoria, observed on 1962 Sept. 16 a luminescence of intensity corresponding to $d = 0.13$ in the light of the calcium H-line; and as his slit trailed over the Moon during exposure, the luminescence was probably widespread at that time (SPINRAD, 1964).

Since that time, a photoelectric scanning spectrometer of optical power comparable to that built by Grainger and Ring was employed in quest of lunar luminescence by the line-depth method by Scarfe at Cambridge (England), who extended this search to other spectral lines than the H-line of Ca II (such as Hα, the sodium D-lines, or a group of Fe I lines near $\lambda$ 5450 Å) in the yellow and red part of the spectrum (SCARFE, 1965). Near $\lambda$ 5450, a time-variable luminescence attaining as much as 30% of intensity of the adjacent continuum was observed on 1963 Oct. 5 - while at the wavelengths of Hα or of the sodium D lines no luminescence exceeding 2% could be detected at the same time. The crater Aristarchus showed the strongest effect, but Copernicus almost as strong, and Kepler only somewhat less. These were the only places where the luminescence was marked. Moreover, normal integrated sunlight (diffused from a plate coated with magnesium oxide) exhibited line profiles virtually identical with those of diffuse moonlight – a fact supporting the idea that the time-independent radiation from the Sun does not induce observable luminescence; and Scarfe concluded that the latter must be related to solar activity. Further evidence for the luminescence of the lunar surface in the neighbourhood of the crater Aristarchus by the line-depth method was since reported by MYRONOVA (1965).

However, the last two months of 1963 (when a solar cycle expiring about that time was in its last throes) had more surprises in store for the students of the Moon. On Oct. 28, at 1.58 UT the largest flare of the year (of class 3) made its appearance on the Sun, and was followed by a week of disturbed Sun, with several minor flares, which affected profoundly the particle density in space at the distance of the Earth (see Figure 22-2). And – behold – not less than two transient luminous phenomena were observed on the Moon during this time which we shall briefly describe.

The first instance occurred on October 30th (between 1:30–1:50 UT – i.e., 48 hours after the flare of Oct. 28), when Greenacre and Barr at Lowell Observatory noted that three distinct spots in the neighbourhood of the crater Aristarchus (in locations marked on Figure 22-3) flared up temporarily in reddish-orange light, intense enough to be seen visually without the aid of a filter, and lasting approximately 25 minutes (cf. GREENACRE, 1963). No photographs or spectra were obtained because of the short duration of the phenomenon; but its intensity was at least comparable with that of scattered sunlight, and moreover, seemed to change from minute to minute.

Less than three days later, Kopal and Rackham at Observatoire du Pic-du-Midi succeeded in securing photographs of another – and much larger – "lunar flare" on the night of November 1–2, which are reproduced on the accompanying Figure 22-4.

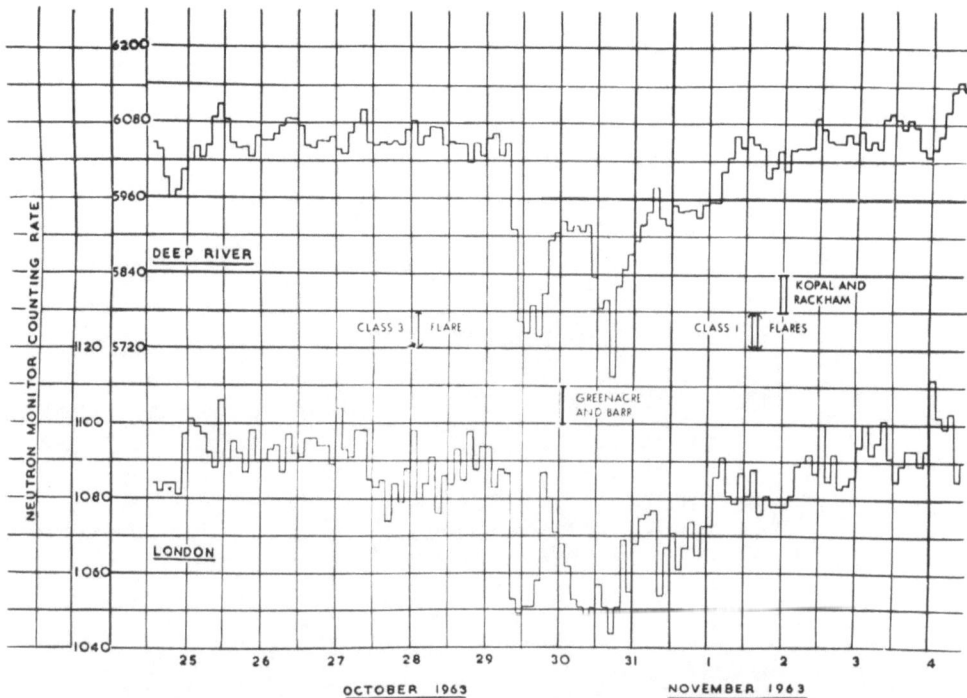

Fig. 22-2. Variation in the terrestrial neutron flux, caused by the solar events at the end of October and beginning of November 1963, as recorded at Deep River (Canada) and the Imperial College in London (after Thambyahpillay). The occurrence of the lunar events are marked on the same time scale.

The first pair of photographs showing it as exposed between 22.35–22.42 UT on November 1st, within the same minute, in the focal plane of the observatory's 24-inch refractor through two different interference filters: one centered at $\lambda$ 6725 Å in the red, of 45 Å half-width; the other centered at $\lambda$ 5450 Å off the green, with 95 Å half-width; photographic emulsions in both cases being the same (Kodak 1-F spectroscopic); and so was their subsequent processing in the dark room. The "red" photograph of a portion of Oceanus Procellarum between the craters Aristarchus, Copernicus, and Kepler showed a striking enhancement of surface brightness of a large area around (and to the north of) the crater Kepler, which is normally of lesser albedo and bluer than the surroundings of Copernicus (to the left on the photographs), and which was totally absent in the green. Shortly after 23.00 UT the red enhancement had virtually disappeared; but it became once more very distinct on four pairs of plates exposed after midnight (between 0.20m to 0.35m UT on Nov. 2nd), one of which is reproduced on Figure 22-4. The red enhancement of the Kepler region occurred, therefore, at least twice that night – each time lasting probably no longer than 15–20 minutes; and no subsequent night for the rest of the month revealed anything unusual. It was, in particular, cloudy on the Pic on November 27, when Greenacre and Barr

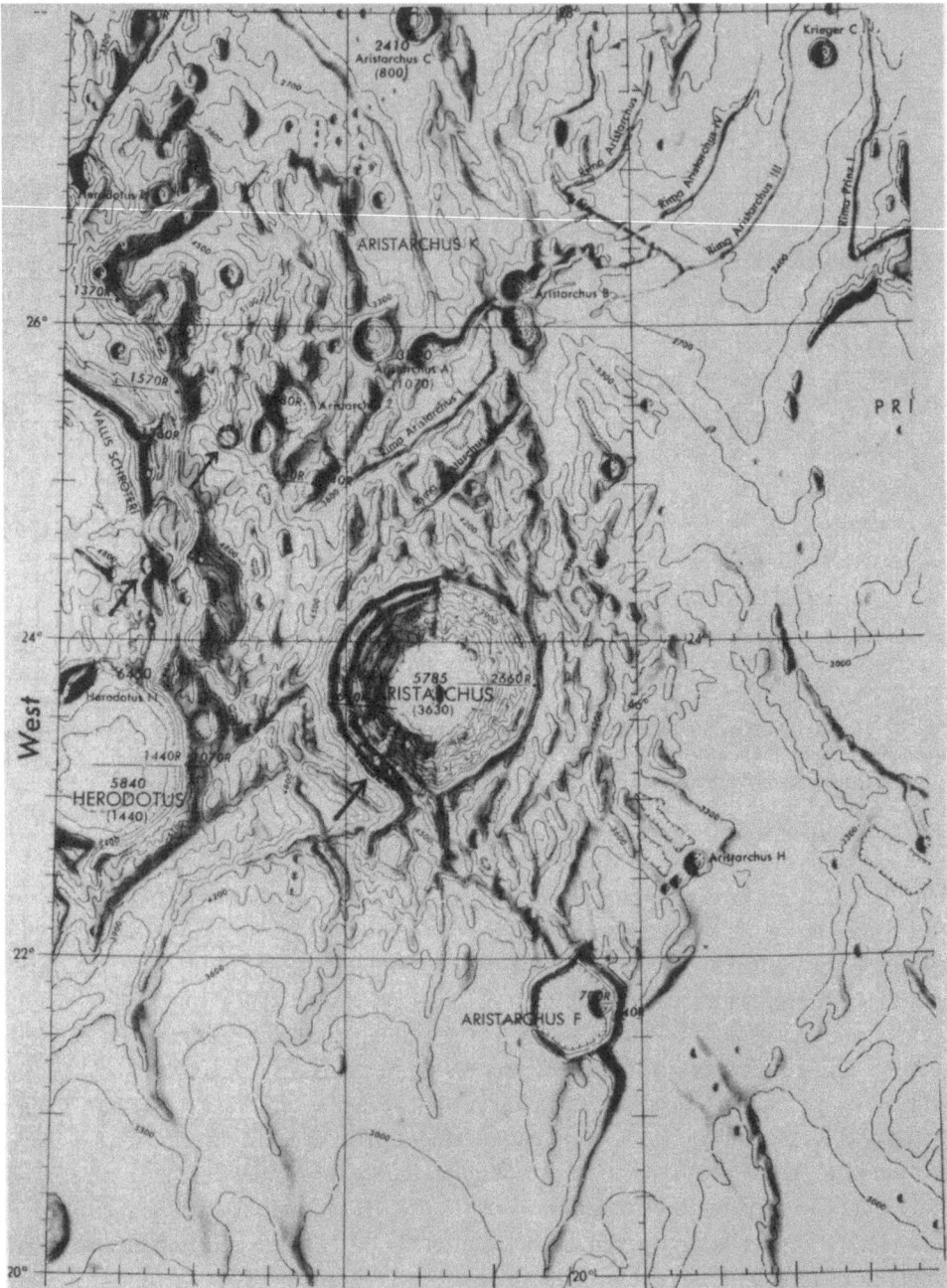

Fig. 22-3. The surroundings of the lunar crater Aristarchus, where Greenacre and Barr have repeatedly observed (in regions indicated by arrows) anomalous red colour phenomena in October and November 1963. (U.S. Air Force Chart of the Moon No. 39; reproduced by courtesy of the Aero Chart and Information Center, St. Louis.)

reported a recurrence of anomalous reddening of other spots in the proximity of Aristarchus, lasting 1¼ hour (GREENACRE, 1964).

Fuller details of Kopal and Rackham's observations of Nov. 1–2nd were adequately reported elsewhere (cf. KOPAL and RACKHAM, 1963, 1964). Some of their plates were calibrated for photometric purposes; and their microdensitometric analysis revealed that, at the peak of the second brightening the relative red enhancement of the Kepler area corresponded to a value of $d = 0.86 \pm 0.03$ – i.e., several times as large as those observed previously by the linedepth method. The "red spots" of Greenacre and Barr may have been equally intense (and possibly more); but the area covered by them – a few square kilometers of the lunar surface – was relatively minute in comparison with the enhanced area as shown on Figure 22-4, which covered more than 60000 square kilometers of lunar ground. The transient enhancement shown on this plate represents the largest and most conspicuous example known of lunar luminous phenomena of this kind; and together with the visual observations at Lowell represents so far the best evidence we possess of the rapid time variations of these phenomena – all secured in the course of one month!

However, a scruting of older literature revealed that phenomena of this type were observed before. The first astronomer who indubitably noticed them on the Moon in the neighbourhood of the crater Aristarchus (and, as far as we can say now, on the same spots as Greenacre and Barr) was William Herschel, who on April 18–19, 1787 noted there the appearance of spots glowing like "slowly burning charcoal thinly covered with ashes" (he thought these were active volcanoes on the Moon); and so certain was he of their existence that he invited his royal patron (King George IIIrd) to come and look at them through his telescope! Now May 1787 was just about the time of the maximum of the solar cycle (and unusually active one); and although no flare observations are of course available for that time, the fact that the Sun must have been greatly disturbed is attested by the reported visibility (on both April 18 and 19, 1787) of polar aurorae as far south as Padua, Italy – an event which occurs scarcely once in a decade (cf. FRITZ, 1873).

More recently, FLAMM and LINGENFELTER (1965) collected evidence that transient luminous phenomena of this kind in the vicinity of Aristarchus were noted by different observers not less than 16 times between 1787 and 1963; and, in addition, the literature contains references to luminescent events in the vicinity of several other lunar features. There seems thus no room for doubt that temporary reddish enhancements represent recurrent phenomena on the lunar surface demanding explanation.

How to account for them? First, let it be stressed that any explanation in terms of ordinary thermal phenomena are utterly out of question; for no known matter could possibly heat up (by whichever process) and cool off on the observed scale in a time of the order of one hour or less. The light emission must obviously be non-thermal; but ordinary thermo-luminescence is likewise ruled out by the fact that (unlike lunar monthly temperature changes) the luminous phenomena are not recurrent each lunation, and occur also at times when local temperatures on the spot (i.e., the height of the Sun above horizon) are very different. The fact that they seem to recur in certain

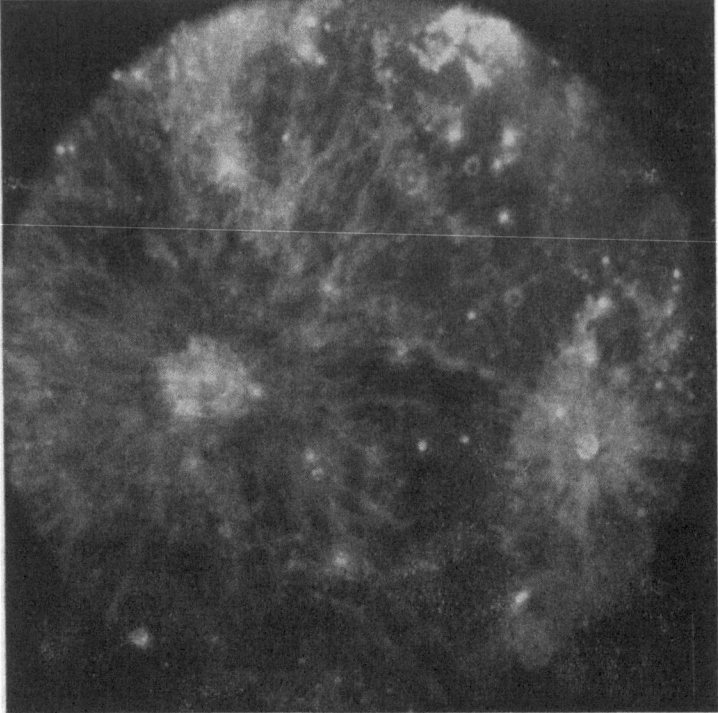

Fig. 22-4.   A comparison of green (left) and red (right) photographs of the Copernicus-Kepler region on the Moon, taken on 1963 November 2, 00.35 UT at the Observatoire Pic-du-Midi by Kopal and Rackham within the same minute. Note a (transient) red enhancement of the area around Kepler.

(and often quite small) places indicates that they favor certain type of ground; moreover, their spectra are obviously not continuous, but emission seems confined to broad bands concentrated in the yellow and red parts of the spectrum.

A possibility of photo-excitation of gas escaping from lunar interior at the spots near Aristarchus exhibiting the anomalous colour phenomena was mentioned recently by Swings, who conjectured that ammonia $NH_3$ is dissociated by sunlight into $NH_2$ and $H_1$ and that the amine radical fluoresces in the red, as is observed in comet tails. The quantitative consequences of such a hypothesis are, however, scarcely any more admissible; for too large a mass of gas would be required to account in this way for the luminous phenomena observed by Greenacre and Barr, and the discharge of such a mass of it into vacuum would be bound to bring about also structural changes of the surrounding landscape which have not been observed. The size of the Kepler enhancement as photographed by Kopal and Rackham makes these difficulties overwhelming; and the fact that such phenomena (particularly around Aristarchus) occur also at night-time – when no sunlight is available to cause photo-dissociation or luminescence – seems to rule out such a hypothesis altogether.

The manifestly transient nature of the luminescent enhancement appears to be

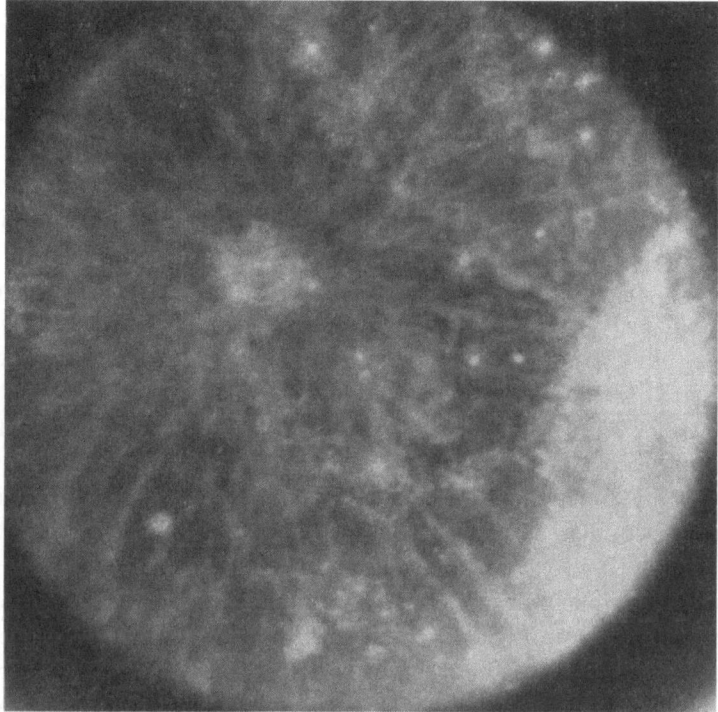

Fig. 22-4.

completely uncorrelated with the lunar thermal calendar; and its origin must, therefore, be obviously sought outside – and what else could control external events than
the Sun? The rhythm of the solar cycle has clearly made itself felt in the residual
brightness of lunar eclipses (Danjon); and it is only natural to ask to what extent may
the "lunar flares" be correlated with other specific aspects of solar activity.

Of these, the one which immediately comes to mind (because of their time-scale)
are solar flares. These are known to be as short-lived (from minutes to hours) as the
observed lunar enhancements; and while they last they emit sufficient amount of energy – both electromagnetic (X-rays) and corpuscular (mainly fast protons) – to disturb
the inner precincts of the solar system for hours and day afterwards. It is, in principle,
quite simple to differentiate between the effects produced by their electromagnetic and
corpuscular emission: while the effects of X-ray emission, propagating with the speed
of light, would reach the Moon at (virtually) the same time as the flare is observed on
Earth – particles emitted from the flare may reach us with a time-lag of many hours
(depending on their speed and directness of their trajectories).

For many years now solar disturbances have been monitored by astronomers with
close to 100% efficiency. We know, in particular, almost exhaustively what the Sun has
been doing from day to day in October – November 1963 – from ground-based as

well as satellite observations – when so many lunar phenomena have been noted by several observers. Thus the occurrence of the "red spots" near Aristarchus, as observed by Greenacre and Barr on October 30th, followed by 48 hours the great flare of October 28th; and the Kopal-Rackham photographs of the enhancement of the Kepler region followed the same flare by 118 hours (and a class 1 double flare of November 1st by $8\frac{1}{2}$ hours). Herschel's observations of April 18–19, 1787 coincided also with a period of highly disturbed Sun. On the other hand, the Greenacre-Barr observations of November 27th were made at a time when the Sun was exceptionally quiet.

Whatever the general case may be, one feature emerges from the foregoing facts quite clearly: namely, the *none* of the lunar phenomena occurred *simultaneously* with any major manifestation of solar activity; and cannot, therefore, be due to electromagnetic (X-ray) excitation. When we add to it the fact that the Herschel "volcanoes" on the Moon of 1787 (as well as, according to Flamm and Lingenfelter, the majority of the past lunar events) occurred on the dark (night) side of the Moon – which could not be reached by direct sunlight – the case against direct electromagnetic excitation becomes overwhelming; and the corpuscular radiation seems to offer the only remaining avenue of approach to the solution of our problem.

This road is feasible in principle; for not only can corpuscular radiation reach the Moon after a time-lag of the observed order of magnitude (i.e., from several hours to a few days), but particles of sufficient energy (spiralling along the respective lines of force) can impinge on any part of the day – or night – hemisphere of the Moon at any time, provided only that the gyro radius $\rho$ of such spirals exceeds that of our satellite. As is well known, this gyro radius is given by the formula

$$\rho = \frac{E}{300\,H}\ \text{cm} \tag{22-2}$$

if $E$ stands for the energy of the respective particles in electron-volts, and $H$ denotes the strength of the interplanetary magnetic field in gauss.

From recent extensive measurements performed aboard Mariner 2 (cf. SMITH *et al.*, 1962) the quantity $H$ is known to fluctuate between 5 and 50$\gamma$, with a value of $10\gamma = 10^{-4}$ gauss representing a fair average. If so, however, the radius $\rho$ of gyration for 1 MeV protons proves to be approximately equal to $3 \times 10^7$ cm; increasing to $3 \times 10^8$ cm (i.e., about twice the radius $a$ of the Moon) for $E = 10$ MeV, and $3 \times 10^9$ cm (i.e., about 17$a$) for 100 MeV protons. Therefore, it is not till for energies close to 10 MeV that $\rho$ attains the dimensions of the lunar globe, and exceeds if for $E > 10$ MeV (i.e., for proton velocities in excess of 50000 km/sec).

However, when we come to consider quantitative aspects of corpuscular stimulation of lunar luminescence, the situation becomes less clear. The 1963 observations by Kopal and Rackham of the large enhancement of the Kepler region which temporarily almost doubled its surface brightness in the red indicated that the energy flux $F$ of energy stimulating luminescent emission should have been of the order of $10^5$ ergs/cm$^2$sec (KOPAL and RACKHAM, 1963); and for typical observations of luminescence by the "line-depth method" it could have been down to $10^3 - 10^4$ ergs/cm$^2$ sec.

Now if protons alone had been responsible for its excitation, the energy balance would require that the ratio

$$\frac{F}{v} = \tfrac{1}{2} m_H v^2 N, \qquad (22\text{-}3)$$

where $m_H$ denotes the mass of a proton; $v$, their velocity; and $N$, their number per ccm. For $m_H = 1.67 \times 10^{-24}$ g and $F = 10^5$ ergs/cm² sec, the foregoing equation requires that the product $Nv \sim 10^{29}$ sec$^{-3}$ for the 1963 Nov. 1–2 events, and about $10^{27}$ sec$^{-3}$ for smaller events observed spectroscopically through narrower passbands.

These numbers seem rather large. For $v = 5000$ km/sec (about the maximum velocity of "slow" protons emitted by solar flares) the corresponding particle density $N$ should be of the order of $10^3$ cm$^{-3}$ for the maximum flux of $10^5$ ergs/cm² sec, and 10–100 per ccm for more moderate events. Now the velocity of the quiet-Sun solar wind (as measured during the flight of Mariner 2 to Venus in 1962, for instance) is only about 400–500 km/sec; and the particle density $N$, only about 0.1 per ccm. During storm conditions both the values of $v$ and $N$ may increase, perhaps, ten times; but even this leaves us with values of $N$ which are 10–100 times too small – a fact which either discloses that the actual solar activity may still have surprises for us in store, or that solar particles may not be the primary source of energy for lunar luminescence emission, but merely trigger its release; we cannot as yet be sure.

If we turn our attention to the lunar night events – when luminescence has to compete in contrast only with earthshine which is (*ceteris paribus*) $10^4$ times less intense than direct sunlight – luminescent glow concentrated in emission bands of 100–1000 Å width and invoked by incident energy flux as low as 1–10 ergs/cm² sec could become visually observable. Only protons with energies in excess of 10 MeV (i.e., velocities larger than 50000 km/sec) can, to be sure, follow interplanetary trajectories which are curved enough to enable them to impinge on the dark side of the Moon, and a class of fast particles emitted by solar flares are known indeed to move with speeds clustering around 150000 km/sec (cf., e.g., ELLISON, 1963). A flux of 1–10 ergs/cm² sec of such particles would (for $v = 1.5 \times 10^5$ km/sec) correspond by (22-3) to a density $N$ of the order of $10^{-7}$ to $10^{-6}$. Now solar-flare proton events in the 10–100 MeV range do exhibit fluxes of $10^3$–$10^4$ particles per cm² per sec – the one following the flare of 1960 November 12 attained almost $10^5$ particles/cm² sec (cf. FREIER and WEBER, 1963) – corresponding to $N \sim 10^{-7}$–$10^{-6}$ particles per ccm or even more. This seems to be indeed of the right order of magnitude to account for the visibility of night-time luminescence on the Moon; but the daytime luminescence still seems anomalously large.

This suspicion is strengthened by some events of the last solar cycle when, following the great flare (class 3+) of 1958 July 7th, Blackwell and Ingham observed at Chacaltaya (in the high Bolivian Andes) a temporary reddening and general increase in brightness of the zodiacal light, and obtained evidence which satisfied them that the extra emission came from the interplanetary dust cloud. This phenomenon could again be scarcely understood otherwise than as luminescence produced by the corpus-

cular output of that flare; but for an assumed particle density of 300 protons per ccm
their velocity had to be in excess of 40000 km/sec to make the luminescent process
less than 100% efficient.

Moreover, BLACKWELL and INGHAM (1961) pointed out that the surface brightness
of the zodiacal cloud – another example of solid matter illuminated by the Sun in
space, and from greater proximity than the Moon – seems correlated with the planetary magnetic index $K_p$, and increased by 40% when $K_p$ was between 8 and 9. The
relevance of the events represented by this number also to lunar luminescence has
recently been emphasised by CAMERON (1964), who pointed out that a daytime luminescence as intense as that photographed by KOPAL and RACKHAM on Nov. 1–2, 1963
could have been caused, not by the direct impact of primary solar particles associated
with a flare, but indirectly by their effect on the terrestrial magnetosphere. The interaction between solar wind and the Earth's magnetosphere is very complex; but it
seems that in solar direction the pressure of this wind compresses the magnetosphere,
and in the antisolar direction causes the formation of a long cavity in which the terrestrial magnetic field can expand. Since the velocity of the solar wind relative to the
Earth is supersonic, a stationary shock wave must be produced beyond the magneto-
pause, which may cause extensive particle acceleration. Or it may be that lunar luminescence is produced by particles trapped in the distant tail of the magnetosphere.
Cameron pointed out that, in all cases of intense daytime luminescence on record, the
Moon was never too far from full – which may be an important factor in the explanation; but it may also be that principal concentrations of suspected luminophors
(around Aristarchus, or Kepler, for instance) are localized in regions which do not
become sunlit till shortly before full-moon; we cannot as yet be sure. But more recent
studies by DODSON and HEDEMAN (1964) indicate the existence of an unexpected correlation between the proton and neutron events of solar origin and the lunar cycle; and
if so, our magnetosphere may indeed have something to do with it; but more work
remains again yet to be done before this suggestion can be placed on a more secure
basis.

In summary, we may say that the existence of a transient luminescence of the lunar
surface – long suspected from several independent lines of evidence – should now be
accepted as an established fact. It seems to recur preferentially in certain parts of the
lunar surface; and its spectrum is not continuous, but confined to certain emission
bands clustering towards the red. A transient emission of intensity amounting to a few
per cent of that of the adjacent continuum in narrow bandwidths seems, moreover,
to be of fairly frequent occurrence. Instances of enhancements amounting from 30 to
80 per cent (and possibly more) are on record; but their frequency can so far only be
guessed at.

Secondly, it is probable that all known aspects of this luminescence seem (directly
or indirectly) related with some activity of the Sun. A correlation between the two
may not be simple; and very probably there is more to it than a mere proportionality
with sunspot numbers. FLAMM and LINGENFELTER (1965) rightly contend that the
majority of past lunar events occurred in the years of low sunspot numbers – the

Greenacre-Barr and Kopal-Rackham events of Oct.–Nov. 1963 are eloquent instances of this fact. But (as was shown first by Danjon) the abrupt change in residual brightness of eclipsed Moon occurs also, not at the maximum, but the *minimum* of solar activity.

The establishment of one-to-one correlation between the corresponding solar and lunar events still represents a goal facing us at some distance in the future; but from an obvious lack of simultaneous occurrence it is clear that luminescence on the Moon is not stimulated by solar electromagnetic (X-ray), but corpuscular, radiation which reaches the Moon after a transit time ranging from several hours to a few days. At daytime – in the absence of any appreciable magnetic field – particles of all energies can impinge on the lunar surface; while at night only energetic particles ($E > 10$ MeV) can do so. Their numbers, as measured by independent experiments, are just about adequate to produce visible luminescence at nighttime; but for daytime events the number of direct solar particles seems deficient by two or three orders of magnitude; so that some secondary accelerating (or storage) facilities may have to be considered. Besides, there are other problems to face: for instance, if the luminescence observed on 1963 October 30 by Greenacre and Barr or on November 1–2 by Kopal and Rackham, was indeed excited by corpuscular emission of the Class 3 flare of October 28, is it possible that these particles could have remained so long compacted in space to give rise to transient phenomena lasting no more than 20–30 minutes after a time-lapse of 48 or 118 hours?

Third, the luminescence of the Moon is observed to recur in certain regions of its surface often limited to a few kilometers in size – suggesting strong localization of the luminophor. Its chemical structure can again so far be only guessed at. Because fluorescent phenomena are the result of electronic transitions rather than molecular or lattice vibrations, they are not related to the mineralogy of surface materials so much as to the presence of minor contaminants or impurities. However, an eventual identification of its mineralogical constituents may, in due course, provide a valuable tool for the petrographic prospecting of the outer crust of the Moon at a distance.

The argument that any appreciable quantum efficiency of this process on the lunar surface would have been quenched by radiation damage over astronomically long intervals of time is belied by the fact that meteorites luminesce in the laboratory when excited at first by artificial corpuscular bombardment (cf., DERHAM and GEAKE, 1964) – in spite of the fact that, while in space, they must have suffered as much radiation damage (from cosmic rays, etc.) as anything that lies exposed on the lunar surface. Like the laboratory scientist, Nature resorts probably to the same restorative of the luminescent power: namely, periodic heating of the sample; and on the Moon this may be taken care of by the diurnal cycle.

Moreover, recent Manchester experiments with corpuscular excitation of luminescing enstatite meteorites by use of a van de Graaf generator have disclosed that the intensity of visible luminescence remains approximately proportional to the energy of the impinging protons up to energies of several MeV's – a fact which reveals that such particles are not trapped at depths from which visible light could no longer reach us. This experiment proved that night-time visual luminescence can occur on the Moon if the particle density and (or) the surface concentration of the luminophor is sufficiently high to make it visible; and observational evidence has already been reviewed which tends to show that this is indeed the case.

Eventual identification of the mineralogical constituents which may luminesce on the Moon might, in due course, provide a workable tool for the geological prospecting of its outer crust at a distance. However, one of the aims of this section has been to introduce to the reader the Moon also in another capacity: namely, as a potential tool for studies of the solar-terrestrial relations. In these days of space exploration – when hundreds of man-made craft are probing various properties of interplanetary space – we should not forget that we have at our doorstep (astronomically speaking) another permanent probe ever present and exposing no small target to all particles which it intercepts on its perpetual journey through space: namely, our Moon. To be sure, the Moon can scarcely as yet be regarded as an instrumented probe, though the day is not far in the future when this may come true. But at least in one sense the Moon deserves this title already to-day; for parts of its surface may constitute a natural *wavelength converter*, transforming high-energy corpuscular radiation from the Sun into visible light through the medium of luminescence. There is indeed little room for doubt that, in the years to come, the "lunar flares" due to this cause will be observed with the same care and completeness as solar flares and phenomena associated with them are watched to-day by international cooperative programs, in order to improve further our understanding of the solar-terrestrial relations.

CHAPTER 23

# STRUCTURE OF THE LUNAR SURFACE

Having completed in the preceding chapter a brief outline of the information which reaches us from the lunar surface by means of the electromagnetic radiation in different parts of the spectrum, a time has come for us to take stock of this knowledge reaching us through these different channels of information to see what it reveals concerning the structure of the lunar surface concealed below the limits of direct telescopic resolution.

First, a few retrospective words on the methods by which the lunar environment can be studied from the Earth at a distance. As we mentioned already in Chapter 16, telescopic observations at optical frequencies – visual or photographic – can provide direct information concerning the features of the lunar surface down to the characteristics of rather less than half a kilometre in size. A 24-inch refractor should (under ideal conditions) resolve on the lunar surface details approximately 400 metres in size; and a 40-inch telescope – the largest aperture used effectively for systematic lunar studies so far – may depress this limit down to somewhat less than 300 metres. Below this limit, however, the outlines of individual lunar objects become blurred in a haze arising from the contributions of diffraction phenomena, unsteadiness of seeing, and photographic plate grain – the action of which is well illustrated on the accompanying Figure 23-1. When the Sun stands very low above the lunar horizon, long shadows cast in its rays may enable us to establish the presence of (preferentially orientated) surface unevenesses on a scale 10–100 times smaller than that necessary for direct resolution (cf. pp. 203 of Chapter 14); but few results of such studies have been reported so far. (cf. KOPAL and RACKHAM, 1963a.)

Actual measurements of local deformations of the lunar surface from the mean selenoid (Chapter 13) carried out in recent years by the shadow method on a large scale (cf., Chapter 14) have revealed the overwhelming part of the lunar surface to be smooth and gently sloping; its average inclination to the horizontal not amounting to more than a few degrees. This appears to be true not only in the maria, but also in the continental regions. Former notions of great ruggedness of the lunar surface, based on telescopic views of the shadows along the terminator, failed to take proper account of the low altitude of the illuminating source.

This statement does not, to be sure, rule out the existence, on the Moon, of occasional slopes considerably steeper than the average. For instance, the inner rims of small craters and crater pits may be considerably more inclined to the horizontal. However – and this should be emphasized – the total area occupied by such slopes represents only a tiny fraction (less than one per cent) of the entire lunar surface; so

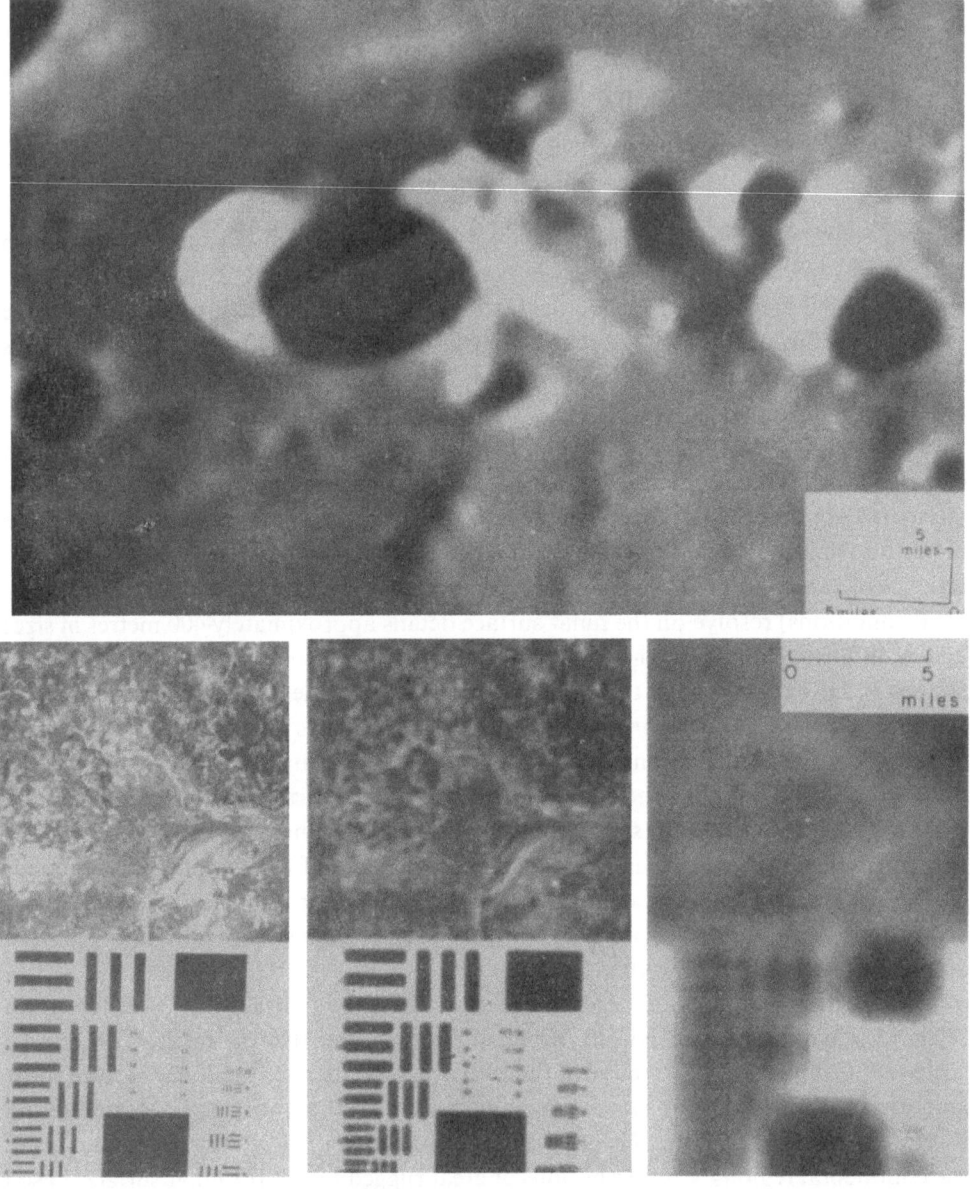

Fig. 23-1.  Photography of the Moon (above) and Washington, D. C. (below) at the same scale, demonstrate that aerial photography of the Earth at a ground resolution corresponding to the best lunar photography is extremely poor. The above photograph of the interior of the crater Clavius, taken with the 200-inch telescope of Mt Wilson and Palomar Observatories represents a section of that already reproduced on Figure 16-5. (After KATZ, 1960.)

that, on the whole, the large-scale nature of the lunar surface must be accepted as smooth. This fact, on the face of it, does not perhaps seem very unusual; for very much the same is true of our Earth (or Mars). However, the principal levelling factors – air and water – on Earth (and, to a lesser extent, on Mars) are totally absent on the Moon and must have been so from time immemorial; so that the reason why the Moon appears today to be so similar to our Earth in this respect must be sought along different lines.

Below the limits of direct optical resolution from the distance of the Earth, a much

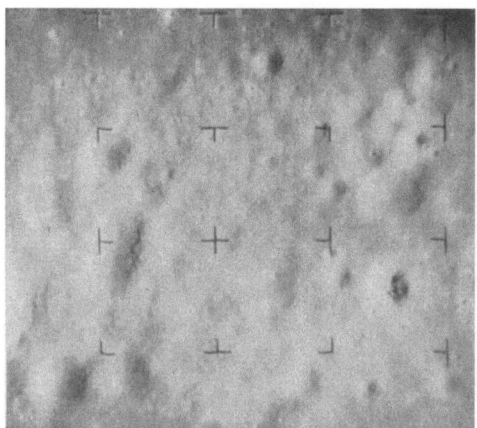
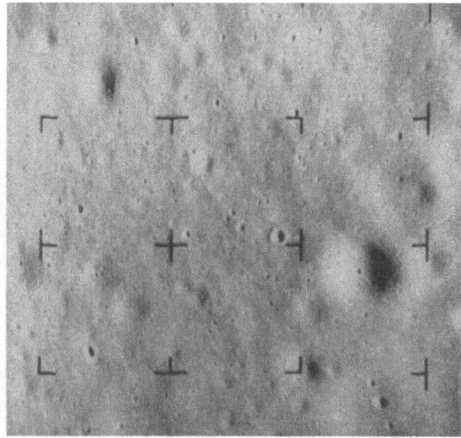

Fig. 23-2. Close-ups of the landing places of Ranger 7 (upper left; $\lambda = -20.7$, $\beta = -10.7$; on July 31, 1964); of Ranger 8 (upper right; $\lambda = 24.7$, $\beta = 2.7$; on February 20, 1965); and of Ranger 9 (lower left; $\lambda = -2.4$, $\beta = -12.9$; on March 24, 1965). Photographed from altitudes 4–6 km above the lunar surface, seconds before the impacts. The size of the individual fields is approximately 2.5 km across; and the smallest details resolved on them are of the order of 1 m for Rangers 7 and 8, and substantially smaller for Ranger 9. The reader may note that, on this resolution, the structure of the surface of all three different regions appear to be essentially the same – a fact suggesting that the forces which shaped it up are not local in nature.

more detailed view of three small regions of the lunar surface has recently been obtained from successful missions of the hard-landing American spacecraft Rangers 7,
8, and 9; the former two attaining towards the end of their flight ground resolutions
of one metre; the last, about 0.3 metre (see Figure 23-2); and even these feats were
since greatly overtaken in surface resolution by the accomplishments of the Russian
soft-lander Luna 9 (Figure 23-3). The total area covered at these resolutions represents, of course, still a very insignificant fraction of the entire lunar surface, and refers
to much the same type of ground. Nevertheless, one of the main outcomes of these
missions has been to confirm the essential smoothness of the lunar surface, inferred

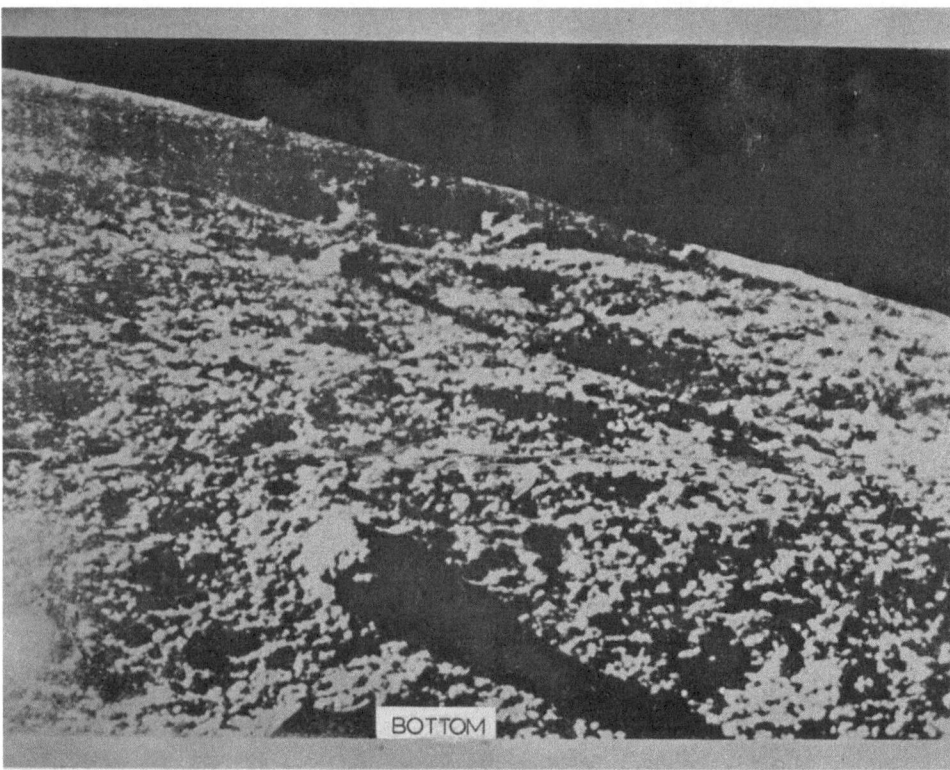

Fig. 23-3. A close-up view of the lunar surface, as televised to the Earth by the Russian soft-lander
Luna 9 on February 4, 1966. The reader can judge for himself the extent to which it confirms our
anticipations in Chapter 21, based on an analysis of high-frequency radar echoes (... a surface smooth
down to a decimetre scale, with a few obstacles here and there larger than the wavelength...). A
stone in the foreground, casting long shadow in the rays of the rising sun, is approximately 15 cm in
size, and lies about 2m from the camera. Other and larger stones are visible at some distance away;
but only a few and far in between. A conspicuous roughness of the surface, inferred previously from
photometric investigations, emerges on the cm-scale, and dominates the surface micro-structure as
far as one can see; pits as small as 2–3 mm are visible in the immediate neighbourhood of the spacecraft. There is nothing in this evidence to rule out the possibility that all this micro-structure is
entirely due to meteoritic abrasion.

previously from Earth-bound work down to a scale of the order of meters for impact areas of the Rangers, and decimeters for Luna 9.

With the exception of these tiny closeups of the lunar surface obtained with the aid of a spacecraft, below the limits of optical resolution from the distance of the Earth all studies of the microstructure of the lunar surface must rely completely on indirect methods, in which the principal link between us and the object of our inquiry is the lunar radiation in the full range of its spectrum. With one quite insignificant exception (i.e., thermal radiation of the Moon due to the leakage of its internal radiogenic heat) all moonlight derives its origin from the Sun – whether this be sunlight falling on the Moon directly, or scattered towards it through the intermediary of our Earth ("earthshine"). All this light must be absorbed or scattered by the lunar surface, in accordance with its local optical properties. A small part of the solar radiation (both electromagnetic and corpuscular) may be absorbed and re-emitted by cascade processes giving rise to fluorescence in the visible part of the spectrum. Energetic corpuscles of the solar wind may even induce the lunar surface to emit "bremsstrahlung" in the X-ray domain, a spectrum of which could reveal to the terrestrial observer above the atmosphere the atomic chemical composition of the lunar crust – just as the studies of lunar luminescent spectra may provide some information about its molecular structure or chemical impurities.

At optical frequencies and in the near infra-red (up to wavelengths of approximately $4-5\mu$) the lunar spectrum is dominated by scattered sunlight; while at wavelengths longer than $5\mu$ – i.e., in the deeper IR or the domain of radiofrequencies – the radiation of our satellite is due almost exclusively to a thermal emission of its globe. The energy sent out as thermal radiation is much greater than that of scattered light; for only about 7 per cent of incident sunlight gets scattered from the lunar surface; the balance being absorbed and re-emitted.

Even as large a balance of incident flux is, however, insufficient to maintain the outermost layer of the lunar surface at a temperature higher than 390 °K at the subsolar point, and less than 90 °K late at night. This means that much of the thermal radiation of the Moon (radiating like a black body at these temperatures) is bound to be emitted at wavelengths which are absorbed by our own terrestrial atmosphere. Fortunately, this atmosphere is fairly transparent in the $8-12\mu$ wavelength window, which should include the region of maximum emissivity of the lunar surface in daytime. The second atmospheric window through which the thermal radiation of the Moon can be observed in the microwave domain of the spectrum (for $\lambda \geqslant 1$ mm) lies already so far on the descending branch of the intensity-distribution of a black body emitter of temperature as low as that of the lunar night time that the energy flux received from it is quite small, but still measurable up to wavelengths of the order of one metre.

The main significance of the measurements of the thermal emission of the lunar globe rests on the fact that, inasmuch as such long-wave radiation originates at an increasing depth below the visible surface, the attenuation of the diurnal heat wave and increase of its phase-lag with increasing wavelength provide us with direct means

to establish the absolute value of the thermal conductivity (and of dielectric properties) of the lunar surface down to an appreciable depth.

The emitted and scattered components of lunar radiation can also be distinguished
by their different polarization properties; for while the scattered part of moonlight
becomes distinctly polarized in the process (and the direction of its plane of polarization rotates with the phase), thermal emission in the near IR (though not in the
microwave domain) remains essentially unpolarized. The same distinction exists also
between the illumination of the Moon by the Sun and the Earth: while the incident
sunlight is unpolarized, the earthlight is already partly polarized by the scattering of
sunlight in our atmosphere.

These are all passive sources of light provided by nature. In addition, it has proved
possible in recent years to reflect from the Moon man-made radar pulses corresponding
to wavelengths ranging from less than one centimetre to almost twenty metres, and
to record with sufficient precision their time profiles modified by the reflecting properties of the lunar surface – a very powerful method of exploration, since long-wave
radar pulses penetrate much deeper in the lunar crust than the diurnal heat wave.
Moreover, quite recently we witness the first use of laser beams (i.e., radar at optical
frequencies) for purposes of lunar topography.

The principal results of the measurements of different properties of moonlight and
its variation with the phase can be summarized as follows:

(1) The intensity of light scattered at optical frequencies varies so rapidly before
and after full Moon (the full phase being approximately 19 times as bright as the first
or last quarter, when due regard is paid to the "opposition effect") as to defy explanation in terms of diffuse reflection from smooth surface of any known natural substance. Moreover, the apparent disk of the Moon exhibits no trace of limb-darkening
in visible light. These phenomena reveal that, at some scale which is large in comparison with wavelength, the lunar surface must become extremely rough, and capable
of an extraordinary amount of back-scattering.

(2) The foregoing statement is, moreover, true not only of the global light of the
apparent lunar disk, but also of any element of it – be it a part of the continental
blocks or maria. Each element attains its maximum brightness at full Moon, regardless of its relative position or angular distance from the Moon's centre.

(3) The reflectivity (albedo) of the Moon varies from place to place within the
range 0.05–0.18 in yellow light ($\lambda = 0.56\mu$) – i.e., much less than for most common
terrestrial rocks – and increases somewhat with the wavelength. The ratio of the
albedo of the brightest and darkest spot optically resolvable on the Moon exceeds,
therefore, scarcely a factor 3; while the continental areas are, on the average, not more
than 1.8 times as bright as the maria. This may, of course, be due to some extent to
limited spatial resolution of photometric studies; cf. again Figure 23-1 illustrating the
way in which differences in ground albedo on Earth tend to vanish with diminishing
resolution of the image.

(4) The light of the Moon as a whole is distinctly redder than the illuminating sunlight, and becomes more so with increasing phase (i.e., away from the full Moon); but

its local colour differs but little from spot to spot. In general, the magnitude of the colour index of any surface detail appears to increase statistically with its albedo.

(5) The scattered moonlight is polarized to the extent of several per cent. At small phase angles $(0 < |\alpha| < 30°)$ the polarization proves to be negative, but it changes sign thereafter and attains a maximum roughly at a phase of 90°; the direction of the plane of polarization being either parallel with, or perpendicular to, the "equation of illumination". The phase at which the phase of polarization rotates so by 90° coincides (for most lunar objects) with the quadratures. The actual amount of polarization is found to increase with diminishing albedo; the maximum polarization of dark maria exceeds 15 per cent.

(6) The intensity distribution of thermal radiation of the Moon in the $8-12\mu$ domain of the spectrum exhibits distinct limb-darkening; and the corresponding light (or, rather, heating or cooling) curves are of strongly local character – indicating that the lunar surface is greatly diversified in its thermal properties.

(7) The intensity of thermal radiation of lunar origin in the 1–1000 mm microwave domain of its spectrum, and its variation during lunation (or eclipse), reveals that the effects of a diurnal heat wave penetrate to a depth of barely a foot below the surface, where a constant temperature close to 240 °K prevails day and night. Moreover, its phase-lag grows with the depth of penetration in such a way as to indicate a far lower coefficient of heat conduction for lunar surface layers than that of any known terrestrial solid rock – a result explainable only on the assumption that lunar surface material consists of loose rubble or dust, in which heat can flow only through the corners of contact between individual elements of the debris.

(8) The power reflection coefficients deduced from the thermal radiation of our satellite in the infrared as well as microwave domain of its spectrum lead, moreover, to a dielectric constant $\varepsilon$ of the lunar crust which is much smaller than that of solid silicate rocks – indicating again a considerable degree of fragmentation of the surface material (the volumetric concentration of the material of not more than 40–50 per cent).

(9) The observations of radar echoes at 30–3000 Mc/sec (10–1000 cm wavelength) reflected from the Moon reveal that approximately 50 per cent of the echo power arises as a result of quasi-specular reflection from a small central region of radius about one-tenth of that of the apparent disk of the Moon. The power reflection coefficients resulting from the observed echo strengths disclose that low dielectric constant $\varepsilon$ (indicating low volumetric concentration of the material), inferred previously from observed properties of thermal radiation of our satellite, extend down to a depth from which 10-metre radar pulses are returned; and this level may be many dozens of metres below the visible surface. It is not till in the domain of decametre waves that the returning echoes point to an increase in the effective value of $\varepsilon$, indicating an increased compression of the material and, eventually, a contact with solid rocks. But the average level at which loose debris gives way to solid rocks would appear to be somewhere between 50–100 metres below the visible surface.

(10) The slope of the trailing edge of the specular part of radar echoes at different wavelengths indicates that the reflecting surface continues to be essentially smooth

(with average gradient of one in ten or twenty), and covered with objects below the limit of radar resolution to no more than 10 per cent of its area down to almost centimetre wavelengths. It is not till for radar waves characterized by $\lambda < 1$ cm that the surface begins to appear quite rough (cf. LYNN *et al.*, 1964), as anticipated previously from its light-scattering properties at optical frequencies. In certain localities (ray craters, for instance) radar methods have indicated surface roughness on a metre scale (cf., PETTENGILL and HENRY, 1962; THOMPSON, 1965); and such anomalies appear to be related with parallel thermal or albedo anomalies. The lunar surface appears to be no more a uniform radar reflector than it is a uniform light reflector or heat conductor; and local anomalies in these respects are as numerous as they are conspicuous.

The ten points just enumerated lend themselves for the following tentative conclusions regarding the nature and the relief of the lunar surface. First, direct telescopic observations from the distance of the Earth have indicated that, on the scale of 1 km and greater, the lunar surface is essentially smooth and its average inclination to the horizontal is close to 4 degrees in the continental areas, and to 1 deg. for the maria (cf. p. 251).

Moreover, recent contributions of the Ranger spacecraft to lunar studies have revealed that (at least in the immediate neighbourhood of their respective points of impact) this essential smoothness of the lunar surface continues down to the scale-length of the order of one metre (cf. Figure 23-2). A glance at these and other photographs secured by Rangers 7–9 show the surface to be smooth and gently undulating, with no more than a few formations looking like boulders interspersed here and there. Their presence is consistent with the studies of radar echoes at different wavelengths; for these can sample only average properties of the reflecting lunar ground; and the limits for possible areal density of obstacles large in comparison with the wavelength as quoted on pp. 373–374 in accordance with EVANS and PETTENGILL (1963) are in accord with a more direct photographic evidence.

Quite recently, a new and dramatic contribution to our knowledge of the structure of the lunar surface was made by the Russian spacecraft Luna 9 – the first soft-lander sent out by human hand to succeed in its mission, and to provide us with an intimate glimpse of the lunar surface down to a cm–mm ground resolution (i.e., at least a hundred times as large as that of the best photographs secured by Ranger 9) in the immediate neighbourhood of the spacecraft; and with progressively diminishing resolution up to the horizon about 1.5 km away (see Figure 23-3).

The view seen from the vantage point of a television camera approximately $1\frac{1}{2}$ m above the surface reveals a panorama of the kind which the investigations summarized in the preceding chapters led us to anticipate. The surface appears indeed to be smooth down to a decimetre scale, with a few larger stones interspersed here and there. However, the greatest contribution of this radically new evidence has been to confirm that *the lunar surface changes over from smooth to extremely rough ground on the cm–mm scale* (see Figure 23-3) at which it acquires its honeycombed and porous structure anticipated from the photometric evidence.

What is the cause of this roughness of the lunar surface on the cm–mm scale, suggested by its reflection properties for high-frequency radar pulses, and confirmed so convincingly by the Luna 9 photographs? There is more than one way by which one can approach this problem; but of these let us consider first the one which must be demonstrably operative on the lunar surface; namely, the downpour of micrometeoritic dust.

As is well known, since time immemorial the lunar surface has been exposed to continuous bombardment by all ingredients of the interplanetary space, ranging in mass from billions of tons for comets or asteroids down to the finest micro-meteoritic dust which has so far been tracked to particles of masses of the order of $10^{-17}$–$10^{-16}$ grams.* The physical characteristics of these particles depend, in turn, largely on their mass. When the latter is measured in kilograms or more, the object is invariably a solid piece of stone (largely silicates), or iron, or a mixture of both. Between masses of approximately $10^2$ and $10^{-8}$, the particles can be broadly described as meteors, and derive largely from the icy nuclei of comets. On Earth (and, to a lesser extent, on Mars) meteor particles of this class invariably spend themselves in the upper atmosphere and never reach the surface; their properties can thus be studied only by optical or radar methods. The population below this mass limit is composed of small particles of the dimensions measured in microns, which can be decelerated by collisions with air molecules without destruction. These are the true micrometeorites; and it is possible to collect them with the aid of high-altitude rockets, or retrieve them as they float down for weeks through the atmospheric molecular sieve to the surface of the Earth.

On the Earth (or any other planet surrounded by an atmosphere) only those particles can come into contact with the actual surface which can withstand atmospheric deceleration without total dispersal of their mass; and this can happen if the deceleration is either very small (i.e., large mass) making the time of flight short, or again if the particles are so small that they can be decelerated already in the upper atmosphere without becoming too hot for vaporization. On the Earth, the prevailing air density tends to segregate such particles in two distinct classes: meteorites weighing $10^3$ grams and more, and micrometeorites of $10^{-7}$ grams or less; particles of the intermediate mass range being effectively excluded from any contact with the terrestrial surface by the shielding effect of our air cushion. On the planet Mars, whose atmosphere is less dense than our own, the range of the masses which are thus forbidden access to the surface is smaller than $10^{10}$, but still very large.

On the Moon, unprotected from the celestial intruders by any atmosphere to speak of, the solid surface is directly exposed to a continued infall of solid particles of all sizes and masses which the Moon intercepts on its perpetual journey through space. The influx of objects intercepted per unit area and time at the mean distance of the Earth from the Sun is now fairly well established for the entire particle population ranging from $10^{17}$ to $10^{-17}$ g in mass. The results based on a recent re-discussion of this subject by HAWKINS (1964) have already been schematically reproduced on Figure 18-2. These results are based on very extensive studies of observational data obtained

---

* Particles smaller still would be expelled from the solar system by solar radiation pressure and the Poynting-Robertson effect.

by various methods in the past; and as many investigators working in this field are in substantial agreement (at least down to masses of the order of $10^{-6}$ g) their results are unlikely to undergo any important changes in the future.

Before coming to estimate the cratering effects which such incessant rain of meteors is bound to produce on the lunar surface we may, however, wish to consider a few additional factors which may influence the outcome. Thus it has been repeatedly suggested that the spatial density of micro-meteorites (estimated so far mainly from the material swept up in the neighbourhood of the Earth) may diminish at greater distances from our planet (cf., e.g., WHIPPLE, 1961). Although the available evidence is still rather inconclusive, the circumterrestrial dust cloud probably exists; and if so, the particle densities as plotted on Figure 18-2 may have to be reduced in deep space. On the other hand, in the proximity of the lunar surface this deficiency may be more than made up by a great number of secondary particles thrown out from the Moon by primary impacts. Recent laboratory simulations by GAULT, SHOEMAKER and MOORE (1963) indicate that the total mass of the secondary ejecta may be ten or even a hundred times as large as that of the intruder; but as (in order to balance the kinetic energy of the primary intruder and the secondary spray) their velocities may be largely below that of escape from the Moon, they will be certain to increase the density of dust in the immediate neighbourhood of the Moon; and this leads us to believe that the cislunar environment may be no less dusty than the circumterrestrial one.

A plot of the cumulative influx rate of dust particles of different mass for the lunar environment, based on a recent work by MCCRACKEN and DUBIN (1964), is shown on the accompanying Figure 23-4. Adopting the results contained in it, let us enquire about the 'etching' of the lunar surface produced by a continuous infall of the particles of such dust. In order to do so, let us assume that primary dust particles of masses listed as ordinates on Figure 23-4 impinge on the lunar surface with a velocity of (say) 10 km/sec, and enquire as to the mass of particles capable of producing on impact micro-craters 1, or 10, mm in size.

If Equation (18-6) is extrapolable for this purpose, the requisite masses turn out to be of the order of $5 \times 10^{-9}$ and $10^{-5}$ g, respectively – corresponding to radio-meteors of essentially cometary origin. Moreover (in accordance with the data of Figure 23-4), the corresponding accretion times should be about one per $mm^2$ per 300–400 years, and one per $cm^2$ per 2–3 million years, respectively – in substantial agreement with our previous estimates in Chapter 18 based on Hawkins's data. This should be the time-scale on which the micro-relief of the lunar surface gets formed, destroyed and re-created by external impacts raining down incessantly on the ground – in much the same way as raindrops on Earth keep checkering the dry dusty surface with their pockmarks.

If this explanation of the roughness of the lunar surface indicated by photometric observations were correct, we should expect that a rather sudden onset of roughness below 1 cm scale-length (indicated by radar reflections at 8.6 mm) should correspond to an increase in spatial frequency of particles capable of producing on impact craterlets of this size. For moderate velocities of primary impacts, this should occur on

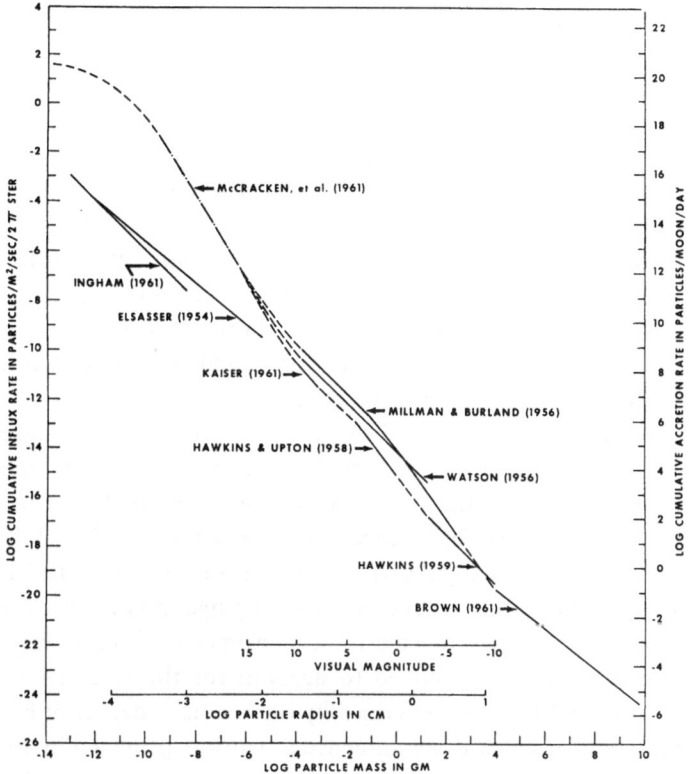

Fig. 23-4. Cumulative mass distribution for interplanetary dust particles in the vicinity of the Moon, as derived from studies of metorites and meteors and from photometric studies of the zodiacal light and the solar F-corona. (The results obtained with rockets and satellites near the Earth are shown for the purposes of comparison.) (After McCracken and Dubin, 1964.)

Figure 23-4 close to an ordinate corresponding to a mass of $10^{-6}$ grams. A glance at our diagram reveals that this appears indeed to be the case, if the data by McCracken *et al.* (1961) are preferred to those of Elsässer (1954) or Ingham (1961). As the former are based on circumterrestrial evidence, while the latter on an extrapolation of estimated density of the zodiacal cloud by a rather wide margin, the former may correspond more closely to the conditions prevailing in the cislunar environment; and the recent close-up photographs of the lunar surface secured by the Russian Luna 9 (cf., Figure 23-3) completely vindicated in this respect our previous surmises based on the interpretation of lunar radar echoes at 8.6 mm wavelength.

Whether or not a sudden jump in roughness of the mean lunar surface below 1 cm, and an upturn of the frequency-curve on Figure 23-4 for interplanetary meteors in the vicinity of the Moon capable of checkering the lunar surface with craterlets of 1 cm and less in size, represents a mere coincidence, only the future can tell. In order to account in this manner for the salient features of lunar photometry, it would be necessary that innumerable craterlets be produced by meteoritic bombardment in the form of rather deep pits, capable of casting shadows on themselves almost immediately after the Sun has ceased to stand directly overhead; for only a deeply vesicular structure could reproduce the observed amount of back-scattering.

It could be argued that, if these micro-craters were scaled-down versions of their macroscopic models, they should be expected to resemble shallow pockmarks rather than the honeycomb surface required by photometric considerations. However, one could counter by saying that the cratering effects of cometary ices – for this is what particles in the requisite mass-range mainly consist of – are different from those produced by more solid particles. Moreover, recent experiments by Gault and his associates at NASA's Ames Research Center have shown (GAULT et al., 1963) that the scaling laws may break down for craters of very small sizes; and that little pits produced by meteorite impacts could indeed be deep.

The fact that the salient features of lunar photometry (i.e., its phase-law and absence of limb-darkening) in the visible part of the spectrum are so different from those exhibited – for instance – by the planet Mars could be attributed to the fact that particles primarily responsible for surface micro-structure – in the $10^{-6}$–$10^{-9}$ g mass range – are (like on Earth) already destroyed by a passage through the Martian atmosphere, and thus filtered out from what eventually impinges on the surface of this planet.

In any comparison of the relative merits of micrometeoritic cratering and any other internal process which can be invoked to account for the vesicular or honeycomb structure of the lunar surface characterized by the requisite degree of back-scattering, a powerful argument for an *external* influence at work is provided by the *photometric homogeneity* of the entire lunar surface, demonstrating that micro-relief required to explain the observed light changes overlies *all* types of the lunar ground – in the maria, continents or bright rays – whatever their location or albedo. *What else but an external influence could impress the same uniform type of micro-relief all over the Moon?*

The data bearing on the rate of meteoritic impact on the Moon and represented graphically on Figure 23-4 lend themselves also for estimates of the total amount of meteoritic material (by integration of the observed influx as a function of the mass) which must have been so deposited on the lunar surface over long intervals of time. The most reliable existing estimate of the total cumulative rate of deposit appears to be that by ÖPIK (1960) estimating the infall of cosmic debris per year at $10^{-8}$ g/cm$^2$ – which, if extrapolated linearly over the entire lunar past, would lead to a total accumulation of some 40–50 grams of material per square centimetre, corresponding (for a mass density of rather less than one gram per ccm) to a layer about one foot in thickness.

That such cosmic dust must have been swept up by the Moon from interplanetary space is obvious, as there is nothing to shield its surface from such impacts; and there seems little room for doubt that *the polarization of scattered moonlight at optical frequencies is due to the action of the top layer of such a dust cover* overlying lower strata. It may (at least, to some extent) be responsible for the observed thermal properties (i.e., low thermal conductivity) of the topmost layer of the lunar crust; but assuredly not for the low value of the dielectric constant down to the level of penetration of long-wave radar pulses. Besides, if this external dust cover were to overlie the entire surface, the visible face of the Moon would have to appear uniformly gray, and show no local differences in albedo or colour. As we stressed already in Chapter 19, the colour of the Moon is indeed very nearly uniform; but the albedo is not, varying as it does from spot to spot by at least a factor of three (and probably more if smoothing effects of limited optical resolution are taken into account).

This fact, which should be obvious to anyone who looks at the Moon with the naked eye and recognizes the difference in brightness between the continents and the maria, reveals that cosmic dust cover on the Moon cannot be uniform. Instead, this dust must be intermingled, or move relative to, the 'native' lunar rocks or debris of local origin*; and it is the degree of their mixture which determines the local reflectivity and colour.

But how could anything mix on the dead and inert landscape of the Moon, or what could stir any dust to motion? One physical agent to provide a continuous gentle stirring is, of course, the temperature giving rise to the Brownian motion of the grains. Let us apply the kinetic theory of gases to such a situation and consider the dust particles in free space as large 'molecules' of very high molecular weight: what are the properties of an 'atmosphere' to which this dust can be stirred (in the absence of any gas molecules) by the temperature prevailing on the Moon? This question admits indeed of a simple answer; and the requisite calculations reveal that if, at $110°$ C (i.e. the temperature of lunar noon in the tropics), the density of our dust atmosphere is to diminish to one-half at the height of 10 cm above the surface, the mean 'molecular weight' of the dust grains should be approximately $7.8 \times 10^{-17}$ grams, corresponding (for a mean density of 3 g/cm$^3$) to a size of $7.9 \times 10^{-6}$ cm. If, moreover, the scale height of this atmosphere were to be increased ten times from 1 cm to one metre, the molecular weight should be further diminished by a factor of 10, and the mean size to $1.7 \times 10^{-6}$ cm or 170 Ångstrom units.

It would, indeed, have to be an extremely fine dust to remain stirred above the surface by its own temperature even to a very small height. No natural process is known which would grind it to such a degree of perfection; and coarser grains would act as very effective 'coolant'. Besides, the polarization of light scattered by the lunar surface as well as the coefficient of thermal conductivity resulting from the microwave measurements are conclusive in indicating that the mean size of the dust grains covering the lunar surface is of the order of a micron or two – i.e., about a hundred times

---

* It may be mentioned that – contrary to the cosmic dust of meteoritic origin in which small particles largely predominate – only a small fraction (by mass) of lunar debris can be fine dust; as no processes producing it appear to be operative on the lunar surface. The oft-repeated argument about thermal cracking of rocks due to temperature changes during a lunation (or eclipse) has now been thoroughly disproved (cf., e.g., RYAN, 1962) on quantitative grounds; and even if it were not, it could operate only with extremely limited efficiency; for solid rocks are barely anywhere exposed on the Moon to direct sunlight; and once only a few millimetres of dust would have formed on their surface, their thermal insulation (cf. Chapter 20) would be almost complete. This would, of course, automatically stop any further thermal damage and no dust could henceforward be produced in this way – unless, of course, the dust formed could move to another place and thus expose the parent rock to continued erosion.

as large. Such dust cannot be stirred by temperature to form an atmosphere of more than millimetre scale height; and so minute a layer cannot, in turn, provide any effective means of transporting dust from one place to another in daytime – let alone at night.

The same conclusion cannot be altered much when we come to consider, not gravitational, but electrostatic forces which dust on the lunar surface is almost certainly bound to accumulate as a result of daytime photo-ionization discussed already in Chapter 21. We concluded earlier that a diurnal space charge formed above the lunar surface is insufficient to interfere with radar communications; and the extent to which dust so charged can be levitated and acquire a limited amount of mobility on account of electrostatic repulsion has been likewise found to be almost negligible (cf., SINGER and WALKER, 1962a, b). Besides, the best proof of the fact that even fine lunar dust cannot be stirred – by whatever action – above the lunar surface is continued visibility (particularly during daytime) of its rough micro-structure on the mm scale. Any dust stirring sufficient to conceal it would at once largely obliterate the back scattering property of the lunar surface, and thus destroy the most conspicuous known characteristic of the lunar light curve.

Since no continuously acting effective mechanism for dust motion on the lunar surface has so far been discovered, we are forced to limit our considerations to occasional stir-ups by mechanical action, when a meteorite hits the lunar surface – with effects discussed already in Chapter 17. We may recall that the explosive action of primary impacts will shock a large volume of adjacent strata into the state of loose debris. A smaller volume of rocks will be ejected from the point of impact to raise the ramparts and cover a large part of the surface of the Moon with ejecta, some of which (thrown out with a velocity in excess of that of escape from the gravitational field of our satellite) may leave the Moon altogether.

A recent study of such a situation in the particular case of the crater Tycho, undertaken recently by KOPAL (1966) indicated that, in this particular case, the volume of the debris ejected outside the ramparts of the crater itself was of the order of $10^3$ km$^3$; and such a volume of ejecta would, if dispersed uniformly over the entire Moon, cover its surface with a layer 3.2 cm in thickness. As, moreover, there are no less than 75 craters on the visible side of the Moon (and many more on the far side) which are equal to, or larger than, Tycho in size, it follows that these alone could, by their cumulative action, have covered the whole Moon with a layer of debris 5–6 metres in depth; and the cumulative effects of many thousands of smaller impact craters could have easily increased this depth ten or twenty fold. *It is the layer of such debris*, consisting of material shattered in situ by a passage of the shocks from individual primary impact points, intermingled with material actually ejected and transferred along ballistic trajectories to different parts of the lunar surface, *which may extend down to the depth of penetration of the radar pulses at decametre waves, and is responsible for the low value of its effective dielectric constant.*

According to this model, we are envisaging the lunar crust to consist largely of 'native' rocks shattered and dislocated by primary impacts, covered on top by a thin layer of cosmic dust (of very much finer degree of pulverization) which may, on the average, extend in depth to as many centimetres as the underlying strata of shattered lunar rocks do in metres.

The proposed model can thus explain in a natural manner not only the observed thermal and electromagnetic properties of the lunar surface discussed in Chapters 20 and 21, but also the observed differences in reflectivity and colour in the visible part of the spectrum; for the latter should clearly be the resultant of the proportion in which the cosmic dust and native rocks have been locally mixed up by mechanical action.

To lay mind, the Moon at night may appear as a sufficiently bright celestial body, adequate to endow the moonlit landscape with a certain degree of poetic charm. Sober astronomers measuring the intensity of moonlight have, however, long been puzzled by the opposite problem: namely, why is the Moon so faint when – according to all reasonable expectations – it should be two to three times brighter? The immediate reason is, of course, its abnormally low reflectivity at optical wavelengths – the mean visual albedo of the lunar disk being close to 0.072 (cf. Equation 19-14) which is two to three times smaller than the average reflectivity of common terrestrial silicate rocks (basalt, granite, etc.,) of densities comparable with the mean density of the lunar globe. For a long time this discrepancy has remained puzzling; for the only known natural rocks on Earth of so low a reflectivity were volcanic lavas or pumice (i.e., substances which are not expected to cover most part of the lunar surface). However, astronomers have realized of late that another sample of solid matter in the solar system exhibits an even lower albedo than that in the visible light – namely, the zodiacal cloud – and its proximity to the Sun pointed the direction in which its low reflectivity was to be sought: namely, the *radiation damage* suffered by any solid surface exposed to solar radiation (mainly corpuscular) for astronomically long intervals of time.

Recent laboratory experiments by HAPKE (1964) simulating radiation damage inflicted on common silicate rocks by proton bombardment under controlled conditions revealed that a gradual surface darkening of silicates from the pristine to lunar conditions will be accomplished by an exposure to the protons of the ordinary 'solar wind' in time intervals of the order of $10^5$ years down to a depth of the order of 10 microns; and subsequent measurements by HAPKE (1964) as well as by DOLLFUS and GEAKE (1965) of the polarization properties of such radiation damaged surfaces revealed that these match the observed polarization of the lunar surface (including negative polarization at small phase angles) almost to perfection. As a result of this work, there hardly remains any doubt now that *the abnormally low reflectivity of most parts of the lunar surface, and their polarization properties in the visible part of the spectrum, are due to solar radiation damage suffered by cosmic dust,* both prior to its infall on the Moon and since, *as well as by the 'native' lunar rocks in the course of their exposure on the surface.*

The length of this exposure may, of course, vary on the Moon from place to place. For instance, in most parts of the plains of the maria – the surface of which (judging from the low density of impact craters per unit area) has not been disturbed greatly by mechanical action since the time of their formation – the cosmic exposure time may be as long as the age of the mare itself (some 2000 million years?); and, hence, the darkening process may have gone to considerable lengths indicated by their relatively low visual albedos (0.05–0.06). On the other hand, any impact craters formed since on their surface (e.g., Copernicus, Kepler or Aristarchus in the Oceanus Procellarum) would have exposed rocks and thrown out debris previously undamaged by radiation, which should appear as bright splashes on the background of older and darker strata; and a glance at such photographs as reproduced on Figure 16-30 reveals that this may indeed be the case. Moreover, the fact that the impact craters of higher than average albedo (Aristarchus, Copernicus, Kepler, Langrenus, Tycho, etc.) have been found to exhibit thermal as well as radar-reflection anomalies pointing to greater compaction and (or) roughness of their surface (on metre rather than millimetre scale) suggests that their original surface torn out into boulders by the impact effects has not yet been levelled off by an overlay of cosmic or other dust; and that the microstructure of such surfaces is still dominated by mechanical effects of impact rather than the gradual and slow-acting meteoritic erosion, working hand in hand with a gradual darkening of the surface by solar radiation damage.

If this picture is indeed correct – and there is little reason to doubt it at the present time – the possibility suggests itself to utilize the observed albedos of the lunar surface as absolute age indicators of the respective type of ground; or, failing that, to use at least the observed albedo differences as indicators of time differences. This latter task we discussed already to some extent in Chapter 18. The former (more difficult) task would call for matching the absolute radiation damage, manifesting itself through altered photometric properties of the surface, of known minerals for a given dose of radiation; and also for an extrapolation of the intensity of the solar wind in the past.

Besides, there are also other problems to face. For example, as we already mentioned, a 10-micron layer of silicate dust will be darkened to the reflectivity of the lunar maria by a proton beam equivalent to solar-wind exposure for $10^5$ years (HAPKE, 1964); while, during the same time, the lunar surface should be turned over by meteor erosion down to a depth about ten times as large (WHIPPLE, 1965). If so, a gradual darkening of the lunar surface by solar radiation would not represent a continuous process, but one in which the requisite dose is accumulated by installments; the exposure being interrupted each time when any particular grain is submerged by mechanical action of micrometeoritic impacts beyond the depth of penetration of the solar protons, and resumed when the grain emerges again on the surface. The speed of this turnover is, unfortunately, more difficult to estimate than the requisite total dose, and may extend the actual time within which the newly ejected materials darken to the level of their substrate from $10^5$ to $10^7$ or $10^8$ years. As we already mentioned in Chapter 18, a gradual darkening of the bright rays on the Moon is apparently a very slow process, yet one operative on a time-scale which must be short in comparison with the age of our satellite (i.e., $10^9$ years); for otherwise there should be many more ray craters on the visible face of the Moon than those which we actually see there.

The meteoritic erosion and abrasion of the lunar surface, which must be expected as a result of its interaction with the full contents of interplanetary space, entails many other interesting consequences. One of them is the fact (made evident by recent cratering experiments in the laboratory; cf., e.g., GAULT *et al.*, 1963) that hypervelocity impacts on the Moon may actually eject from the lunar surface an amount of debris weighing a great many times the mass of the primary cosmic intruder – to eject it at much lower velocity, to be sure (as required for the energy balance), but the some of it still sufficiently fast for escape from the gravitational field of the Moon (i.e., with a velocity $u > \sqrt{ga}$ of Chapter 17). It appears, therefore, that *the secular effect of meteoritic bombardment is likely to be a removal of matter from the Moon rather than net accretion of it*; and a slow abrasion of its surface, in the course of which the newly crushed material gets intermingled with the swept-up interplanetary dust.

That this is indeed the case may also be indicated by the effects of the yearly bombardment of the lunar surface by energetic meteor showers – like the November Leonids or August Perseids. The former are known to impinge on the lunar surface with a relative velocity close to 70 km/sec; and such is apparently the havoc caused by this bombardment that at least some gets ejected from the Moon and is captured by the Earth; for the artificial Earth satellites equipped with micrometeorite detectors registered a secondary maximum of impact frequency following the maximum of the Leonid shower after an interval equal to the time of free fall from the Moon to the Earth; and the delayed particles exhibit velocities of about 11 km/sec (i.e. one of free fall to the Earth) rather than 70 km/sec of the original Leonids. There seems indeed but little room for doubt that this secondary maximum signifies an infall of lunar dust stirred by the energetic Leonids; and a realization that particles so small can apparently knock off an overspill of dust on to the Earth gives us some idea of the violence which must descend on the Moon during such spells.

## CHAPTER 24

# ORIGIN OF LARGE-SCALE FORMATIONS ON
# THE LUNAR SURFACE

In the preceding chapter our aim has been to reconstruct an average micro-structure of the lunar surface from information obtained through diverse channels of mainly photometric evidence. In what follows, we wish to conclude this discussion – and this entire volume – by a few retrospective considerations concerning the probable structure of the lunar surface on a larger scale. In earlier parts of this volume we have stressed repeatedly that the sculpture of the lunar surface, on any scale, can be regarded as the boundary condition of all internal processes that have been going on beneath the surface of our satellite since the time of its formation as an astronomical body, plus an impact counter of all external events recording in stone all collisions which it had suffered with the full range of the mass contents (from asteroids or comets to the protons and electrons of the solar wind) of interplanetary space. We wish now to return here once more to this *leitmotiv*, to examine from broader perspective some of its implications concerning the internal or external origin of some of the salient features of the lunar surface – such as the 'craters' in the widest sense of the word, or the difference between the 'continental' and 'mare' ground.

In doing so we shall also depart from our earlier treatment of this subject in Chapter 16–18 in another respect. In those parts of our text we set out (or, at least, tried) to answer all questions which had arisen in the course of our enquiry. In order to gain more freedom we shall abandon now this attitude, and shall not avoid raising more questions than can be answered at the present time – as a landmark of the present state of our ignorance, and an expression of confidence in the future.

To begin with, in Chapter 8 of this book we stressed that – almost regardless of how the Moon initially came into being – such traces of spontaneously radioactive elements, which its primordial mass must have contained, must by now have raised the present mean internal temperature to a level somewhere between 1000°–2000 °K. The nascent Moon could not have been very hot to begin with (at most a few hundred degrees K); for otherwise its relatively small mass could not have coalesced into its body as we know it at the present time. If, however, our new-born satellite contained an above-average abundance of shortlived radioactive elements like $Al^{26}$, with half-lives of the order of a few million years, it could have theoretically been melted by the radiogenic heat released by their spontaneous decay during this period of time but – and this is significant – it would have cooled off again (by convection) and re-solidified in a comparable time-span. If, on the other hand, the new-born Moon was endowed with an above-average abundance of long-lived radioactive elements like $K^{40}$ or $U^{238}$ (with half-lives of the order of the age of the solar system), their action would have kept warming up the Moon continuously, so that the interior of the Moon would now be at a higher temperature than ever before, still secularly warming up (cf., e.g., Ko-PAL, 1962a). Therefore, the known properties of radioactive elements from the periodic table hold a theoretical possibility for a melting of the Moon either at the earliest stage of its evolution, or at present (or in the future); but – and this is significant – *at no time in between*.

This fact which, we repeat, goes back to the properties of known elements of the periodic table, rules out the possibility of the Moon becoming molten by natural causes at any intermediate stage of its life; and this fact has an obvious bearing on any theories of the origin of the lunar craters and the maria.

To take up first the problem of the lunar maria, we pointed out already in Chapter 17 that their origin has so far been sought along two different lines: namely they have been regarded (at the risk of some schematization) as 'lava flows' or 'dust reservoirs'; and as both theories still keep contending for recognition, in what follows we wish to discuss and place in proper perspective their respective salient features.

Both alternative hypotheses possess a long pedigree. The dust hypothesis has been considered – among others – as far back as 1910 by that colourful American astronomer T. J. J. SEE – and in more recent years has been revived and re-emphasized, in particular, by GOLD (1955). A gist of these views can be briefly summarized as follows. According to present density of impact craters per unit area of different parts of the lunar surface, the 'continents' represent an older type of ground than the maria. On the other hand, their higher albedo points to a surface which has suffered smaller damage by protracted exposure to the solar wind. The only way in which greater age hypothesis based on higher crater density would seem to be compatible with the existence of a surface which has suffered less radiation damage would be by an assumption that this continental surface is constantly being *eroded* by meteor abrasion lying bare strata previously undamaged – and where else would such dust be removed but into reservoirs of the maria?

This view – attractive as it may be in certain respects – suffers on a closer scrutiny, from a number of objections (some of which should have been obvious by 1955, and others have emerged since) the ensemble of which almost invalidates it altogether – certainly in its more extreme form. Thus, to begin with, it is a commonly proven fact that fine powders of silicate rocks are much *lighter* than their solid parent material – and the finer the dust, the lighter the albedo. This would require dusty maria to be brighter than the rocky continents – unless (which is dubious) this situation has been completely reversed by radiation damage en route.

Secondly, the floors of the majority of large craters in the continental areas (as well as post-mare craters in the maria) show no evidence of either the accumulation of a darker material inside their walls, or the flattening of their interior. If migrating dust due to the continental abrasion filled in the hypothetical mare basins, why did it not fill, by the same kind of migration, smaller bowls of many craters on the way? Many maria (e.g., Mare Imbrium, Humorum, etc.) are bordered by rille systems running roughly parallel with their shores. If dust drifts into maria from outside, why did that sliding down the Apennines not fill the rille pralleling their course?

Besides, do we know that any large scale systematic motion of dust can occur on the lunar surface over long intervals of time? Contrary to certain qualitative conclusions by BERG (1964) the Brownian motion stirred by monthly temperature cycle is utterly insufficient to provide such a mechanism in the laboratory; and by itself it could not account for any appreciable lateral transfer of mass.

Could a motive for lateral dust motion be provided by gravity differences? This idea, entertained by Gold, was effectively dispelled by more recent findings (cf. Chapter 13) that a notion of the continental regions as 'highlands' and of maria as 'lowlands' is only indifferently true – and over relatively large parts of the lunar surface the opposite appears to be the case. If a difference in gravity had been the motivating force of migration, many high lying maria (for instance Sinus Medii and Oceanus Procellarum south of the Apennines) should be largely devoid of dust – an expectation patently belied by the lower-than-average albedo of these plains.

If any further arguments were needed to weaken still more the case for the dust nature of lunar maria, let us recall that the small but relatively abrupt changes in colour of the lunar maria (see Figure 19-3), suggestive of fairly sharp discontinuities in surface composition (Whitaker), which are scarcely compatible with a hypothesis of shifting dust. Let us recall, furthermore, that according to recent measurements of the thermal radiation of the Moon in the microwave domain ($\lambda = 3$ mm) by GARY, STACEY and DRAKE (1965) the mean temperature of the lunar maria is by 3 °K higher than that of the continental regions; while the difference in the mean albedo (causing the darker maria to absorb more sunshine) should account for a difference of barely $1°.5$ K. The balance may be due to internal heat; but if so, one should expect areas thickly covered by dust to be less efficient conductors – and, therefore, cooler – than the regions which communicate thermally with greater ease with the layer beneath. Under these conditions, the photographic evidence provided by the Russian Luna 9 spacecraft on February 3rd, 1966 (see Figure 23-4) can be regarded as a 'coup de grace' to any possibility of dust flow on the lunar surface, and thus to Gold's hypothesis regarding the lunar maria as deep dust basins.

In the face of such a situation, the alternative hypothesis of *lava flows* suggests itself for closer examination. This latter hypothesis has also found several champions among contemporary scientists – in particular, Kuiper and (in a somewhat different form) Urey. The latter proposed (following Gilbert) that, for example, the Mare Imbrium (and other circular maria at least, like Mare Serenitatis, Crisium, etc.) was the result of a flow of lava consisting of local material melted by the impact of a planetesimal of kinetic energy of the order of $10^{30}$–$10^{31}$ ergs. This argument is supported by the fact that the kinetic energy of an object impinging on the lunar surface with a velocity of 2.38 km/sec (equal to a minimum speed with which external bodies can impact on the Moon) is known to be about 2800 joules/g; whereas only 2000 joules/g are sufficient to heat most terrestrial silicates to their melting point.

However, this view runs also against many objections, some of which were already discussed in Chapter 16 and others may be added in this place. First, it is a fact that in no terrestrial laboratory experiments impacts with velocities of 2.4 km/sec and higher produced any melting of the material (silicate or other) whatsoever; nor is there any evidence that melting occurred in the terrestrial meteor craters. No trace of molten rocks has ever been found in the Arizona meteor crater, or at any other comparable terrestrial sites.

Secondly, the cosmic ages of all lunar maria (as inferred from the areal density of post-mare impacts) appear to be comparable, and by at least several hundred million years younger than those of the continents (cf. Chapter 18). It would seem indeed unlikely that, after most of the intense crater-forming activity which shaped the landscape surrounding the lunar south pole had subsided, a whole group of large planetesimals suddenly appeared within a relatively short interval of time to mutilate the lunar face by the greatest impacts of all – after which the situation became again very quiet.

Thus, in spite of certain attractive features, the collision hypothesis does not seem to be very promising for large-scale production of lava flows on the lunar surface; and may not be directly relevant to the phenomena under consideration. This does not mean, of course, that the entire hypothesis of lava flow would have to be abandoned; but only a specific way of mechanical lava production; and not necessarily in the more general (and more vague) way in which it has, in recent years, been defended, e.g., by KUIPER (1954, 1955, 1959).

It is true that any references to natural radioactivity as a source of heat for the hypothetical lava flows (made on occasions by Kuiper) are probably ill-founded; for our present knowledge of cosmochemistry and of the contents of the periodic table of natural elements lends no support for any hypothesis of the occurrence of a large-scale melting of the lunar surface at an intermediate stage of the Moon's life by radiogenic heat. This does not again dispose altogether of the hypothesis; for, as Kuiper and others have eloquently pointed out on many recent occasions, the features observable on the surface of the lunar maria – with their wrinkle ridges, submerged 'ghost' craters (cf. Chapter 16 and 17), etc. – suggest manifestations of an arrested flow of viscous magma so vividly that its possibility cannot be dismissed out of hand at this stage. In other words – argues Kuiper – the circumstantial evidence in favour of some kind of lava flows on the lunar surface is so strong that it should be considered even if we do not yet know what caused them. This is a defensible attitude, which has proven its worth on previous occasions in the annals of science; for it may happen that the cause which is still obscure to us now may be discovered in the future; and with these reservations Kuiper's view may find many a sympathetic ear. However, it should be kept in mind at the same time that, at present, the postulated cause of hypothetical large-scale lava flows on the lunar surface is still unknown (for neither impacts, nor natural radioactivity offer any satisfactory solutions or alternatives); and that Kuiper's views on this matter are less vulnerable than Urey's mainly because they are more vague.

Suppose, however, for the sake of argument that lunar mare ground does represent solidified lava flow: what kind of microscopic structure should such a surface possess?

Some indication can be gathered from recent laboratory experiments on solidification of molten silicates in vacuum (DOBAR *et al.*, 1964, 1966; cf., also WARREN, 1963a, b) showing structures not unlike that disclosed recently by close-up photographs of the lunar surface taken by Luna 9 (cf., Figure 23-3). It is impossible, of course, to decide from available evidence whether or not such surface actually obtains in the immediate neighbourhood of the landing place of Luna 9 to the exclusion of other possibilities discussed elsewhere in this chapter; but neither can it be dismissed.

There exists another large-scale characteristic of the lunar maria which, when properly understood, may bear vitally on the problem of their origin: and that is a conspicuous disparity in their distribution on the two hemispheres of the lunar globe. Ever since the first Russian photographs of the far side of the Moon, taken by Lunik 3 on October 7, 1959 (see Figure 16-2), confirmed since by improved photographs by Zond 6 (cf., LIPSKI, 1965), became available, it has been evident that *all lunar maria are largely situated on the hemisphere facing the Earth*: and with a slight rotation of the principal axes of inertia of the lunar globe this disparity could be made almost complete.

What could be the cause of this very fundamental and striking phenomenon? Without wishing to enter this subject in more than a preliminary manner, the author of this book would like to point out a possible relevance, in this connection, of certain dynamical considerations discussed already in Chapter 6. There we learned that retrograde extrapolation of the tidal evolution of the Earth-Moon system into the past makes it probable that, about 2000 million years ago, the Moon may have come so close to the Earth as to dip temporarily within the terrestrial "Roche limit". According to both GERSTENKORN (1955) and MACDONALD (1964) this dip was only grazing and – which may be significant in this connection – did not engulf (even temporarily) the entire lunar globe, but only its front hemisphere (facing the Earth) seems to have entered this danger zone.

The mechanical damage which the lunar surface may have suffered as a penalty for entering this zone (cf., ALFVÉN, 1963) was already mentioned in Chapter 6. May we, at this stage, suggest that a conspicuous asymmetry in the distribution of the maria over the lunar globe appears to be consistent with an assumption that *the mare plains may constitute 'scars' of mechanical damage, produced on the lunar hemisphere which entered the Roche limit by the disruptive tides of the Earth?*

At the first sight, such a hypothesis would again seem to require the maria to be 'lowlands' lying below the level of the continental areas, which (as we mentioned already) is not found to be generally the case. However, it is also possible that the strain to which the lunar surface may then have been exposed would have led to a cracking of the crust, and lava could have been exuded at the surface to fill up the original 'scars'.

As to the origin of such lava, where would the heat have come from to melt it? It may be mentioned that one process by which this could have been accomplished at (or near) that time would be through viscous dissipation of bodily tides raised in the mass of the Moon by terrestrial attraction which, at the time of the Moon's closest approach to the Earth, should have been about ten thousand times as high as at the present time. The actual mechanism by which viscous dissipation of tidal motion can generate internal heat we discussed already in Chapter 10, and found it barely significant at the present time. However, some 2000 million years ago it could have been a very different story; and may be the present lunar maria represent reminders of that time.

To return to a further confrontation of internal versus external influences operative on the lunar surface, while doubts continue to exist concerning the possibility of large-scale surface melting on the Moon in the past* a more localized activity which could be loosely termed 'volcanic' is not only possible, but probable. The Moon's mass – like that of the Earth – possesses a certain (albeit low) abundance of radioactive elements; and their chance concentration in accidental sub-surface pockets may give rise to phenomena which would be termed volcanic on Earth. The maar-type of formations on the floor of Alphonsus (see Figure 24-1) or hill-top craters like Regiomontanus A (Figures 16-6 or 17-9) represent examples of the cases in point. Moreover, one could go one step further and regard all such lunar surface formations as rilles, domes or wrinkle ridges (described in Chapter 16) as of internal (sub-surface) rather than external origin; at least it is very difficult to conceive of any external processes which could have given rise to them. But how about the lunar craters in general?

In view of the old vintage of some theories of internal crater origin, as well as of its recent revival as reflected, e.g., in GREEN et al. (1965), let it be emphasized that all information we now possess on the particulate contents of interplanetary space (summarized diagramatically on Figures 18-2 or 23-4) renders it a virtual certainty that a large majority of lunar craters must be of external origin; for there is nothing to shield the lunar surface from impacts of such bodies as we know are moving through space in the neighbourhood of the Moon over long intervals of time.

The size-frequency distribution of lunar craters on the visible side of the Moon has been exhaustively investigated (cf., e.g., HARTMANN, 1965; SHOEMAKER, 1965). Moreover, the argument summarized on Figure 18-3 discloses not only that the present density of solid particles in interplanetary space – deduced by methods which are totally unrelated to studies of the lunar surface – is adequate to account for the accumulation of a large majority of known craters by impacts in the past $4\frac{1}{2}$ thousand million years; but also that the majority of such impacts occurred with relative velocities of a few km/sec only (implying that, prior to their collisions with the Moon, the impinging bodies revolved around the Sun in direct orbits of moderate eccentricity).

Statistical arguments of this nature cannot, of course, assign particular origin to any single specific formation; and if only one per cent of lunar craters of dimensions in excess of 1 km on the visible face alone were of non-impact origin, the total number of such formations could well exceed one thousand. A number of specific craters

---

* The volume-to-surface ratio for the Earth being four times as large as for the Moon, the former could have accumulated a greater store of internal heat. Yet, in spite of it, no lava flows comparable in size with the lunar maria have been found anywhere on the Earth; and, for this reason, their lunar likelihood is correspondingly lessened.

which we hold to be of other than impact origin were mentioned already on pp. 298–299; and these included some of the largest surface formations on the Moon like Wargentin or Ptolemy and Alphonsus; and alternative possibilities of their origin may perhaps call for a few additional explanatory comments.

Suppose, for the sake of argument, that a limited part of the lunar surface became (for whatever cause) molten at some time and subsequently cooled off by radiation. Since the main heat loss then occurs at the outer boundary, the cooling of the lava should have proceeded most rapidly at the top, with eventual solidification commencing from above. However, since the density of solid silicates exceeds that of the molten phase, the solidifying pieces could not (unlike ice on water) float on top of the layer which may still be liquid, but would be bound to sink to the bottom – thus giving rise to a hydrodynamical flow of certain pattern.

Could such a pattern assume the quasi-hexagonal form manifested by many of the large crater formations (Figure 17-7) or circular maria (Figure 17-8) on the lunar surface, the walls of which could perhaps be regarded as material slowly pressed upwards by their sinking solid floor? Such a mechanism would safeguard automatically the validity of Schröter's rule (p. 250); and other characteristics could also find their natural explanation in this way.

Consider, for example, certain details of the ramparts of the crater Alphonsus, as revealed by the recent photographs secured by Ranger 9 (Figure 24-1). From the vantage point of approaching spacecraft, the ramparts of Alphonsus proved to consist of a rather loose system of quasi-parallel ridges, the gentle slopes of which show evidence of much less celestial bombardment than the rugged and darkened floor of the crater – suggesting that the oldest landmarks visible on the crater floor are very much older than the walls.

Moreover, in a few places patches of flat ground can be seen in between the individual rampart ridges and – which is interesting – their ground appears to be just as dark and heavily pockmarked as that of the main crater floor. This suggests that the flat patches in between the ramparts are generically related with the floor, and may once have been a part of it before the ramparts were raised. Moreover, the fact that the surface structure of patches, later blocked off by ramparts, was not altered appreciably by this event suggests that *the raising of the ramparts was a gradual, rather than cataclysmic, process*; for the latter – such as an impact – would have certainly buried pre-existing surface structure under a thick layer of debris.

This argument strengthens our belief that the crater Alphonsus (and, by implication, other formations akin to it) are not of external, but internal origin. *Its floor*, as revealed to us by Ranger 9 photographs, *represents probably the oldest type of solid ground in the solar system within easy observational reach*, containing in its sculpture a continuous record of events going back to the oldest chapters of the geological history of our satellite. Its ramparts are probably much younger; but not as young as the surface of the adjacent plains of Mare Nubium to the west; for the latter are at a higher level than the crater floor; and their crater density very much less. It must have been the walls around the crater which protected its floor from being submerged by what-

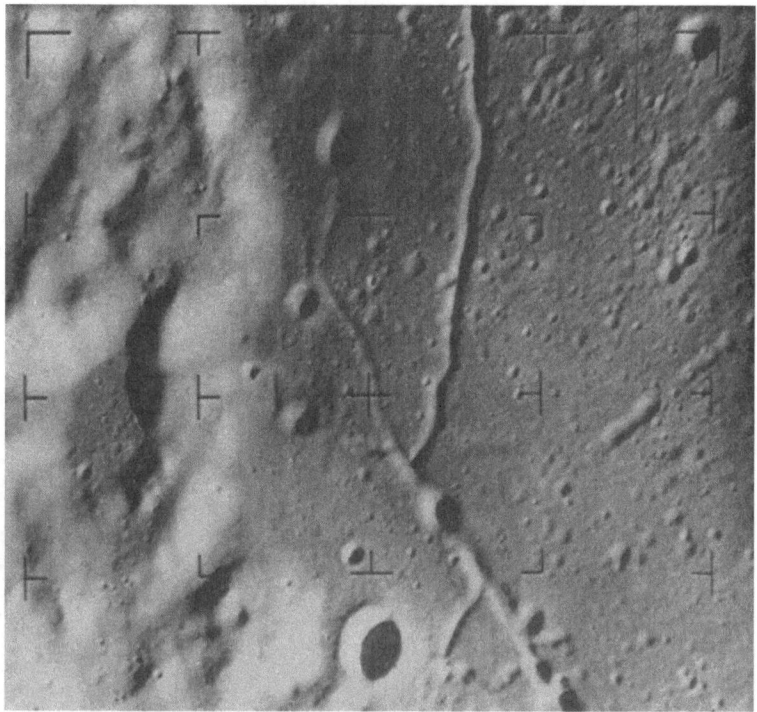

Fig. 24-1. Maar-like formations inside the crater Alphonsus, as photographed by Ranger 9 on March 24, 1965, from an altitude of 184 km above the lunar surface. A part of the ramparts of Alphonsus can be seen on the left. Note the greatly diminished reflectivity of the ground immediately surrounding the maar formation; and also the fact that the rille on which it lies is partly filled by ejecta on both sides of the crater.

ever covered up the mare later; and the crater density in the respective types of ground, together with their differences in altitude which permit us to decipher their stratification.

But – apart from large formations of the type of Ptolemy or Alphonsus, which are difficult to separate generically from small quasi-circular maria – the majority of the craters we see on the Moon must be of impact origin, from large formations of the size of Copernicus, Langrenus, Theophilus or Tycho down to the smallest pits resolvable on Ranger or Luna photographs. This cannot be otherwise; for the density of solid particles of requisite mass now floating in interstellar space is sufficient to account for most of them by chance encounters with the Moon; and there is nothing we know of that could have prevented them from doing so in the course of time. In other words, the cratering by impacts must have been demonstrably operative on the lunar surface in the course of time – as it has been on the Earth* or on Mars, as evidenced on recent

* The ground resolution of the Mariner photographs would correspond, on the lunar surface, to an optical resolution of a 4-inch telescope from the distance of the Earth.

close-up photographs of its surface secured by the American Mariner 4 in the course of its historic fly-by on July 14, 1965 (one of which is reproduced on the accompanying Figure 24-2). On the other hand, the operation of any large-scale volcanic processes on the lunar surface at any time of its long past remains still not proven; and the principal arguments advanced in favour of this view are based largely on analogies.

A comparison of the lunar and Martian landscape on the same scale* should be instructive, in so far as a proportion of the particles of different origin is likely to be different: namely, because of the location of its orbit, Mars is likely to suffer many more impacts of asteroidal debris than the Moon, which must again show a greater

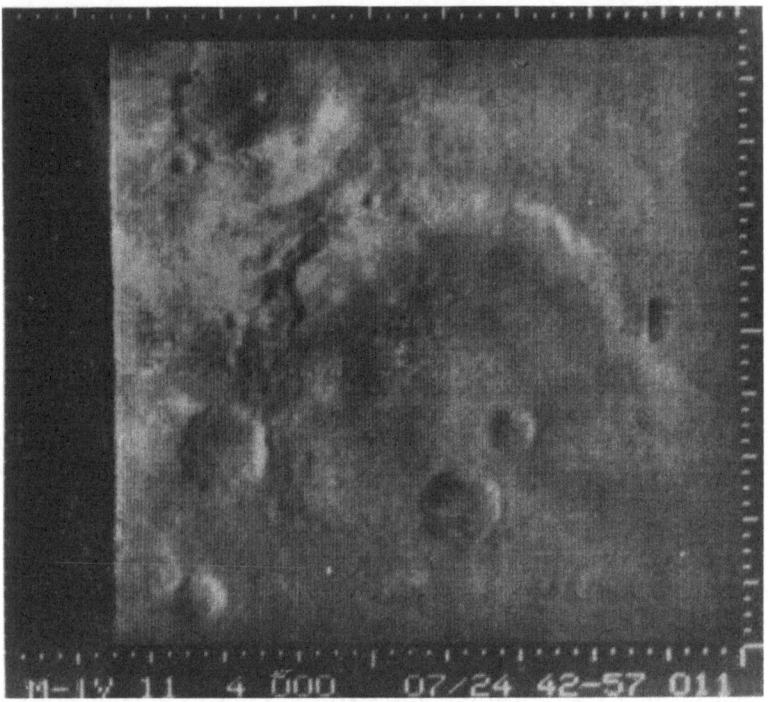

Fig. 24-2. A photograph of the Martian surface obtained by the U.S. spacecraft Mariner 4 on July 14, 1965, during its historic fly-by around this planet, from a distance of approximately 12400 km. The area in view is close to 270 × 240 km in size, and is centred on 31° south areographic latitude and 197° east longitude (including the region of the Atlantis, between Mare Sirenum and Cimerium). North is on top; and the Sun was 47° from the local zenith. The distance between individual scan lines corresponds to approximately 3 km on the Martian surface.

Note a great number of craters resembling very much those on the Moon (the one in the upper left-hand corner of the field of view possesses a central peak). The largest visible formation is about 270 km across (i.e., considerably larger than the largest known crater – Clavius – on the Moon). By courtesy of the Jet Propulsion Laboratory, California Institute of Technology.

* Cf., e.g., KRINOV, BEALS *et al.*, or DIETZ in *The Moon, Meteors and Comets* (ed. by B. M. MIDDLE-HURST and G. P. KUIPER), Univ. of Chicago Press, 1963; pp. 183-300.

proportion of cometary impacts. However, to separate these by their morphological characteristics represents clearly a task for the future.

All these miscellaneous arguments should not, however, distract our attention from the fact that none of them helps us as yet to resolve the principal dilemma met at the commencement of this chapter: namely, the one concerning the relative ages of the lunar continents and the maria. If an increasing density of impact formations per unit area of the surface is a genuine indication of the age of the respective ground, the continents should be obviously much older than the maria. If, on the other hand, the smallness of the albedo is taken as an indication of the age during which the respective ground suffered radiation damage by cumulative exposure to the solar wind, it is the maria which should be older. Both these views are difficult to gainsay, and yet appear to be mutually contradictory, thus posing the greatest single challenge met in the field of the lunar studies today.

How to reconcile them, when moving dust seems no longer to offer legitimate avenue of escape? Is it, perchance, a greater perishability of the craters in the maria – by subsidence, perhaps, into a softer ground (with the 'ghost' craters gradually sinking rather than being submerged)? Or is it the fact that lunar continental land masses are less prone to be darkened by radiation than the mare material (but if so, how about the gradual perishability of the bright ray systems?). Is there a large-scale erosion operating on the continents (where the average slope of the ground is indeed more inclined to the horizontal than in the maria); and is the dust or debris transported down the inclined slopes by a cumulative action of innumerable moon-quakes triggered by each major primary impact? But if so, where does it come to rest?

At the commencement of this chapter we promised eventually to raise more questions than can be satisfactorily answered at the present time, and we have now got entangled in their net. This, however, is tantamount to an admission that somewhere on the way along which we have taken the reader through this book we have, at least temporarily, lost our way.

A fuller understanding of the structure of the lunar surface – on both large and small scale – is still obscured by our inability to reconcile the two different ways of ground dating, based on the crater density per unit area on one hand, and its degree of darkening by radiation damage on the other. Whether or not the conflicting consequences of these two methods of approach to the stratigraphy of the lunar surface can be resolved without a fundamental revision of some of the beliefs and conclusions which we regarded as safely established in earlier parts of this book, cannot be foreseen. The history of scientific progress should teach us to keep an open mind.

We do not need to be unduly concerned as to whether our attempts to explore the nature of the lunar surface at a distance brought us to anything like the final truth; for that will come out in the near future, after men will have landed there and brought back both the results of more detailed topographic surveys and sample materials. In the meantime, however, in the course of this introduction we have learned something of the varied interests involved. We have traced, in particular, the pattern in which interactions of the lunar surface with particles of the largest (asteroids, comets) down

to the smallest (protons) are interwoven in its composite fabric. Our partial successes
in accounting for observational results revealed by recent spacecraft operating in the
proximity of the Moon encourage us to think that we are not far from the right track
in understanding the principal features of the structure of our satellite. Moreover, the
current spectacular advances in astronautics can fill us with confidence that most
remaining problems facing us across the intervening gap of space may soon be settled
by the geologist's hammer on the spot.

## POSTSCRIPT*

While the text of this book has gone through the press, lunar research – chiefly by
means of spacecraft – has continued successfully to make great strides; and the aim
of this postscript will be to acquaint the reader with the principal results of such investigations carried out in the meantime.

Following the first successful soft landing of the Russian Luna 9 on February 4,

Fig. 24-3.   Sunset over the lunar surface in Oceanus Procellarum, as televised by Surveyor 1 on June
12, 1966 (reproduced by courtesy of NASA and JPL). Note the flat horizontal panorama, and a lonely
small crater – obviously of impact origin – some three meters in size, at a distance of about 26 metres
from the spacecraft.

* September 1st, 1966.

1966, the second soft landing by the American Surveyor 1 on June 2nd in the Oceanus Procellarum near the crater Flamsteed (for its location, see the enclosed fold-out map) proved to be a brilliant success, and furnished us with a large number of televised images of lunar environment, which are technically much superior to those furnished by Luna 9 (as reproduced on Figure 23-3).

A view of the lunar landscape at the time of the first sunset over Surveyor 1 on June 13th has been reproduced on the frontispiece to this volume; and two others are shown on the accompanying Figures 24-3 and 24-4. The landscape panorama as revealed on Figure 24-3 confirmed further our anticipations based on the evidence furnished previously by Luna 9, as well as on all indirect evidence discussed in previous chapters of the last part of this book; and so did the detailed view of surface structure as shown on Figure 24-4. In particular, no evidence was revealed of any loose dust on the surface – in agreement with the conclusions drawn in the early part of Chapter 24. Direct measurements of the diurnal temperature variation, performed likewise by Surveyor 1, are in close accord with the results of the terrestrial measurements of

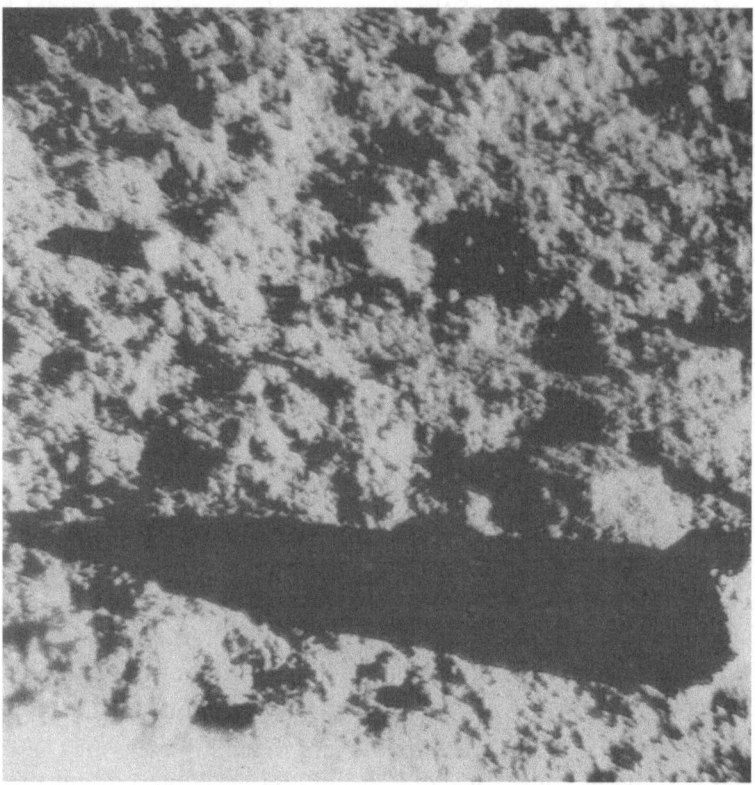

Fig. 24-4.  A closeup view of a portion of the lunar surface at the time of sunset, as televised by Surveyor 1 on June 13, 1966 (reproduced by courtesy of NASA and JPL). Note the long shadow cast by a small boulder in the immediate neighbourhood of the spacecraft; smallest details discernible on the photograph are only millimeters in size.

lunar infrared flux as discussed in Chapter 20. Lastly, the bearing strength of the lunar surface in the region of the Surveyor landing proved to be more than adequate for a support of the spacecraft which is to land men on the Moon in the near future.

In addition to the soft-landers, not less than three satellites sent out by human hand were successfully injected in circum-lunar orbits this spring and the summer (Luna 10, Orbiter 1, and Luna 11); and appropriate analyses of their trajectories have furnished valuable direct information on the properties of the lunar gravitational field. In particular, the values of the zonal harmonic coefficients $J_{0,2}$ and $J_{2,2}$ in the expansion of lunar gravitational potential proved to be in good agreement with the values of the ratios of the lunar moments of inertia (cf. Equations 4-120) deduced from the Moon's physical librations. The value of $J_{0,3}$ proved, however, to be *negative* and absolutely half as large as $J_{0,2}$; while $J_{0,4}$ turned out also to be negative and about half as large as $J_{0,3}$. These facts dispose once for all of the possibility that the underlying deformations of the lunar globe could have anything to do with the tidal action of the Earth or of any other body – now or at any time in the past; for if so, all $J_{0,j}$'s should have to be of the *same* algebraic sign, and much more rapidly convergent (a ratio of one-half of the successive $J_{0,j}$'s. would, on tidal hypothesis, correspond to a centre of gravity of the tide-generating body to be only two lunar radii away from the Moon's centre).

Moreover, the absolute magnitudes as well as the signs of different $J_{i,j}$'s furnished by the Orbiter came out to be in remarkable accord with their prediction by GOUDAS (1966a), based on an assumption that the gravitational potential of the Moon is essentially that of a homogeneous body filling up the observed form of the lunar surface (cf. Chapter 13). Small departures from complete homogeneity, as evidenced by the observed lunar profiles (see p. 186) do not, therefore, probably extend far below the surface.

Photographs of the Moon's far side secured by the American Orbiter 1 this August confirmed (within the limits of their coverage) the earlier Russian findings on the virtual absence of the maria on the Moon's far side, and thus provided further post-mortem for Gold's hypothesis of the maria as being receptacles for migrating dust abraded from the continental land masses. This hypothesis we buried already on pp. 421–422 for several reasons, to which the following can be added: should a hypothetical migration of dust on the lunar hemisphere facing us have filled its low-lying basins, why had not the same occurred on the far side (which is equally exposed to all abrasive influences)?

Another major step forward in understanding the face of the Moon and its stratigraphy has come out this summer from the Jet Propulsion Laboratory, California Institute of Technology: namely, through a discovery by D. B. Nash (private information) that the rate at which silicate rocks are darkened as a result of protracted exposure to corpuscular radiation simulating solar wind depends sensitively on small admixtures of certain chemical impurities (such as, for instance, hydro-carbons). Experiments on this subjects are still in progress; but if their basic conjecture is confirmed, a way would be found out of the dilemma discussed on pp. 421 or 429 of

Chapter 24: namely, a conflict between the results of the relative dating of lunar surface features – such as the maria – by their crater density and reflectivity. If the extent to which the lunar surface layer can be darkened by a continued action of the solar wind is sensitive to chemical impurities, the relatively smooth and pockmark-free maria may well be of later origin than the continental land masses in spite of their lower albedo; for the latter may be due to a small difference in chemical composition of their material from that of the continents. If so, however, a dilemma which on page 429 we called "the greatest single challenge in the field of lunar studies today" may have found a natural explanation earlier than we hoped.

# BIBLIOGRAPHICAL NOTES

## Chapter 19

For determinations of the absolute magnitude of full Moon in different photometric systems cf. MARTYNOV (1959), FESSENKOV (1962; containing a comprehensive survey of earlier literature on the subject); or, more recently, GALLOUET (1963) or GEHRELS et al. (1964).

The law of "photometric homogeneity" of the lunar face at optical frequencies was developed by BARABASHEV (1923, 1924) and MARKOV (1924); cf. also BARABASHEV and MARKOV (1925, 1926). A recent discovery of the "opposition effect" is due to GEHRELS, COFFEEN and OWINGS (1964), followed by VAN DIGGELEN (1965), OETKING (1965) and others; while the equally anomalous photometric "new Moon" effect was pointed out by DANJON (1933).

Photometric investigations of lunar light changes, and of the reflectivity of different parts of the lunar surface, that have been published in the past forty years are so numerous that only a brief reference to most can be noted in this place. Of more significant contributions to this literature cf., e.g., BARABASHEV (1923, 1924, 1927); BARABASHEV and CHEKIRDA (1945, 1948, 1959); BENNETT (1938); AKIMOV (1963, 1965); BORISOVA (1953); FEDORETS (1952); FESSENKOV, STAUDE and PARENAGO (1928); FESSENKOV (1929); FESSENKOV and PARENAGO (1929); FESSENKOV (1960, 1962); GRAFF (1949); KING (1922); MARKOV (1924, 1927a, b, 1948, 1952); MARKOV and BARABASHEV (1925, 1926); NIKONOVA (1949); ÖPIK (1924); ORLOVA (1941, 1952, 1954, 1955, 1956, 1957, 1958, 1962); ROSENBERG (1921); SCHOENBERG (1925); SHARONOV (1934, 1936, 1940, 1965); SYTINSKAYA (1953a, b, 1957, 1963, 1965); TCHUNKO (1949); VAN DIGGELEN (1959); and others. In looking over this rather extensive literature the reader cannot fail to be impressed with the fact that, in the four decades between 1920–1960, the progress of work in the field of lunar photometry (and colorimetry) was largely in the hands of the Russian astronomers working at the Universities of Kharkov and Leningrad – to which contributions as distinguished as they were numerous were made by Russian lady astronomers, of the calibre of Borisova, Fedorets, Nikonova, Orlova, Petrova, Radlova (Mrs. Levin) and Sytinskaya (Mrs. Sharonov) – proving, perhaps, a subtle influence on scientific inclinations of the ancient female deity of the sky.

Almost all photometric work contained in the sources referred to above was visual or photographic. Photoelectric photometry of the Moon – going back to STEBBINS and BROWN (1907) – did not really commence systematically until the work of ROUGIER (1933, 1934a, b, 1936a, b); followed in more recent years by GEHRELS, COFFEEN and OWINGS (1964), or WILDEY and POHN (1964).

Theoretical work aiming to account for peculiar features of the lunar light changes in terms of the structural characteristics of the surface of our satellite goes back at least to BARABASHEV (1924); ÖPIK (1924); FESSENKOV (1928); SCHOENBERG (1930); and BENNETT (1938), who all tried (with a limited success) to explain the observed facts by a varying degree of surface roughness. The limitations of such an approach were in recent years discussed (and refined proposals made) by VAN DIGGELEN (1959); MINNAERT (1961); HAPKE and VAN HORN (1963), or GEHRELS et al. (1964). For other investigations of the theory of lunar light changes cf. MINNAERT (1941); HAPKE (1963) or WILDEY (1963).

As far as the spectral composition of the scattered components of moonlight is concerned, the first indications of the fact that the colour of selected lunar regions depends on the phase go back to the work of MIETHE and SEEGERT (1911, 1914); WOOD (1910, 1912), or WRIGHT (1929). Of other investigations concerned with the colorimetry of the lunar surface cf. e.g., WILSING and SCHEINER (1909, 1921); BARABASHEV (1924a, 1951, 1953); BARABASHEV and CHEKIRDA (1953, 1954, 1955, 1956); FESSENKOV (1929b); HARGREAVES (1924); KEENAN (1931); LIPSKI (1959); MARKOV (1950); PLATT (1958); RADLOVA (1941, 1943, 1957, 1960, 1962); RADLOVA and SHARONOV (1958); SHARONOV (1953, 1955, 1956a, b); SYTINSKAYA and SHARONOV (1952 a, b); TEIFEL (1957, 1958, 1959, 1960); VIGROUX (1956); YEZERSKI and FEDORETS (1955, 1956); PETROVA (1965); and others.

For extensive recent three-colour photoelectric photometry of selected regions of the Moon in the course of a lunation cf., in particular, WILDEY and POHN (1963) or GEHRELS et al. (1964); also VAN DEN BERGH (1962) or COYNE (1963, 1965).

For a determination of the lunar reflectivity in the ultraviolet part of the spectrum cf., e.g., O'BRIEN and O'BRIAN (1931) or STAIR and JOHNSTON (1953) and, more recently, HEDDLE (1963); cf. also HEDDLE et al. (1962). For infrared albedos of the Moon determined recently from Stratoscope II cf. WATTSON and DANIELSON (1965); while for infrared photography from ground cf. KUPREVICH (1962a, b) or RACKHAM (1964). For infrared spectra of the Moon cf. also ADEL (1946).

The polarization of moonlight, detected first by ARAGO as far back as 1811, and Lord ROSSE (who first established the variation of its amount with the phase), was subsequently studied by SECCHI (1859, 1860a, b) or LANDERER (1889, 1890, 1910); cf also SALET (1922). Fundamental contributions to this field have been made by LYOT (1924 and, in particular, 1929). Of other contributions cf., e.g., BARABASHEV (1926); DOLLFUS (1952, 1962); FESSENKOV and KRAMER (1943); HEILES (1963); SYTINSKAYA (1956); TURNER (1957, 1958) or WRIGHT (1927). Photoelectric polarimetry of the Moon has in recent years been carried out by DZHAPIASHVILI (1957); DZHAPIASHVILI and KSANFOMALITI (1962); KOHAN (1962, 1964, 1965); CLARKE (1962, 1963, 1965) and, in particular, by GEHRELS (1960) or GEHRELS et al. (1964).

For recent studies of the radiation damage to the lunar surface due to the solar wind, and its photometric consequences, cf. HAPKE (1964, 1965) or DOLLFUS and GEAKE (1965).

For the photometry of the "ashen light" of the Moon, the fundamental contributions are those of DANJON (1928, 1936); cf. also ÖPIK (1924), GRIMM (1931) or DUBOIS (1943). A connection between the brightness of the ashen light of the Moon and the solar activity was pointed out by DUBOIS (1944, 1955); while LYOT and DOLLFUS (1949) measured the polarization of the ashen light.

The photometry of lunar eclipses constitutes so extensive a subject that only a merest outline of it could have been given in this chapter. Its student is, however, well served by existing literature, and a full account of it can be found in the recent fundamental contributions by LINK (1956, 1962, 1963) and BARBIER (1961). For studies of a dependence of the residual brightness of lunar eclipses on solar activity cf. DANJON (1920); DE VAUCOULEURS (1944); BAKHAREV (1953); or, most recently, BELL and WOLBACH (1965).

## Chapter 20

Thermal radiation of the Moon in the infrared domain of the spectrum was detected first by Lord ROSSE (1869, 1870, 1872) with the aid of his 3-foot telescope at Parsontown, followed by LANGLEY (1884, 1887); LANGLEY and VERY (1889); VERY (1898, 1906), and COBLENTZ (1906).

Modern work on this subject commenced with PETTIT and NICHOLSON (1929, 1930) and was continued by PETTIT (1935, 1940) at the Mt. Wilson Observatory. In more recent years, important contributions on the observational side have been made by SINTON (1959, 1960); GEOFFRION, KRAMER and SINTON (1960); SHORTHILL, BOROUGH and CONLEY (1960); SHORTHILL and SAARI (1961); SAARI and SHORTHILL (1963); MURRAY and WILDEY (1963, 1964); SAARI (1964); RYADOV, FURASHOV and SHARONOV (1964); LOW (1965); SAARI and SHORTHILL (1965); BLIZARD (1965); CHISTYAKOV (1965); and others.

Observations of the thermal emission of the lunar globe in the microwave domain of the Moon's radio spectrum, commencing with the work of PIDDINGTON and MINNETT in 1949, are too numerous to be reviewed here in full; the reader will find the necessary references in Table 20-2. Of particular interest, in this connection, should be the recent radiometric measurements of GARY, STACEY and DRAKE (1965) at 3.3 mm wavelength, revealing the sub-surface temperature of lunar maria to remain warmer over a lunation by more than one would expect from their lower albedo.

The inward propagation of diurnal heat wave (or eclipse cold wave) in the lunar crust as a problem of conductive energy transfer has been considered by many investigators, with the aim of determining from the observations the thermal-conduction properties of the lunar crust. Following the pioneer work of EPSTEIN (1929) and WESSELINK (1948), the conductive heat transfer in the lunar crust was investigated further by JAEGER and HARPER (1950); JAEGER (1953a, b); SINTON (1956); MEZGER and STRASSL (1959); FREMLIN (1959); JAEGER (1959); MUNCEY (1962); and, most recently and extensively, by KROTIKOV and SHCHUKO (1963); while for a discussion of errors of lunar temperature determination cf. BURNS and LYON (1964) or MERTZ (1965); also BLIZARD (1965). Radiative transfer – of both infrared and microwave thermal radiation – in the lunar crust was considered first by PIDDINGTON and MINNETT (1949) and, most recently and extensively, by KOPAL (1964a) whose treatment of the subject has been largely followed in this chapter.

Of previous summarizing discussions of the thermal radiation of the Moon, and its transmission through lunar surface layers cf., e.g., GARSTANG (1958); GILVARRY (1958); SINTON (1962); TROITSKI (1962); or KROTIKOV and TROITSKI (1962) etc.

Of other recent literature concerned with thermal conduction properties of porous materials which may be present on the lunar surface cf., e.g., WOODSIDE and MESMER (1961); BERNETT *et al.* (1963); JAFFE (1964); LIU and DOBAR (1964); WECHSLER and GLASER (1964); DOBAR (1965); GLASER, WECHSLER and GERMELES (1965); GLASER and WECHSLER (1965); WECHSLER and GLASER (1965); etc.

For studies of the polarization of thermal emission in the microwave domain of the spectrum cf. SOBOLEVA (1962) or HEILES and DRAKE (1963).

## *Chapter 21*

Following the pioneer work of BAY (1946, and DEWITT and STODOLA (1949), in Hungary and the United States, systematic observations of lunar radar echoes commenced in Australia and England with KERR, SHAIN and HIGGINS (1949); MURRAY and HARGREAVES (1954); BROWNE, EVANS, HARGREAVES and MURRAY (1956); EVANS (1956, 1957a, b); TREXLER (1958); HEY and HUGHES (1959); LEADABRAND *et al.* (1960); and was continued by PETTENGILL (1960); EVANS and PETTENGILL (1963a); MILLMAN and ROSE (1963); LYNN, SOHIGIAN and CROCKER (1964); MUHLEMAN (1964); DAVIS and ROHLFS (1964); KLEMPERER, W. K. (1965); etc., on frequencies corresponding to wavelengths from 8.6 mm (LYNN *et al.*) to the domain of decameter wavelengths (Davis and Rohlfs).

In the meantime the lunar echo theory (involving the diffraction theory of radiowave scattering and radar reflections from rough surfaces (was gradually developed by KERR and SHAIN (1951, 1957); HARGREAVES (1959); BRIGGS (1960); BROWN (1960); DANIELS (1960, 1961, 1962, 1963); HUGHES (1960, 1961, 1962a, b); LEADABRAND *et al.* (1960); HAGFORS (1961); (1964), SIEGEL and SENIOR (1959, 1960, 1961, 1962); HAYRE (1961, 1964); BRAMLEY (1962); EVANS (1963); EVANS and PETTENGILL (1963a); EVANS and HAGFORS (1964); FUNG (1964); FUNG and MOORE (1964); REA, HETHERINGTON and MIFFLIN (1964); BECKMANN (1965); HAGFOTS (1966); KATZ (1966); and others.

It should be noted, in this connection, that Equation (21-10) expressing the frequency-dependence of the power reflection coefficient $R$ on the conductivity was given incorrectly by SIEGEL and SENIOR (1959, 1960), due to a mistake in separating the real and imaginary parts of the complex refraction coefficient (21-1) to the desired degree of approximation; and their incorrect result was subsequently quoted by EVANS (1962).

The effects of lunar librations and Doppler shifts on radar echoes have been considered by BROWNE *et al.* (1956); EVANS (1957a); EVANS, EVANS and THOMSON (1959) or PETTENGILL (1960). The range-frequency analysis of lunar echoes was initiated by PETTENGILL and HENRY (1962), and carried on by THOMPSON (1965).

Of summarising articles on lunar radar echoes, and their use for prospecting the lunar crust at a distance cf., e.g., EVANS (1962, 1966) or EVANS and PETTENGILL (1963b); also EVANS and HAGFORS (1964). For laser echoes, cf. SMULLIN and FIOCCO (1962).

Concerning the literature on the occultations of radio sources by the Moon, cf. ELSMORE and WHITFIELD (1955); LINK (1956); COSTAIN, ELSMORE and WHITFIELD (1956); or ELSMORE (1957).

Recent investigations of different properties of the lunar ionosphere and surface charge include ÖPIK and SINGER (1960); HERRING and LICHT (1959, 1960); BRANDT (1960); SINGER (1961); WEIL and BARASH (1963); BERNSTEIN *et al.* (1963); HINTON and TAEUSCH (1964), etc. Electrostatic erosion and possibility of a dust transport on the lunar surface has recently been considered by KOPAL (1960c, pp. 82–84); GRANNIS (1961); WALKER (1962, 1963, 1964); SINGER and WALKER (1962); GOLD (1962); COFFMAN (1964); and others.

## *Chapter 22*

In view of the interest shown in recent years in transient luminous phenomena on the lunar surface attributed now to luminescence, let it be recalled that the first observer who noted and described such phenomena was nobody lesser than Sir William HERSCHEL, in 1783, then at the beginning of his great career. Apparently he saw first a red glow in (or near) Aristarchus on May 4, 1783; and again 1787. An account of his earlier observations was never published; but when his collected works (*The Scientific Papers of Sir William Herschel*, ed. by J. L. E. Dreyer, London 1912) appeared, their editor's preface contained extensive extracts from Herschel's correspondence; and among them was a letter to J. H. de Magellan (a Portugese scientist then resident in London) who had asked Herschel for details of the reported "lunar volcano" of 1783.

As quoted by Dreyer, a part of Herschel's reply reported that, ... "May 4, 1783. I perceived in the dark part of the Moon a luminous spot. It had the appearance of a red star of about the 4th magnitude. It was situated in the place of Hevelii Mons Porphyrites (i.e., Aristarchus); the instrument with which I saw it was a 10-feet Newtonian reflector of 9 inches aperture. Dr. Lind's lady who looked in the telescope immediately saw it, tho' no person had mentioned it, and compared it to a star. Dr. Lind tried to see it in an achromatic of $3\frac{1}{2}$ feet of Dollond's but could not perceive it, tho' with difficulty perceive it in the refractor."

Herschel saw such phenomena again in April 1887; and in a note entitled 'An account of Three Volcanoes on the Moon' (HERSCHEL, 1787) he reported that ...

"April 19, 1787, 10 h 36 m Sidereal Time: I perceive three volcanoes in different places of the dark part of the new moon. Two of them are either already nearly extinct, or otherwise in a state of going to break out; which perhaps may be decided next lunation. The third shows an actual eruption of fire, or luminous matter. I measured the distance of the crater from the northern limb of the Moon, and found it to be $3'\ 57''.3$. Its light is much brighter than the nucleus of the comet which M. Méchain discovered at Paris the 10th".

"April 20, 1787, 10 h 36 m Sidereal Time: The volcano burns with greater violence than last night. I believe its diameter cannot be less than $3''$, by comparing it with that of the Georgian planet [i.e., Uranus]. As Jupiter was near at hand, I turned the telescope to his third satellite, and estimated the diameter of the burning part of the volcano to be equal to at least twice that of the satellite. Hence we may compute that the shining or burning matter must be above three miles in diameter. It is of an irregular round figure, and very sharply defined on the edges. The other two volcanoes are much farther towards the center of the Moon, and resemble large, pretty faint nebulae, that are gradually much brighter in the middle; but no well defined luminous spot can be discerned in them. These three spots are plainly to be distinguished from the rest of the marks upon the Moon; for the reflection of the sun's rays from the earth is, in its present situation, sufficiently bright, with a ten-feet reflector, to show the Moon's spots, even the darkest of them: nor did I perceive any similar phenomena last lunation, though I then viewed the same places with the same instrument."

"The appearance of what I have called the actual fire or eruption of a volcano, exactly resembled a small piece of burning charcoal, when it is covered by a very thin coat of white ashes; and it had a degree of brightness about as strong as that which such a coal would be seen to glow in faint daylight.

"All the adjacent parts of the volcanic mountain seemed to be faintly illuminated by the eruption and were gradually more obscure as they lay at a greater distance from the crater".

Moreover, in a letter of April 26th, 1787, to Sir Joseph Banks (then President of the Royal Society) accompanying his above communication, Herschel stated "... Inclosed I have sent you an account of three volcanoes in the moon, one of which is now actually burning with great violence, and probably disgorging an immense lava" (cf., C. A. LUBBOCK, *The Herschel Chronicle*, Cambridge 1933, pp. 217–218). So convinced was he of the reality of this phenomenon that, on May 20th, he sent the following message to a page of his royal patron, King George III: "Last month I discovered three volcanos on the Moon, and saw the actual eruption, or fire, of one of them, yesterday I examined the same place again and found that one of these volcanos is not yet quite extinct. Will you do me the favour to acquaint the King with these circumstances; and if his Majesty would wish to see the Moon, the best time for viewing the Crater, which continues still to be considerably luminous, will be this evening between 9 and 10 o'clock. I will be at Windsor in good time to see the King's ten-foot telescope brought out and prepared, if it should please his Majesty to have it done".

We do not know what his Majesty's pleasure may have been in this matter; but more important to us may be what the contemporary scientific opinion thought of Herschel's discovery. His accuracy as an observer was held already by then in such high regard that the existence of these volcanoes was not immediately disputed; but neither was it readily accepted. Lalande probably spoke for many when he wrote to Herschel from Paris that "... the volcano on the Moon has been visible these last few days; but there are astronomers who are inclined to believe that Mount Aristarchus, which is naturally very brilliant, might very well reflect the light of the Earth in such a manner as to produce this bright appearance across the ashen light of the Moon" (cf. again LUBBOCK, 1933, op. cit.). How Herschel took on to this suggestion we do not, unfortunately, know; the two astronomers may have discussed it when Lalande visited Herschel at Slough in August of the same year. At least Lalande's missive located for us the region on the Moon where the volcano was supposed to glow in May 1787. Herschel himself never identified the active region with Aristarchus itself, but described its position to be located 3′57″.3 from the northern limb of the Moon (the center of Aristarchus being only 3′35″ so distant). If we assume that this was in a direction perpendicular to the limb for the libration at that time, it coincides pretty well with the limb distance of the active spots which were located there more recently by Greenacre and Barr (GREENACRE, 1963) and are marked on Figure 22-3.

In fact, the modern reader may readily discern that a simple change of a few expressions and turns of the text should enable us paraphrase, in Herschel's language, the more recent observations of Greenacre and Barr ("a burning charcoal thinly covered by ashes, three miles across on the Moon") as well as those of Kopal and Rackham ("large, pretty faint nebulae, that are gradually brighter in the middle, but no well-defined luminous spot can be discerned in them"). As none of the observations which Herschel reported to have made were ever found wrong in any essential respect – though many had to await posthumous verification – we have no reason to believe that the same should not be true of his lunar observations as described in the preceding paragraphs; and if so, there seems but little doubt that the first indications of the luminescence on the dark side of the Moon were noted already by Herschel not less than 180 years ago.

Of subsequent observers who recorded from time to time the occurrence of similar phenomena (for appropriate references cf., e.g., FLAMM and LINGENFELTER, 1965) it should be merely noted that stellar magnitudes assigned by them to such bright spots possess little or no quantitative meaning; for the definition of the stellar magnitude as we know it to-day (i.e., corresponding to an intensity ratio of the fifth root of 100) was not proposed till 1850 by Pogson; and further decades had yet to elapse before such magnitudes could be properly measured. To Herschel, the term "stellar magnitude" carried still but little more quantitative meaning than it did to Hipparchos two thousand years before.

In order to appreciate the fact that Herschel's reference to a luminous spot on the dark side of the Moon in 1783 having the appearance of a red star of about the 4th magnitude" must have grossly exaggerated the situation, suffice it to recall that one square second of earthlit lunar ground corresponds in brightness to a star of $+ 13.5$ visual magnitude. Accordingly, (as one square second at the mean distance of the Moon corresponds approximately to one square mile of its surface) the doubling of the normal surface brightness of each square second would give it the brightness of a star of only $13.5 - 0.75 = 12.7$ vis. magn.; and ten square miles of such ground would then possess an integrated magnitude of $12.7 - 2.5 = 10.2$.

In more recent times, the subject of the lunar luminescence as such was originally opened up by KHAN (1946), KAPLAN (1946) and HERZBERG (1946), seeking luminescence of such traces of tenuous gas as may surround the lunar globe; but it was KOZYREV (1956) who first realized clearly that the virtual absence of any such atmosphere is bound to relegate the lunar "auroral zone" on to the solid surface.

Kozyrev's quest for the indications of such luminescence by the line-depth method commenced in the autumn of 1955, with the aid of a three-prism spectrograph attached to the 50-inch reflector of the Crimean Astrophysical Observatory, giving a dispersion of 50 Å/mm at Hγ, with a spectral resolution of about 2.5 Å; and his 0.05 mm slit-width intercepted on the lunar surface a strip of little less than 4 km across. In order to secure accurate observations of luminescent intensity, he restricted his measurements to the H and K lines of Ca II – two of the strongest absorption lines in that part of the solar spectrum where the intensity of the adjacent continuum rapidly falls off. In this way he studied the spectrum of the moonlight scattered from the plains of Mare Serenitatis and Imbrium; from the craters Aristarchus, Copernicus, Plato, Schickard and Wood's spot. Only Aristarchus (together with the neighbouring Herodotus) gave positive results, as described in the text.

The only other specific observations aimed at investigation of lunar luminescence by the line-depth method before 1960 were those of DUBOIS (1957, 1959) – in point of fact, DUBOIS (1959, p. 20) stated that his first observations of this effect were undertaken between 1949 November – 1952 March; and that a report (unpublished) based upon them was read before the Société Française de Physique (section du Sud-Ouest) on November 21, 1952. His best observations were made subsequently with the high-dispersion solar spectrographs at Arcetri and Utrecht. He employed slit widths as small as 0.2 Å, which allowed him to measure the depths of many more lines than Kozyrev; but in his publications he quoted luminescence measurements only on the Fraunhofer lines D, F, E, G, and Hα. The characteristics of his spectrographs prevented him from observing too far in the violet – a fact which rendered comparison with Kozyrev's results difficult. However, Dubois's observations referred to some 90 different points of the lunar surface; and he claimed that about half of these showed luminescence at the time of observation, of fractional intensity ranging from 3 to 25 per cent.

Unlike Kozyrev, GRAINGER (1962) used a grating spectrograph giving a dispersion of 5 Å/mm and spectral resolution of more than 0.1 Å; the registration was photoelectric, with a separate channel to compensate the seeing. The size of the slit intercepted on the lunar surface a rectangular area of approximately 1 × 10 miles in size. The photoelectric spectrograph used by Scarfe was described by GRIFFIN (1961).

For photometric evidence of lunar luminescence from lunar eclipse studies cf., DANJON (1920); DUBOIS (1944, 1955); LINK (1946); LINK and ŠIROKÝ (1951); BAKHAREV (1953); CIMINO and FORTINI (1953); CIMINO, FORTINI and GIANUZZI (1955); FORTINI (1954, 1955); CIMINO (1957); CIMINO and FRESA (1958); MATSUSHIMA (1965); COHEN (1965); or SANDULEAK and STOCK (1965).

For polarimetric evidence of lunar luminescence cf., GEHRELS (1964) or GEHRELS, COFFEEN and OWINGS (1964).

For transient colour changes which may be attributable to luminescence, cf., e.g., BLACKWELL and INGHAM (1961), KOPAL and RACKHAM (1963; 1964a, b), or GREENACRE (1963, 1965). Cf. also KOPAL (1964b); MIDDLEHURST (1964), BOTLEY (1964), or FLAMM and LINGENFELTER (1965) for luminous phenomena on the lunar night side.

For spectroscopic investigations of lunar luminescence by the line-depth method, cf. LINK (1951); KOZYREV (1956); DUBOIS (1956, 1957, 1959); GRAINGER (1962); GRAINGER and RING (1962); SPINRAD (1964); SCARFE (1965) or MYRONOVA (1965).

Solar-lunar particle relations underlying lunar luminescence have recently been studied by CAMERON (1964), ANAND, OSTER and SOFIA (1964); DODSON and HEDEMAN (1964); BIGG (1963, 1964); or ARONOWITZ and MILFORD (1965) who considered the magnetic shielding of the lunar surface from the solar wind as a function of the lunar magnetic moment; while the magneto-hydrodynamic wake of the Moon in solar wind was recently investigated by NESS (1965).

## *Chapter 23*

For a previous treatment of a temperature-stirred Brownian dust atmosphere above the lunar surface cf. KOPAL (1960c, p. 81; cf. also 1965c, pp. 60–61). BERG's qualitative discussion (1964) of this problem failed to appreciate the smallness of the grains which could be levitated by this mechanism, and misled its author to erroneous conclusions.

Of other recent investigations of lunar surface dust cf., e.g., WALKER (1964) or COFFMANN (1964).

# REFERENCES

ADEL, A.: 1946, *Astrophys. J.* **103**, 19.
AKIMOV, A. A.: 1965, *Věstn. Kharkov Univ.*, No. 4, pp. 43–61.
AKIMOV, L. A.: 1963, *Circ. Astron. Observ. Kharkov Univ.*, No. 26, pp. 14–19.
ALFVÉN, H.: 1963, *Icarus* **1**, 357.
ALLAN, D. W.: 1956, *Am. Math. Monthly* **63**, 315.
ALTER, D.: 1955, *Publ. Astron. Soc. Pacific* **67**, 237.
ALTER, D.: 1956, *Publ. Astron. Soc. Pacific* **68**, 404; 437.
ALTER, D.: 1957, *Publ. Astron. Soc. Pacific* **69**, 411; 533.
ALTER, D.: 1958, *Publ. Astron. Soc. Pacific* **70**, 416; 489.
ALTERMAN, Z., JAROSCH, H., and PEKERIS, C. L.: 1959, *Proc. Roy. Soc. (A)* **242**, 80.
ANAND, S. P. S., OSTER, L., and SOFIA, S.: 1964, *Nature* **202**, 1097.
ANDERSON, J. D., NULL, G. W., and THORNTON, C. T.: 1964, in *AIAA Series on Progress in Astronautics and Aeronautics*, Vol. 14 (ed. V. SZEBEHELY), Acad. Press, New York.
ARNOLD, J. R.: 1965, *Astrophys. J.* **141**, 1548.
ARONOVITZ, L. and MILFORD, S. N.: 1965, *J. Geophys. Res.* **70**, 227.
ARTHUR, D. W. G.: 1954, *J. Brit. Astron. Assoc.* **64**, 127.
ARTHUR, D. W. G., KUIPER, G. P., and WHITAKER, E. A.: 1961, *Orthographic Atlas of the Moon*, Univ. of Arizona Press.
ARTHUR, D. W. G.: 1962, in *The Moon* (I.A.U. Symp. No. 14, ed. by Z. KOPAL and Z. K. MIKHAILOV), Acad. Press, New York and London, pp. 317–324.
ARTHUR, D. W. G.: 1963, 'Selenography' in *The Moon, Meteorites and Comets* (ed. by B. M. MIDDLEHURST and G. P. KUIPER), Univ. of Chicago Press, pp. 57–89.
ASHWORTH, D.: 1964, Ph. D. Thesis, Univ. of Manchester (unpublished).
ATKINSON, R. d'E.: 1951, *Monthly Not. Roy. Astron. Soc.* **111**, 448.
BAKER, G.: 1960, *Nature* **185**, 291.
BAKHAREV, A. M.: 1953, *Bull. All-Union Astron.-Geodet. Soc.*, No. 14, pp. 50–51.
BALDWIN, J. E.: 1961, *Monthly Not. Roy. Astron. Soc.* **122**, 513.
BALDWIN, R. B.: 1949, *The Face of the Moon*, Univ. of Chicago Press.
BALDWIN, R. B.: 1963, *The Measure of the Moon*, Univ. of Chicago Press.
BALDWIN, R. B.: 1964, *Astron. J.* **69**, 377.
BALDWIN, R. B.: 1965, *Astron. J.* **70**, 545.
BANACHIEWICZ, TH.: 1955, *Trans. I.A.U.* **7**, 174.
BARABASHEV, N. P.: 1923, *Astron. Nachr.* **217**, 445.
BARABASHEV, N. P.: 1924, *Astron. Nachr.* **221**, 289.
BARABASHEV, N. P.: 1924a, *Astron. Zh.* **1**, 44.
BARABASHEV, N. P.: 1926, *Astron. Nachr.* **229**, 7.
BARABASHEV, N. P.: 1927, *Publ. Kharkov Astron. Obs.* **1**, 35.
BARABASHEV, N. P.: 1948, *Publ. Kharkov Astron. Obs.* **8**, 29.
BARABASHEV, N. P.: 1951, *Astron. Circ. USSR Acad. Sci.*, No. 127, pp. 9–10.
BARABASHEV, N. P.: 1953, *Circ. Kharkov Astron. Obs.*, No. 2.
BARABASHEV, N. P.: 1954, *Trudy Kharkov Univ.* **3**, 13.
BARABASHEV, N. P.: 1955, *Circ. Kharkov Astron. Obs.*, No. 13, pp. 3–13.
BARABASHEV, N. P.: 1956, *Astron. Zh.* **33**, 549.
BARABASHEV, N. P.: 1959, *Astron. Zh.* **36**, 851.
BARABASHEV, N. P., MIKHAILOV, A. A., and LIPSKI, YU. N.: 1960, *Atlas of the Far Side of the Moon*, U.S.S.R. Acad. Sci., Moscow (English edition by the Pergamon Press, London, 1961).
BARABASHEV, N. P. and TCHEKIRDA, A. T.: 1945, *Astron. Zh.* **22**, 11.

BARBIER, D.: 1961, in *Planets and Satellites* (ed. by G. P. KUIPER and B. MIDDLEHURST), Univ. of Chicago Press, pp. 249–271.

BARNES, V. E.: 1958, *Geochim. Cosmochim. Acta* **14**, 267.

BARNES, V. E., KOPAL, Z., and UREY, H. C.: 1958, *Nature* **181**, 1457.

BAY, Z.: 1946, *Hungarian Acta Phys.* **1**, 1.

BECKMAN, P.: 1965, *J. Geophys. Res.* **76**, 2345.

BEER, W. and MÄDLER, J. H.: 1837, *Der Mond*, Simon Schropp, Berlin.

BELKOVICH, I. V.: 1936, *Bull. Engelhardt Obs. Kazan*, No. 10.

BELKOVICH, I. V.: 1948, *Astron. Circ. USSR Acad. Sci.*, No. 81.

BELKOVICH, I. V.: 1949, *Izv. Engelhardt Obs. Kazan*, No. 24.

BELL, B. and WOLBACH, J. G.: 1965, *Icarus* **4**, 409.

BENARD, H.: 1900, *Rev. Gen. Sci.*, **11**, 1261; 1309.

BENARD, H.: 1901, *Ann. Chim. Phys.* (7) **23**, 62.

BENNETT, A.: 1965, *Icarus* **4**, 177.

BENNETT, A. L.: 1938, *Astrophys. J.* **88**, 1.

BENNETT, E. C., WOOD, H. L., JAFFE, L. D., and MARTENS, H. E.: 1963, *Aeron. Astronaut. J.* (6) **1**, 1402.

BERG, CH. A.: 1964, *Nature* **204**, 461.

BERGH, S. van den: 1962, *Astron. J.* **67**, 147.

BERNSTEIN, W., FREDRICKS, R. W., VOGL, J. L., and FOWLER, W. A.: 1963, *Icarus* **2**, 233.

BESSEL, F. W.: 1839, *Astron. Nachr.* **16**, 257.

BESSEL, F. W.: 1841, *Astron. Untersuchungen Königsberg* **1**, pp. 1ff.

BIGG, E. K.: 1963, *J. Geophys. Res.* **68**, 1409; 4099.

BIGG, E. K.: 1964, *J. Geophys. Res.* **69**, 4971.

BIRCH, F.: 1952, *J. Geophys. Res.* **57**, 227.

BLACKWELL, D. E. and INGHAM, M. F.: 1961, *Monthly Not. Roy. Astron. Soc.* **122**, 143.

BLAGG, M. A.: 1929, *J. Brit. Astron. Assoc.* **39**, 328.

BLAGG, M. A. and MÜLLER, K.: 1935, *Named Lunar Formations*, Vol. 1 (Catalogue) and 2 (Maps), London.

BLIZARD, J. B.: 1965, *Trans. Am. Geophys. Union* **46**, 132.

BOETTCHER, C. J. E.: 1952, *Theory of Electric Polarization*, John Wiley and Sons, New York.

BOLT, B.: 1960, *Nature* **188**, 1176.

BONACINI, C.: 1931, *Coelum* **1**, 82.

BORISOVA, A. P.: 1953, *Věstn. Univ. Leningrad* **8**, 89.

BORN, M. and WOLF, E.: 1959, *Optics*, Pergamon Press, London, Chapter 13.

BOTLEY, C. M.: 1964, *Icarus* **3**, 502.

BRAMLEY, E. N.: 1962, *Proc. Phys. Soc. London* **80**, 1128.

BRANDT, J. C.: 1959, *Science* **131**, 1606; 1671.

BREECE, S., HARDY, M., and MARCHANT, M. Q.: 1964, 'Horizontal and Vertical Control for Lunar Mapping (Part II: AMS Selenodetic Control System)', U.S. Army Map Service, Tech. Rep. No. 29.

BREIDO, I. I. and SHCHEGOLEV, D. YA.: 1962, in *The Moon* (I.A.U. Sympos. No. 14, ed. by Z. KOPAL and Z. K. MIKHAILOV), Acad. Press, New York and London, pp. 25–38.

BRIDGMAN, P.: 1948, *Proc. Am. Acad. Arts Sci.* **76**, 55.

BRIGGS, B. H.: 1960, *Nature* **187**, 490.

BRIGGS, M. H.: 1962, *J. Brit. Interplanet. Soc.* **18**, 386.

BRINKMANN, R. T.: 1966, *J. Geophys. Res.* **71**, 340.

BROCKAMP, B.: 1960, *Z. Geophys.* **26**, 271.

BROCKHAUS, K. and JOKSCH, H. C.: 1960, *Z. Geophys.* **26**, 9.

BROOKS, S. A.: 1963, M.Sc. Thesis, Univ. of Manchester (unpublished).

BROUWER, D. and CLEMENCE, G. M.: 1961, in *Planets and Satellites* (ed. by G. P. KUIPER and B. MIDDLEHURST), Univ. of Chicago Press, p. 59.

BROUWER, D. and CLEMENCE, G. M.: 1961a, *Methods of Celestial Mechanics*, Academic Press, London and New York.

BROUWER, D. and HORI, G. I.: 1962, 'The Motion of the Moon in Space', in *Physics and Astronomy of the Moon* (ed. by Z. KOPAL), Acad. Press, New York and London, pp. 1–26.

BROWN, E. W.: 1896, *An Introduction to the Lunar Theory*, Cambr. Univ. Press.

BROWN, S. HARRISON: 1960, *J. Geophys. Res.* **65**, 1679.
BROWN, S. HARRISON: 1961, *J. Geophys. Res.* **66**, 1316.
BROWN, W. E. 1960, *J. Geophys. Res.* **65**, 3087.
BROWNE, I. C., EVANS, J. V., HARGREAVES, J. K., and MURRAY, W. A. S.: 1956, *Proc. London Phys. Soc.* **69 B**, 901.
BRUNGART, D. L.: 1965, *Trans. Am. Geophys. Union* **46**, 139.
BRUTON, R. H., CRAIG, K. J., and YAPLEE, B. S.: 1959, *Astron. J.* **64**, 325.
BUETTNER, K. J. K.: 1952, *Publ. Astron. Soc. Pacific* **64**, 11.
BULLEN, K. E.: 1947, *Introduction to the Theory of Seismology*, Cambr. Univ. Press.
BURNS, E. A. and LYON, R. J. P.: 1964, *J. Geophys. Res.* **69**, 3771.
BYSTROV, N. F.: 1962, *Astron. Zh.* **39**, 527.
CAMERON, A. G. W.: 1962, *Icarus* **1**, 13.
CAMERON, A. G. W.: 1964, *Nature* **202**, 785.
CAMERON, W. S.: 1964, *J. Geophys. Res.* **69**, 2423.
CARR, R. E. and KOVACH, R. L.: 1962, *Icarus* **1**, 75.
CARLSON, D. H. and JONES, G. D.: 1965, *J. Geophys. Res.* **70**, 1897.
CARSLAW, H. S. and JAEGER, J. C.: 1959, *Conduction of Heat in Solids* (sec. edition), Oxford Univ. Press.
CARSON, D., DAVIDSON, M., GOUDAS, C. L., KOPAL, Z., and STODDARD, L. G.: 1966, *Icarus* **5**, 334.
CHANDON, E.: 1941, *Compt. Rend. Acad. Paris* **212**, 1026.
CHANDRASEKHAR, S.: 1939, *An Introduction to the Study of Stellar Structure*, Univ. of Chicago Press, Chapter IV.
CHANDRASEKHAR, S.: 1952, *Phil. Mag.* (7) **43**, 1317.
CHANDRASEKHAR, S.: 1953, *Phil. Mag.* (7) **44**, 233 and 1129.
CHANDRASEKHAR, S.: 1961, *Hydrodynamic and Hydromagnetic Stability*, Oxford Univ. Press, Chapter II.
CHAPMAN, D. R.: 1960, *Nature* **188**, 353.
CHAPMAN, D. R.: 1961, *Lunar-Planetary Exploration Colloquium* (North-American Aviation, Downey, Cal.) **2** (No. 4), p. 37.
CHAPMAN, D. R. and LARSON, H. K.: 1963, *J. Geophys. Res.* **68**, 4305.
CHEVALIER, S.: 1917, *Bull. Astron.* **34**, 5, 161.
CHISTYAKOV, N. YU.: 1965, *Izva. Pulkovo Obs.*, **24**, No. 2, pp. 175–181.
CIMINO, M.: 1957, *J. Internat. Lunar Soc.* **1**, 9.
CIMINO, M. and FORTINI, T.: 1953, *Rend. Accad. Naz. Lincei* (8) **14**, 619.
CIMINO, M., FORTINI, T., and GIANUZZI, M. A.: 1955, *Rend. Accad. Naz. Lincei* (8) **18**, 173.
CIMINO, M. and FRESA, A.: 1958, *Rend. Accad. Naz. Lincei* (8) **25**, 58.
CLARK, S. P.: 1956, *Bull. Geol. Soc. Am.* **67**, 1123.
CLARK, S. P.: 1957, *Trans. Am. Geophys. Union* **38**, 931.
CLARK, S. P.: 1957a, *Am. Mineralogist* **42**, 732.
CLARKE, D.: 1962, *Astron. Contrib. Univ. of Manchester*, Ser. III, No. 93.
CLARKE, D.: 1963, Ph. D. Thesis, Univ. of Manchester (unpublished).
CLARKE, D.: 1965, *Monthly Not. Roy. Astron. Soc.* **130**, 83.
COBLENTZ, W. W.: 1906, *Phys. Rev.* **23**, 247.
COFFMAN, M. L.: 1964, *J. Geophys. Res.* **69**, 567.
COHEN, A. J.: 1964, *Nature* **201**, 1015.
COHEN, A. J.: 1965, *Astron. J.* **70**, 135.
COURANT, R. and FRIEDRICHS, K. O.: 1948, *Supersonic Flow and Shock Waves*, Interscience Publishers, New York.
CONTOPOULOS, G.: 1965, *Astrophys. J.* **142**, 802.
COSTAIN, C. H., ELSMORE, B., and WHITFIELD, G. R.: 1956, *Monthly Not. Roy. Astron. Soc.* **116**, 380.
COYNE, G. V.: 1963, *Astron. J.* **68**, 49.
COYNE, G. V.: 1965, *Astron. J.* **70**, 115.
CROSS, C. A.: 1954, *J. Brit. Astron. Assoc.* **64**, 167.
CROSS, C. A.: 1955, *J. Brit. Astron. Assoc.* **65**, 72.
CROSS, C. A.: 1965, *J. Brit. Astron. Assoc.* **75**, 199.
DALE, E. D.: 1962, M. Sc. Thesis, Univ. of Manchester (unpublished).
DALY, R. A.: 1946, *Proc. Am. Phil. Soc.* **90**, 104.

DANBY, J. M. A.: 1962, *Fundamentals of Celestial Mechanics*, Macmillan, New York, 1962, Chapter 14.
DANBY, J. M. A.: 1964, *Astron. J.* **69**, 165.
DANIELS, F. B.: 1960, *Nature* **187**, 399.
DANIELS, F. B.: 1961, *J. Geophys. Res.* **66**, 1781.
DANIELS, F. B.: 1962, *J. Geophys. Res.* **67**, 895.
DANIELS, F. B.: 1963, *J. Geophys. Res.* **68**, 449; 2864.
DANJON, A.: 1920, *Compt. Rend. Acad. Paris* **171**, 127; 1207.
DANJON, A.: 1928, *Compt. Rend. Acad. Paris* **187**, 336.
DANJON, A.: 1933, *Ann. Observ. Strasbourg* (3) **2**, 139.
DANJON, A.: 1936, *Ann. Observ. Strasbourg* (3) **3**, 139–180.
DARWIN, G. H.: 1879, *Phil. Trans. Roy. Soc.* **170**, 1–35; 447–550.
DARWIN, G. H.: 1880, *Phil. Trans. Roy. Soc.* **171**, 713–891.
DAVIDSON, M.: 1963, M. Sc. Thesis, University of Manchester (unpublished).
DAVIS, J. R. and ROHLFS, D. C.: 1964, *J. Geophys. Res.* **69**, 3257.
DELANO, F.: 1950, *Astron. J.* **55**, 129.
DERHAM, C. J. and GEAKE, J. E.: 1964, *Nature* **201**, 62.
DEVADAS, P.: 1962, *J. Brit. Astron. Assoc.* **72**, 380.
DeWITT, J. H. and STODOLA, E. K.: 1949, *Proc. Inst. Radio Engrs.* **37**, 229.
DICKE, R. H. and BERINGER, R.: 1946, *Astrophys. J.* **103**, 275.
DIETZ, R. S.: 1946, *J. Geol.* **54**, 359.
DIGGELEN, J. van: 1951, *Bull. Astron. Inst. Neth.* **11**, 283.
DIGGELEN, J. van: 1959, *Recherches Obs. Utrecht* **14**, 1–114.
DIGGELEN, J. van: 1965, *Plan. Space Sci.* **13**, 271.
DOBAR, W. I.: 1965, in *Geological Problems of Lunar Exploration* (ed. by J. GREEN), *Ann. N.Y. Acad. Sci.* **123**, 495–515.
DOBAR, W. I.: 1966, *Icarus* **5**, 399.
DOBAR, W. I., TIFFANY, O. L., and GNAEDINGER, J. P.: 1964, *Icarus* **3**, 323.
DODD, R. T., SALISBURY, J. W., and SMALLEY, V. G.: 1963, *Icarus* **2**, 466.
DODSON, H. W. and HEDEMAN, E. R.: 1964, *J. Geophys. Res.* **69**, 3965.
DOLGINOV, S. S., YEROSHENKO, E. G., ZHUZGOV, L. I., PUSHKOV, N. V., and TYURMINA, L. O.: 1960, *Iskustvennye Sputniki Zemli* **5**, 149.
DOLGINOV, S. S., YEROSHENKO, E. G., ZHUZGOV, L. I., and PUSHKOV, N. V.: 1962, in *The Moon* (I.A.U. Sympos. No. 14, ed. by Z. KOPAL and Z. K. MIKHAILOV), Acad. Press, New York and London, pp. 45–62.
DOLLFUS, A.: 1952, *Compt. Rend. Acad. Paris* **234**, 2046.
DOLLFUS, A.: 1956, *Ann. Astrophys.* **19**, 71.
DOLLFUS, A.: 1962, in *The Physics and Astronomy of the Moon* (ed. by Z. KOPAL), Acad. Press, New York and London, pp. 131–159.
DOLLFUS, A. and GEAKE, J. E.: 1965, *Compt. Rend. Acad. Paris* **260**, 4921.
DOMMANGET, J.: 1962, *Comm. Obs. Roy. Belgique*, No. 208.
DREYER, J. L. E.: 1906, *History of the Planetary Systems*, Cambr. Univ. Press.
DROFA, V. K.: 1962, *Publ. Astron. Obs. Univ. Kiev*, No. 10.
DUBOIS, J.: 1943, *Ciel et Terre* **59**, 375.
DUBOIS, J.: 1944, *L'Astronomie* **58**, 136.
DUBOIS, J.: 1955, *L'Astronomie* **69**, 242.
DUBOIS, J.: 1956, *L'Astronomie* (Bull. de la Société Astronomique de France) **70**, 225, 297.
DUBOIS, J.: 1957, *J. Phys. Radium* **18**, 13S.
DUBOIS, J.: 1959, *Rozpravy Czech. Acad. Sci.* **69**, part 6 (pp. 1–44).
DUHAMEL, J. M. C.: 1837, *J. Ecole Polytechnique* **15**, 1–57.
DZHAPIASHVILI, V. P.: 1956, *Astron. Circ. USSR Acad. Sci.*, No. 167, pp. 16–19.
DZHAPIASHVILI, V. P.: 1957, *Bull. Abastumani Obs.*, No. 21.
DZHAPIASHVILI, V. P. and KSANFOMALITI, L. V.: 1962, in *The Moon* (I.A.U. Sympos. No. 14, ed. by Z. KOPAL and Z. K. MIKHAILOV), Acad. Press, New York and London, pp. 463–467.
ECKERT, W. J.: 1965, *Astron. J.* **70**, 787.
ECKHARDT, D. H.: 1965, *Astron. J.* **70**, 466.
EDWARDS, W. F. and BORST, L. B.: 1958, *Science* **127**, 325.

EGGLETON, R. E. and MARSHALL, C. H.: 1963, *Trans. Am. Geophys. Union* **43**, 464.

ELGER, T. G.: 1895, *The Moon*, George Philip and Son, London.

ELLISON, M. A.: 1963, *Plan. Space Sci.* **11**, 597.

ELSÄSSER, H.: 1954, *Z. Astrophys.* **33**, 274.

ELSMORE, B.: 1957, *Phil. Mag.* (8) **2**, 1040.

ELSMORE, B. and WHITFIELD, G. R.: 1955, *Nature* **176**, 457.

EPSTEIN, E. E. and STACEY, J. M.: 1964, *Astron. J.* **69**, 541.

EPSTEIN, P. S.: 1929, *Phys. Rev.* **33**, 269.

EVANS, J. V.: 1956, *Proc. Phys. Soc. London* **69B**, 953.

EVANS, J. V.: 1957a, *Proc. Phys. Soc. London* **70B**, 1105.

EVANS, J. V.: 1957b, *J. Atmos. Terr. Phys.* **11**, 259.

EVANS, J. V.: 1962, in *Physics and Astronomy of the Moon* (ed. by Z. KOPAL), Acad. Press, New York and London, pp. 429–479.

EVANS, J. V.: 1963, *J. Res. Nat. Bur. Stand.* **67D**, 1.

EVANS, J. V., EVANS, S., and THOMSON, J. H.: 1959, in *Paris Symposium on Radio Astronomy* (I.A.U. Sympos. No. 9, ed. by R. N. BRACEWELL), Stanford Univ. Press, pp. 8–12.

EVANS, J. V. and HAGFORS, T.: 1964, *Icarus* **3**, 151.

EVANS, J. V. and PETTENGILL, G.: 1963a, *J. Geophys. Res.* **68**, 423; 5098.

EVANS, J. V. and PETTENGILL, G.: 1963b, in *The Moon, Meteorites and Comets* (ed. by B. M. MIDDLEHURST and G. P. KUIPER), Univ. of Chicago Press, pp. 129–161.

FAUTH, PH.: 1964, *Mondatlas*, Olbers-Gesellschaft, Bremen.

FEDORETS, V. A.: 1952, *Uch. Zap. Kharkov Univ.* **42**, 49.

FENSLER, W. E., KNOTT, E. F., OLTE, A., and SIEGEL, K. M.: 1962, in *The Moon* (I.A.U. Sympos. No. 14, ed. by Z. KOPAL and Z. K. MIKHAILOV), Acad. Press, New York and London, pp. 545–565.

FESSENKOV, V. G.: 1929a, *Astron. Zh.* **5**, 219.

FESSENKOV, V. G.: 1929b, *Astron. Nachr.* **236**, 7.

FESSENKOV, V. G.: 1943, *Dokl. USSR Acad. Sci.* **39**, 275; *Astron. Zh.* **20**, 212.

FESSENKOV, V. G.: 1960, *Astron. Zh.* **37**, 496.

FESSENKOV, V. G.: 1962, in *Physics and Astronomy of the Moon* (ed. by Z. KOPAL), Acad. Press, New York and London, pp. 99–130.

FESSENKOV, V. G. and KRAMER, O. P.: 1943, *Dokl. USSR Acad. Sci.* **40**, 152.

FESSENKOV, V. G. and PARENAGO, P. P.: 1929, *Astron. Zh.* **6**, 279.

FESSENKOV, V. G., STAUDE, N. M., and PARENAGO, P. P.: 1928, *Trudy Sternberg State Astron. Inst. Moscow* **4**, No. 1, pp. 1–90.

FIELD, G. B.: 1963, *Am. Scientist* **51**, 349.

FIELDER, G.: 1958, *Monthly Not. Roy. Astron. Soc.* **118**, 547.

FIELDER, G.: 1960, *Sky and Telescope* **19**, 334.

FIELDER, G.: 1961, *Astrophys. J.* **134**, 425.

FIELDER, G.: 1962, *Astrophys. J.* **135**, 632.

FIELDER, G.: 1962a, *J. Brit. Astron. Assoc.* **72**, 24, 119.

FIELDER, G.: 1963, *Nature* **198**, 1256.

FIELDER, G.: 1965, *Monthly Not. Roy. Astron. Soc.* **129**, 351.

FIRSOFF, V. A.: 1959, *Science* **131**, 1669.

FISCHER, I.: 1962, *Astron. J.* **67**, 373.

FLAMM, E. J. and LINGENFELTER, R. E.: 1965, *Nature* **205**, 1301.

FORSYTH, A. R.: 1929, *Treatise on Differential Equations* (6th edition), Macmillan and Co., London, § 193, p. 401.

FORTINI, T.: 1954, *Rend. Accad. Naz. Lincei* (8) **17**, 209.

FORTINI, T.: 1955, *Rend. Accad. Naz. Lincei* (8) **18**, 65.

FOWLER, W. A., GREENSTEIN, J. and HOYLE, F.: 1962, *Geophys. J. Roy. Astron. Soc.* **6**, 148.

FRANZ, J.: 1887, *Astron. Nachr.* **116**, 1.

FRANZ, J.: 1899, *Königsberg Beob.*, No. 38.

FRANZ, J.: 1901, *Mitt. Sternw. Breslau* **1**, 1–48.

FREIER, P. S. and WEBBER, W. R.: 1963, *J. Geophys. Res.* **68**, 1605.

FREMLIN, J. H.: 1959, *Nature* **183**, 1317.

FRIDLAND, M. V.: 1959, *Bull. Inst. Theor. Astron. Leningrad* **7**, 293.

FRIDLAND, M. V.: 1961, *Bull. Inst. Theor. Astron. Leningrad* **8**, 225.

FRIESEN, D. and HAWKINS, G. S.: 1964, *Plan. Space Sci.* **12**, 318.
FRITZ, H.: 1873, *Verzeichnis der Beobachteten Nordlichter*, K.u.K. Akad. Wiss., Wien.
FUJINAMI, S.: 1952, *Publ. Astron. Soc. Japan* **4**, 115.
FUJINAMI, S., INA, T., and KAWAI, S.: 1954, *Publ. Astron. Soc. Japan* **6**, 67.
FUNG, A. K.: 1964, *J. Geophys. Res.* **69**, 1063.
FUNG, A. K. and MOORE, R. K.: 1964, *J. Geophys. Res.* **69**, 1075.
GALILEI, Galileo: 1610, *Nuncius Sidereus*, Bartoluzzi, Padua.
GALILEI, Galileo: 1632, *Dialogue on the Great World Systems*, Florence.
GALLOUET, L.: 1963, *Compt. Rend. Acad. Paris* **256**, 4593; Thèse (Paris).
GARSTANG, R. H.: 1958, *J. Brit. Astron. Assoc.* **68**, 155.
GARY, B., STACEY, J. and DRAKE, F. D.: 1965, *Astrophys. J. Suppl.*, No. 108.
GAULT, D. E., SHOEMAKER, E. M., and MOORE, H. J.: 1963, NASA Tech. Note D-1767.
GAVRILOV, I. V.: 1959, *Astron. Circ. USSR Acad. Sci.*, No. 206.
GAVRILOV, I. V.: 1961, *Izv. Centr. Astron. Obs. Ukrainian Acad. Sci.* **3**, 68.
GEHRELS, T.: 1960, *Astron. J.* **65**, 470.
GEHRELS, T.: 1964, *Astron. J.* **69**, 542.
GEHRELS, T., COFFEEN, T., and OWINGS, D.: 1964, *Astron. J.* **69**, 826.
GEOFFRION, A. R., KORNER, M., and SINTON, W. M.: 1960, *Lowell Obs. Bull.* **5**, 1.
GERSTENKORN, H.: 1955, *Z. Astrophys.* **36**, 245.
GIAMBONI, L. A.: 1959, *Astrophys. J.* **130**, 324.
GIBSON, J. E.: 1958, *Proc. Inst. Radio Engrs.* **46**, 280.
GILBERT, G. K.: 1893, *Bull. Phil. Soc. Washington* **12**, 241.
GILVARRY, J. J.: 1957, *Nature* **180**, 911.
GILVARRY, J. J.: 1958, *Astrophys. J.* **127**, 751.
GILVARRY, J. J.: 1960, *Nature* **188**, 886.
GILVARRY, J. J.: 1964, *J. Theoret. Biol.* **6**, 325.
GILVARRY, J. J.: 1964a, *Icarus* **3**, 121.
GILVARRY, J. J.: 1964b, *Publ. Astron. Soc. Pacific* **76**, 245.
GILVARRY, J. J.: 1965a, *Ann. N.Y. Acad. Sci.* **123**, 1061.
GILVARRY, J. J.: 1965b, *Icarus* **4**, 317.
GILVARRY, J. J. and HILL, J. E.: 1965a, *Publ. Astron. Soc. Pacific* **68**, 224.
GILVARRY, J. J. and HILL, J. E.: 1956b, *Astrophys. J.* **124**, 610.
GLASER, P. E. and WECHSLER, A. E.: 1965, *Icarus* **4**, 104.
GLASER, P. E., WECHSLER, A. E., and GERMELES, A. E.: 1965, in *Geological Problems in Lunar Research* (ed. by J. GREEN), *Ann. N.Y. Acad. Sci.* **123**, 656–670.
GOLD, L.: 1960, *J. Astronaut. Sci.* **7**, No. 1.
GOLD, L.: 1961, *J. Geophys. Res.* **66**, 2531.
GOLD, T.: 1955, *Monthly Not. Roy. Astron. Soc.* **115**, 585.
GOLD, T.: 1962, in *The Moon* (I.A.U. Sympos. No. 14, ed. by Z. KOPAL and Z. K. MIKHAILOV), Acad. Press, New York and London, pp. 433–439.
GOLD, T.: 1962a, *Astron. J.* **67**, 577.
GOLD, T.: 1965, *Trans. Am. Geophys. Union* **46**, 139.
GOLDREICH, P.: 1963, *Monthly Not. Roy. Astron. Soc.* **126**, 257.
GOLDSTEIN, R. M. and CARPENTER, R. L.: 1963, *Science* **139**, 910.
GOODACRE, W.: 1931, *The Moon*, Bournemouth.
GORYNIA, A. A.: 1960, *Izv. Central Astron. Observ. Ukrain. Acad. Sci.* **3**, 23.
GORYNIA, A. A.: 1962, *Bull. Astron. Obs. Kiev* **4**, 35.
GORYNIA, A. A. and DROFA, V. K.: 1962, *Relief of the Limb Regions of the Moon*, Publ. House Ukrainian Acad. Sci., Kiev, pp. 163.
GOUDAS, C. L.: 1963, *Icarus* **2**, 423.
GOUDAS, C. L.: 1964a, *Icarus* **3**, 168.
GOUDAS, C. L.: 1964b, *Icarus* **3**, 273.
GOUDAS, C. L.: 1964c, *Icarus* **3**, 375.
GOUDAS, C. L.: 1964d, *Icarus* **3**, 476.
GOUDAS, C. L.: 1965a, *Icarus* **4**, 188.
GOUDAS, C. L.: 1965b, *Icarus* **4**, 218.
GOUDAS, C. L.: 1965c, *Icarus* **4**, 528.

GOUDAS, C. L.: 1966a, 'The Figure and Gravity Field of the Moon' in *Advances in Astronomy and Astrophysics* (ed. by Z. KOPAL), Academic Press, New York and London, Vol. 4, pp. 27–151.

GRAFF, K.: 1901, *Veröff. König. Rechen-Instituts Berlin*, No. 14.

GRAFF, K.: 1949, *Sitzungsber. Österr. Akad. Wiss. (Abt IIa)* **157**, 17.

GRAINGER, J. F.: 1962, Ph.D. Thesis, University of Manchester (unpublished).

GRAINGER, J. F. and RING, J.: 1962, in *Space Research* (Proc. of 3rd Internat. Space Science Symposium, Washington), pp. 989–996.

GRANNIS, P. D.: 1961, *J. Geophys. Res.* **66**, 4293.

GREEN, J. et al.: 1965, *Geological Problems in Lunar Research* (ed. by J. GREEN), *Ann. N.Y. Acad. Sci.* **123**, pp. 367–1257.

GREENACRE, J. A.: 1963, *Sky and Telescope* **26**, 316.

GREENACRE, J. C.: 1965, in *Geological Problems in Lunar Research* (ed. by J. GREEN), *Ann. N.Y. Acad. Sci.* **123**, pp. 811–815.

GREENLAND, L. and LOVERING, J. F.: 1963, *Geochim. Cosmochim. Acta* **27**, 249.

GRIEG, D. D., METZGER, S., and WAER, R.: 1948, *Proc. Inst. Radio Engrs.* **36**, 652.

GRIFFIN, R. F.: 1961, *Monthly. Not. Roy. Astron. Soc.* **122**, 181.

GRIMM, H.: 1931, *Z. Geophys.* **7**, 92.

GROVES, G.: 1960, *Monthly. Not. Roy. Astron. Soc.* **121**, 497.

GROVES, G.: 1962, 'Dynamics of the Earth-Moon System' in *Physics and Astronomy of the Moon* (ed. by Z. KOPAL), Acad. Press, New York and London, Chapter 3, pp. 61–98.

GRUSHINSKII, N. P. and SAGITOV, M. U.: 1962, *Soviet Astron.* **6**, 113.

HABIBULLIN, SH. T.: 1958, *Physical Libration of the Moon*, Kazan, pp. 182.

HABIBULLIN, SH. T.: 1961, *Astron. Zh.* **35**, 669.

HACKER, S. G. and STEWART, J. Q.: 1935, *Astrophys. J.* **81**, 37.

HAGFORS, T.: 1961, *J. Geophys. Res.* **66**, 777.

HAGFORS, T.: 1964, *J. Geophys. Res.* **69**, 3779.

HAGFORS, T.: 1966, *J. Geophys. Res.* **71**, 379.

HAPKE, B. W.: 1963, *J. Geophys. Res.* **68**, 4571.

HAPKE, B. W.: 1964, *Trans. Am. Geophys. Union* **45**, 347.

HAPKE, B. W.: 1965, *Ann. N.Y. Acad. Sci.* **123**, 711.

HAPKE, B. W. and GOLDBERG, L. S.: 1965, *Trans. Amer. Geophys. Union*, **46**, 139.

HAPKE, B. W. and HORN, H. VAN: 1963, *J. Geophys. Res.* **68**, 4545.

HARGREAVES, F. J.: 1924, *J. Brit. Astron. Assoc.* **34**, 243.

HARGREAVES, J. K.: 1959, *Proc. Phys. Soc. London* **73B**, 536.

HARTMANN, W. K.: 1965, *Icarus* **4**, 157, 207.

HARTWIG, E.: 1880, *Die Konstanten der physischen Libration des Mondes aus Beobachtungen am Strassburger Heliometer*, Diss. Karlsruhe.

HARTWIG, E.: 1881, *Monthly Not. Roy. Astron. Soc.* **41**, 375.

HAWKINS, G. S.: 1959, *Astron. J.* **64**, 450.

HAWKINS, G. S.: 1960, *Astron. J.* **65**, 318.

HAWKINS, G. S.: 1963, *Nature* **197**, 781.

HAWKINS, G. S.: 1963a, *Astron. Contr. Boston Univ.*, Ser. II, No. 36.

HAWKINS, G. S.: 1963b, *J. Geophys. Res.* **68**, 895.

HAWKINS, G. S.: 1964, in *Ann. Rev. Astronomy Astrophys.* (ed. by A. J. DEUTSCH, L. GOLDBERG and D. LAYZER), **2**, 149–164.

HAWKINS, G. S. and UPTON, E. K. L.: 1958, *Astrophys. J.* **128**, 727.

HAYN, F.: 1902, 'Selenographische Koordinaten', *Abh. König. Sachs. Gesell. Wiss.* **27**, 861.

HAYN, F.: 1904, *Abh. König. Sachs. Gesell. Wiss.* **29**, pp. 1–83.

HAYN, F.: 1907, *Abh. König. Sachs. Gesell. Wiss.* **30**, pp. 1–105.

HAYN, F.: 1914a, *Abh. König. Sachs. Gesell. Wiss.* **32**, pp. 1–113.

HAYN, F.: 1914b, *Astron. Nachr.* **199**, 261.

HAYN, F.: 1920, *Astron. Nachr.* **211**, 311.

HAYN, F.: 1923, 'Die Rotation des Mondes' in *Enzykl. Math. Wiss.* **6**, 1020–1043.

HAYRE, H. S.: 1961, *Proc. Inst. Radio Engrs.* **49**, 1433.

HAYRE, H. S.: 1964, *J. Franklin Inst.* **277**, 197.

HEDDLE, D. W. O.: 1963, *Observatory* **83**, 225.

HEDDLE, D. W. O. et al.: 1962, *Nature* **193**, 861.

HEILES, C. E.: 1963, *Icarus* **2**, 350.

HEILES, C. E. and DRAKE, F. D.: 1963, *Icarus* **2**, 281.

HERRING, J. R. and LICHT, A. L.: 1959, *Science* **130**, 266.

HERRING, J. R. and LICHT, A. L.: 1960, in *Space Research* (Proc. COSPAR Symposium, Nice; ed. by H. KALLMANN-BIJL), North-Holland Publ. Co., Amsterdam, pp. 1132–1145.

HERSCHEL, W.: 1787, *Phil. Trans. Roy. Soc.* **77**, 229.

HERZBERG, G.: 1946, *Popular Astronomy* **54**, 414.

HESS, W. N. and NORDYKE, M. D.: 1961, *J. Geophys. Res.* **66**, 3405.

HEY, J. S. and HUGHES, V. A.: 1959, in *Paris Symposium on Radio Astronomy* (I.A.U. Sympos. No. 9, ed. by R. N. BRACEWELL), Stanford Univ. Press, pp. 13–18.

HILL, G. H.: 1878, *Amer. Journ. Math.* **1**, 5, 129, 245.

HILL, J. E. and GILVARRY, J. J.: 1956, *J. Geophys. Res.* **61**, 501.

HINTON, F. L. and TAEUSCH, D. R.: 1964, *J. Geophys. Res.* **69**, 1341.

HIPPEL, A. VON: 1954, *Dielectric Materials and Applications*, John Wiley, New York, p. 314.

HOFSOMMER, D. J., POTTERS-ALLEDA, G. C. F. E., and POTTERS, M. L.: 1959, *Report Math. Center of Amsterdam*, No. 344.

HOLMES, A.: 1959, *Trans. Edinburgh Geol. Soc.* **17**, 183.

HOPMANN, J.: 1963, *Sitzungsber. Österr. Akad. Wiss.* (*Math.-Naturwiss. Klasse*) **172**, 103.

HOPMANN, J.: 1964, *Anzeiger Österr. Akad. Wiss.* (*Math.-Naturwiss. Klasse*) **12**, 367.

HUGHES, D. S. and MCQUEEN, R. G.: 1958, *Trans. Am. Geophys. Union* **39**, 959.

HUGHES, V. A.: 1960, *Nature* **186**, 873.

HUGHES, V. A.: 1961, *Proc. Phys. Soc. London* **78**, 988.

HUGHES, V. A.: 1962a, *Proc. Phys. Soc. London* **80**, 1117.

HUGHES, V. A.: 1962b, *J. Geophys. Res.* **67**, 892.

INGHAM, M. F.: 1961, *Monthly Not. Roy. Astron. Soc.* **122**, 157.

JAEGER, J. C.: 1953a, *Australian J. of Phys.* **6**, 10.

JAEGER, J. C.: 1953b, *Proc. Cambridge Phil. Soc.* **49**, 355.

JAEGER, J. C.: 1959, *Nature* **183**, 1316.

JAEGER, J. C. and HARPER, A. F. A.: 1950, *Nature* **166**, 1026.

JAFFE, L. D.: 1964, in *The Lunar Surface Layer* (ed. by J. W. SALISBURY and P. E. GLASER), Acad. Press, New York and London, pp. 355–380.

JAFFE, L. D.: 1966, *J. Geophys. Res.* **71**, 1095.

JEAN-PIERRE, J.: 1964, *J. Roy. Astron. Soc. Canada* **58**, 276.

JEANS, J. H.: 1918, *Problems of Cosmogony and Stellar Dynamics*, Cambr. Univ. Press.

JEFFREYS, H.: 1929, *Proc. Cambridge Phil. Soc.* **26**, 101.

JEFFREYS, H.: 1930, *Monthly Not. Roy. Astron. Soc.* **91**, 169.

JEFFREYS, H.: 1947, *Monthly Not. Roy. Astron. Soc.* **107**, 260.

JEFFREYS, H.: 1957, *Monthly Not. Roy. Astron. Soc.* **117**, 475.

JEFFREYS, H.: 1961a, *Monthly Not. Roy. Astron. Soc.* **122**, 339.

JEFFREYS, H.: 1961b, *Monthly Not. Roy. Astron. Soc.* **122**, 421.

JEFFREYS, H.: 1962, *The Earth* (4th edition), Cambr. Univ. Press.

JEFFREYS, H. and BLAND, M. E. M.: 1951, *Monthly Not. Roy. Astron. Soc.* (*Geophys. Suppl.*) **6**, 148, 272.

JOKSCH, H. C.: 1957, *Z. Geophys.* **23**, 250.

JONES, M. T.: 1965, M. Sc. Thesis, University of Manchester (unpublished).

JÖNSSON, A.: 1917, *Lund Medd.*, Ser. II, No. 15.

JUDAY, R. D.: 1965, *Trans. Am. Geophys. Union* **46**, 142.

KAISER, T. R.: 1961, *Monthly Not. Roy. Astron. Soc.* **123**, 265.

KALINYAK, A. A. and KAMIONKO, L. A.: 1962, in *The Moon* (I.A.U. Sympos. No. 14, ed. by Z. KOPAL and Z. K. MIKHAILOV), Acad. Press, New York and London, pp. 273–287.

KAPLAN, J.: 1946, *Popular Astron.* **54**, 313.

KATZ, A.: 1960, *Proc. Lunar and Planet. Explor. Colloq.* **2**, 27.

KATZ, I.: 1966, *J. Geophys. Res.* **71**, 361.

KAULA, W. M.: 1963, *J. Geophys. Res.* **68**, 4957.

KAULA, W. M.: 1964, *Rev. Geophys.* **2**, 661.

KEENAN, P. C.: 1931, *Publ. Astron. Soc. Pacific* **42**, 58.

KERR, F. J. and SHAIN, C. A.: 1951, *Proc. Inst. Radio Engrs.* **39**, 230.

KERR, F. J. and SHAIN, C. A.: 1957, *Nature* **179**, 433.

KERR, F. J., SHAIN, C. A., and HIGGINS, C. S.: 1949, *Nature* **163**, 310.

KHAN, M. A. R.: 1946, *Popular Astron.* **54**, 312.

KING, E. S.: 1922, *Popular Astron.* **30**, 617.

KLEMPERER, W. K.: 1965, *J. Geophys. Res.* **70**, 3798.

KOHAN, E. K.: 1962, in *The Moon* (I.A.U. Sympos. No. 14, ed. by Z. KOPAL and Z. K. MIKHAILOV), Acad. Press, New York and London, pp. 453–461.

KOHAN, E. K.: 1964, *Izv. Pulkovo Obs.* **23**, No. 5, pp. 93–107.

KOHAN, E. K.: 1965, *Izv. Pulkovo Obs.* **24**, No. 2, pp. 182–201.

KONDURAR, V. T.: 1963, *Astron. Zh.* **40**, 757.

KOPAL, Z.: 1959, *Close Binary Systems*, Chapman-Hall and John Wiley, London and New York, Secs. II, 1–2.

KOPAL, Z.: 1959a, *Nature* **183**, 169, 737.

KOPAL, Z.: 1959b, *Space J.* **3**, 6.

KOPAL, Z.: 1959c, *Am. Scientist* **47**, 505.

KOPAL, Z.: 1960a, in Proc. 1st COSPAR Space Sci. Sympos. Nice (ed. by H. K. KALLMANN-BIJL), Amsterdam, pp. 1123–1131.

KOPAL, Z.: 1960b, *Figures of Equilibrium of Celestial Bodies*, Univ. of Wisconsin Press, Madison, Wis.

KOPAL, Z.: 1960c, *The Moon – our Nearest Celestial Neighbour* (1st edition), Chapman-Hall and Acad. Press, London and New York.

KOPAL, Z.: 1961, Jet Propulsion Lab., Calif. Inst. of Technology, Tech. Report 32–108.

KOPAL, Z.: 1961a, *Discovery* **22**, 143.

KOPAL, Z.: 1961b, *Rev. Intern. Council Sci. Unions* **3**, 173.

KOPAL, Z.: 1961c, *Discovery* **22**, 334.

KOPAL, Z.: 1962, in *Physics and Astronomy of the Moon*, Acad. Press, New York and London, pp. 231–282.

KOPAL, Z.: 1962a, Jet Prop. Lab. Tech. Rep. No. 32-225.

KOPAL, Z.: 1962b, Jet Prop. Lab. Tech. Rep. No. 32-276.

KOPAL, Z.: 1962c, *Plan. Space Sci.* **9**, 625.

KOPAL, Z.: 1962d, in *Proc. of the Conference on Lunar Exploration, Blacksburg, Virginia*, Part A, paper No. 5.

KOPAL, Z.: 1963a, *Icarus* **1**, 391.

KOPAL, Z.: 1963b, *Icarus* **1**, 412.

KOPAL, Z.: 1963c, *Icarus* **2**, 376.

KOPAL, Z.: 1963d, *Mem. Soc. Roy. Sci. Liège* (5) **7**, 29.

KOPAL, Z.: 1963e, *The Moon – our Nearest Celestial Neighbour* (2nd edition), Chapman-Hall, London, pp. 24–25 and 50–51.

KOPAL, Z.: 1963f, in *Advances in the Astronautical Sciences*, Plenum Press, New York, **8**, 537–548.

KOPAL, Z.: 1964a, *Icarus* **3**, 8.

KOPAL, Z.: 1964b, *Icarus* **3**, 78.

KOPAL, Z.: 1965a, *Icarus* **4**, 166.

KOPAL, Z.: 1965b, *Icarus* **4**, 173.

KOPAL, Z.: 1965c, in Introduction to the *Photographic Atlas of the Moon*, Acad. Press, New York and London, pp. 60–61.

KOPAL, Z.: 1966, *Icarus* **5**, 201.

KOPAL, Z. et al.: 1961, 'Studies in Lunar Topography', *Air Force Cambridge Research Laboratories, Research Note* No. 67, pp. 1–188.

KOPAL, Z. and LYTTLETON, R. A.: 1963, *Icarus*, **1**, 455.

KOPAL, Z. and MILLS, G. A.: 1965, Report under Contract AF 61(052)-524, Manchester.

KOPAL, Z. and RACKHAM, T. W.: 1962, in *The Moon* (I.A.U. Sympos. No. 14, ed. by Z. KOPAL and Z. K. MIKHAILOV), Acad. Press, New York and London, pp. 343–360.

KOPAL, Z. and RACKHAM, T. W.: 1963, *Icarus* **2**, 481.

KOPAL, Z. and RACKHAM, T. W.: 1963a, *Icarus* **2**, 329.

KOPAL, Z. and RACKHAM, T. W.: 1964, *Sky and Telescope* **27**, 140.

KOZIEL, K.: 1948, *Acta Astron. Cracoviae* (a) **4**, 61–139.

KOZIEL, K.: 1949, *Acta Astron. Cracoviae* (a) **4**, 153–193.

KOZIEL, K.: 1957, *Acta Astron. Cracoviae* **7**, 228.

KOZIEL, K.: 1962, in *Physics and Astronomy of the Moon* (ed. by Z. KOPAL), Acad. Press, New York and London, pp. 27–59.

KOZIEL, K.: 1964, *I.A.U. Draft Report*, p. 217.

KOZYREV, N. A.: 1956, *Izv. Crimean Astron. Obs.* **16**, 148.

KOZYREV, N. A.: 1959, *Priroda*, No. 3.

KOZYREV, N. A.: 1962, in *Physics and Astronomy of the Moon* (ed. by Z. KOPAL), Acad. Press, New York and London, pp. 361–383.

KOZYREV, N. A.: 1962a, in *The Moon* (I.A.U. Sympos. No. 14, ed. by Z. KOPAL and Z. K. MIK-HAILOV), Acad. Press, New York and London, pp. 263–271.

KOZYREV, N. A.: 1963, *Nature* **198**, 979.

KREITER, T. J.: 1960, *Publ. Astron. Soc. Pacific* **72**, 393.

KRISTENSON, H.: 1954, *Arkiv Astron.* **1**, 411.

KROTIKOV, V. D. and TROITSKY, V. S.: 1962, *Astron. Zh.* **39**, 1089.

KROTIKOV, V. D. and SHCHUKO, O. B.: 1963, *Astron. Zh.* **40**, 297.

KUIPER, G. P.: 1954, *Proc. U. S. Nat. Acad. Sci.* **40**, 1096.

KUIPER, G. P.: 1955, *Proc. U. S. Nat. Acad. Sci.* **41**, 820.

KUIPER, G. P.: 1959, *J. Geophys. Res.* **64**, 1713.

KUIPER, G. P.: 1965, in 'Ranger VII, Part II: Experimenters' Analyses and Interpretations', Jet Propulsion Laboratory, Calif. Inst. of Tech., Tech. Report No. 32-700, p. 49.

KUPREVICH, N. F.: 1962a, *Astron. Zh.* **39**, 1136.

KUPREVICH, N. F.: 1962b, in *The Moon* (I.A.U. Symposium No. 14, ed. by Z. KOPAL and Z. K. MIKHAILOV), Acad. Press, New York and London, pp. 361-368.

KVÍZ, Z.: 1964, *Bull. Astron. Inst. Czech.* **15**, 227.

LANDERER, J. J.: 1889, *Compt. Rend. Acad. Paris* **109**, 360.

LANDERER, J. J.: 1890, *Compt. Rend. Acad. Paris* **111**, 210.

LANDERER, J. J.: 1910, *Compt. Rend. Acad. Paris* **150**, 116.

LANGLEY, S. P.: 1884, *Mem. U. S. Nat. Acad. Sci.* **3**, 3.

LANGLEY, S. P.: 1887, *Mem. U. S. Nat. Acad. Sci.* **4**, 107.

LANGLEY, S. P. and VERY, F. W.: 1889, *Am. J. Sci.* (3) **38**, 421.

LAPWOOD, E. R.: 1952, *Monthly Not. Roy. Astron. Soc., Geophys. Suppl.* **6**, 402.

LEADABRAND, R. L., DYCE, R. B., FREDRIKSEN, A., PRESNELL, R. I., and SCHLOBOHM, J. C.: 1960, *Proc. Inst. Radio Engrs.* **48**, 932.

LEADABRAND, R. L., DYCE, R. B., FREDRIKSEN, A., PRESNELL, R. I., and SCHLOBOHM, J. C.: 1960a, *J. Geophys. Res.* **65**, 3071.

LENHAM, A. P.: 1955, *J. Brit. Astron. Assoc.* **65**, 241.

LEVIN, B. YA.: 1962, in *The Moon* (I.A.U. Sympos. No. 14, ed. by Z. KOPAL and Z. K. MIKHAILOV), Acad. Press, New York and London, pp. 157–167.

LEVIN, B. YA.: 1963, *Astron. Zh.* **40**, 1071.

LEVIN, B. YA.: 1964, *Proc. 13th Internat. Astronaut. Congress*, Springer Verlag, Wien and New York, pp. 11–20.

LEVIN, B. YA. and MAYEVA, S. V.: 1960, *Dokl. USSR Acad. Sci.* **133**, 44.

LEVIN, B. YA. and MAYEVA, S. V.: 1964, *Astron. Zh.* **41**, 997.

LEVIN, L.: 1954, *Theory of Modern Waveguides*, State Publ. Office of Foreign Lit., Moscow.

LIN, N. C. and DOBAR, W. I.: 1964, in *The Lunar Surface Layer* (ed. by J. W. SALISBURY and P. E. GLASER), Acad. Press, New York and London, pp. 381–387.

LINK, F.: 1946, *Compt. Rend. Acad. Paris* **223**, 976.

LINK, F.: 1951, *Bull. Astron. Inst. Czech.* **2**, 131.

LINK, F.: 1956, *Die Mondfinsternisse*, Akad. Verlagsgesellschaft, Leipzig.

LINK, F.: 1956a, *Bull. Astron. Inst. Czech.* **7**, 1.

LINK, F.: 1962, in *Physics and Astronomy of the Moon* (ed. by Z. KOPAL), Acad. Press, New York and London, pp. 161–229.

LINK, F.: 1963, in *Advances in Astronomy and Astrophysics* (ed. by Z. KOPAL), Acad. Press, New York and London, **2**, 87–198.

LINK, F. and ŠIROKÝ, J.: 1951, *Bull. Astron. Inst. Czech.* **2**, 86.

LIPSKI, YU. N.: 1959, *Astron. Zh.* **36**, 322.

LIPSKI, YU. N.: 1960, *Astron. Zh.* **37**, 1043.

LIPSKI, YU. N.: 1962, in *The Moon* (I.A.U. Sympos. No. 14, ed. by Z. KOPAL and Z. K. MIKHAILOV), Acad. Press, New York and London, pp. 7–23.

LIPSKI, YU. N.: 1963, in *The Moon, Meteorites and Comets* (ed. by B. M. MIDDLEHURST and G. P. KUIPER), Univ. of Chicago Press, pp. 90–122.

LIPSKI, YU. N.: 1965, *Sky and Telescope* **30**, 338.

LOVE, A. E. H.: 1927, *Treatise on the Mathematical Theory of Elasticity* (4th edition), Cambr. Univ. Press.

LOW, F. J.: 1964, *Astron. J.* **69**, 143.

LOW, F. J.: 1965, *Astrophys. J.* **142**, 806.

LOW, F. J. and DAVIDSON, A. W.: 1965, *Astrophys. J.* **142**, 1278.

LOWAN, A. N.: 1933, *Phys. Rev.* **44**, 769.

LOWAN, A. N.: 1934, *Am. J. Math.* **56**, 254.

LOWAN, A. N.: 1935, *Am. J. Math.* **57**, 174.

LOWMAN, P. D.: 1962, *Geochim. Cosmochim. Acta* **26**, 561.

LYNN, V. L., SOHIGIAN, M. D., and CROCKER, A. E.: 1964, *J. Geophys. Res.* **69**, 781.

LYOT, B.: 1924, *Compt. Rend. Acad. Paris* **178**, 1796.

LYOT, B.: 1929, *Ann. Obs. Meudon*, **8**, fasc. 1, pp. ix–161.

LYOT, B. and DOLLFUS, A.: 1949a, *Compt. Rend. Acad. Paris* **229**, 1277.

LYOT, B. and DOLLFUS, A.: (1949b) *Compt. Rend. Acad. Paris* **228**, 1773.

LUBBOCK, C. A.: 1933, *The Herschel Chronicle* Cambr. Univ. Press.

LYTTLETON, R. A.: 1963, Jet Propulsion Laboratory, Calif. Inst. of Tech., Tech. Rep. No. 32-522.

MAFFEI, P.: 1963, *Carte Lunari di ieri e di oggi*, Inst. Geog. Militare Firenze.

MACDONALD, G. J. F.: 1959, *J. Geophys. Res.* **64**, 1967.

MACDONALD, G. J. F.: 1960, *Plan. Space Sci.* **2**, 249.

MACDONALD, G. J. F.: 1962, *J. Geophys. Res.* **67**, 2945.

MACDONALD, G. J. F.: 1963, *Space Sci. Rev.* **2**, 473.

MACDONALD, G. J. F.: 1964, *Rev. Geophys.* **2**, 467.

MACDONALD, G. J. F.: 1964a, *Science* **145**, 881.

MACDONALD, T. L.: 1929, *J. Brit. Astron. Assoc.* **39**, 314.

MACDONALD, T. L.: 1931, *J. Brit. Astron. Assoc.* **41**, 172, 228, 288, 367.

MACDONALD, T. L.: 1932, *J. Brit. Astron. Assoc.* **42**, 291.

MACDONALD, T. L.: 1940, *J. Brit. Astron. Assoc.* **50**, 160.

MACDONALD, T. L.: 1942, *J. Roy. Astron. Soc. Canada* **36**, 159.

MACDONALD, T. L.: 1951, *J. Brit. Astron. Assoc.* **62**, 36.

MAINKA, C.: 1901, *Mitt. Sternwarte Breslau*, **1**, 55–70.

MAKOVER, S. G.: 1962, *Bull. Inst. Theor. Astron. Leningrad* **8**, 249.

MALVASIA, C.: 1662, *Ephemerides novissimae, ecc.*, Modena.

MANNO, F. F., SAUERMANN, G. O., and ENGELMAN, A.: 1965, *Trans. Am. Geophys. Union* **46**, 130.

MARCUS, A. H.: 1964, *Icarus* **3**, 460.

MARCUS, A. H.: 1965, *Icarus* **4**, 267.

MARCUS, A. H.: 1966, *Monthly Not. Roy. Astron. Soc.* **132**, (in press).

MARKOV, A. V.: 1924, *Astron. Nachr.* **221**, 65.

MARKOV, A. V.: 1927a, *Astron. Nachr.* **231**, 57.

MARKOV, A. V.: 1927b, *Astron. Zh.* **4**, 60.

MARKOV, A. V.: 1948, *Astron. Zh.* **25**, 172.

MARKOV, A. V.: 1950, *Bull. Abastumani Obs.* **11**, 107.

MARKOV, A. V.: 1952, *Izv. Pulkovo Obs.* **19**, No. 149, pp. 64–80.

MARKOV, A. V.: 1958, *Izv. Pulkovo Obs.* **20**, No. 158, pp. 138–155.

MARKOV, A. V.: 1962, in *The Moon* (I.A.U. Sympos. No. 14, ed. by Z. KOPAL and Z. K. MIKHAILOV), Acad. Press, New York and London, pp. 39–44.

MARKOV, A. V. and BARABASHEV, N. P.: 1925, *Astron. Nachr.* **226**, 129.

MARKOV, A. V. and BARABASHEV, N. P.: 1926, *Astron. Zh.* **3**, 55.

MARKOV, A. V. and SHCHEGOLEV, D. YA.: 1963, *Plan. Space Sci.* **11**, 549.

MARSHALL, R.: 1962, *Icarus* **1**, 95.

MARTYNOV, D. YA.: 1959, *Astron. Zh.* **36**, 648.

MASLOWSKI, J. and MIETELSKI, J.: 1963, *Acta Astron. Cracoviae* **13**, 135.

MATSUSHIMA, S.: 1965, *Astron. J.* **70**, 326.

McCracken, C. W., Alexander, W. M., and Dubin, M.: 1961, *Nature* **192**, 441.

McCracken, C. W. and Dubin, M.: 1964, in *The Lunar Surface Layer* (ed. by J. W. Salisbury and P. E. Glaser), Acad. Press, New York and London, pp. 179–214.

McGillem, C. D. and Miller, B. P.: 1962, *J. Geophys. Res.* **67**, 4787.

McMath, R. R., Petrie, R. M. and Sawyer, H. E.: 1937, *Publ. Univ. Obs. Michigan* **6**, 67.

Mertz, L.: 1965, *J. Geophys. Res.* **70**, 999.

Meyer, D. L. and Ruffin, B. W.: 1965, *Icarus* **4**, 513.

Middlehurst, B.: 1964, *Sky and Telescope* **28**, 83.

Miethe, A. and Seegert, B.: 1911, *Astron. Nachr.* **188**, 9, 239, 371.

Miethe, A. and Seegert, B.: 1914, *Astron. Nachr.* **198**, 121.

Mikhailov, A. A.: 1965, *Astron. Zh.* **42**, 1062.

Miller, B. P.: 1965, *J. Geophys. Res.* **70**, 2265.

Millman, G. H. and Rose, F. L.: 1963, *J. Res. Nat. Bur. Stand.* **67D**, 107.

Millman, P. M. and Burland, M. S.: 1956, *Sky and Telescope* **16**, 222.

Minnaert, M. J. G.: 1941, *Astrophys. J.* **93**, 403.

Minnaert, M. J. G.: 1961, in *Planets and Satellites* (ed. by G. P. Kuiper and B. M. Middlehurst), Univ. of Chicago Press, pp. 213–248.

Mofenson, J.: 1946, *Electronics* **19**, 92.

Moulton, F. R.: 1914, *An Introduction to Celestial Mechanics* (2nd edition), Macmillan, New York, Chapter VIII.

Moulton, F. R.: 1926, *New Methods in Exterior Ballistics*, Univ. of Chicago Press.

Muhleman, D. O.: 1964, *Astron. J.* **69**, 34.

Muhleman, D. O., Holdridge, D. B., and Block, N.: 1962, *Astron. J.* **67**, 191.

Muncey, R.: 1962, *Australian J. Phys.* **16**, 24.

Munk, W. and MacDonald, G. J. F.: 1960, *Rotation of the Earth*, Cambr. Univ. Press, Chapter 11.

Murray, B. C. and Wildey, R. L.: 1963, *Astrophys. J.* **137**, 692.

Murray, B. C. and Wildey, R. L.: 1964, *Astrophys. J.* **139**, 734.

Murray, W. A. S. and Hargreaves, J. K.: 1954, *Nature* **173**, 944.

Myronova, M. M.: 1965, *Dokl. Akad. Nauk. Ukrainian SSR*, No. 4, pp. 455–459.

Nakada, M. P. and Mihalov, J. D.: 1962, *J. Geophys. Res.* **67**, 1670.

Nash, D. B.: 1963, *Icarus* **1**, 372.

Nasmyth, J. and Carpenter, J.: 1885, *The Moon*, Scribner and Welford, London.

Naumann, H.: 1939, *Veröff. Sternwarte Leipzig*, Heft 7.

Nechvíle, V.: 1917, *Časopis pro pěstování matematiky a fysiky (Praha)* **46**, 123.

Nechvíle, V.: 1926, *Compt. Rend. Acad. Paris* **182**, 310.

Nefediev, A. A.: 1951, *Izv. Engelhardt Obs. Kazan*, No. 26.

Nefediev, A. A.: 1955, *Izv. Engelhardt Obs. Kazan*, No. 29.

Nefediev, A. A.: 1957, *Bull. Engelhardt Obs. Kazan*, No. 30.

Nefediev, A. A.: 1958, 'Maps of the Relief of the Moon', *Izv. Engelhardt Obs. Kazan*, No. 30.

Nefediev, A. A.: 1963, *Bull. Engelhardt Obs. Kazan*, No. 36.

Neison, E.: 1876, *The Moon*, Longmans, Green and Co., London.

Ness, N. F.: 1965, *J. Geophys. Res.* **70**, 2989.

Neugebauer, M.: 1960, *Phys. Rev. Letters* **4**, 6.

Newcomb, S.: 1895, *Astron. Papers Am. Ephem. Naut. Almanac* **5**, 398.

Nikonova, E.: 1949, *Izv. Crimean Astrophys. Obs.* **4**, 114.

Nininger, H. H.: 1943, *Sky and Telescope* **2**, No. 4 (pp. 12–15) and No. 5 (pp. 8–9).

Nininger, H. H.: 1952, *Out of the Sky*, Denver Univ. Press, Denver, Colorado; p. 302.

Nininger, H. H.: 1947, *Chips from the Moon*, Desert Press, El Centro, California.

Nordyke, M. D.: 1962, *J. Geophys. Res.* **67**, 1965.

O'Brien, B. and O'Brian, E. D.: 1931, *Phys. Rev.* **37**, 1012.

Odelevsky, V. I.: 1951, *Zh. Tekhn. Fiz.* **21**, 667.

Oetking, P.: 1965, *Trans. Am. Geophys. Union* **46**, 131.

O'Keefe, J. A.: 1957, *Astrophys. J.* **126**, 466.

O'Keefe, J. A.: 1959, *Science* **130**, 97.

O'Keefe, J. A.: 1961, *Science* **133**, 562.

O'Keefe, J. A.: 1963, in *The Tektites* (ed. by J. A. O'Keefe), Univ. of Chicago Press, pp. 167–188.

O'KEEFE, J. A. and ANDERSON, J. P.: 1952, *Astron. J.* **57**, 108.
O'KEEFE, J. A. and CAMERON, W. S.: 1962, *Icarus* **1**, 271.
ÖPIK, E. J.: 1924, *Publ. Obs. Tartu* **26**, No. 1.
ÖPIK, E. J.: 1955, *Irish Astron. J.* **3**, 137.
ÖPIK, E. J.: 1956, *Irish Astron. J.* **4**, 84.
ÖPIK, E. J.: 1958, *Irish Astron. J.* **5**, 34.
ÖPIK, E. J.: 1960, *Monthly Not. Roy. Astron. Soc.* **120**, 404.
ÖPIK, E. J.: 1961, *Astron. J.* **66**, 60.
ÖPIK, E. J.: 1962, *Plan. Space Sci.* **9**, 211.
ÖPIK, E. J.: 1966, in *Advances in Astronomy and Astrophysics* (ed. by Z. KOPAL), Acad. Press, New York and London, vol. 4, pp. 301–336 (cf., in particular, pp. 322–328).
ÖPIK, E. J. and SINGER, S. F.: 1960, *J. Geophys. Res.* **65**, 3065.
ÖPIK, E. J. and SINGER, S. F.: 1961, *Science* **133**, 1419.
OPPOLZER, TH. V.: 1887, *Canon der Finsternisse*, Wien. English translation 1962 (ed. by O. GINGE-RICH), Dover Publications, New York.
ORLOVA, N. S.: 1941, *Uch. Zap. Univ. Leningrad*, No. 82.
ORLOVA, N. S.: 1952, *Uch. Zap. Univ. Leningrad*, No. 153, pp. 166–193.
ORLOVA, N. S.: 1954, *Věstn. Univ. Leningrad* **9**, 77.
ORLOVA, N. S.: 1955, *Astr. Circ. USSR Acad. Sci.*, No. 156, pp. 19–21.
ORLOVA, N. S.: 1956, *Astron. Zh.* **33**, 93.
ORLOVA, N. S.: 1957, *Věstn. Univ. Leningrad* **12**, 152.
ORLOVA, N. S.: 1958, *Astr. Circ. USSR Acad. Sci.*, No. 192, pp. 20–21.
ORLOVA, N. S.: 1962, in *The Moon* (I.A.U. Sympos. No. 14, ed. by Z. KOPAL and Z. K. MIKHAILOV), Acad. Press, New York and London, pp. 411–414.
ORNATSKAYA, O. I.: 1964, *Astron. Zh.* **41**, 995.
ORO, J., WIKSTRÖM, S. A. *et al.*: 1965, *Proc. COSPAR Sympos. Mar del Plata, Argentina* (in press).
OVENDEN, M. W. and ROY, A. E.: 1961, *Monthly Not. Roy. Astron. Soc.* **123**, 1.
PALM, A. and STROM, R. G.: 1962, *Publ. Astron. Soc. Pacific* **74**, 316.
PALM, A. and STROM, R. G.: 1963, *Plan. Space Sci.* **11**, 125.
PARTRIDGE, W. S. and VAN FLEET, H. B.: 1958, *Astrophys. J.* **128**, 416.
PEKERIS, C. L.: 1935, *Monthly Not. Roy. Astron. Soc., Geophys. Suppl.* **3**, 343.
PETROVA, N. N.: 1965, *Izv. Pulkovo Obs.* **24**, No. 2 (pp. 168–174).
PETTENGILL, G.: 1960, *Proc. Inst. Radio Engrs.* **48**, 933.
PETTENGILL, G. H. and HENRY, J. C.: 1962, *J. Geophys. Res.* **67**, 4881.
PETTENGILL, G. H. and HENRY, J. C.: 1962a, in *The Moon* (I.A.U. Sympos. No. 14, ed. by Z. KOPAL and Z. K. MIKHAILOV), Acad. Press, New York and London, pp. 519–525.
PETTIT, E.: 1926, *Publ. Astron. Soc. Pacific* **38**, 242.
PETTIT, E.: 1935, *Astrophys. J.* **81**, 17.
PETTIT, E.: 1940, *Astrophys. J.* **91**, 408.
PETTIT, E.: 1945, *Astrophys. J.* **102**, 14.
PETTIT, E. and NICHOLSON, S. B.: 1927, *Publ. Astron. Soc. Pacific* **39**, 227.
PETTIT, E. and NICHOLSON, S. B.: 1929, *Publ. Astron. Soc. Pacific* **41**, 257.
PETTIT, E. and NICHOLSON, S. B.: 1930, *Astrophys. J.* **71**, 102.
PIDDINGTON, J. H.: 1962, *Plan. Space Sci.* **9**, 305.
PIDDINGTON, J. H. and MINNETT, H. C.: 1949, *Australian J. Sci. Res.* (*A*) **2**, 63.
PIOTROWSKI, S. L.: 1953, *Acta Astron. Cracoviae* **5**, 115.
PITHER, C. M.: 1964, *J. Brit. Astron. Assoc.* **74**, 318.
PLATT, J. R.: 1958, *Science* **127**, 1502.
PLUMMER, H. C.: 1918, *Introduction to Dynamical Astronomy*, Cambr. Univ. Press., Chapter 23.
POHN, H. A.: 1963, *Publ. Astron. Soc. Pacific* **75**, 186.
POHN, H. A., MURRAY, B., and BROWN, H.: 1962, *Publ. Astron. Soc. Pacific* **73**, 333.
POTTER, H. I.: 1962, in *The Moon* (I.A.U. Sympos. No. 14, ed. by Z. KOPAL and Z. K. MIKHAILOV), Acad. Press, New York and London, pp. 63–66.
POTTER, H. I. and BYSTROV, N. F.: 1962, *Astron. Zh.* **38**, 946.
PREY, A.: 1922, *Abhandl. Gesell. Wiss. Göttingen* (*Math.-Physik. Kl.*) **2**, 1.
PRZYBYLLOK, E. G. H.: 1905, *Astron. Nachr.* **168**, 310.
PUISEUX, P. H.: 1906, *Bull. Soc. Astron. de France* **20**, 465.

PUISEUX, P. H.: 1908, *La Terre et la Lune, forme Extérieure et Structure Interieure*, Gauthier-Villars, Paris.

PUISEUX, P. H.: 1910, *Comp. Rend. Acad. Paris* **151**, 133.

PUISEUX, P. H.: 1916, *Comp. Rend. Acad. Paris* **163**, 341.

PUISEUX, P. H.: 1925, *Ann. Observ. Paris (Mémoires)*, vol. 32.

QUAIDE, W.: 1965, *Icarus* **4**, 374.

RACKHAM, T. W.: 1962, *Astron. Contr. Univ. Manchester*, Ser. III, No. 96.

RACKHAM, T. W.: 1964, *Icarus* **3**, 45.

RADLOVA, L. N.: 1941, *Uch. Zap. Univ. Leningrad*, No. 82, pp. 99–129.

RADLOVA, L. N.: 1943, *Astron. Zh.* **20**, 1.

RADLOVA, L. N.: 1957, *Astron. Circ. USSR Acad. Sci.*, No. 179.

RADLOVA, L. N.: 1960, *Astron. Zh.* **37**, 1053.

RADLOVA, L. N.: 1962, in *The Moon* (I.A.U. Sympos. No. 14, ed. by Z. KOPAL and Z. K. MIKHAILOV), Acad. Press, New York and London, pp. 409–410.

RADLOVA, L. N. and SHARONOV, V. V.: 1958, *Astron. Zh.* **35**, 788.

RAE, W. L.: 1963, *J. Brit. Astron. Assoc.* **73**, 169.

RAYLEIGH, J. W.: 1916, *Phil. Mag.* **32**, 529.

REA, D. G., HETHERINGTON, N., and MIFFLIN, R.: 1964, *J. Geophys. Res.* **69**, 5217 (cf. also **70**, 1565).

RITTER, H.: 1934, *Astron. Nachr.* **252**, 157.

ROBERTS, W. A.: 1964, *Icarus* **3**, 348.

ROBERTS, W. and FULMER, C. V.: 1963, *Icarus* **2**, 452.

RONCA, L. B.: 1965, *Icarus* **4**, 390.

ROSENBERG, H.: 1921, *Astron. Nachr.* **214**, 137.

ROSSE, (W. PARSONS), Earl of: 1869, *Proc. Roy. Soc. London* **17**, 436.

ROSSE, (W. PARSONS), Earl of: 1870, *Proc. Roy. Soc. London* **19**, 9.

ROSSE, (W. PARSONS), Earl of: 1872, *Proc. Roy. Soc. London* **21**, 24.

ROSSE, (W. PARSONS), Earl of: 1878, *Proc. Roy. Soc. Dublin* **1**, 19.

ROUGIER, G.: 1933, *Ann. Observ. Strasbourg* (3) **2**, 203–339.

ROUGIER, G.: 1934a, *J. Phys. Radium* (7) **5**, 25S.

ROUGIER, G.: 1934b, *L'Astronomie* **48**, 220, 281.

ROUGIER, G.: 1936a, *J. Phys. Radium* (7) **7**, 156S.

ROUGIER, G.: 1936b, *Comp. Rend. Acad. Paris* **202**, 463.

RUNCORN, S. K.: 1962, *Nature* **195**, 1150.

RUNCORN, S. K.: 1963, Jet Propulsion Lab., Calif. Inst. of Technology, Technical Rep. No. 32-529.

RUSKOL, E. L.: 1960, *Astron. Zh.* **37**, 690.

RUSKOL, E. L.: 1962, 'The Origin of the Moon' in *The Moon* (I.A.U. Sympos. No. 14, ed. by Z. KOPAL and Z. K. MIKHAILOV), Acad. Press, New York and London, pp. 149–155.

RUSKOL, E. L.: 1963a, *Astron. Zh.* **40**, 288.

RUSKOL, E. L.: 1963b, *Izv. USSR Acad. Sci. (Geophys. ser.)* No. 2, pp. 216–222.

RUSKOL, E. L.: 1966, *Icarus* **5**, 221.

RYADOV, V. YA., FURASHOV, N. I., and SHARONOV, V. A.: 1964, *Soviet Astron.* **8**, 82.

RYAN, J. A.: 1962, *J. Geophys. Res.* **67**, 2549.

SAARI, J. M.: 1964, *Icarus* **3**, 161.

SAARI, J. M. and SHORTHILL, R. W.: 1963, *Icarus* **2**, 115.

SAARI, J. M. and SHORTHILL, R. W.: 1965, *Nature* **205**, 964.

SABANEYEV, P. F.: 1962, in *The Moon* (I.A.U. Sympos. No. 14, ed. by Z. KOPAL and Z. K. MIKHAILOV), Acad. Press, New York and London, pp. 419–431.

SAGAN, C.: 1960, *Proc. U. S. Nat. Acad. Sci.* **46**, 393.

SALET, P.: 1922, *L' Astronomie* **36**, 406.

SALISBURY, J. W.: 1961, *Astrophys. J.* **134**, 126.

SALISBURY, J. W., SMALLEY, V. G., and RONCA, L.: 1965, *Nature* **206**, 385; *Trans. Am. Geophys. Union* **46**, 139.

SANDULEAK, N. and STOCK, J.: 1965, *Publ. Astron. Soc. Pacific* **77**, 237.

SOROKIN, N. A.: 1965, *Astron. Zh.* **42**, 1070.

SAUNDER, S. A.: 1900, *Monthly Not. Roy. Astron. Soc.* **60**, 174.

SAUNDER, S. A.: 1901, *Monthly Not. Roy. Astron. Soc.* **62**, 41; **63**, 190, 432.

SAUNDER, S. A.: 1905, *Monthly Not. Roy. Astron. Soc.* **65**, 458.

SAUNDER, S. A.: 1907, *Mem. Roy. Astron. Soc.* **57**, 1.
SAUNDER, S. A.: 1911, *Mem. Roy. Astron. Soc.* **60**, 1.
SCARFE, C. D.: 1965, *Monthly. Not. Roy. Astron. Soc.* **130**, 19.
SCHMIDT, J. F. J.: 1878, *Die Charte der Gebirge des Mondes*, Dietrich Reimer, Berlin.
SCHOENBERG, E.: 1925, *Acta Soc. Sci. Fennicae*, **50**, No. 9 (pp. 1–70).
SCHOENBERG, E.: 1929, in *Handb. der Astrophysik* (ed. by G. EBERHARD, A. KOHLSCHÜTTER, and
    H. LUDENDORFF), Julius Springer, Berlin, II 1, pp. 1–280.
SCHRUTKA-RECHTENSTAMM, G. v.: 1958, *Sitzungsber. Österr. Akad. Wiss. (Math.-Naturwiss. Klasse)*
    **165**, 97–126.
SCHRUTKA-RECHTENSTAMM, G. v.: 1958, *Sitzungsber. Österr. Akad. Wiss. (Math.-Naturwiss. Klasse)*
    **167**, 71–106.
SCHURMEIER, H. M. *et al.*: 1964, Jet Propulsion Laboratory, Technical Report No. 32-605.
SCHWARZSCHILD, K.: 1906, *Göttingen Nachr.* **5**, 41–53.
SECCHI, A.: 1859, *Monthly Not. Roy. Astron. Soc.* **19**, 289.
SECCHI, A.: 1860a, *Monthly Not. Roy. Astron. Soc.* **20**, 70.
SECCHI, A.: 1860b, *Astron. Nachr.* **52**, 91.
SEE, T. J. J.: 1910, *Publ. Astron. Soc. Pacific* **22**, 15; *Popular Astron.* **18**, 137.
SENIOR, T. B. A. and SIEGEL, K. M.: 1959, in *Paris Symposium on Radio Astronomy* (I.A.U. Sym-
    posium No. 9, ed. by R. N. BRACEWELL), Stanford Univ. Press, Stanford, Cal.
SENIOR, T. B. A. and SIEGEL, K. M.: 1960, *J. Res. Nat. Bur. Stand.* **64D**, 217.
SENIOR, T. B. A. and SIEGEL, K. M.: 1961, *Proc. Inst. Radio Engrs.* **49**, 1944.
SENIOR, T. B. A., SIEGEL, K. M., and GIRAUD, A.: 1962, in *The Moon* (I.A.U. Sympos. No. 14, ed.
    by Z. KOPAL and Z. K. MIKHAILOV), Acad. Press, New York and London, pp. 533–543.
SHAKIROV, K. S.: 1963, *Bull. Engelhardt Obs. Kazan*, No. 34.
SHARONOV, V. V.: 1934, *Astron. Zh.* **11**, 225.
SHARONOV, V. V.: 1936, *Uch. Zap. Univ. Leningrad*, No. 1, pp. 26–33.
SHARONOV, V. V.: 1940, *Uch. Zap. Univ. Leningrad*, No. 31, pp. 28–60.
SHARONOV, V. V.: 1953, *Astron. Circ. USSR Acad. Sci.* No. 137, p. 7.
SHARONOV, V. V.: 1955, *Věstn. Univ. Leningrad*, No. 11, pp. 113–120.
SHARONOV, V. V.: 1956a, *Věstn. Univ. Leningrad*, No. 1, pp. 155–167.
SHARONOV, V. V.: 1956b, *Astron. Circ. USSR Acad. Sci.*, No. 166, pp. 9–11.
SHARONOV, V. V.: 1962, *Astron. Zh.* **39**, 87.
SHARONOV, V. V.: 1962a, in *The Moon* (I.A.U. Sympos. No. 14, ed. by Z. KOPAL and Z. K. MIKHAI-
    LOV), Acad. Press, New York and London, pp. 385–390.
SHARONOV, V. V.: 1965, *Astron. Zh.* **42**, 136.
SHARONOV, V. V.: 1965a, in *Geological Problems in Lunar Research* (ed. by J. GREEN), *Ann. N.Y.
    Acad. Sci.* **123**, 740–750.
SHOEMAKER, E. M.: 1962, in *Physics and Astronomy of the Moon* (ed. by Z. KOPAL), Acad. Press,
    New York and London, pp. 283–359.
SHOEMAKER, E. M.: 1965, in Jet Propulsion Lab. Tech. Rep. No. 32-700, pp. 75–134.
SHOEMAKER, E. M. and HACKMAN, R. J.: 1962, in *The Moon* (I.A.U. Sympos. No. 14, ed. by Z. KOPAL
    and Z. K. MIKHAILOV), Acad. Press, New York and London, pp. 301–316.
SHOEMAKER, E. M., HACKMAN, R. J., and EGGLETON, R. E.: 1961, in *Advances in the Astronautical
    Sciences*, Plenum Press, New York **8**, 70–89.
SHORTHILL, R. W., BOROUGH, H. C., and CONLEY, J. M.: 1960, *Publ. Astron. Soc. Pacific* **72**, 481.
SHORTHILL, R. W. and SAARI, J. M.: 1961, *Publ. Astron. Soc. Pacific* **73**, 335.
SHORTHILL, R. W. and SAARI, J. M.: 1965, *Science* **150**, 210.
SIDA, D. W.: 1965, *Observatory* **85**, 42.
SIEGEL, K. M. and SENIOR, T. B. A.: 1962, *J. Res. Nat. Bur. Stand.* **66D**, 227.
SINGER, S. F.: 1961, *Astronaut. Acta* **7**, 135.
SINGER, S. F. and WALKER, E. H.: 1961 *J. Geophys. Res.* **66**, 2561.
SINGER, S. F. and WALKER, E. H.: 1962, *Icarus* **1**, 7, 112.
SINGER, S. F. and WENTWORTH, R. C.: 1959, *J. Geophys. Res.* **64**, 1807, 1959.
SINTON, W. M.: 1955, *J. Opt. Soc. Am.* **45**, 975.
SINTON, W. M.: 1956, *Astrophys. J.* **123**, 325.
SINTON, W. M.: 1959, *Lowell Observ. Bull.* **4**, 260.
SINTON, W. M.: 1960, *Lowell Observ. Bull.* **5**, 23.

SINTON, W. M.: 1960a, *Publ. Astron. Soc. Pacific* **72**, 362.

SINTON, W. M.: 1962, in *Physics and Astronomy of the Moon* (ed. by Z. KOPAL), Acad. Press, New York and London, pp. 407–427.

SITTER, W. DE: 1917, *Monthly Not. Roy. Astron. Soc.* **77**, 155.

SMITH, E. J., DAVIS, L., Jr., COLEMAN, P. J., Jr., and SONNETT, C. P.: 1962, *Science* **138**, 1099.

SMITH, E. J., DAVIS, L., Jr., COLEMAN, P. J., Jr., and SONNETT, C. P.: 1963, *Science* **139**, 909.

SPENCER JONES, H.: 1941, *Mem. Roy. Astron. Soc.* **66**, part 2.

SPINRAD, H.: 1964, *Icarus* **3**, 500.

SPITZER, L.: 1949, in *The Atmospheres of the Earth and Planets* (ed. by G. P. KUIPER), Univ. of Chicago Press, Chapter VII.

SOBOLEVA, N. S.: 1962, *Astron. Zh.* **39**, 1124.

STAELIN, D. H., BARRETT, A. H., and KUSSE, B. R.: 1964, *Astron. J.* **69**, 69.

STAIR, R. and JOHNSTON, R.: 1953, *Nat. Bur. Stand. J. Res.* **51**, 81.

STANYUKOVICH, K. P. and BRONSHTEN, V. A.: 1962, in *The Moon* (I.A.U. Sympos. No. 14, ed. by Z. KOPAL and Z. K. MIKHAILOV), Acad. Press, New York and London, pp. 415–418.

STANYUKOVICH, K. P. and FEDYNSKI, V. V.: 1947, *Dokl. USSR Acad. Sci.* **70**, 129.

STEBBINS, J. and BROWN, F. C.: 1907, *Astrophys. J.* **26**, 326.

STERNE, T. E., GUTHE, K. F., and ROBERTS, W. O.: 1940, *Proc. U.S. Nat. Acad. Sci.* **26**, 399.

STRATTON, F. J. M.: 1909, *Mem. Roy. Astron. Soc.* **60**, part 4.

STROM, R. G. and PALM, A.: 1963, *Nature* **199**, 1052.

SUCKSDORF, E.: 1956, *Geophysica* **5**, 95.

SUDBURY, P. V.: 1965, M. Sc. Thesis, University of Manchester (unpublished).

SUTTON, G. H., MEIDELL, N. S., and KOVACH, R. L.: 1963, *J. Geophys. Res.*, **68**, 4261.

SYTINSKAYA, N. N.: 1953a, *Astron. Circ. USSR Acad. Sci.*, No. 144, pp. 11–12.

SYTINSKAYA, N. N.: 1953b, *Astron. Zh.* **30**, 295.

SYTINSKAYA, N. N.: 1954, *Astron. Circ. USSR Acad. Sci.*, No. 153, pp. 17–18.

SYTINSKAYA, N. N.: 1956, *Astron. Circ. USSR Acad. Sci.*, No. 168, p. 18.

SYTINSKAYA, N. N.: 1957, *Uch. Zap. Univ. Leningrad*, No. 190, pp. 74–87.

SYTINSKAYA, N. N.: 1959, *Astron. Zh.* **36**, 315.

SYTINSKAYA, N. N.: 1962, in *The Moon* (I.A.U. Sympos. No. 14, ed. by Z. KOPAL and Z. K. MIKHAILOV), Acad. Press, New York and London, pp. 391–394.

SYTINSKAYA, N. N.: 1963, *Astron. Zh.* **40**, 1083.

SYTINSKAYA, N. N.: 1965, in *Geological Problems in Lunar Research* (ed. by J. GREEN), *Ann. N.Y. Acad. Sci.* **123**, 756–767.

SYTINSKAYA, N. N. and SHARONOV, V. V.: 1952a, *Uch. Zap. Univ. Leningrad*, No. 153, pp. 114–154.

SYTINSKAYA, N. N. and SHARONOV, V. V.: 1952b, *Věstn. Univ. Leningrad*, No. 9, pp. 97–109.

SZEBEHELY, V. and GIACAGLIA, G. E. O.: 1964, *Astron. J.* **69**, 230.

TAKEUCHI, H., SAITO, M., and KOBAYASHI, N.: 1961, *J. Geophys. Res.* **66**, 3895.

TCHEKIRDA, A. T.: 1946, *Bull. Kharkov Astr. Obs.*, No. 1, pp. 5–8.

TCHUNKO, H. P. A.: 1949, *Z. Astrophys.* **26**, 279.

TEYFEL, V. G.: 1957, *Astron. Circ. USSR Acad. Sci.*, No. 79, pp. 8–10.

TEYFEL, V. G.: 1958, *Astron. Circ. USSR Acad. Sci.*, Nos. 192, pp. 21–23; 194, pp. 11–13; 196, pp. 5–6.

TEYFEL, V. G.: 1959, *Astron. Zh.* **36**, 114, 1041.

TEYFEL, V. G.: 1960, *Astron. Zh.* **37**, 703.

TEYFEL, V. G.: 1962, in *The Moon* (I.A.U. Sympos. No. 14, ed. by Z. KOPAL and Z. K. MIKHAILOV), Acad. Press, New York and London, pp. 399–407.

THOMPSON, T. W.: 1965, Ph.D. Thesis, Cornell University (unpublished).

TISSERAND, F.: 1891, *Mécanique Celeste,* Gauthier-Villars, Paris, vol. 2, Chapter 28, pp. 444–475.

TISSERAND, F.: 1894, *Mécanique Celeste,* Gauthier-Villars, Paris, vol. 3.

TOMASHEK, R.: 1957, 'Tides in the Solid Earth' in *Handb. d. Phys.* **48** (ed. by J. BARTELS), pp. 775–845.

TREXLER, J. H.: 1958, *Proc. Inst. Radio Engrs.* **46**, 286.

TROITSKY, V. S.: 1954, *Astron. Zh.* **31**, 511.

TROITSKY, V. S.: 1962, in *The Moon* (I.A.U. Sympos. No. 14, ed. by Z. KOPAL and Z. K. MIKHAILOV), Acad. Press, New York and London, pp. 475–489.

TURNER, G.: 1957, *J. Brit. Astron. Assoc.* **67**, 185.

TURNER, G.: 1958, *J. Brit. Astron. Assoc.* **68**, 251.
TURNER, G.: 1959, *Astron. Contr. Univ. Manchester*, Ser. III, Nos. 72 and 73.
TURSKI, W.: 1963, *Icarus* **1**, 170.
UREY, H. C.: 1952, *The Planets*, Yale Univ. Press, New Haven, Conn.
UREY, H. C.: 1952a, *Geochim. et Cosmochim. Acta*, **2**, 263.
UREY, H. C.: 1955, *Proc. U. S. Nat. Acad. Sci.* **41**, 27, 423.
UREY, H. C.: 1956a, *Proc. U. S. Nat. Acad. Sci.* **42**, 889.
UREY, H. C.: 1956b, in *Vistas in Astronomy* (ed. by A. BEER), Pergamon Press, London, **2**, 1667–1680.
UREY, H. C.: 1956c, *Sky and Telescope*, **15**, 108, 160.
UREY, H. C.: 1956d, *Observatory* **76**, 232.
UREY, H. C.: 1957, in *Progress in Physics and Chemistry of the Earth*, Pergamon Press, London, **2**, 46–76.
UREY, H. C.: 1957a, *Nature* **179**, 556.
UREY, H. C.: 1958, *Proc. Chem. Soc. London*, March, pp. 67–78.
UREY, H. C.: 1958a, *Zeit. Phys. Chem.* **16**, 346.
UREY, H. C.: 1959, *J. Geophys. Res.* **64**, 1731.
UREY, H. C.: 1960, *Astrophys. J.* **132**, 502.
UREY, H. C.: 1960a, *J. Geophys. Res.* **65**, 358.
UREY, H. C.: 1960b, *J. Geophys. Res.* **65**, 2529.
UREY, H. C.: 1961, *Astrophys. J.* **134**, 268.
UREY, H. C.: 1962, in *Physics and Astronomy of the Moon* (ed. by Z. KOPAL), Acad. Press, New York and London, Chapter 13 (pp. 481–523).
UREY, H. C.: 1962a, in *The Moon* (I.A.U. Sympos. No. 14, ed. by Z. KOPAL and Z. K. MIKHAILOV), Acad. Press, New York and London, pp. 133–148.
UREY, H. C.: 1962b, in *Space Research* (Proc. COSPAR Symposium, Nice; ed. by H. KALLMANN-BIJL), North-Holland Publ. Co., Amsterdam, pp. 1114–1122.
UREY, H. C.: 1962c, *Nature* **193**, 1119.
UREY, H. C.: 1965, *Science*, **147**, 1262.
UREY, H. C.: 1965a, in 'Ranger VII, Part II: Experimenters' Analyses and Interpretations', Jet Prop. Lab., Cal. Tech., Techn. Report No. 32-700, pp. 135–136.
UREY, H. C.: 1966, *Science* **151**, 157.
UREY, H. C.: 1966a, *Monthly Not. Roy. Astron. Soc.* **131**, 199.
VAND, V.: 1945, *J. Brit. Astron. Assoc.* **55**, 47.
VARSAVSKY, C. M.: 1958, *Geochim. Cosmochim. Acta* **14**, 291.
VAUCOULEURS, G. DE: 1944, *Compt. Rend. Acad. Paris* **218**, 655; 805.
VENING MEINESZ, F. A.: 1959, *Proc. Koninkl. Ned. Akad. Wetenschap.* (*Ser. B*) **62**, 115.
VERY, F.: 1898, *Astrophys. J.* **8**, 265.
VERY, F.: 1906, *Astrophys. J.* **24**, 351.
VIGROUX, E.: 1956, *J. Observateurs* **39**, 134.
VÖLKEL, M.: 1908, 'Beitrag zur Bestimmung der physischen Libration des Mondes', *Trudy Astron. Inst. Univ. Kazan*, No. 17.
WALKER, E. H.: 1962, *J. Geophys. Res.* **66**, 4293.
WALKER, E. H.: 1964, *J. Geophys. Res.* **69**, 566.
WARNER, B.: 1961, *Plan. Space Sci.* **5**, 321.
WARNER, B.: 1961a, *Plan. Space Sci.* **5**, 283.
WARNER, B.: 1961b, *Publ. Astron. Soc. Pacific* **73**, 349.
WARREN, C. R.: 1963a, *Science* **140**, 188.
WARREN, C. R.: 1963b, Art. 39 in *U. S. Geol. Survey Profess. Paper* 475-B, pp. B148–152.
WASIUTYNSKI, J.: 1946, *Astrophys. Norwegica*, **4**, pp. 195–205.
WATSON, F. G.: 1956, *Between the Planets* (sec. ed.), Harv. Univ. Press, Cambridge.
WATSON, G. N.: 1958, *Theory of Bessel Functions* (2nd edition) Cambr. Univ. Press.
WATSON, K., MURRAY, B., and BROWN, H.: 1961, *J. Geophys. Res.,* **66**, 1598, 3033.
WATTS, C. B.: 1963, 'The Marginal Zone of the Moon', *Astron. Papers Am. Ephem. Nautical Almanac*, vol. 17.
WATTSON, R. B. and DANIELSON, R. E.: 1965, *Astrophys. J.* **142**, 16.
WEBB, D. H.: 1946, *Sky and Telescope* **5**, No. 3.

WECHSLER, A. E. and GLASER, P. E.: 1964, in *The Lunar Surface Layer* (ed. by J. W. SALISBURY and P. E. GLASER), Acad. Press, New York and London, pp. 389–410.

WECHSLER, A. E. and GLASER, P. E.: 1965, *Icarus* **4**, 335.

WEHNER, G. K., KENKNIGHT, C. E., and ROSENBERG, D.: 1963, *Plan. Space Sci.* **11**, 1257.

WEIL, H. and BARASH, M. L.: 1963, *Icarus* **1**, 346.

WEIMER, TH.: 1952, *Atlas des Profils Lunaires*, Publ. de l'Observ. de Paris.

WEIMER, TH.: 1954, *Bull. Astron.* **17**, 271.

WELLS, J. W.: 1963, *Nature* **197**, 948.

WESSELINK, A. J.: 1948, *Bull. Astron. Inst. Neth.* **10**, 351.

WESSELINK, A. J.: 1954, *Observatory* **74**, 215.

WHIPPLE, F. L.: 1961, *Nature* **189**, 127.

WHITAKER, E. A.: 1963, in *The Moon, Meteorites and Comets* (ed. by B. M. MIDDLEHURST and G. P. KUIPER), Univ. of Chicago Press, pp. 123–128.

WHITWELL, T.: 1929, *J. Brit. Astron. Assoc.* **39**, 255.

WICHMANN, M.: 1846, *Astron. Nachr.* **26**, 289.

WICHMANN, M.: 1847, *Astron. Nachr.* **27**, 53, 81, 97, 211.

WILDEY, R. L.: 1963, *Nature* **200**, 1056.

WILDEY, R. L.: 1964, *Icarus* **3**, 136.

WILDEY, R. L. and POHN, H. A.: 1964, *Astron. J.* **69**, 619.

WILSING, J. and SCHEINER, J.: 1909, *Publ. Astrophys. Obs. Potsdam* **20**, No. 61, pp. 1–68.

WILSING, J. and SCHEINER, J.: 1921, *Publ. Astrophys. Obs. Potsdam* **24**, No. 77, pp. 3–25.

WILSON, A. T.: 1962, *Nature* **196**, 11.

WINTER, D. F.: 1962, *J. Res. Nat. Bur. Stand.* **66D**, 215.

WISLICENUS, W. F.: 1902, *Bull. Soc. Belge d'Astron.* **7**, 39.

WOOD, R. W.: 1910, *Monthly Not. Roy. Astron. Soc.* **70**, 226.

WOOD, R. W.: 1912, *Astrophys. J.* **36**, 75.

WOODSIDE, W. and MESSMER, J. H.: 1961, *J. Appl. Phys.* **32**, 1688.

WRIGHT, F. E.: 1927, *Proc. U. S. Nat. Acad. Sci.* **13**, 535.

WRIGHT, F. E.: 1929, *Publ. Astron. Soc. Pacific* **41**, 125.

YAKOVKIN, A. A.: 1923, *Astron. Nachr.* **219**, 61.

YAKOVKIN, A. A.: 1928, *Izv. Engelhardt Obs. Kazan*, No. 13.

YAKOVKIN, A. A.: 1934, *Bull. Astron. Inst. Engelhardt*, No. 4.

YAKOVKIN, A. A.: 1939, *Izv. Engelhardt Obs. Kazan*, No. 21.

YAKOVKIN, A. A.: 1945, *Izv. Engelhardt Obs. Kazan*, No. 23.

YAKOVKIN, A. A.: 1950, *Kiev Obs. Publ.*, No. 3.

YAKOVKIN, A. A.: 1952, *Trans. I.A.U.* **8**, 231.

YAKOVKIN, A. A. and BELKOVICH, I. V.: 1935, *Astron. Nachr.* **256**, 305.

YAPLEE, B. S., BRUTON, R. H., CRAIG, K. J., and ROMAN, N. G.: 1958, *Proc. Inst. Radio Engrs.* **46**, 293.

YEZERSKI, V. I. and FEDORETS, V. A.: 1955, *Astron. Circ. USSR Acad. Sci.*, No. 159, pp. 18–20.

YEZERSKI, V. I. and FEDORETS, V. A.: 1956, *Circ. Kharkov Astr. Obs.*, No. 15, pp. 17–20.

ZELTSER, M. S.: 1959, in *The Moon* (ed. by A. V. MARKOV), Moscow, pp. 174–201.

ZHARKOV, V. N. and ULINICH, F. R.: 1960, *Trudy Inst. Phys. Zemli* **11**, 61.

# INDEX OF PERSONS